CANDID SCIENCE VI
More Conversations with Famous Scientists

Some other books by István (IH) and Magdolna (MH) Hargittai

IH, *The Martians of Science: Five Physicists Who Changed the Twentieth Century*, Oxford University Press, 2006.
IH (with B. Hargittai), *Candid Science V: Conversations with Famous Scientists*, Imperial College Press, 2005.
MH, IH, *Candid Science IV: Conversations with Famous Physicists*, Imperial College Press, 2004.
IH, *Our Lives: Encounters of a Scientist*, Akadémiai Kiadó, Budapest, 2004.
IH, *Candid Science III: More Conversations with Famous Chemists*, edited by MH, Imperial College Press, 2003.
IH, *The Road to Stockholm: Nobel Prizes, Science, and Scientists*, Oxford University Press, 2002; 2003.
IH, *Candid Science II: Conversations with Famous Biochemical Scientists*, edited by MH, Imperial College Press, 2002.
IH, *Candid Science: Conversations with Famous Chemists*, edited by MH, Imperial College Press, 2000.
IH, MH *In Our Own Image: Personal Symmetry in Discovery*, Kluwer/Plenum, New York, 2000.
IH, MH, *Symmetry through the Eyes of a Chemist*, Second edition, Plenum, New York, 1995.
IH, MH, *Symmetry: A Unifying Concept*, Shelter Publications, Bolinas, CA, 1994.
IH (with R. J. Gillespie), *The VSEPR Model of Molecular Geometry*, Allyn & Bacon, Boston, 1991.

Edited books

IH (with A. Domenicano), *Strength from Weakness: Structural Consequences of Weak Interactions in Molecules, Supermolecules, and Crystals*, Kluwer, Dordrecht, 2002.
IH (with T. C. Laurent), *Symmetry 2000*, Vols. I–II, Portland Press, London, 2002.
MH, IH, *Advances in Molecular Structure Research*, Vols. 1–6. JAI Press, Greenwich, CT, 1995–2000.
IH (with C. A. Pickover), *Spiral Symmetry*, World Scientific, Singapore, 1992.
IH, *Fivefold Symmetry*, World Scientific, Singapore, 1992.
IH (with A. Domenicano), *Accurate Molecular Structures*, Oxford University Press, 1992.
IH, *Quasicrystals, Networks, and Molecules of Fivefold Symmetry*, VCH, New York, 1990.
IH, *Symmetry 2: Unifying Human Understanding*, Pergamon Press, Oxford, 1989.
IH, MH, *Stereochemical Applications of Gas-Phase Electron Diffraction*, Vols. A–B, VCH Publishers, New York, 1988.
IH (with B. K. Vainshtein), *Crystal Symmetries, Shubnikov Centennial Papers*, Pergamon Press, Oxford, 1988.
IH, *Symmetry: Unifying Human Understanding*, Pergamon Press, Oxford, 1986.

CANDID SCIENCE VI
More Conversations with Famous Scientists

István Hargittai

Magdolna Hargittai

Imperial College Press

Published by

Imperial College Press
57 Shelton Street
Covent Garden
London WC2H 9HE

Distributed by

World Scientific Publishing Co. Pte. Ltd.
5 Toh Tuck Link, Singapore 596224
USA office: 27 Warren Street, Suite 401-402, Hackensack, NJ 07601
UK office: 57 Shelton Street, Covent Garden, London WC2H 9HE

István Hargittai
Budapest University of Technology and Economics
Eötvös University and Hungarian Academy of Sciences
H-1521 Budapest, Pf. 91, Hungary

Magdolna Hargittai
Eötvös University and Hungarian Academy of Sciences
H-1518 Budapest, Pf. 32, Hungary

Library of Congress Cataloging-in-Publication Data
Hargittai, István.
    Candid science VI : more conversations with famous scientists / István
Hargittai, Magdolna Hargittai.
      p. cm.
    Includes indexes.
    ISBN 1-86904-693-3 (alk. paper) -- ISBN 1-86094-694-1 (pbk. : alk. paper)
    1. Scientists--Interviews. 2. Scientists--Biography. 3. Scientists--History--20th century.
    I. Title: Candid science six. II. Title: Candid science 6. III. Hargittai, Magdolna. IV. Title.

Q141 .H264 2006
509.2'5--dc22

                                                                                                  2006048576

British Library Cataloguing-in-Publication Data
A catalogue record for this book is available from the British Library.

Copyright © 2006 by István Hargittai and Magdolna Hargittai

All rights reserved. This book, or parts thereof, may not be reproduced in any form or by any means, electronic or mechanical, including photocopying, recording or any information storage and retrieval system now known or to be invented, without written permission from the Publisher.

For photocopying of material in this volume, please pay a copying fee through the Copyright Clearance Center, Inc., 222 Rosewood Drive, Danvers, MA 01923, USA. In this case permission to photocopy is not required from the publisher.

Printed by Mainland Press Pte Ltd

Foreword

This is the sixth and final volume of a remarkable series of interviews carried out by Magdolna and István Hargittai with some of the most accomplished scientists of the late twentieth and early twenty first centuries. These volumes provide an unusual insight not only into some extraordinary science but also into the minds of the scientists who carried out that science. Reading these conversations brings you closer to understanding what makes scientists do science and what they feel while they are doing it. They allow the reader to experience the curiosity which drives scientists to find out how the natural world works, or said another way, the passion of just wanting to know, the satisfaction of successfully completing the near impossible experiment, the concerns many scientists feel about the implications of their work for society, and the comradeship and collegiality which marks many scientific endeavors. Gaining access to scientific minds working at this level of achievement is difficult but these interviews often get you close.

Why is it important to understand how science and scientists work? The characteristics of science, respect for observation and experimentation, reliance upon refutable hypotheses, consistency across different fields of inquiry, critical skepticism, all contribute to making science the most reliable approach for gaining knowledge about the natural world. Advances in science have spectacularly changed our understanding of the world and of what it is to be human, and have also underpinned many new practices and technologies which have improved the quality of human life. This has happened to such an extent that it is now impossible to imagine society without the benefits that science has provided. Although science has this universal relevance, scientists are trained specialists and are relatively few in number, and there is always a danger that they can easily become remote from the rest of society. The conversations reproduced in this and the earlier volumes help bridge that gap, and we should all be grateful for the vision and fortitude of István Hargittai in chronicling so many of the stories of science and scientists that mark the present age.

New York City, June 2006   Paul Nurse

PREFACE

This sixth volume of the *Candid Science* series is its concluding volume. Once again, we included interviews with scientists from all three major areas of the scope of the series, biomedical sciences, chemistry, and physics, and the entries follow loosely this order. Our usual procedure of recording of interviews and the follow-up work have been described in prefaces of previous volumes. Each of us contributed roughly half of the interviews, with István focusing mostly on the biomedical sciences and Magdi on physics, and there were four joint interviews as well.

When one of us (IH) started this interviews project back in 1994, it was not envisioned to become so big. Eventually, Magdi joined in, and after well over two hundred interviews we can look back to the past decade with good feelings. These feelings include gratitude toward our interviewees; not only did they graciously share their experience and thoughts with us, but acted as patient teachers in many fields that we had been unfamiliar with.

Many people keep asking me about the inception of the project, which, again, has been described in previous volumes, and many have suggested that I should include an interview by someone with me. I must admit that I played with this idea, but decided against it because we have followed rather strict criteria for selecting interviewees, and it would have been imprudent to include ourselves among them. What I would like to suggest instead is a book that I wrote about my encounters with famous scientists, including a considerable amount of autobiographical material: I. Hargittai, *Our Lives: Encounters of a Scientist*, Akadémiai Kiadó, Budapest, 2004. This book is in English (for a review, see, *Nature*, November 15, 2004) and its German version will have been published by the time the present volume appears.

Acknowledgments

We would like to express our thanks to Val L. Fitch (Princeton, New Jersey); Robin K. Harris (Durham, England); E. Thomas Strom (Dallas, Texas); Lars Strother (Charlottesville, Virginia) and Martin Veltman (Bilthoven, The Netherlands) for their kind assistance in improving the presentation of the material communicated in this volume.

As we are completing the *Candid Science* series, we would also like to express our sincere thanks to Imperial College Press (London) and World Scientific (Singapore) for bringing out this series. Our special appreciation goes to Senior Editor Ms. Ying Oi Chiew for her expertise and thoughtful dedication, guiding the last three volumes through production.

We are grateful to the Budapest University of Technology and Economics, the Hungarian Academy of Sciences, and Eötvös University as well as to the Hungarian National Scientific Research Funds for their support of our research activities in structural chemistry.

Budapest, February 2006   István Hargittai

Contents

Foreword	v
Preface	vii
Francis H. C. Crick	2
Sydney Brenner	20
Matthew Meselson	40
Paul M. Nurse	62
Richard T. Hunt	88
Seymour Benzer	114
Christiane Nüsslein-Volhard	134
Werner Arber	152
David Baltimore	164
J. Michael Bishop	182
Harold E. Varmus	200
Peter Mansfield	216
Avram Hershko	238
Aaron Ciechanover	258
Irwin Rose	304
Alexander Varshavsky	310

Osamu Hayaishi	360
Ada Yonath	388
Isabella L. Karle	402
Jerome Karle	422
Yuan T. Lee	438
Darleane C. Hoffman	458
Richard L. Garwin	480
Donald A. Glaser	518
Nicholas Kurti	554
Herbert Kroemer	566
James W. Cronin	586
Wolfgang K. H. Panofsky	600
Burton Richter	630
Samuel C. C. Ting	654
Martin L. Perl	668
Carlo Rubbia	680
Simon van der Meer	698
Douglas D. Osheroff	710
Jack Steinberger	732
Masatoshi Koshiba	752
Riccardo Giacconi	762
Brian D. Josephson	772
Ivar Giaever	786
Vitaly L. Ginzburg	808
David J. Gross	838
Frank Wilczek	856
Name Index	871
Cumulative Index of Interviewees	883

Francis H. C. Crick, 2004 (photograph by I. and M. Hargittai).

1

FRANCIS H. C. CRICK

Francis Harry Compton Crick (1916, Northampton, England – 2004, La Jolla, California) at the time of his death was a distinguished research professor of the Salk Institute for Biological Studies in La Jolla, California. He was also a former president of the Institute. He was co-recipient of the Nobel Prize in Physiology or Medicine for 1962 together with James D. Watson[1] (b. 1928) and Maurice Wilkins (1916–2005) "for their discoveries concerning the molecular structure of nuclear [*sic*] acids and its significance for information transfer in living material," in short, for the discovery of the double helix structure of DNA.

Francis Crick attended Northampton Grammar School, then Mill Hill School in North London. He studied physics at University College London and received his B.Sc. degree in 1937. He continued his studies for a doctorate, doing research on high-stress failure in engineering materials. His studies were interrupted by the war, and he did war-related research for the British Admiralty during World War II. In 1947, he started doing research for the Medical Research Council and from 1949 he worked at the Cavendish Laboratory of Cambridge University. His cooperation with James Watson started in 1951, and led to the discovery of the double helix in 1953. During the subsequent years he continued working in molecular biology, primarily involved in the understanding of protein synthesis and the genetic code. He left Great Britain and joined the Salk Institute in 1976 and changed his research fields for understanding the brain and the nature of consciousness.

Francis Crick was a Fellow of the Royal Society (London); a Foreign Member of the National Academy of Sciences of the U.S.A. (1969), and other learned societies. His many honors included the Lasker Award (1960), the Gairdner Award (1962), and the Prix Charles Leopold Meyer of the French Academy of Sciences. In 1991, he was named a Member of the Order of Merit of the United Kingdom, whose membership is restricted to 24.

This account is about a conversation with Francis Crick and his wife Odile Crick a few months before his death and about my (IH) correspondence with him over the years.

The Visit

My wife, Magdi and I visited Francis and Odile Crick in their home in La Jolla, California, on February 8, 2004. It was a full two hours of conversation over lunch. I had written to Crick about our forthcoming visit to Pasadena, and he had written to me: "My health is still poor but it would be a pity for us to be so close and yet not meet. So I suggest you come to lunch on Saturday, February 7th at my house in La Jolla, arriving between 12:30 and 1:00 p.m. Attached is a crude map to help you find our house." This was our first and only personal meeting. When we rang the Cricks' bell, Francis Crick opened the door, Magdi stepped in first, and Francis stretched out his hand and introduced himself, "I'm Francis Crick." There was also Odile right away, and the two of them made the atmosphere so light and pleasant that I could not help telling them at once about what I had just read in Maurice Wilkins's new book, *The Third Man of the Double Helix*. Wilkins writes about his trying hard to find someone to marry and hoped that Crick and Odile would bring some nice young woman with them. Wilkins adds that he had no wish to separate Odile from Francis. Francis and Odile laughed heartily at Wilkins's "magnanimity".

Our conversation covered many topics. Erwin Chargaff's name came up and Francis found it strange that Chargaff did not discover base-pairing in the light of his observations on the base ratios in DNA. However, this may be more surprising in hindsight. Just looking at the data, there is much fluctuation, about 10 per cent, about the 1 to 1 ratios. So even recognizing and suggesting the 1 to 1 ratio was a sharp observation. Francis then added that Chargaff's mind might have not wandered towards pairing because he, that is, Chargaff was thinking in terms of one chain rather than two. In a single chain, nothing would prompt one's thinking towards pairing. Once you know that two chains need be considered, pairing enters one's thinking more naturally. Francis seemed careful not to use the word helix at this point, as if placing himself into Chargaff's position. Even if Chargaff was thinking about the significance of the 1 to 1 ratios, about the possible meaning of such a ratio in a nucleotide chain, it's not surprising that nothing suggested itself for a reasonable solution.

Once the idea of two chains or helices came up, base-pairing was more probable to be thought of. That it was not trivial is witnessed by the fact that Crick and Watson did not think of base-pairing either until the very last moment in the story of the discovery. It was very late in the story that complementarity came up. Even when Watson was pairing bases, first he was seeking correspondence between like bases. Francis gently reminded me that solving the problem was less straightforward than I might have thought. Complementarity could have been accomplished between like bases if the two like bases would not be turning toward each other with the same end. When they started thinking about pairing bases though, whether like to like or between different ones, the solution was found relatively quickly. As Crick was talking about finding base-pairing, he distinctly spoke about "our" and not just Watson's findings. Watson did not even want to believe in base pairs initially.

Another feature of the double helix structure that we talked about in detail was C_2 symmetry. Here, Francis said that Jim did not understand its significance. This is in accord with my own experience when we were in Cold Spring Harbor in 2002 and I talked with Jim about it. My impression was that even almost 50 years after the discovery, he still underestimated the significance of C_2 symmetry in the DNA structure and in particular, in the story of the original discovery. This symmetry is the most unambiguous indication of the complementarity of the two helices. Francis added that Rosalind Franklin did not quite recognize the significance of C_2 symmetry either in solving the DNA structure. Although Franklin was a crystallographer — and more so than Watson — she had never solved a structure before. She did not have extensive experience in structure analysis and even less with organic systems and polymeric molecules at that. Of course, nobody else had much experience with solving organic polymeric structures either, at that time. Crick thought that Wilkins speaks about base-pairing in his book as if he knew more about it than he could have and did at that time. Crick was sure that Wilkins did not have the idea at the time.

He told us that one of his most important findings had never been written up and was recorded only in a manuscript for a lecture which seems to have been lost. This was about the one-dimensional sequence of the amino acids in proteins and the importance of the three-dimensional structures of the proteins. The sequence, which was one-dimensional, determined the folding, which produced the three-dimensional structure. In terms of replication, the three-dimensional structure could not replicate itself; the only part that was capable of replication was the surface of the three-dimensional structure. The essence of the idea was that for replication it sufficed to repeat the sequence. Crick told us about this idea when I raised the question about who was the one that for the first time brought up the idea that the nucleic acids code for the proteins. Then he said that actually he had this idea, but, he added, then others also had this idea.

When we talked about the connection between nucleic acids and proteins, Francis said that Jim and he were definitely thinking about that in the spring of 1953. When they announced on that fateful day in the pub, The Eagle, that they solved the secret of life, they could make such an announcement only because they realized that there was a connection. For calling it the secret of life, the double helix structure of DNA would not have sufficed. They understood the implication of the double helix so quickly because they had thought about the question of information transfer.

Actually, Crick told us, he had had this idea even before he had met Jim Watson. This is fascinating as we may try to delineate their contributions to the story of the double helix. When they worked together, they talked a lot to each other, so it is hardly possible to delineate their contributions. In raising this question about how nucleic acids code for proteins, it is very difficult to delineate their shares. However, this idea about the importance of sequence in replication, whether it is the sequence in nucleic acids or proteins, was Crick's idea alone.

We also talked about Jim and Francis enumerating the 20 naturally occurring amino acids. This was in connection with the notion that sometimes it happens that important findings do not appear so important at the time of their being made. Today, this enumeration is there in every textbook, but seldom is it associated with Crick and Watson.

A good part of our conversation concerned religion. I started with the general notion by Jim Watson that he was not happy that Crick seemed to have been moving from a more radical position towards the center. According to Jim, Francis was in a better situation to criticize religion

Jacob (Yakov) Varshavsky and Francis Crick in Moscow at the Fifth International Congress of Biochemistry, 1961 (courtesy of Alex Varshavsky).

than he was himself. This was because Francis was not the head of a major organization and he was not involved in fundraising as he, Jim, was. Crick said that to fight religion at the present time produces only frustration. First we have to understand how the brain operates and after that it will be much easier to convince people that religion is meaningless. We also talked about the recent changes in the views of the Catholic Church regarding evolution, for example. Francis stressed, however, that the Catholic Church seems to want to solve all the problems of religion within its own framework and without the involvement of science.

Francis had high hopes for the success of the new book about the mind, *The Quest for Consciousness*, by Christof Koch.[2] He told us that the book was the result of their joint work, but Koch was the sole author; Francis authored only the introduction. We talked about the importance of book titles and we learned that Odile had something to do with the final title of this book. She suggested replacing the initially used "search" by "quest".

I told Francis that we very often ask the question in our interviews about heroes. Of our contemporaries Francis Crick is mentioned more often than anybody else is (of the non-living, Albert Einstein is mentioned most often). Francis was not shy about this and appeared pleased and genuinely interested in what the reason might be. We mentioned one example, that of Fred Sanger. Sanger liked Crick's style of lecturing, the ease with which he did it, and the jokes he inserted into his lectures. At one point Sanger decided to emulate Crick. He had carefully prepared his presentation in Crick's style, but he came away disappointed because all his jokes came out flat and were received with silence. Francis enjoyed the story and I understood what it meant when people described him as roaring with laughter. He said that it did not work this way. He never prepared his jokes specifically although they came out of his reservoir; but they came out — or appeared coming out — spontaneously during the talk. Once he was asked to give a lecture in Paris but he was asked to give it in French. His French was not that good, so he wrote it up and Odile corrected and translated it. The first thing that had to be left out was the jokes; it would have been difficult to have his old jokes in French, besides, planning them in advance — so that they could be translated — made it impossible for them to appear as spontaneous jokes.

When I asked Francis about his heroes, he mentioned Linus Pauling. He also added that he was a latecomer in science, and people "worship" heroes when they are younger. Francis was about 30 years old when he started in science.

I knew that Francis was gravely ill, yet there was no impression of illness during our meeting. It was an unforgettable encounter for us and it gives me a good feeling that Francis and Odile visibly enjoyed our visit. Maybe it meant a brief break away from gloomier times that in less than half a year would end in Francis's death. I cherish all that I learned from him in our conversation, but we have what he told us through the filter of our memory. What he wrote me in our correspondence is of course a more reliable record.

The Correspondence

Over the past few years, I sent a few letters to Francis Crick with questions that I thought only he could answer. I wrote him on July 24, 2000, asking him about the story of the introduction of the isomorphous replacement

Maurice Wilkins, Max Perutz, Francis Crick, John Steinbeck, James D. Watson, and John Kendrew in Stockholm, at the Nobel Prize ceremony, 1962 (courtesy of the late Lars Ernster).

method into protein crystallography. In our book with Magdi, *In Our Own Image*,[3] we quoted Max Perutz who — with reference to the hemoglobin structure — told me a few years before the following: "In 1953, I discovered that it could be solved by the method of isomorphous replacement by comparing the X-ray diffraction pattern from a crystal of pure hemoglobin to one from hemoglobin to which I attached two mercury atoms." At the time of writing our book, we were a little puzzled that he did not mention J. M. Robertson, David Harker, and J. M. Bijvoet, who originated the technique of isomorphous replacement although they never applied it to proteins.

In Crick's book, *What Mad Pursuit*,[4] I noticed that he might have been the first at the Cavendish Laboratory to suggest the use of the isomorphous replacement method in protein structure analysis. Crick was unambiguous in his response saying that he and others might have made such suggestions, but it was Perutz who carried out the tremendous amount of work that was involved and the credit should be his alone.

I wrote Crick again in early spring of 2001. I had just completed the manuscript of *The Road to Stockholm*.[5] I wondered about the missing Nobel Prize for Sydney Brenner. I knew that Brenner and Crick used to work together in Cambridge and that they had a sizzling intellectual interaction for years. However, I felt that from the point of view of the Nobel Prize, theirs was an asymmetric relationship. Whereas Crick had already had his Nobel Prize, the assignment of any major research achievement to Brenner might have been hindered by his close relationship with Crick. So I asked Crick about this. Here is what he wrote me on April 13, 2001:

> Although Sydney Brenner and I shared an office for 20 years, for most of that time I worked in the office (not always the same office) whereas Sydney worked mainly in the lab. However we did talk together for an hour or more on most days.
>
> The adaptor hypothesis was my idea, but Sydney coined the name for it. Sydney had the idea that acridine mutants were probably the addition or subtraction of bases. I did all the initial work on the phase-shift mutants, but Sydney designed the special genetic cross to show that +++ mutants were like wild-type. I worked out that shifts to the left were different from shifts to the right. Sydney did almost all the work to establish the stop-chain codons. Sydney realized that the Volkin–Astrachan DNA was really messenger RNA, though I immediately saw it too. Sydney, with Meselson and Jacob, established the existence of mRNA experimentally. Sydney (and another group) established experimentally the co-linearity of gene and protein. My recollection is that all this is fairly accurately described in Horace Judson's book, *The Eighth Day of Creation*.
>
> All the initial work on the nematode was conceived and carried out by Sydney, and he organized the study of its cell lineage and its detailed neuroanatomy.
>
> In my opinion Sydney ought to have the Nobel Prize but although he has done a vast amount of important work it is difficult to select just one particular discovery that would attract a Nobel Prize.
>
> However Sydney's work is widely recognized by everyone. In fact he has received every other important award other than the Nobel — many more than I have!
>
> I hope this is of some help. Incidentally, *What Mad Pursuit* is mostly about the mistakes I made.

This letter was only 18 months before Brenner's long-awaited Nobel Prize was announced.

In my letter dated April 26, 2001, I asked Crick about what Jim Watson told me in my interview with him[1]: "Francis Crick gave a provocative lecture in 1968 at University College London where he said you should only be declared alive two days after birth. Later I have been mistakenly accused of that remark — Watson continued — ascribing to me Hitler-like motivations. Francis also then said the state should not spend any money for medical care about people above 80. Now that he is 84, he would probably disagree. He said this when he was 52." Crick responded on June 28, 2001, and I quote from his letter:

> My apologies for not replying sooner to your letter of April 26th. I did indeed give a provocative lecture in 1968 (or thereabouts) at University College London, but I'm not sure that I still have a copy of it.
>
> To reply to your two questions I would indeed modify my suggestions today. In the old days doctors quickly let a very deformed or handicapped baby die, rather than make exceptional efforts, as they often do now, to keep the baby alive. I now realize that it would be impossible, at least in this country, to count life as starting after the first two days of a baby's life because so many religious people believe life effectively starts much earlier, even at conception. In other words one has to consider not just the feelings of the baby (who hardly has any) but also the feelings of the parents, and of other members of society, however silly one may think them to be. But I do believe that doctors should not make exceptional efforts to keep a very handicapped baby alive.
>
> As to the age limit, people now live longer than they did in the sixties, so I think such an age might be a little higher, but I doubt if a rigid rule would be acceptable. Again I think very expensive treatments, or ones that have only a limited availability, should be allocated in some sensible way. I've heard that the State of Oregon is trying out such a scheme.
>
> If I were to give such a lecture again (which is unlikely) I would instead stress the right of a person who is incurably ill to terminate his own life. I believe this is being tried out in Holland.

In my next letter of July 27, 2001, I asked him whether there were any scientists that could be considered directly as his pupils. I found this question of interest because by then I had experienced that other famous scientists named Crick more than anybody else among living scientists as their hero. Again, Crick's response of August 1, 2001, is of interest in full:

> To reply to your question, I don't think there is anyone whom I could call my pupil. I only supervised a graduate student for a year, but after that year someone else took him over. I think I deliberately avoided such tasks.
>
> On the other hand I have had several close collaborators. The major ones have been Jim Watson, Sydney Brenner and (more recently) Christof Koch. Others I have had more than transient collaborations with are Aaron Klug, Beatrice Magdoff, Leslie Orgel and Graeme Mitchison. In all these collaborations we have published papers together. These collaborators (except possibly for Magdoff and Mitchison) have each had many pupils of their own.
>
> I think I work best, not entirely by myself, but with one other close collaborator. Sydney Brenner and I shared an office for 20 years. At the moment my close collaborator is Christof Koch, a neuroscientist at Caltech.
>
> Of course I have interacted for most of my scientific life with a very large number of scientists and over the years have given lectures in many different places. Some people have told me that they were strongly influenced by a lecture of mine they heard. I think I must have been rather a good lecturer, because at meetings no one liked to have to lecture after me!

Finally, in my letter of August 8, 2003, I asked Crick several questions. During the summer of 2003, I was preparing a talk "Success in Science" for the Ph.D. students of the Karolinska Institute at their annual retreat in the Stockholm Archipelago in September. During the preparation, I had several questions that I decided to pose to Crick. I need to describe my questions in some detail because Crick would refer to them in making his responses succinct.

1. George Gamow. I have had the impression that the molecular biologists did not quite appreciate his ideas for the genetic code. On the other hand, the backs of the photographs I had received from the University of Colorado

Francis Crick, Alex Rich, George Gamow, James D. Watson, and Melvin Calvin at the Cold Spring Harbor Laboratory (courtesy of J. D. Watson).

on Igor Gamow's behalf indicated Gamow as "the originator of the triplet code". In my conversation with Arno Penzias, when we considered Gamow's place in science history, he told me that Gamow was a better scientist than Galileo.[6]

2. Religion. Gunther Stent told me[7] that he wrote an essay for the 20th anniversary of the double helix in which he presented a linguistic analysis of some of Crick's writings. Stent substituted the word "God" wherever Crick had used the word "nature". According to Stent, the substitution did not change the essential meaning of the text. Privately Crick let him know that he didn't like what Stent implied. I also asked Crick for comment in connection with the Church's recent acceptance of the idea of evolution. I mentioned Wigner's position that physics did not endeavor to explain nature; it only endeavored to explain the regularities in the behavior of objects.

3. When Crick was searching for a research area, he seemed very lucky to have the possibility of consulting always top people, like the Nobel laureate A. V. Hill. I asked him how did this happen.

4. I once again returned to the question about the evaluation of Perutz's contribution to protein crystallography.

5. I asked Crick about J. D. Bernal who could have been included in the 1962 Nobel Prize in Chemistry that went to Perutz and John Kendrew or in the 1964 Nobel Prize in Chemistry that went to Dorothy Hodgkin although nobody protested that Bernal was not included. Bernal was also a pioneer in protein crystallography, if only considering his taking the first ever X-ray pictures of a protein with its mother liquor. My impression was that although Bernal might have had seminal contributions, he did not live up to his own enormous potentials. His communist politics might have also hurt him when recognition (and research support) was concerned.

Francis Crick and James D. Watson in Cold Spring Harbor Laboratory, 1990 (courtesy of J. D. Watson).

6. My next question concerned the importance of C_2 symmetry in describing the double helix structure of DNA. In one of my conversations with Jim Watson, I had suggested an alternative description of the double helix using a more technical language of symmetry than the original description of the Watson–Crick announcement in *Nature*. Jim had dismissed it saying that the knowledge that the structure had C_2 symmetry was not essential. I also asked Crick whether Rosalind Franklin was ever told during the last five years of her life after the double helix discovery that Watson and Crick had had access to her data.

7. Finally, I asked Crick whether he had any comment on the topic of "Success in Science".

Here is Crick's response dated August 29, 2003:

> I am in poor health so I will respond only briefly to the long list of questions in your letter of August 8th.
>
> About Gamow, I liked him very much, especially as he was very kind to two such junior scientists as Jim and me. He did not originate THE Triplet Code — his triplet code was completely wrong. I am not sure that he was the first to introduce the idea of triplets. (Sir Cyril) Hinshelwood had a silly argument for pairs, but it's possible Dounce had earlier suggested triplets.
>
> I would rank Galileo far above Gamow, because he was the first real scientist (with the possible exception of one or two Greeks, such as Archimedes). That is, he both did experiments as well as mathematics (or quantitative thinking) as opposed to thinking in words, as for example Aristotle did. Aristotle made many perceptive observations (not all completely correct, however) but he never did an <u>experiment</u> to test his ideas. When Newton said he was "standing on the shoulders of giants" one of the people he had in mind was Galileo. Galileo's trouble with the Catholic Church has been exaggerated, and was mainly due to the Inquisition, a quite inexcusable institution.
>
> I will not otherwise comment on Gamow's place in modern physics. I certainly think he was original.
>
> Gunther Stent, as usual, has produced an entertaining mixture of sense and nonsense. I will only say that my position is that

I am an agnostic, with strong inclination towards atheism. For Gunther's term "nature" I would prefer "The Entire Universe". I agree with Wigner that a little modesty would not be out of place.

I will not comment on the so-called "religious" views of Einstein and Bohr. Gunther's remarks about Babylon, etc. miss the point, which is that Darwin effectively discredited "The Argument from Design" which before him seemed unanswerable. Enough of my old friend Gunther! Incidentally he wrote an excellent review of the six reviews of Jim's book in the Norton edition of it.

By "The Church" I presume you meant the Catholic Church. All the "religions of the book" (the Bible) differ substantially among themselves. All three have both extensive sects and sub-sects. A recent encyclical by the present Pope said evolution must now be regarded as a fact, though it disapproves of what I do now. But in the U.S.A. millions of, say, Southern Baptists think evolution is quite wrong, that the earth is less than 10,000 years old, etc.

As to A. V. Hill, the MRC and so on, Bob Olby (who is writing my biography) has recently covered this in his draft. You could consult him about it.

About Max Perutz. He was certainly not the first to suggest the method of isomorphic replacement for proteins, but he was the first to make it work and this transformed the field. I don't think he was especially quick in applying it. This is because it is not easy to do. We had taken on Vernon Ingram to develop chemical methods to do this, so it was lucky Max got a supply of sickle cell hemoglobin AND the Hg worked. He did a wonderful job of sticking with hemoglobin till he had proved his theory of its action correct. Also in running the LMB so well and so smoothly. About his and Kendrew's Nobel Prize, I have always suspected that the major influence was Tiselius, but wait and see!

I agree with your comments on Bernal.

About C_2 symmetry and DNA. I enclose a letter I wrote to Mark Bretschler about this. Whatever Jim may say, a few days before he discovered the base-pairs he was still building models with parallel DNA chains and the bases paired like with like.

It was surely obvious (to us and to her) that Rosalind knew all the facts we might be expected to know, since she gave them in her 1951 seminar that Jim attended — though Jim forgot them all, including the C_2 symmetry. The striking picture of the B form which she had put on one side for six months, and which I myself never saw till later, certainly excited Jim and made it easier for him to persuade Bragg to let us build models again.

Much later she told Aaron Klug that one thing she regretted in it all was missing the implications of the C_2 symmetry. Incidentally Klug recently gave an accurate review of it all, and is writing it up for publication. You should always follow Klug about Rosalind. Brenda Maddox's interesting book is not scientifically accurate.

Success in science can take many distinct forms. I think I once said, "It was partly a matter of luck, and partly good judgment, inspiration and persistent application."

Odile and Francis Crick and István Hargittai in the Cricks' home in La Jolla, California, 2004 (photograph by M. Hargittai).

Odile and Francis Crick and Magdolna Hargittai in the Cricks' home in La Jolla, California, 2004 (photograph by I. Hargittai).

Best of luck for your lecture at the Karolinska Retreat. If you would like you could send me your impressions of it, but please don't expect me to reply again.

Then came our meeting on February 8, 2004, in La Jolla.

References and Notes

1. In 2000–2002 we recorded several conversations with Jim Watson. Excerpts from the first one appeared in Hargittai I., *Candid Science II: Conversations with Famous Biomedical Scientists*, edited by M. Hargittai. Imperial College Press, London, 2002, pp. 2–15.
2. Koch, C. *The Quest for Consciousness: A Neurobiological Approach*. Roberts and Co., 2004.
3. Hargittai, I.; Hargittai, M. *In Our Own Image: Personal Symmetry in Discovery*. Kluwer/Plenum, New York, 2000.
4. Crick, F. *What Mad Pursuit: A Personal View of Scientific Discovery*. Basic Books, 1988, pp. 49–51.
5. Hargittai, I. *The Road to Stockholm: Nobel Prizes, Science, and Scientists*. Oxford University Press, 2002.

6. Hargittai, M.; Hargittai, I. *Candid Science IV: Conversations with Famous Physicists.* Imperial College Press, London, 2004, pp. 272–285. Penzias was co-discoverer of the residual heat in the Universe that gave proof for Gamow's Big Bang theory of the origin of the Universe. Penzias became a Nobel laureate whereas Gamow never received this award.
7. Hargittai, B.; Hargittai, I. *Candid Science V: Conversations with Famous Scientists.* Imperial College Press, London, 2005.

Sydney Brenner, 2003 (photograph by I. Hargittai).

2

SYDNEY BRENNER

Sydney Brenner (b. 1927 in Germiston, South Africa; British national) is Distinguished Research Professor at The Salk Institute in La Jolla, California. He studied at and received his first degrees from the University of Witwatersrand in South Africa. He received his D.Phil. degree from Oxford University in 1954. After having spent some time in the United States and South Africa, he returned to England in 1956 where he joined the predecessor of the Medical Research Council (MRC) Laboratory of Molecular Biology (LMB) in Cambridge. He was director of the MRC LMB 1979–1986 and then directed the MRC Molecular Genetics Unit, also in Cambridge, 1986–1991. In 1995, he founded The Molecular Sciences Institute in Berkeley, California, from which he retired in 2000. He has been with the Salk Institute since 2001.

Dr. Brenner was awarded the Nobel Prize in Physiology or Medicine in 2002 together with H. Robert Horvitz and John E. Sulston[1] "for their discoveries concerning genetic regulation of organ development and programmed cell death." He is Fellow of the Royal Society (London, 1965); foreign member of the National Academy of Sciences of the U.S.A. (1977) and of the French Academy of Sciences (1992); and has been a member of many other learned societies. His numerous awards include the Albert Lasker Medical Research Award (1971), the Albert Lasker Award for Special Achievement in Medical Science (2000), the Royal Medal of the Royal Society (1974), the Gairdner Foundation Award (Canada, 1978, 1991), the Krebs Medal of the Federation of European Biochemical Societies (1980), the Kyoto Prize (1990), the King

Faisal International Prize for Science (1992), among many others. We recorded our conversation at King's College in Cambridge on April 22, 2003.*

If you had received a Nobel Prize, say, 30 years ago, as you might have, would you have had the same research career during the past 30 years as you have had?

Nothing would've been different. In fact, to me, this is my second Nobel Prize. I just failed to get the first one.

For what?

For all the molecular biology; for messenger RNA, the Code, but that is what I got the first Lasker Award. So the first Nobel Prize Seymour Benzer and I should have shared for our work in molecular genetics. Seymour has not gotten his prize either.

He received the Crafoord Prize in the biosciences from the Royal Swedish Academy of Sciences.

Yah, yah, but it's not the same.

It is not the same but once someone gets the Crafoord Prize, it's unlikely that the same person would get the Nobel Prize although, of course, it is not officially stipulated so.

Quite.

Ideally there should be a specific discovery for the Nobel Prize, so what would have been the discovery for which you would have received your first Nobel Prize?

The messenger RNA, that was an important discovery, the proof that messenger RNA existed.

I had an exchange of letters with Francis Crick when I asked about his cooperation with you while you were together at the Laboratory of Molecular Biology in Cambridge. My impression was, as I wrote to him,

---

*István Hargittai conducted the interview.

that you are one of those scientists who are more interested in advancing science than themselves and although we should not measure success by Nobel Prizes alone, it was conspicuous that you never received one. Crick said that you have discovered so many things that it would be difficult to select the one for which the prize should be given.

One of the difficulties about the messenger RNA is that we delayed our paper for five months. In the meantime Jacob and Monod produced their review; the discovery was assigned to them; and they received the Nobel Prize. This is my theoretical reconstruction of that story.

Crick said that you were the one who noticed the co-linearity of the gene and the proteins.

We proved that.

The first who cracked the genetic code was Marshall Nirenberg, but his name is often not mentioned as if people tended to forget him.

He got his Nobel Prize. He and Khorana did it by chemical methods. That was the proof of it. We, Francis and I, showed that it was triplet in nature and we showed it by genetics. The prize should be on genetics, not on the code.

In this respect, what was Gamow's contribution?

Gamow defined the problem although Jim and Francis had thought about it and I had thought that it was a one-dimensional sequence that could be translated into a three-dimensional structure.

Gamow wrote his first letter to Jim about what was to become the genetic code soon after the announcement of the double helix, in June 1953.

That's right. However, I went to see Francis in April 1953, before their paper appeared. We were already talking about what came out in their second paper, which appeared in May. We talked about some way to translate the DNA information into the amino acid sequence. What Gamow did was to propose a form of the code. He introduced a kind of terminology with which one could begin to discuss it. In fact, everything what he did was wrong.

What specifically was wrong? That he defined the problem that was correct.

He defined the problem; he took the view that the amino acids were assembled directly on the DNA in what he called the diamond-shaped cavities. That was his physical model, but the big mistake about this was that he did not realize that DNA has a polarity, it has a chemical polarity that reads in one direction. There is only one message because the second strand will be derived from the first by the rules of complementation. Gamow thought that you could read DNA equivalently in either direction. That was one of his degeneracies.

Alan Mackay, who was one of J. D. Bernal's disciples, has told me that Bernal recognized early on, in the 1930s, that the genetic code could not be three-dimensional, it could not even be two-dimensional, it had to be one-dimensional, and it had to have two-fold symmetry.

About DNA?

It was not yet known whether it was DNA or proteins, it was referring to the genetic material. In fact, he thought it was protein.

The question is why did he say it should have two-fold symmetry?

For complementarity, I suppose.

Pauling and Delbrück also wrote about complementarity in 1940, but then forgot about it. The ideas about complementarity in replication were around. Pauling had forgotten about it and Watson probably never read it and I had not either. Neither had Watson read von Neumann's paper on self-reproducing machines, which I did. I read that before I came to England. Anyway, that doesn't matter anymore.

It's interesting.

It's very interesting and a lot of it is discussed in the book by Lily Kay. She was a historian and died very recently of cancer. The book is called *Who Wrote the Book of Life?*

You said something to the effect that it is not enough to say that the future of an organism is written in its genes somehow, we should know how. Has it been solved yet?

It hasn't actually because we can't read it. All we know is this linear script and we have known it for some time and we know that regions of it are translated into amino acid sequence. We also know that some other regions carry information for products, which are themselves nucleic acids. We know about transfer RNA and we know about lots of other small RNAs, which have been discovered, but which we don't know so much about. Then, of course, we know about the ribosome and several other entities in the cell. We also know that in some way the regulation is written there. But we don't have a lexicon and we can't interpret it. So if you would ask, Could we compute organisms from their DNA, the answer is No. Not now.

Von Neumann said that, if you can't compute it, you don't understand it.

Von Neumann said that there were two ways of explaining complex things. One was to explain them in terms of essentially the level above, that is, in a matter of language, in other words. Then he said, certain things were so complex that effectively we could not give an explanation of them and we had to define a prescription for constructing objects that perform the same behavior; in other words, to give an algorithmic explanation of them. He accepted that as a scientific explanation. As an example he quoted pattern recognition. Now, how to explain pattern recognition itself; what you can do is describe the essential features of pattern recognition, to describe an object with this internal structure. Usually now it is a computer program rather than just the solution. So the answer is, I believe, the following. We can describe everything in the Universe today, we have the power to give atom by atom a description of everything, but that's just data, that's just description. What science depends on is taking not the "morest" data but the "leastest" best and predicting the remainder from some other information. In other words, it is the classic technique of science to effectively form a theory from the facts as ascertained and then you can predict it. For explaining such things as behavior of organisms, we could essentially make a description of how an organism behaves under all circumstances, but that is description. The best thing, I believe, is to know what generates the behavior, the machine, the structure, and then we can predict the behavior. Once you have that you have the explanation of it.

This is more ambitious than what Eugene Wigner claimed as the objectives of physics. According to him, the great success of physics is due to a restriction of its objectives. It does not endeavor to explain nature; it only endeavors to explain the regularities in the behavior of objects.

Physics is, of course, very different from biology. Biology is the art of the satisfactory rather than mathematics being the art of the perfect. In biology, if something works, it will survive. In order to try to explain, we do what I call doing it by simulation. But I would like to stress the difference between simulation and imitation. It's very easy to write something that copies and imitates a human being, imitates his behavior. But it's very difficult to write something that generates the same behavior of a human being or, for that matter, a worm. I think that's the task of biology, that is, simulation rather than imitation.

How do you define yourself?

I am a biologist. I do a little theory and a lot of experiments.

Coming back to your having shared an office with Francis Crick for 20 years, wasn't there a psychological barrier for you to overcome in discussing high-level theory with Crick and then going back to the lab and doing meticulous experiments for years?

No, it was not quite like that. It was just an exaggerated description of how we worked. It is also an exaggerated view that one works out the theory and formulates an experiment and then goes to the lab to carry out the experiments. This is the old Popperian idea. The other extreme is what people are now doing working on experiments, collecting natural facts. It doesn't work like that. You start by collecting facts, do a little bit of theory, then you come back and collect more and so on. It's a play between two things, how history of science views this. What I think the most important thing to do in the lab is to interpret the experiment correctly that does not work. Many people today have no capacity, many of the younger people, to analyze why an experiment did not work.

Is it the lack of patience?

I don't think it is patience; they just have no way of doing this, rather, they go and get another set and try again. Understanding why experiments don't work is just as important as understanding why they do work. One often learns from failed experiments.

Working with Hinshelwood, did you have a training in physical chemistry?

I had no training in physical chemistry. I worked with Hinshelwood, but I worked in biology; I was not trained in physical chemistry.

You can now see that this can make it essentially concentration-independent because you have only half of the amount of protein or you introduce a change in the induction period, but the output, provided it's not saturated, remains the same. This system is very much buffered against fluctuations in the concentrations of the actual molecules participating, which does this essentially by counting. Once I just sketched the data in, but now there is an enormous simplification because we can simply say, this transforms into this and this transforms into that, and we have many such things. Sometimes we can convert the front levels of concentrations into frequencies of pulses. It might be easier for a cell rather than measuring ten per cent change in a concentration, which, as you know, is very difficult to make a discrimination like this, to measure the difference between ten and eleven pulses in a fixed amount of time. Understanding the nature of signaling inside the cells immediately gives you a means of simplifying it. That's the answer to not to be defeated by complexity. A lot of the complexity is something we can't think about. We can't think about twenty thousand things going one and the same time, but the cell has the means of making this difference.

I've written a paper on this and I once explained it to Gödel, about the act of center of enzymes, the fact that enzymes and products can coexist in the same cell in solution. In other words, you don't have to send the product through a pipe to the next enzyme at the scale of bacterium collisions. Of course, collisions are highly frequent events, so that basically at most of the time a molecule is hitting the wrong protein and most of the time it hits the right one it hits at the wrong place. But all other things are ignored and the system has not to worry about anything. When I told about this to Gödel, he looked at me and he said, "That is the end of vitalism." It was an interesting remark.

It was at a late stage ...

... of his life.

Not only that but at a late stage in the history of vitalism.

Of history, yes, but the fact is that you could not grasp all this order; this order was obtained by using a trick, by using molecular collisions.

It may be a big jump, but what is your relationship to religion?

I am an atheist and I just have no use about religion. I don't mind it, I just don't believe in it basically.

According to Gunther Stent, although Francis Crick denies to be religious, if you substitute the word nature by God in his writings, you get a religious description of things.

Maybe he uses the word nature for what other people may use God, but Francis is against the supernatural. I don't use it. My story is that I was phoned by the Gallup Poll and they asked me a lot of questions, and then they asked me, What's your religion? I said I'm an atheist. There was a long pause and finally the man said, I can't seem to find it amongst these religions. So I had to say that I had no religion. That's my view.

I've read that in your youth you used to earn your living as a tenth man in the Synagogue.

I was taught by being simply forced to go to the Synagogue, to Hebrew school, but I quit when I was about five or even four. I had to go through a very rough part of the town and I was terrified. There were always gangs of Afrikaner and blacks, the usual things, and they decided to beat me up. As I stood there I said, "Shema Yisrael, Adonai Elohenu, Adonai Ehad," but nothing came. I got beaten up, nobody helped me, and I said, forget it. That sort of things stuck in my mind. To me it was just a lot of nonsense basically.

But religion is a very powerful concept.

Sure. You can explain everything with it.

Many scientists claim that they are religious.

Young Sydney Brenner (courtesy of the Laboratory of Molecular Biology, Cambridge, England).

There's a funny kind of thing, which I find very practical about Christian religions. They say there is no reward on Earth, wait till you get to Heaven, and you've got to do all these things, and, of course, there is the punishment also.

But can this be an incentive for scientists?

It's neither an incentive for nor an incentive against. I'm not trying to disprove the existence or non-existence of God; it's not my problem. It may be someone else's problem. For me, I don't have to take it into the compass of what I'm thinking about.

Aaron Klug was also beaten up as a child in South Africa.

Probably. But he stayed religious or, rather, he went back and became more religious in his older age.

Is he more part of the Establishment than you are?

Now, yes. Before, neither of us were.

At one time you were director of the MRC Laboratory of Molecular Biology in Cambridge, but then did not continue.

I didn't want to. I decided I really wanted to get back to the lab. I was tired of arguing with my colleagues.

You seem to have always found it important to coin names. Was it a conscious effort?

I'm very interested in words; that's something that I do.

A few names come to mind, such as molecular genetics, adaptor hypothesis, messenger RNA, codon.

A good name can carry a lot of message. Most of these names have just evolved. I thought, let's be a little bit sophisticated about it. I have introduced another word, instantiation. It's a tough word, meaning an example. A gene becomes instantiated because what we call a gene is now no longer one gene; it can have many different modes of expression, and genes carry much more information than just the amino acid sequence. So we talk about a gene being instantiated in five different ways; they are five different instantiations. That encompasses when it is expressed, in which cells it is expressed at, and where it goes in the cell. Genes carry little addresses with them. Once you realize that all this has been encoded, we better go and find out what all that is, because that's the key to putting this into a computational form.

I would like to ask you about your relationship with others. I have received a photograph from LMB in which Fred Sanger is being congratulated on the occasion of his second Nobel Prize. You are there putting on not a very friendly face.

I was making a speech, giving him something, I don't remember what. But I had a very good relationship with Fred. He is a marvelous person. He's a different kind of scientist from everybody else. He was a good friend.

How about Max Perutz?

I would say it was a reasonable relationship. Max had two levels. There was Max Perutz the scientist and there was the Archduke Maximilian Ferdinand from old Vienna expecting a lot of things done for him. I would not say our relationship was negative. When I took over the finances of LMB in 1977, I had to do a lot of repairing because Max had a vision of the

Frederick Sanger and Sydney Brenner at an impromptu celebration of Sanger's second Nobel Prize, 1980, in the canteen of the Laboratory of Molecular Biology (courtesy of the Laboratory of Molecular Biology, Cambridge, England).

Lab, which may not have been entirely corresponding to reality. I would not consider him a mentor; he was then parallel. The person I worked a lot with was John Kendrew and, of course, Francis. They were my biggest connections in the Lab.

Do you have heroes?

I think Francis is a hero for me, and Fred is another kind of hero. Fred is a hero for just going about doing things. He is a craftsman of science. Francis has a more brilliant mind, but Fred has the craft of doing things. He is a good chemist; that's what Fred is, a superb chemist.

What will be your legacy? Of course, it is early to ask but we don't meet every day.

I just wrote my biography for the Nobel site and I think my legacy is the people that have worked with me. It's a kind of living legacy. It corresponds much to the idea of a Jew. You gain immortality through your children, which is an old Judaic idea, and my legacy will be through the people that have worked for me.

Do you write about your children in your Nobel biography?

Not very much. I'm afraid my family had to endure my scientific life. I married only once; we have three children and a stepson.

You write beautifully about Szilard. He could not be a very easy person.

I met him in America in 1954. I always took him for a schemer, not a schemer in a conspiratorial sense, but he always had schemes for doing things. I used to have long discussions with him about various schemes, about reforming the scientific literature and so on. I liked him a lot. I also thought his schemes of dieting were ridiculous. He came to my house. When he came to a Pugwash [Conference on Science and World Affairs] meeting here in Cambridge, he was put up in a rooming house with a working class family. When the man came home from work and found out that Szilard was Hungarian, he wanted him to leave because he didn't want any communist in his house. The man assumed that anybody foreign must be a communist. So Szilard came to stay with us and we put him up. He came down in the morning for breakfast; he had his little pills for his coffee, his sugar substitute, he then proceeded — there was a sugar bowl on the table — and quite unconsciously he took a teaspoon and he began to eat the sugar from the bowl, spoon after spoon. My children, who were never allowed to do this, very rapidly started to do this themselves. I just watched this because I knew that Szilard was trying to reduce his weight. Szilard often would go to a restaurant and would pick out everything very carefully for eating low calorie dishes, going to the kitchen and explaining to the cooks. Then he would leave the place but if it had a confectionary, he would buy a dozen éclairs to eat. Once we were in Washington and I, after having watched him doing this, asked him, Why did you go through this performance? His answer was, I like it.

Szilard was a great man. He had all these wonderful ideas. But he never had any possessions. Never kept anything. When he finished a book, he threw it out of the window. Once he was talking to Mel Cohen in Paris; I was sitting there. Szilard had a notebook and he was writing down every single word Mel Cohen said, and he came to the end of the notebook and he threw it into the rubbish bin. He took out a second notebook and went on writing into it. He was a marvelous sort of character.

* * * * *

On April 26, 2003, Sydney Brenner gave the closing talk at the one-day program of lectures at the MRC Laboratory of Molecular Biology devoted to the 50th anniversary of the discovery of the double helix. Various alumni talked about their work since they had left the LMB or reminisced about their days at the LMB. Below I summarize Brenner's remarks in my words in sketchy notes. He started by noting that he did not prepare a Powerpoint presentation because "one good phrase is worth a thousand power points."

LMB was not only ideas and experiments; it was also developing the necessary tools to carry out research, such as X-ray diffraction apparatus, apparatus for electron crystallography, the experimental set-up for sequencing, etc. People found collaborative efforts very useful but they had to form collaboration themselves; there was no institutionalized effort to this end. Francis Crick said, "You can't make people collaborate but surely you can stop them."

There were two DNA revolutions. One was the discovery of the double helix structure because it explained the function of DNA. The second revolution was DNA cloning and sequencing. It revolutionized not only genetics but also biochemistry. People have learned to say, don't worry,

Present and past luminaries of the Laboratory of Molecular Biology: from left to right, Max Perutz; John Kendrew; Aaron Klug; César Milstein; James D. Watson; Frederick Sanger; Hugh Huxley; and Sydney Brenner (courtesy of the Laboratory of Molecular Biology, Cambridge, England).

if there is a problem, there'll be an enzyme to solve it. We are no longer at the mercy of laboratory animals; we can make ourselves the proteins we need. Solving protein structures used to take five years, nowadays it takes five hours.

What will be the next question we should attempt to solve? It will be to answer the question, How do the genes build up the organisms?

DNA technology is unique in that it has a linear sequence of information, leading to the three-dimensional structures of proteins.

Yesterday Jim Watson made some remarks [at the conclusion of the Symposium commemorating the discovery of the double helix]. I agree with some and disagree with some others. I agree that the thing to do is to get into a new research area from the very beginning (of course, you have to know when is the beginning and not get into it before). I disagree, however, with the notion that we'll have to have only big labs in the future. Some of those big labs, like CERN, are just like factories. In a big lab, the members may be working on details without knowing

Sydney Brenner, Richard Roberts, and Frederick Sanger, Cambridge, England, 2003 (photograph by I. Hargittai).

Sydney Brenner and István Hargittai at King's College, Cambridge, England, 2003 (photograph by M. Hargittai).

what the whole problem is. In physics it is conceivable, in biology it is not.

One of LMB's secrets was that it recognized the power of computers early on. I remember a story when I was not yet director of LMB but was already handling its finances. I was lying in hospital with a broken leg and Fred Sanger came for a visit. He asked for money to buy a disk drive and explained to me why he needed it. Of course, I agreed. The computer saved us in handling complex information.

The question is often asked whether after the Human Genome, will our life be the same? Yes, it will be because the Human Genome is only a telephone book; but try to find out from a city telephone directory how the city works. There is need for a theory. It is the wrong approach to collect the data, stuff them into the computer, and wait for the *emerging* results. This is not artificial intelligence, rather, it is artificial stupidity. We have to have a theory.

Before molecular biology, chemistry dealt with matter and energy. Molecular biology has brought information into the picture; it made information also into a chemical problem. Sequence is information and it is chemical information (it may also be considered a low energy physics problem). DNA gave a framework to think about information.

The next level of challenge is organization, to understand how inventory is organized, we should find out about the organization of the cells, how they work and interact, etc., and we have to find out how the genome maps human behavior. But we should still be concerned with causation. We should ask questions like: What causes things? What is the chain of causation? Knowing causation would simplify representation.

Biology is different from many other fields because in biology we have the possibility to interfere whereas we can't change the weather or the origin of the Universe. At the same time biology should remain a predictive science and this is why we need to worry about causality.

We have to continue collecting data; we have to collect a lot of data, but, remember, when you are collecting data, you are collecting a lot of noise too. Nonetheless, we need to get back from the hangover of the Human Genome to experimentation.

Reference

1. An interview with John Sulston appeared in Hargittai, B.; Hargittai, I. *Candid Science V: Conversations with Famous Scientists*. Imperial College Press, London, 2005, pp. 528–549.

Matthew Meselson, 2004 (photograph by I. Hargittai).

3

MATTHEW MESELSON

Matthew Meselson (b. 1930 in Denver, Colorado) is the Thomas Dudley Cabot Professor of the Natural Sciences in the Department of Molecular and Cellular Biology of Harvard University in Cambridge, Massachusetts. He studied chemistry at the University of Chicago and did his graduate work with Linus Pauling at the California Institute of Technology. While Meselson was still a graduate student, he and Franklin Stahl (b. 1929 in Boston and currently a Professor of Molecular Biology at the University of Oregon in Eugene) devised an experiment in which they provided proof for the semi-conservative replication of DNA in 1957, published in 1958. Their experiment was a new technique — invented by them — using centrifugal force to separate molecules based on their densities. It is called the density gradient centrifugation. Their joint work is described in the late Frederick L. Holmes's book *Meselson, Stahl, and the Replication of DNA: A History of "The Most Beautiful Experiment in Biology"* (Yale University Press, 2001).

Matthew Meselson completed his Ph.D. dissertation at Caltech in 1957 and stayed on as Assistant Professor. In 1960, he moved to Harvard University as Assistant Professor and has stayed at Harvard ever since. He is a member of the National Academy of Sciences of the U.S.A. (1968), of the American Academy of Arts and Sciences (1962), and of the American Philosophical Society (1981).

In addition to his scientific research, Dr. Meselson has been concerned about the use of chemical and biological weapons since 1963. He has been very active in writing and consulting on this topic as well as in directing various studies related to the protection from and elimination of chemical and biological weapons. We recorded our conversation with Matthew Meselson in his summer home in Woods Hole, Massachusetts, on August 8, 2004.*

A few days before we held this conversation, Francis Crick died, so he came up at the beginning of our recording.

The last time I met with Francis Crick was the day when they captured Saddam Hussein. As usual, Francis was in charge of the conversation. I brought him cookies because Francis loves cookies. He was a very kind person to me. Although it was not quite often, but whenever I would do a nice experiment, I'd get a little friendly note from Francis. I loved talking with him. He was the intellectual leader of solving the problems that were posed by the DNA molecule. Unlike most molecules, the double helix structure of this molecule told you everything. It told you what the problems were. When you look at other molecules, even using the best computational possibilities of today, their structures don't tell you what these molecules do. You look at DNA and you see it right away how it is going to duplicate, how it keeps information, and how it's going to mutate. The DNA structure left a lot of problems open and Francis was the leader of solving them. They included the transfer RNA idea, the solution of the code, and the brilliant experiment he and Sydney [Brenner] did. When I teach I always say that the high point of molecular biology is that one experiment. It showed that the code is a triplet code and it defined the reading frame.

How would you describe George Gamow's role in the code problem?

Gamow did very little. Gamow, being a physicist, saw that it must be information in a four-letter code. I still remember listening to Gamow in the summer of 1954. He always wore a white suit and drove a white convertible car, and usually had a lot of whiskey. He tried to build a model in which the amino acids would fit into the grooves of the DNA molecule. The diamond-shaped grooves had sides which are in common. That means that the base pair that forms one side of a diamond-shaped cavity also forms another side of the previous one. Then he imposed constraints of which amino acids can follow which ones. Eventually, it was understood

---

*István Hargittai conducted the interview.

that there were no restrictions. Any amino acid could follow any other amino acid. Gamow's whole idea was just not very biological. Nonetheless, Gamow had the ability of creating ebullience, that summer anyway and later too. He was a lovely fellow and lovely to talk with.

Whose idea was it that DNA would code for proteins? Today it's trivial, but somebody must have thought of it first.

It was so obvious. Once you see that molecule, the double helix, you know that the sequence of the nucleotides is information. It must be. If the storage of genetic information is there, the sequence of the bases is the only thing there that is not perfectly regular. The backbone is just monotonous. I don't think it is a concept that takes somebody to have. It's totally obvious. The same applies to replication. There are two chains and it's obvious that they come apart and make templates for new chains.

But then your experiments were needed nevertheless to convince people.

You used the right words, to convince people. Late that summer of 1954, there were two kinds of people. There were those who said it was too simple to be true and those who said it was too simple to be wrong. For a lot of people biology seemed to be extraordinarily complex, with all kinds of different interactions, different proteins and colloids. If you had that mindset, you might say that DNA as information storage for information, as self-replicating entity, might seem too simple. There were a lot of people like that. When Frank [Stahl] and I did this simple experiment, it had an educational value, like you said.

There is a big book about your experiment, Meselson, Stahl, and the Replication of DNA: A History of "The Most Beautiful Experiment in Biology" *[Yale University Press 2001].*

Fred Holmes double and triple checked everything in our laboratory, read our notebooks, went back and forth to Francis and Jim [Watson] to see if he could shed any light or correct any memories. Finally, he knew more about it than Frank and I. Unfortunately, he died two years ago. Coming back to Francis, he defined the problems and pointed the way. Of course, at that time the problems were almost linear in their sequence. Each answer led to the next question. Some of Francis's brilliant ideas were never published. The idea of adaptor molecules was one. His ideas just became part of the lore. I use them when I teach. Now, it's a network of questions and answers and it's not clear that there is any one road ahead.

Did you find your experiment so beautiful when you were actually working on it?

It was beautiful when we saw it.

For me it seems that its importance was in showing that replication of DNA was semi-conservative and it also gave a new technique. How did it come about?

I was a student of Linus Pauling and all of his students took his course "The Nature of the Chemical Bond". At one point he told us how from the quantum mechanical zero-point energy we could calculate the strength of a hydrogen bond. Even better, he told us how we could calculate the ratio of the strength of a deuterium bond to a hydrogen bond. It was a two-line proof and I found it astonishing. I got interested in deuterium. I wanted to work with Linus not only because of who he was, but also because I wanted to get into biology. I just didn't know how. I thought the best way was to study molecular structure. I hoped that this architectural chemistry would point my way into biology. I never liked organic chemistry, but I liked physical chemistry and I liked physics. At that time we didn't know anything about DNA. As I became interested in deuterium, I began reading the literature. I read about a man at Columbia University who was feeding a mouse with deuterated water trying to create a deuterated mouse. The mouse didn't survive because there was some heavy-metal contamination in the water.

Otherwise, heavy water should be all right if expensive to feed the mouse.

It would be a heavy mouse and they would need a special mouse trap for it. Anyway, I was thinking about it when Jacques Monod came to Caltech and gave a lecture. In those days people were asking, when you induce beta-galactosidase, is it a *de novo* synthesis of a new enzyme or is it that the enzyme is sitting there, inactive, and you twig it, maybe methylate it or acetylate it, do something to it, and suddenly it became an active enzyme? The experiment itself detects only the activity of the enzyme. Monod said that maybe this could be determined by measuring the osmotic pressure at equilibrium. I'm sitting there, a young student, and I think this is crazy. There are so many variables that may impact the permeability of the membrane. I thought it was a terrible idea, but it got me thinking.

As I was thinking about deuterium — and don't ask me why it worked this way — it seemed to me that the way to do the experiment was to grow the bacteria in D_2O. Then you centrifuge it down, re-suspend it in ordinary water and at the same time induce beta-galactosidase with IPTG [isopropyl beta-D-thiogalactoside]. Then you look to see if the beta-galactosidase activity will either sink or float in a solution whose density is adjusted to be just in between the calculated density of deuterated or hydrogen beta-galactosidase, and you do this in a centrifuge. It's a wrong idea but it's like a good idea. I went to Linus and told him that we could do this experiment. He said it was good but told me to finish my crystal structure first.

Then I came to Cold Spring Harbor as a laboratory assistant for Jim. He wanted to solve the structure of RNA. It was in 1954. After the double helix, Jim spent a whole year at Caltech. He worked with Max Delbrück. He lived in the Atheneum and I lived in the Atheneum. We talked occasionally and I got to know him. I told him about this experiment and he said, don't do it here, do it in Sweden. He said there were two reasons for this. One, because the centrifuges were in Sweden. The other was that there were no girls at Caltech, but there were lots of girls in Sweden. When I came to Cold Spring Harbor, I was doing experiments for Jim. It was to titrate RNA, which Alex Rich has made, supposedly very carefully, without denaturing it.

In 1947, a great British physical chemist, Galant — I think he was great — titrated DNA to a quite strong acid. He observed that at pH 2 it became a very good buffer. If you back-titrated it with an alkaline, it didn't back-titrate with the same curve. Instead, it released a lot of groups that could ionize. The same thing happened, if you took it to a very high pH, say, pH 10. It doesn't matter whether you had exposed it to a strong base or a strong acid, from then on, it titrates along the same curve. What it means is that there must be a lot of amino groups and hydroxo groups, which are blocked, but when you go to a high enough acid or a high enough base, they become available. He concluded that there must be hydrogen bonds forming a systematic network and until you get to a pH where they dissociate, they're not available for titration. After you hit that pH, the whole molecule reorganizes. Then he studied the viscosity and when the amino groups and the hydroxo groups become available for titration, the viscosity collapses. He concluded that there was some kind of a structure, either intramolecular or between molecules, maybe even a micelle.

This was a very simple way also to ask if RNA would have such a structure. So I was to do this experiment, which I did, and it showed that RNA was not like DNA. This was in Cold Spring Harbor. Francis Crick was here and Sydney Brenner was here. I was in the red brick building and Jim was looking out of the window and he saw a guy sitting on the grass by a tree and he says, see, that guy, that's Frank Stahl. He thinks he's pretty good. Let's give him the Hershey–Chase experiment to do it all by himself and see if he's really good. So I went to talk to him. He was sitting there with a big bottle of gin and several bottles of tonic and he was selling gin and tonic. He didn't have a lot of money so this is what he was doing. He was sitting under a tree, which we now call the gin-and-tonic tree. As he was selling gin and tonic, he was also trying to work out a problem in phage radiation genetics. It involved some integrals, and he didn't really know how to solve the problem and I didn't know anything about phages, but I knew something about calculus. So I solved the equations for Frank and he educated me a little bit about bacteria phages and bacteria. Now he probably knows more about mathematics than I do. We got to talking and I told him about the experiment I wanted to do and asked him if he would join me? He said yes. At that time he was a student at Rochester, but the next year he was going to come to Caltech. When he came, sooner or later we rented a house together. I didn't want to finish my X-ray crystallography and I could hardly wait to start my experiment with Frank. But they kept telling me that it would be bad for my character not to finish my X-ray crystallography.

Who said that?

Frank as well as Linus. Finally, I finished it and we started the experiment. We started it by estimating the density of DNA and finding something in which it would float. We went to the periodic chart and, of course, you couldn't use a divalent or a trivalent salt because they make a precipitate with DNA. We had to use a monovalent salt. First we made a saturated solution of rubidium chloride which the Caltech stockroom had. It was no good. We had estimated the density of DNA to be about 1.7. So we ordered cesium chloride. It was a byproduct of atomic bomb production. We were first doing this not with DNA but with T4 phages. It was a very bad way of doing it because it'd all fall apart. There would be all kinds of intermediates, and we didn't know that a gradient would form. We were using a preparative ultracentrifuge in Renato Dulbecco's laboratory. After each run we would stop the ultracentrifuge, poke a hole in the plastic tube, take out drops and assay the phages to see if we could make them go up or go down, what the right density would be. It was a mess. The phages would die and when we were making optical density measurements, there were bands all over the place. Caltech had an ultracentrifuge, but it was an air turbine and people said, you could kill yourself. It might blow up. Jerry Vinograd, who was in the Chemistry Department had one of the very first ultracentrifuges built by the Beckman company. We asked Jerry whether we could use his ultracentrifuge because there we could see what was happening. It had quartz windows in the cell and as the rotor goes around, we could take pictures. With ultraviolet light, we could see where the DNA was. We put together cesium chloride solution and the phages into the cell and turned on the ultracentrifuge. There is a device called the Schlieren, which measures the refractive index. As we turned on the centrifuge, the Schlieren line kept tilting, which indicated for us that the cesium was redistributing. I'd read the big book by The Svedberg, but I didn't know that The was a Swedish first name.

It stands for Theodor.

I didn't know that at the time but as we measured the sedimentation velocity in Svedberg units, I thought that the book was about how to measure sedimentation velocity. In that book, there was a chapter by Anderson on the measurement of molecular weights of simple salts by equilibrium density gradient centrifugation. It was in German and my German was very bad. In it he said that you had to centrifuge for seven *Stunde* before reaching equilibrium. My intuition told me that cesium would take maybe weeks, maybe months to reach equilibrium. I saw in the dictionary that *Stunde* meant hours but I thought that it was a colloquial expression of longer time periods. I was totally unbelieving that you could make little molecules do something in seven hours. I asked Max Delbrück and he said that it was hours and I asked him whether it might have any colloquial meaning and he said he didn't know. Anyway, pretty soon we figured out that that was what was happening. A gradient was forming. Then Frank and I embarked on quite a large and time-consuming effort to calculate what would be the bandwidth of the bands. The question was, would we have enough resolution to distinguish heavy phages from hybrid phages from light phages. We calculated that we would have enough resolution if we used 5-uracil instead of deuterium. But 5-uracil you can't substitute uniformly, so you would get broad bands and we still had many bands. We realized that we were getting nowhere. Finally we decided to switch from phages to bacteria and from heavy hydrogen to something that just appeared on the market or it was the first time that we learned of it, nitrogen-15. That experiment worked the first time out. But I mislabeled the tubes apparently. We succeeded in experiment two, but we re-numbered our experiments because we could never figure out what we had done in experiment one. So experiment two became experiment one. Then we did experiment three and we renumbered it to experiment two. This is what we published, experiments one and two although they were really two and three. It worked like a charm. That's the story.

Quite a story.

I have a crazy memory that it was New Year's Eve, but in reality it was in October. In my memory there was a New Year's Eve party right across the street and I went there and told everyone about this result. Maybe it's an imagination.

Was there anybody who might have remembered you coming over and telling about the experiment?

I never tried to find out. I don't know. They might have all been quite drunk. If there was a party, they wouldn't remember.

Wasn't Frank around?

Frank was interviewing for a job in Missouri.

Measuring densities by buoyancy for isotopic ratios, Michael Polanyi's experiment comes to mind.

Sure. I know the son better than I knew the father. I remember Michael Polanyi's buoyancy experiment dimly. I never read about it, only heard about it.

It was in the early 1930s, but it had no effect on you.

No. It had no effect on me.

Jacob (Yakov) Varshavsky and Matthew Meselson at the Fifth International Congress of Biochemistry, Moscow, 1961 (courtesy of Alex Varshavsky).

Changing topics, as I understand you have been interested in bioterrorism since 1963 and that Linus Pauling taught you that first you should establish yourself as a scientist and only then turn to such questions.

That's right. What happened was — and I only learned about this later — there was this new organization, the General Advisory Committee of the United States Arms Control and Disarmament Agency. They had too much money in their first year and they didn't want to give it back because in the next year maybe they would not be given as much money by the office of the budget. They were a brand new agency and they were expanding so it was wise of them not to have a surplus. They got the idea to invite six academics for the summer and pay them. My officemate was Freeman Dyson, which was wonderful for me because he encouraged me to trust my intuition.

They said that we should work on the European nuclear theater — nuclear weapons arms control. For about a week or two I tried that, but it became quickly obvious that I didn't know anything about that, which is OK, but a lot of other people, like Henry Kissinger, had been experts in this.

Kissinger had written a book about the necessity of choice. So I asked my boss if instead I could work on biological and chemical arms control. My boss was Frank Lawn, a wonderful chemist, a professor at Cornell University in Ithaca. He said, sure, I could work on them. He also told me that there was a man before we came who worked on biological weapons arms control, but he became so depressed that he killed himself. Lawn offered me his desk. He also said that they were all going to Moscow to negotiate the nuclear test ban treaty so he gave me complete freedom to do what I wanted.

I decided to start with biological weapons because chemical and biological weapons were too big a subject. I went to the CIA to see what other countries were doing at that time. The answer was that we don't know much. It's probably still true. I also went to find out what we were doing. We had a big, offensive biological weapons program. A very nice man named Leroy Fathergo was my guide. I had to get all kinds of security clearances. I asked him why we were doing this, working on biological weapons. He told me that it was much cheaper than all kinds of nuclear weapons. I was thinking about what he told me. It puzzled me because we had all those nuclear weapons. Why would we want to pioneer weapons of mass destruction so cheap that everybody could have it? I found this crazy. We should be absolutely opposed to biological weapons. We should be the last one to make such weapons. If we set an example, everybody else is going to make them. Rather, we should try to make them illegal; we should get rid of them.

First I didn't say much in public. I didn't write because it was better if nobody talked about such things. You don't want to get people interested in it. But then the Vietnam War came and people were accusing us with chemical warfare in Vietnam. There was more talk about it. There were some accidents of various kinds that came to public notice. I became more public about it. Then President Nixon was elected and he chose Henry Kissinger to be his National Security Advisor; I knew Henry because his office building was next to the Biological Laboratories at Harvard University. I went to his arms control seminar and I would also see him occasionally at lunch. For some reason we did become friendly. We were both going through a divorce at the same time. For us both, Harvard was an emotionally frigid environment. I remember when Henry had come back from his second trip to Vietnam — he was not yet in government, it was under President Johnson, and he went there at the request of the American Ambassador,

Cabot Lodge — he was very tired, we were sitting on the sofa in his office and had some sherry. He said, now I know how the good Germans felt. This really touched my heart. I thought that the Vietnam War was a mistake and he wouldn't say that.

Did he mean by good Germans those who were against Hitler?

He meant by good Germans who saw what was going on. It didn't occur to me until you just said it that it could have been his implication. But I think he meant only the Germans who saw that their country was doing something wrong.

They saw and did nothing.

I don't know how to answer that. He clearly wasn't very fond of the Vietnam War. That was the implication I drew from it. After Henry became the National Security Advisor, I bumped into him literally on the ramp at the airport in Boston because I was going up and he was going down or the other way around. We collided. He knew that I'd been interested for several years in biological weapons and that I thought that the United States should stop its program. He said, Matt, what should we do about your thing? And I said that I should write him a paper. So I wrote him some papers about how it was counterproductive for a rich country like the United States to pioneer something so dangerous and cheap and readily made. Maybe they are not so readily made but much easier than nuclear weapons. That was the argument I made and that the United States should ratify the Geneva Protocol, which it had not done. Many years later we finally did, which is a no first use agreement about not using chemical and biological weapons. I wrote some more papers for him. Now I feel that there is not much that can be done. Things have gone crazy.

Somewhere you are saying that if intent is recognized, the perpetrators should be destroyed. Would you sanction a preventive war in such a scenario?

No.

So what can be done?

We can do a number of things. First of all you better be sure that you know what you are talking about. In the case of Iraq, we made a big mistake. Their weapons program had been stopped in 1991. A lot of people understood that, but unfortunately, not enough people. To have such a doctrine, it would be misused. You can't start wars over things like that. We have to go far back in such a discussion. Where shall I begin?

First you have to create an understanding in the world that biological weapons are a threat to all humanity. To those who make them, to those who use them, and to those against whom they are used. Biological weapons are enemies of all humanity. That's recognized in the biological weapons convention, which prohibits its development and manufacture.

The terrorists of today don't care about their own safety or anybody else's.

You entered a huge subject.

I'll try to be more specific.

We need to find out whether these terrorists are still led by a leadership. It may be too late. It may be that we delayed too long that there is no leadership anymore. But if there is still a leadership, bin Laden for example, or someone else, still in charge, we should try to make a deal with them. That's the way you deal with terrorists. Part of dealing with them is that if they behave properly, we will do certain things, and they will sell their followers down the river. This is the way you deal with terrorists. There is no way to keep these terrorists from killing an American businessman once a month or an American student once a month. To say that what they're up for is to destroy freedom is stupid. The last war they were running was against the Soviet Union and the Soviet Union was not a bastion of freedom. It was a miserable dictatorship. These guys are not out to destroy freedom. Bin Laden has written what he wants. You could say that what he wants is wrong. That's different. Maybe we shouldn't change our policy. But what's motivating at least their leadership is policy. They are against our policy of tilting too strongly to Israel, the policy of being present where they have their holy places in Mecca and Medina in Saudi Arabia. He's written this again and again. The third policy that he doesn't like is our supporting corrupt and dictatorial Arab governments. If I were in charge, but it's probably too late, I would send out some feelers through a third party or a fourth party or a fifth party, so that nobody could trace it. The feelers could say that maybe we could reduce our presence in Saudi Arabia, but we want to see some things on your side too. At the end of a long process if what they really want is to share some power somewhere, they could do that. That's what the American revolutionaries did and that's what at least the more sane Irish terrorists do, they want to share power. They don't want to be terrorists forever. The real danger is that once they create their terrorist organizations, they trap all kinds of lunatics who enjoy just killing and destruction. Once that happens it is too late to make a deal because these guys are on the loose. It may already be that it's too late. Does it sound too radical to you?

It does.

It's the way the British used to be too. You buy them off.

What I read by you, it didn't impress me that you were a radical person.

I'm not a radical person. I'm very worried about the present president because he takes big risks. He gambles. It's too dangerous.

Do you believe in strong defense?

I believe strongly in a good defense. I certainly advised our Defense Department on defense matters, how to defend against chemical and biological weapons. The problem is that it's very hard to defend against terrorists. What I've done though is together with a friend, we've written a treaty. There are treaties that create universal jurisdiction for certain crimes, namely, airline hijacking, torture, and so on. They do set up special courts. The problem with the Milosevic thing is that it takes so many months and even years to create a new court with new rules and new justices whereas we already have a legal system in this country and almost all other countries. The treaties I mentioned give jurisdiction to national courts for certain crimes.

For example, a man named Yunis takes an airplane from Beirut. He wants to go to Tunisia, but the Tunisians don't let him land there. So he has to go back to Lebanon and he lets all the people out and he destroys the airplane. He blows it up and disappears. The FBI tries to find him and they find him. They make a phony drug deal offer to him. He thinks it's real. They get him to come on to a boat in the harbor in Beirut. It's very foolish for him because the boat goes to an American destroyer, which takes him to Bari, Italy. They fly him to Washington and now he is in court. His attorney says to the judge that this court has no jurisdiction over my client. He's not an American, he committed no crime on American soil, and he didn't hurt any American property or American citizen. But the judge explains that the American court does have jurisdiction because the United States is a party to the 1971 airline sabotage treaty. So Yunis is in jail because of this treaty.

There was then the Pinochet case.

That was another example. How could've such a man be brought to a British court when he is not a British subject and tortured no Brits? We need that, and it has other aspects to it. It sets up cooperation between police departments in different countries. Today, I imagine that if someone is making biological weapons in, say, Argentina, it might not be clear whom do you call in Argentina. This treaty that we've written, we require every state party to designate an officer whom you contact in such a case. The British government publicly declared their approval of our suggestion although they haven't done anything yet. So such a treaty would tighten things up. Then we would have to get rid — as much as we can — of secrecy. We have to demand that each state tells where they're doing bio-defense work. Even if they wouldn't tell us what they are doing they should tell us at least where. That would be that if we found that something was going on elsewhere, that might be suspicious. There is no reason why we shouldn't tell where. That should not be secret. Of course, this is no perfect solution.

Jumping again in topics, I would like to bring up the speech Leo Szilard gave in 1954 about what he called the sensitive minority in science.

I have a long story about that. He gave that speech at Caltech at the Ambassador Hotel. He was visiting Max Delbrück first. Leo used to rehearse his speeches to an empty room with maybe a friend or two. In addition, he didn't know how to drive an automobile. So I was his chauffeur and I was also the kid who was sitting in the room and listening to this wonderful talk.

Now I have this question. He mentions that when he found out that there was this possibility of building the hydrogen bomb, and Edward Teller was working on it…

Don't even tell, because if the communists find out they'll accuse Teller of being a communist — they will have a way of doing this — and he will not be able to work on it anymore. It's just like Szilard to figure out something like that.

But this tells me that Szilard was concerned that something might have disturbed the development of the hydrogen bomb.

That I don't know.

That's how I read this.

You could read it that way but you could read it also to mean that he was trying to give people a joke. At that time we had McCarthy and Szilard may have been just trying to show how crazy McCarthy had become.

When he said that he was terrified that only one person was working on the hydrogen bomb at the time…

That's right. It was to a White House person Szilard had told that there was only one man working on the hydrogen bomb and the White House person told Szilard not to say who he was because if the communists found out, they would see to it that he be destroyed.

That's a Szilardian twist. But from my point of view it meant that Szilard was concerned that the hydrogen bomb might not have been built in America.

That could be.

But that is not the perception people usually have about Szilard.

No. But I would hesitate to say anything about Leo now that he's dead. I only have my stories about Leo that I know. But what you're saying is conceivable. He was not a peacenik. He was definitely not a peacenik.

My impression is that some people may misunderstand him. In my reading, he was a great believer in American democracy.

Yes.

And he understood the Soviet system.

Yes.

And he didn't want the Soviet system.

No.

So my impression is that he wanted the American hydrogen bomb to be developed not because he wanted to destroy the world or it to be used, but he wanted to have a balance between the two superpowers.

I don't think I would have anything to say on that. He tried very hard to make better relations with the Soviet Union. He went to see Khrushchev. It's a famous story. Nobody was there except the Soviets. You know the story about the raiser and shaving. Khrushchev said to Szilard that he would stop shaving when the war started. I wonder if it's true.

I don't think that Szilard lied.

I don't think so either. It's such a story.

It sounds like Khrushchev.

Folksy. I loved Szilard. I was with him the day before he died in La Jolla, in his hotel. He was OK. He died peacefully in his sleep as far as I know.

Who are your heroes?

François Jacob. In a quite different way, I would say that people who are willing to think outside the box. George Kistiakowsky, Linus [Pauling].

I didn't find anything about your origins.

My grandfather and grandmother on my mother's side came from Russia and on my father's side from Romania, Jewish. My father was born in Denver, Colorado. My mother was born in New York City. The family came to Denver when she was a young child. Her father became a farmer. We had what was called the Homestead Act in order to populate the West. If you built a house on a segment of land of sixty acres within one year, you could own the land. My grandfather cultivated sugar beats. Then when they wanted to have their children get a good education, they moved to Denver. I was born in Denver.

When I was two years old, my father and mother moved to Los Angeles and I grew up in Los Angeles. I had a very happy time there. There were open fields near the place where we lived, so I could run and do

Matthew Meselson during the interview in Woods Hole (photograph by M. Hargittai).

things. I had a laboratory in our basement beginning very early, maybe when I was seven years old. By the time I was in high school I was purifying rare earth elements and selling them in order to buy more laboratory equipment. I finished high school early and wanted to go to college. The high school people told me that I needed to stay another year and take gymnastics — physical education for a year and a half because a California law said that you couldn't get a high school degree without a form of physical education for three years. That surprised me, and I started asking people what can I do and they said that I could go to the University of Chicago because they took kids who were young. I was sixteen.

I went to the University of Chicago thinking that I could get a degree in chemistry. But Robert M. Hutchins had abolished undergraduate degrees in specialized subjects like chemistry. Everyone took liberal arts, the classics. In retrospect, I'm very glad now. I would've never read Aristotle and Homer and Sophocles and Euripides and all that stuff. Hutchins is another hero for me. But that form of education is almost dead now. After that, I knew I wanted to get back to biology somehow. I spent a year in Europe trying to find out what to do with myself. That's when I went to Budapest.

Then I came back and went to Caltech as a freshman for one year. I didn't like it; I was much older; I'd lived in Paris. There were no girls at Caltech and I didn't like the way they taught. It seemed too much memory. So I went back to the University of Chicago, and got a letter, which said that we didn't give bachelor's degrees in chemistry but if we did, we would've given one to Meselson. That letter got me into graduate school in physics in Berkeley. I stayed for one year, but Berkeley was so big that I didn't like it. The classes were huge. My advisor said, go back to Caltech, but I decided not to go back to Caltech, but to go to Chicago to Nicholas Reshevsky. I never did go back because one day before going back to Chicago, Linus's children — I knew Linda and Peter — had a party at their swimming pool. I'm in the water and Linus comes out — he knew me from my previous one year at Caltech — and asked me, what're you going to do next year young man? I said that I was going to Nicholas Reshevsky in Chicago. He looked astonished and he said, that's a lot of baloney, why don't you come to Caltech and be my graduate student? I looked up from the water and said, OK, I will. That was it. It was a very lucky thing for me. I became his graduate student. The first time he gave me a problem; he took a rock and he put it on my desk and we had a conversation, which went something like this:

— Matt, do you know about tellurium?
— Yes, Professor Pauling.
— It's under selenium and sulfur in the periodic chart of the elements.
— Yes, Professor Pauling.
— Have you ever smelled hydrogen sulfide?
— Yes, Professor Pauling.
— It smells bad, doesn't it?
— Yes, Professor Pauling.
— Have you ever smelled hydrogen selenide?
— No, Professor Pauling.
— Well, it's much worse.
— I see, Professor Pauling.
— Now, Matt, have you ever smelled hydrogen telluride, you probably have not.
— No, Professor Pauling.
— It's much worse than hydrogen selenide as hydrogen selenide is much worse than hydrogen sulfide.
— I see, Professor Pauling.

— The reason I'm telling you this, Matt, is that I want you to work with the crystal structures of some salts of tellurium, but I want you to be very careful because some chemists had gotten tellurium into their system and they acquired something called tellurium breath. It'd isolate you from society because it's so bad. Some of these people have committed suicide. But I know you'd be very careful. So what do you think?

— I'd like to go home and think about it.

When I came back, I told him that I was really interested in biology and I asked him to give me a molecule with some carbons and hydrogens and nitrogens. He laughed. I think he may have pulled this trick on other students too. I don't think he was serious, but I don't really know. So I did a crystal structure with two amide groups to prove what he'd already proven that the peptide bond is planar because of resonance. He was a wonderful teacher. This is one of my many stories about Linus.

Who else was there in Pauling's group when you were there?

Martin Karplus was there, Jack Dunitz was there, Jim Watson came for a year. I knew Jim's mother before I knew Jim. She was the admissions officer at the University of Chicago. She was a wonderful woman, very warm. I also knew his sister Betty because she was in college with me. At Caltech, I was in Chemistry and the chemistry boys felt that the biology boys were inferior. But I wanted to meet Max Delbrück. I finally got up my courage — because people said that he was intimidating — and I went and sat down in his office. The first thing he did, he said, what do you think of these papers by Watson and Crick? This must have been around June of 1953. I said, I never heard of these papers. He threw at me some reprints and he said, get out and don't come back until you read them. To my ear he said, come back. So I read them. Then, soon after that, Jim was coming to Caltech or maybe he was already there, I just didn't know about him. That was my first introduction to Jim, a violent introduction by having his papers thrown at me. The reason I got interested in the replication of DNA was that Max was writing a paper with Gunther Stent in which they were considering semi-conservative, conservative, and dispersive replication of DNA. Max didn't like the unwinding of the chains. He thought it was geometrically too difficult. I'd planned this experiment with beta-galactosidase and as soon as Max showed me this paper that he was writing with Gunther I realized that it was DNA I wanted to do this experiment with.

Lately, you have posed this question, "Is sex necessary?"

It's an ancient question. It's August Weismann[a] who first reformulated it in a modern way. He published a paper in English in *Nature* and nobody remembered it for a long time. It was a mere paragraph in which he stated his thesis. He said that given the fact that asexual species could reproduce at twice the rate — all else being equal — as compared with sexual species, sexual reproduction must confer some great advantage of life. I believe that this advantage is in the generation of variability upon which natural selection may act. This is very remarkable and true because no two children are the same and everybody remembers this statement. But very few people remember the first statement that sexual reproduction is very expensive. You have to produce males; it makes you vulnerable to venereal disease, and it has other problems. So there are many costs. Also, once you achieve a favorable combination of genes, and the environment is stable, that's the best combination. If you recombine them with some other good combination, you get a mess. You get a combination that's not fit. So there are lots of reasons why people have wondered why sex exists? I started wondering about it because of a silly reason. I was teaching for many years an undergraduate genetics course and it dawned on me that nature didn't make sex just for professors to give students questions about genetic crosses. So I asked myself, why does sex exist? I went to the library and ran across a paper that posed the same question that Weismann had, but by an Englishman named Turner, around 1969. It was titled something like, "Why does the genome not congeal?" That got me interested and I would ask people what they thought about this question. One day I asked an old man, a wonderful man, G. E. Lyn Hutchinson at Yale. I was visiting him; he was a friend. He asked me whether I knew about the bdelloid rotifers, and I never heard of them, but that's why we started working with these creatures which have been known from Leeuwenhoek's time. No one has ever seen a male, no one has ever seen a hermaphrodite; there're 370 described species, and they're all females. We've been working

---

[a]August Weismann (1834–1914), German medical doctor turned biologist formulated in 1892 that all living organisms contain a hereditary substance, the germ plasm, and that changes to the body do not cause alteration of the genetic material.

with them in a molecular way and we've begun to wonder whether the reason why sex exists is defense against retrotransposals rather than for generating variability. Retrotransposals are transposable genetic elements of a particular kind. There are parasites in our genome. Our genome may be forty per cent of this junk.

Paul M. Nurse, 2001 (photograph by I. Hargittai).

4

PAUL M. NURSE

Paul M. Nurse (b. 1949 in London) is President of the Rockefeller University. He shared the Nobel Prize in Physiology or Medicine in 2001 while he was at the Imperial Cancer Research Fund, London, with Leland H. Hartwell (of the Fred Hutchinson Cancer Research Center, Seattle) and R. Timothy Hunt (also of the Imperial Cancer Research Fund in London) "for their discoveries of key regulators of the cell cycle."

Paul Nurse received his B.Sc. degree from the University of Birmingham in 1970 and his Ph.D. from the University of East Anglia in 1973. Before he joined the Rockefeller University, he was Director-General of the Imperial Cancer Research Fund and Head of its Cell Cycle Laboratory. He was elected Fellow of the Royal Society (London) in 1989, received the Gairdner Award (Canada) in 1992, became a foreign associate of the National Academy of Sciences of the U.S.A. in 1995, was awarded the General Motors Cancer Research Foundation Alfred Sloan Prize and Medal in 1997, and the Lasker Award in 1998, among others. He has been knighted and he is Sir Paul Nurse.

We used the occasion of Paul Nurse's award of the Honorary Doctorate from the Budapest University of Technology and Economics to record a conversation in the author's office at the same University on March 1, 2003. At that time, Dr. Nurse was still at the Imperial Cancer Research Fund, but had already accepted the presidency of the Rockefeller University.*

---

*István Hargittai conducted the interview.

Could you please give a brief summary of cell biology before you entered the field and how did you change it?

When I was a graduate student I thought it was important to tackle a big problem. A big problem to me was the distinction between living and non-living things. One of the characteristics of all living things is how they reproduce themselves. This is seen in the simplest form with the division of cells, cells being the basic unit of life. As I was interested in the characteristics of life, I focused on the cell as the basic unit, and for many years I focused on the reproduction of the cell. What changed with the work of my colleagues and myself was the understanding of the mechanism, which controls the reproduction of the cell from one to two. This was the main thing that emerged from our work and that this was universal in all organisms from yeast to human beings. This was not really expected.

It was not expected?

Surprisingly it was not expected. Nobody thought that it would be the same in something so complicated as our cells and in simple organisms like yeast.

When I was in school, 50 years ago, we learned about cell division and it was never even hinted that cell division might be organism-specific. We just learned about one kind of cell division.

Theodore Schwann had certainly been struck by the universality, but in the 1960s and 1970s, just before this work, there was a lot of interest in the differences rather than in the similarities. Animal cells were seen as being different from plants and fungi cells. It was not thought to be regulated in the same way at all. I think that it had more to do with culture, more with the fact that those who worked with mammalian cells because they were interested in cancer were not very interested by the genetic model systems like flies and fungi and the yeasts. This cultural difference perhaps overemphasized the distinction. When you teach, you look at the universal features of the same mitosis, you teach all the names of its different stages. It would also be fair to say that when you come to the very simple microbes, they do not have the same obvious stages of mitosis. Another reason, perhaps, was that at that time the control could be reduced to a simple story. Also, that was the era only one or two decades after the metabolic pathways had been produced in which the biological systems appeared to be very complicated with lots and lots of components. The idea that it was a single switch that regulates the cell reproduction seems very simple. Something else comes to mind, and this is appropriate to do in this 50th anniversary year of the double helix. We can think of the structure of DNA, which carries the means by which it reproduces itself and also the means by which it encodes information. Our work on the cell cycle, how it is regulated, is focused on what regulates the replication of DNA and what then segregates the replicated DNA. So it extends that problem concerned with the reproduction of DNA, and translates it to the cellular level.

In between, there was the genetic code.

Indeed that was the information part.

Marshall Nirenberg told me that when he realized that the genetic code was universal it had a profound impact on him of almost mystical significance. This is the same universality.

This is the same universality, of course, and should not be so surprising because of common descent in evolution, the fact that the range of living organisms comes from common ancestors.

Were you surprised or impressed at all by this revelation?

By the time my group actually established this, which was in the late 1980s, I had really begun to think that it was universal. So it was less of a surprise when it happened, but it was very satisfying when it did happen. For me, perhaps, what was more important was the way in which it was done. The key was to try to show that the genes that controlled cell division in yeasts were the same in human beings. So we had to find a similar gene. We applied a trick: how can you, amongst, what we now know, 25 thousand genes in humans, find that particular gene, how do you know if such a gene exists? And once you find the gene, how will you know it is the gene you are looking for given the evolutionary diversions between yeasts and humans.

Is there only one such gene?

Yes, there is. The way we did it was to use a gene library from humans and to introduce it into the yeast cells that were defective for this particular gene and simply look for the human gene that would rescue that defect. It was an assay based on function, to jump that evolutionary distance. Once we had the gene, it was clear there had to be not only a gene of similar structure, but similar function, and therefore all things were controlled in the same way.

What does this gene exactly do?

We called this gene CDC2 and it encodes an enzyme called the protein kinase. Protein kinases phosphorylate other proteins and this modification can change the way in which the protein operates. By putting a negative charge on the protein it changes the way it folds, for example. It can act as a switch for a variety of other proteins if you phosphorylate them. In very general terms, it's now known to phosphorylate a variety of other proteins to bring about the events of mitosis. This is a complicated choreography of chromosomes condensing; lining up, separating them, all initiated by phosphorylating a set of other proteins. That's its second function. Its first function is to initiate DNA replication and it's still not fully understood how it does that. It does it in a similar way, again, by phosphorylating a set of proteins to bring about DNA replication. In yeast, this CDC2 activity starts at the beginning of the cycle rather low and at the end of the cycle it is high. At the low level it initiates DNA replication, at an intermediate level it actually blocks further DNA replication and then it initiates mitosis. At the end of mitosis, to get out of it, you have to kill the enzyme completely to zero. So it does a simple monotonic rise and fall. What you can see is a very primitive control, which brings about an orderly progression through the events, keeps them in the right sequence, and then starts the next cycle. That reflects a basic mechanism that presumably was invented with the origin of eukaryotic cells, that is, cells with nuclei, maybe fifteen hundred million years ago. Once that was invented, then everything locked into this control mechanism. That's the reason why it's so highly conserved.

How can you be sure that there is only one gene that is common?

What we can be sure of is that the major control is this gene because we can activate the whole process just by turning on this gene. What we now know is that there is only one gene in humans and frogs or flies, which can be rescued in this way. There are other genes that look similar, but they won't work. Of course, one can never be certain, but we've shown that this gene works and other close relative genes don't. It's this functional assay, which is so important to show that that one is important.

Is there any computational component of your work?

I've always been interested in the mathematical-computational aspects of biology. I was an advisory editor of the *Journal of Theoretical Biology* for twenty years, which is unusual for a practicing molecular cell biologist. I'm not a mathematician, but I do think theoretically quite a lot. It helps to clarify the experiments. I would like to give you an example why I don't use computation but do try to think in a more abstract way as geneticists often do, of course. What influenced me greatly in looking for means to study cycle control was the so-called thought experiments. I tried to imagine if there is such a thing as the rate-limiting step in cell cycle progression. To understand this I read a lot about metabolic pathways and how you control flux through pathways. It's getting close to physical chemistry. Understanding flux through pathways you realize that you can distribute control amongst a number of steps or you can focus on fewer steps depending on how the system is set up. It made me realize that if control was invested in one or two steps then it could be found, because you could speed up the reaction. If you speeded up the reaction, you would end up advancing the cells into mitosis and cell division early. So although it was not computation, it was theoretical work in its approach. I myself have never been involved in computation, not because I don't think it's useful, because I think it is as it clearly helps to clarify your thinking. If you have to reduce your thinking to some equations, you have to really know what you're dealing with. Biologists are often a bit sloppy in their thinking, but when you're forced to think about equations, it actually makes you think more precisely about the problem. Half the interest in theoretical-computational biology is getting the equations set up. In the future, we will have to do more than computational work like solving differential equations and like logical representations of the elements within our networks. That way we'll be able to describe a complex network in terms of how it processes information through logical representation rather than flux and quantitation of flux. It will still be computation but a different type.

I would like to ask you about the division of work among the three Nobel laureates and about Yoshio Masui, who was not among the three awardees but was one of the three Lasker awardees along with you and

Hartwell. I remember Tim Hunt saying something self-deprecating in Stockholm in 2001 to the effect that he was not in the same class as the other two laureates.

Tim can be self-deprecating indeed. The first contributions in this area were by Masui and then by Hartwell. Masui was working with frog eggs and he did some quite simple physiological experiments with eggs. He came to the conclusion that there were factors that were important for bringing about particularly mitosis. He did this very early, in the 1960s up to 1971. That work defined this area. It did not move very far during the next decade and a half and Masui himself did not take it beyond establishing the initial controls.

Hartwell took a completely genetic approach at the same time. He had two major contributions. One was in the 1970s when he showed that you could use mutants to analyze cell cycle progression. He isolated the first type of this kind of mutant in a eukaryot after he had isolated one

Paul Nurse receiving the Nobel Prize from the King of Sweden, 2001 (photograph by M. Hargittai).

in bacteria a few years before. I know that you are going to talk with Sydney Brenner, who did something like that in the mid-1960s in connection with DNA replication. Hartwell was the first person to do that systematically and the first person to use that information to try to think about how overall cell cycle progression was regulated. That was his first major contribution in the early 1970s. His next major contribution was in the late 1980s when he devised this idea of checkpoints, which was a derivative from his original work. The idea was that in a sequence of events that lead to the completion of cell division, the cell might check whether the previous events had been completed and if they hadn't, it'd stop further progress. These were Lee's major contributions, all with budding yeasts.

My contributions came probably next. In the mid-1970s I used exactly the same approaches, which I copied from Lee in a different yeast to define the cell cycle genes. What I added to it was to think about what might have been overlooked in what I had already said and to look for mutants that would be altered in rate-limiting steps. Eventually I identified the CDC2 gene, which I then cloned and showed the molecular basis of, that it was a kinase and then I showed that it was also in humans. So I connected yeasts with humans at the end of that.

What Tim did was, he was working again with the frog, not unlike Masui, but he was a biochemist and he showed that certain proteins become degraded as they go through this cycle. He came up with the idea that these proteins can regulate cell cycle progression. Tim identified a protein called cyclin, which turns out to bind the CDC2 protein. That was his contribution. These were the major elements, which went from defining the initial controls to the molecular basis of them and then showing that they are basically the same in yeasts, in frogs, in humans, and other things in between.

Masui got the Lasker award and Tim got the Nobel Prize. It's always incredibly difficult with those sorts of prizes to work out which contributions really mattered and which people made them. I sit on the Lasker Committee and have become familiar with the process there and it is very difficult to make awards of this sort. I know that the Nobel Prize has a lot of difficulty trying to get combinations of people because they have this rule limiting the co-laureates at three. It's very difficult and to some extent artificial and a lot depend on luck. They have to put a combination together, which looks good and which people like. It doesn't always reward necessarily the best people or cuts the best sciences. It will be very difficult, for example,

The three new Nobel laureates in Physiology or Medicine, in 2001, in Stockholm at a reception, with Hans Wigzell of the Karolinska Institute (photograph by I. Hargittai).

with the genome project where they will have to choose from among five or six or more scientists. The artificiality of prizes sometimes can distort those scientists and that work which gets rewarded.

I see an unfair watershed effect; some people become semi-gods and others disappear in oblivion.

It isn't fair. We have to be aware of these weaknesses.

The Lasker procedure is different from the Nobel evaluation and selection process.

There's also the Gairdner award in biomedicine of very high prestige. Masui, Hartwell and myself also won the Gairdner award in the early 1990s. About one in four of the Gairdner awardees go on to win the Nobel Prize in biomedicine. For the Lasker the proportion is even higher, about half to two thirds. The Lasker jury does its work in one committee meeting. What happens is that people get nominated, a jury vote on them, and a shortlist of eight or ten groups of people are selected. Each person on the jury is given two or three projects to evaluate in depth a month or two before the committee gets together. On the day the committee gets together, we go through all the candidates and two or three people talk about each set of people, and then we vote on it. What is interesting about it is that when you get the combination of people that works then it becomes quite straightforward. There may be two or three groups competing and one makes it one year and the other the next year. A lot of effort goes into the combination of people who should receive the prize, but then at some point it clicks. It's not so much about what area and what contributions, but about getting the right combination of people and this is where the artificiality is. The Nobel Committee, as far as I know, does a lot of in-depth evaluation, and uses lots of people outside the committee to help it. It's extremely thorough. I can't say anything but praise about how it operates. It is striking how the decisions of these two committees overlap.

The Lasker is quite American-oriented, isn't it?

Yes, it is. I haven't done a statistical analysis, but most people on the committee are American, only one or two who is not American. I certainly feel some responsibility of thinking about the rest of the world when I am on the committee. It's not a bias; it's just that people tend to think of other people whom they know better.

Have you talked with Masui since the Nobel Prize?

I haven't, only sent some letters to him; last time I visited him in Toronto was prior to the Nobel Prize.

I would like to talk a little about your background. As you write about yourself, you came from humble origins and now you are Sir Paul. How did it happen?

I come from a working class family, not an academic family at all. I had three brothers and sisters who all left school at fifteen. We were quite poor. I don't know what combination of circumstances led to me being academic and staying in school and then going to university and then continuing. My parents and the rest of the family were very encouraging; they were never discouraging; it just was a different world. My mother was a cleaner and my father worked in a factory; he repaired machines. They never thought I left school, their impression was that I continued school at university.

They were right.

My parents could never understand what I could possibly be doing by still learning in my mid-thirties. My father only died a few months ago in his nineties; my mother died ten years ago. It was very nice for me to see my father see that I won the Nobel Prize. He didn't go to Stockholm though because he was too infirm, but he saw it. He was very pleased about that and also about the knighthood, which, in some ways, was more accessible for him. The Nobel Prize is so extreme that was almost unbelievable whereas the knighthood was a bit more close to home.

Don't these distinctions separate you from the rest of your family?

A little bit.

Did you have to overcome any adversity in your life and career?

I can't say that there was a huge barrier that I had to fight against. When I was at school, at grammar school, from eleven to eighteen, I found it quite difficult because most people came from middle class homes and had been exposed to more books and so on. I was just learning so much all the time. I didn't do very well at examinations, but I gradually got better and better. What this meant was, I got used to failing things and not getting too agitated about it. Sometimes people who are overachievers they get used to everything working for them; then when something doesn't work for them in their research, they just collapse. Whereas I had had quite a lot of time when things didn't work for me, not because there was a big hurdle, but just because I wasn't well prepared, I got quite used to things not working, sometimes failing, and I was more relaxed about that. I see more of a chance when things are working. When I was a research worker in my twenties and thirties, I didn't have particularly good financial support, so that was sometimes a struggle. Having said that, I always had a salary, I was never out of a job, I did always get on with my experiments although it was not always easy to compete with the Americans, for example, what I had to do much in the 1980s. I can't say though that there were major problems.

The only adversity, and this is too strong a word, was that I wasn't working in the mainstream, while most of the world was interested in something, I was working on something else. When I gave a talk, people were polite, they thought it was OK and I got published in good journals, I was not being kept out of good journals, but I was not in the center of attention. It wasn't my work that was of interest to others. It then came in a big rush in the mid-1980s when suddenly people realized that this was interesting. By then, I was eighty per cent there. In some ways, I quite liked not being the center of attention because I had room to breathe and to think without worrying about what my competitors were doing because there weren't that many competitors. I'm afraid I can't really describe adversity to you. For other people it might have meant a big barrier seeing that nobody was interested in their work, but I look upon it as an advantage.

Looking back, you didn't go to the best schools, you didn't go to Oxford or Cambridge, you didn't work in the Laboratory of Molecular Biology; this is not adversity, but it didn't point to a great career.

It didn't and I was always relaxed about that. I always felt that it's very easy to say that great science only comes from a few places, but that's not true, it can come from other places, and we should be aware of that when we're thinking about policy. No, it's right, I didn't go through a traditional trajectory, and my early path was not so easily done, I could've not got into Oxford or Cambridge because I didn't do well enough in school.

But you still got into academic research rather than going to teach or working for a pharmaceutical company. Was this ambition, drive, determination?

I liked teaching; you just don't have enough time to do everything. In the second half of my life I had a lot of managerial and administrative responsibilities too.

When was the turning point?

Probably when I returned to London in about 1990, to run research in the Imperial Cancer Fund (ICF); it was a major shift. I was a Professor in Oxford and I didn't have a big administrative load there. I would like to tell you what kept me in research because it is important. I'm driven by curiosity, very strongly; I'm curious about the world around me and I'm very interested in working out the answers. To me doing research was a pleasure.

Was it curiosity that turned you to science?

I think so. Even when I was in primary school, I was particularly interested in the natural world, natural history, the stars, clouds, just everything around me.

Was there any particular teacher or book?

Not initially, just looking on the world. I had a very good teacher in my secondary school that pushed me towards biology. There was then a second aspect, a wish to do something myself. I ended up doing biology. One reason I didn't go to physics or chemistry was because it seemed to me that there were more unknowns and more questions in biology than in physics and chemistry. I felt chemistry was more tightly understood and it seemed to me that biology was something where I could actively contribute to it myself. So it was not only a question of curiosity, but also the ability to do something myself. Not just reading books about it, but doing experiments and finding something out that perhaps nobody had found out before. These two aspects were very important to me.

Any mentor?

I had several. I had a teacher at school, a man called Keith Neal, a very young man; he was a biologist and he encouraged this interest in observation of the world around me and to do experiments. I had various mentors; I had a teacher at the university, Jack Cohen, who was very idiosyncratic. Much of what he believed was unusual, but because he always challenged what was established knowledge, it was a very good lesson to be exposed to him.

Didn't the school mind it?

It was at university and I think a good university should always have one or two people on their faculty who challenges the traditional teachings in some ways. Even if it doesn't make sense, it's good for the students to see that there are different ways for looking at things. Then I had very good mentors during my early postdoctoral work in Edinburgh, Murdoch Mitchison in particular; he featured in the double helix and knew Jim Watson in the 1950s. He was very generous and made me think very biologically. Other mentors, who influenced me were my peer colleagues, the people I worked with and talked with directly. This was an especially positive part in my own development.

Did any venue impact your research career in a decisive way?

I did my Ph.D. in a very new university, two or three years old, in Norwich. I went there partly because I was born there. I found it exciting that it was so new. I had difficulties with my Ph.D., but there was something very exciting about being in that new environment. I did my first degree in Birmingham; this is a high quality redbrick university in England. With my working-class background, it would have been difficult for me at Oxford or Cambridge.

Would it have been difficult to get accepted or to survive?

Both. The Oxbridge colleges were a bit overbearing. I felt more at home in a redbrick university. Then I went to a plate-glass brand new university.

Jim Watson has told me about the advantage he felt about having done his doctoral studies at Indiana University rather than Harvard or Caltech.

We had exactly the same conversation because he originally also came through this other route, from Chicago, and he would've found it difficult in the so-called Ivy League places.

It is now 50 years since the discovery of the double helix. What are your thoughts about this anniversary?

It was a fantastic discovery because from the structure of the molecule so much was immediately understood. It's one of these major leaps that scientists don't often make. Biologists like particulars, they like details, specifics; what we found about a protein regulating cell cycle is a bit of a detail; although it has some general significance, but it's still a detail. The structure of DNA and the genetic code are not details; these are great discoveries. The double helix and the subsequent code were the real great days of this area of biology. We are now more into details, of which my work is an example.

Some trivial questions and I apologize.

Of course.

Genetically modified food? There seems to be a difference in public opinion in the U.K. and the U.S.

The problems with genetically modified food may be ecological problems by which I mean that genetically modified plants may have some potential for environmental damage and this needs to be researched.

It may not only happen with genetically modified (GM) plants. We have seen some examples in Hawaii.

Problems can arise when plants are introduced from one ecological region into another. I think this is an interesting example how scientific debate can go wrong. In the United States, there was very little debate about GM crops; it just happened. In the U.K., it was reported in a very dramatic way in the newspapers. We shouldn't blame the newspapers for it; it simply happened. Then there were people who were very much against GM food, not only for issues to do with safety, but more to do with economic reasons, why should the big agro-industry dominate in this way? The debate is completely mixed up with arguments about capitalism, arguments about domination by big companies overlooking the needs and wishes of the community; arguments about whether this is of any use whatsoever for the community or simply to companies. This is all mixed in with safety issues and people's concern about genetics. I'm involved in a number of dialog initiatives with the public about science. One thing that we discovered is that the reason that many people were frightened about GM crops is because they thought that the food had genes in them. It was a misunderstanding; of course, all plants have genes; they thought that GM plants had genes and natural plants did not. Had scientists known about that beforehand, they ought to have dealt with the arguments with the public better. But because the scientists were not in good contact with the ordinary people somehow, the debate went all wrong. The media and some non-governmental organizations, like Greenpeace, were better informed about what worried ordinary people than scientists were. It was a good example of how not to have a good debate about complex scientific issues.

On the other hand, there seems to be more concerns about other issues in the U.S., like stem cells.

I think this comes from the strong religious beliefs of a minority, but influential religious groups who have certain views and opinions about when life begins and ends. They have very powerful lobbies. Although there is little debate about GM crops, there are debates about stem cells and they are just as distorted in the U.S. as the debates about GM crops in the U.K. People with very clear agendas distort the debate about stem cells very strongly. They use every way they can to set it. In the U.K. and in Europe generally, these things are handled rather better; the stem cell debate in the U.K. was good compared with the GM debate.

May I ask you about religion?

I was brought up as a Baptist.

Different from the mainstream.

The mainstream U.K. is Anglican, but the U.K. is not a very religious country; most people do not have religious beliefs. I moved away from religion in my early teens, largely because of Darwin and evolution and the conflicts there.

You are moving to the U.S. later this year to become President of Rockefeller University. Do you anticipate any difficulty in the U.S. because of your lacking religious beliefs? You will have to be doing a lot of fundraising.

I don't anticipate any difficulty in New York, which may not be typical of the U.S.A. As for fundraising, the Rockefeller presidency differs from other presidencies. The Rockefeller University is more like a big department rather than a big university. There are only 75 faculty members, powerful and high quality, but small in number. The environment is very creative. What's great about the presidency of the Rockefeller is that it's expected that you also carry on your own scholarly research activities, not just a fig leaf. I'll be able to spend one third, maybe one half of my time still doing my own research. That would be impossible in any normal presidency, rectorship or vice-chancellorship of a university.

The President of Cold Spring Harbor Laboratory has fundraising as his main activity.

Yes, but Cold Spring Harbor is not so well endowed; if you like, Cold Spring Harbor is new money, the Rockefeller is old money.

As I understood Rockefeller used to support all its research from its endowment and frowned upon people seeking external support, but that has drastically changed.

That's true and today they raise a lot of money from federal funding, government agencies, and there are many Howard Hughes investigators. But there is still this massive endowment, which provides a lot of flexibility and a buffer. Having said this, there is a major fundraising role, but it is significantly smaller than the one I've been doing for the last seven years. I was responsible for raising much more money in England for the organization I worked for. It's also a somewhat different kind of fundraising from what academic people are used to. I run a cancer research organization and we have to raise all the money each year to support the research and this is around EUR 400 million each year. I'm responsible for it but that doesn't mean that I do it alone; rather, I distribute it among many other people. In the Rockefeller, I'll have to raise money too, but that will be mainly from private donors, which is a different sort of fundraising, but I have to do that also in England, so I'm used to this approach.

I doubt you would answer my next question, but do you plan to introduce any change at the Rockefeller?

I'm very excited about going to Rockefeller; it's a great academic institution. I've worked for the cancer research organization for many years and I would prefer to work for a university with a much wider academic interest. I like the Rockefeller also because it's not just biomedical, but has physics, chemistry, math, and computation, about ten groups, which work in the interface with biomedicine. I'm very excited by the possibility of these interdisciplinary approaches.

There have been attempts to expand the Rockefeller in the direction of these other fields but my impression was that there was some hesitation about spreading themselves thinner.

I'll have to speak about this with the faculty members as we develop an academic plan in the coming year when I appear there. But my thoughts are that we have real strengths, that we should further strengthen, which are, of course, cellular-molecular biology, neurobiology, and then the third area, which is interdisciplinarity involving physics, chemistry, and math. They will help the two other areas understand the complexity of their networks to give rise to the phenomena of biology, the sort of higher-order phenomena, and how you can understand them in terms of the chemistry of networks of information flow that work in practice. I'm interested in that transition to higher-order problems of biology like how to generate biological form and how to generate purposeful behavior. I'm interested in how to generate them from chemistry and networks. You can see these problems in neurobiology, in cell biology, and physicists, chemists, and mathematicians can help to think about them.

So you have a research agenda for Rockefeller. But there are strong individuals working as group leaders.

Of course. If I sounded like giving a research agenda, that may be too strong a word for that because one should never have a top down research agenda. The main task of any research leader is to provide an environment in which others thrive. There are already many strong people, highly creative individuals and basically they should be let get on with their job. In addition, to create the proper environment for them, the job is to provide the interactions among them that might spark off new things. I'm very interested in this and am excited by the possibility.

Being currently in a very high leadership position in England, were you looking for a further challenge?

First I was the research director at the Imperial Cancer Fund for three or four years. Then I was responsible for the whole operation for another seven years. I merged two research charities in the last eighteen months to make a big single biomedical cancer research organization with an annual budget of EUR 400 million. This was a heavy administrative job, probably heavier than the Rockefeller will be. I felt it was time to move on because you can spend too long doing those sorts of jobs; in such a position you either go mad or bad. I did want to change. I wasn't quite sure what change to have. In the middle of all this I got the Nobel Prize, which is like having another job. I did think about just running my own laboratory and that was an attractive possibility and I did think about trying to combine that with some research leadership role. I'm rather hoping that the Rockefeller would provide that because it's a great institution, but it's not too big an institution. It doesn't have undergraduates, it doesn't have medical students, and it just has 175 graduate students. It's a great and interesting place in Manhattan and I can still do my own research work. I rather hope it will give me the best of all worlds.

The Rockefeller has had over twenty Nobel laureates; would that intimidate you? You would be measured against such a past performance in which, of course, it helps that you are a Nobel laureate yourself.

I know, of course, but it does not intimidate me; it's a challenge, certainly. I look upon it as a mark of quality, which means that it will be a good place to work. There are great colleagues; it has done extremely well in the past; it will be very good to work there in the future.

I wonder about your social position in England. On the one hand, there are your humble origins and, on the other hand, you are Sir Paul. Do you feel that you are part of the Establishment?

Yes, I am part of the Establishment; I am on the committee that advises Blair and the Cabinet on science, for example. This is a group of about ten scientists and industrialists. I play an important role in the Royal Society. I am one of the great and the good in the Establishment although perhaps not quite as conventional as some.

Do you live a sizzling intellectual life?

I would say, it's not bad, but a lot of these activities are not particularly intellectual; they are important policy of the power, but not always very interesting and very exciting intellectually. My intellectual stimulation comes mostly from my own laboratory and my own interactions with graduate students and postdocs and my colleagues rather than the great and the good interactions. I come across very interesting people, but we are talking about other things, it's not that exciting. I have wide interests; I go to talks about lots of other things and I find that intellectually exciting, but it's more of a private thing; it's not the great and the good committees. If anything, they're a little dull.

Do you anticipate your New York life more interesting?

I think I might. My problem is that in the United Kingdom I do quite a lot of things on the media, television, radio; I sit on lots of committees; I'm chairman of EMBO [European Molecular Biology Organization]; I'm asked to do lots of things. I'm finding it a little bit too much. Even though I'm saying no to most things, it's still overwhelming. In North America, I will not be part of the Establishment in the same way; I will be in one institution and it will take several years to assimilate me into the Establishment. So, for a while, I will be more free.

I have looked up some of your other interviews and you have sounded strongly European. I have heard voices that you are now abandoning Europe. I personally thought this Rockefeller position was recognition for European science too.

I also looked upon it as recognition of European science and was proud to be asked by the Rockefeller, coming from Europe. I believe I will be the first non-American-based President of Rockefeller. I won't abandon Europe; I'd done my best work in Europe; I'm very supportive of European science and the cultural differences that lead to a different sort of science compared with the U.S. Many people start part of their younger career working in the U.S.; I didn't do that. I'm doing that at the other end of my life. I hope it does not seem to be abandoning, but people have said that to me quite a lot. But there is a really big difference. When people abandon Europe or abandon the U.K. to go to the U.S., they're normally very critical of where they come from. They say they can get better resources, better pay, can do their work better; that is not the case for me. I've been very well supported; I've been personally well paid; I'm very happy with how I've operated. It's just a somewhat different challenge and I consider it as flag-waving for European science. I don't see it as part of the brain drain. Besides, the Rockefeller is rather an international institution; that's another attractive feature of it.

I saw you on TV in Stockholm last December [2002]. You are a busy person, yet you went back to Stockholm one year after your own Nobel Prize [2001].

Why did I go?

Paul Nurse at the Nobel Prize award ceremony, 2001. To the right is the 2001 literature Nobel laureate, Sir Vidiadhar Surajprasad Naipaul; behind Nurse to the right is Burton Richter, and further to the right Sam Ting; behind Richter to the left is Martin Veltman and to the right Richard Taylor (photograph by M. Hargittai).

Of course, in December 2001 you were not so much in the center of action because your Nobel Prize coincided with the centennial celebrations.

I was simply asked by the television, Swedish, BBC and U.S. Public Broadcasting System, to do a debate associated with the Nobel Prize. I knew the people, who made the program, so I did it as a favor to them. The debate was among four Nobel laureates about science and society; this is the sort of thing I do in the U.K. In the end, although I was the only biologist, it became very biomedical oriented. I found the award ceremony in 2002 much more relaxing than the previous year, when it was very hectic. In 2001, I felt myself to be too much in the center of attention and this time I was happy to be in the background. I had been to another Nobel ceremony about ten years ago when I was invited to a Nobel Symposium, so this was the third occasion that I could be there.

Could you summarize your position on science and society?

I feel quite strongly that we do have to have higher quality dialogs about scientific issues within our communities for the health of democracies because so many decisions involve some influence of science. I was getting worried about the quality of debate and the way in which scientists can be viewed as witch doctors, like people who are important but who do things nobody understands and who are feared as much as listened to. I'm interested in having new ways of dialogs with the public; find out the issues that bother them, to find ways of presenting science to the public. If the public gets suspicious of science, we will end up with regulations and controls that hinder doing our work.

Do you have heroes?

In my own area, both Crick and Brenner; both were a generation before me; both worked in the U.K., both did brilliant experiments, and when I was an undergraduate, I knew about all their work, and was very impressed and influenced by their work. Particularly the genetic code information transfer was what impressed me. What caught my attention was the approach by Crick and Brenner, their abstract way of thinking. I met them both when I was quite young and I enjoyed their lectures. Continuing with my heroes, I cannot avoid Darwin, his thinking, and the wonderful idea there. In his case it was his writings that strongly and positively influenced me. These are the people whom I very much respected during my formative years.

How did you feel getting the Nobel Prize before Brenner did?

I felt embarrassed. So I felt great satisfaction when I heard the announcement in 2002 and that was another reason why I went back to Stockholm. It was Sydney and it was also John Sulston, who is a friend of mine. Kurt Wüthrich, one of the chemistry laureates in 2002 is also a friend of mine.

I have heard you lecturing several times and I appreciate your sense of humor. Is it all spontaneous or well planned? I'm asking this because your jokes seem to come out so well timed, just when the audience starts slipping.

I do have some wit and like to play with words. I do try to use humor because any lecture is a performance and I want to engage the audience. Humor is one way of engaging the audience. It means that they will pay more attention. Science is difficult and some things that I talked about yesterday [at the Budapest University of Technology and Economics] were quite difficult, not all of it, but some of it. If it is relentlessly difficult with no lightening, then it becomes too much. It's like Shakespeare, even in *Hamlet* there is humor with the gravediggers. You need to lighten it just to keep the brains boiling. I never let these things to spontaneity, I have to construct them very carefully; I do prepare quite hard for my lectures.

Manners and style in science. Some people say if one scientist does not do something, another will. It's very different from artistic creations.

I always find this a little difficult to deal with. There is some truth in the fact that the discovery of something is inevitable to happen; if A won't do it then B will do it. But I would like to make a few observations about it. First of all, the manner in which a discovery is done, the experiments and the thinking, can be very different. The manner in which the discovery is unraveled, reflects creativity, different approaches, personalities in the same way as the work of great literature may be revealing some truth about human nature. That truth, that universal message may be also in another work of literature of a completely different form, but what resonates with you is the message. Of course, while a piece of literature is completely unique and different, but still the universal messages may have great similarities. Science is not quite as unique, but although the underlying story to be discovered may be similar and different people may be out to approach it, the way in which they do it and the beauty with which it is done, with the risk of sounding pompous, differs. Then there is the question of timing, and we shouldn't forget about it; there are discoveries, which can be made now or can be made in thirty or fifty years. Often things come together at once because of the set of conditions that lead to certain problems being addressed and having the methodologies to do them. But that is not always the case. Some of the problems that we think about now, I remember that we discussed them as graduate students thirty years ago. We thought a lot about them, did some work, could not solve them and now we're coming back to them. It is true that there are some basic structures out there to be discovered, but it's more complicated than that. The ways with which we approach them, the timing by which they are sorted out depend on the individuality of the scientists involved.

Would you try to predict the next set of Nobel laureates?

It's funny that I actually did predict last year's Nobel laureates and I'm proud of it. I do have my guesses, but I would not like to mention names because there's nothing worse than being spoken about before you may get the prize.

Some people don't mind that at all.

I know I minded a lot. I really disliked it because what would happen is that it was known that I was shortlisted, people had written letters. I started hearing about it maybe seven years before it did happen. It gets distressing; every year comes round and it doesn't happen, or does happen. So I tried not to speculate about it, certainly not in public.

There are some discoverers who, not having necessarily completed their education, go directly to the frontiers of science, make a discovery, and disappear in oblivion or in a very profitable biotech company. You had your painstaking education step by step.

I did and it took a long time. My career has been quite purposeful; I used genetics to find out what elements were important for control; then I translated that into a molecular role, and then I showed the validity of my findings for other organisms. It was a rather clear pathway through. I wasn't propelled to the front. Other people are propelled to the front. They're working always at the leading edge, in the limelight. It's partly because they keep an eye on the moving front and go to it. There are then others who are in the forefront some times; they are working somewhere else and find themselves suddenly propelled to the forefront. There are different way of doing things and different scientific tastes.

What are your ambitions?

Apart from the Rockefeller, my ambition is still with my own work. I'm particularly interested at this moment in form and how to generate spatial organization.

Morphology?

It's morphology, but I'm doing it at the level of the cell. I still have my interest in the cell cycle and in genomics too. But my main interest is cell form. All living organisms have their characteristic forms but it's not an easy problem to understand how these forms are generated. I'm interested in how overall shape is generated.

About twenty years ago Aaron Klug concluded his Nobel lecture by charting the future that would be characterized by studies which would try to connect the cellular and the molecular. Among the more complex systems to study he mentioned specifically the mitotic apparatus and he called their study a formidable task for future generations. You have greatly contributed to solving this formidable task.

Some part of it.

How would you formulate the next task for the next generation?

I still think that it is understanding the global characteristics of living things, and I used form as an example of it, and knowing how you can generate these general characteristics, like form, like organization in time, like the ability to respond, that is, purposeful behavior, and so on; how that purposeful behavior emerges from the chemistry and simple networks.

When you say form, do you mean macroscopic form?

Yes, action at a distance. What you have is order over distances of micrometers or even millimeters, which is derived initially from direct chemical intermolecular interactions; so it's based on chemistry. You can make a

Nurse lecturing at the Budapest University of Technology, 2003.

ribosome or phage head by direct protein interaction and I'm not interested in that. I'm interested in how chemistry generates form way beyond local interaction. The same principle applies in other ways as well. If you look at a network of information, for example, signaling pathways, how do they generate the overall purposeful behavior of a living system? You don't rely on the chemistry, you rely on the information and the way in which information is processed and transferred.

I haven't asked you yet about your family.

I have a wife and two children. I met Anne when I was at university. She studied social sciences. We have two daughters. One taught English for six months in Budapest and is now working as a TV producer, making soccer programs for British television. She is a woman on the big editorial staff of the sports department. My other daughter is doing a Ph.D. in theoretical physics at the University of Manchester, but currently spending two years at Fermilab near Chicago.

Is there any question that I might have asked and didn't?

I'm happy with the interview as was done.

Richard Timothy (Tim) Hunt, 2005 (photograph by I. Hargittai).

5

RICHARD TIMOTHY HUNT

Richard Timothy Hunt (known as Tim, b. 1943 in Neston in the Wirral, England) is Principal Scientist at the Cancer Research U.K., Clare Hall Laboratories. He received his education in Dragon School, Oxford (1951–1956), Magdalen College, Oxford (1956–1960), and Clare College, Cambridge (1961–1964). He received his B.A. degree in 1964 and his Ph.D. degree at the Department of Biochemistry, Cambridge University, in 1968. He did postdoctoral work at the Albert Einstein College of Medicine in 1968–1970, held various positions at the Department of Biochemistry of Cambridge University through 1990, and has been with Cancer Research U.K. since 1991.

Dr. Hunt shared the Nobel Prize in Physiology or Medicine in 2001 with Leland H. Hartwell (b. 1939) and Paul M. Nurse (b. 1949) "for their discoveries of key regulators of the cell cycle." He was elected Fellow of the Royal Society in 1991, Member of the Academia Europaea in 1998, Foreign Associate of the National Academy of Sciences of the U.S.A. in 1999, and member of other learned societies. Among his other distinctions, he is Officier of the French Légion d'Honneur (2002), and he gave the Croonian Lecture of the Royal Society in 2003. We recorded our conversation in London, at the headquarters of the Royal Society, on January 10, 2005.*

---

*István Hargittai conducted the interview.

I understand from your autobiography that your father was in Intelligence and you did not know about it until after he died.

That's right. I never liked to ask what he did in the war. The Second World War was very important to both my parents because my mother was quite close to Liverpool where I was born in 1943 and there was a lot of bombing. I never knew what dad did; I knew that he hadn't served in the Army because that would've been manifest. I only discovered it after they had been both dead that he spent a lot of the time — when I was very little — working in London. Even his brother didn't know that. He must have signed the official secrets act and he was the kind of person that if he once signed such an act, he would've felt bound by it to the grave. In any case, he was a man of very few words. When a man named R. V. Jones published a book about his work at Bletchley, cracking the codes, a very amusing book, it was probably in contravention of the official secrets act.

How did you find out about your father?

Mum used to say that she thought that he was something in Intelligence. But I don't know why I never asked him. It is a little symbolic of our relationship. I wasn't that close to him to ask him that kind of a question. He never mentioned it so I understood that he didn't want to bring it

Young Tim Hunt, 1970 (all photographs are courtesy of Tim Hunt unless indicated otherwise).

up. After they were both dead I went through a lot of abandoned correspondence in the bottom of a big old drawer and there were letters there from him to my mother, postmarked from Bush House, which is just down the road here, saying, "How is our little one doing?" So it was perfectly clear that he was not at home and that he was working in London.

Do you think that his intelligence work was related to his professional background?

It must have been because he had a reasonable knowledge of things Italian and ancient manuscripts. Also, he'd worked in Germany as a young student, in Munich, for a while. That was also something that he never ever spoke about, about his experiences in Nazi Germany although he must have seen a lot of that. It was such a taboo subject after the war. It's only very recently, for example, that I began to read biographies of Hitler to see what actually went on. Hitler was a bogy man and they didn't try to explain it. I didn't even understand what Hitler stood for; part of that reveals my own ignorance and my political naivety.

Did your father ever try to show you the beauty of his work?

If you asked him a frivolous question about his work, he would not take it seriously, but if he saw somebody who had a reason to know about what he was doing, that was something different. I remember very vividly just after my mother died, two friends who were both art historians came to visit. They went to the Bodleian Library and he very proudly pulled out a thousand-year-old manuscript from the shelf and showed them. He liked showing people his stuff.

But not very much to you.

Not so much to us. I was rotten at Latin and Greek and he was very good at it, but he just accepted that. He was perfectly proud that I was destined to become a scientist from an early age, who showed no particular interest in his interest. He died in 1979; he must have been 71 or 72. I have two younger brothers and both are in music. I am not very musical; I used to sing in the Cambridge University Choir, but that was about it. I tried to learn the French horn but I was very bad at it by comparison with my brothers.

You lived in Oxford but you went to study in Cambridge.

There were two very good reasons for that. One was to get away from home because up to the age of 18 I lived in Oxford at home. The other reason was that it was clear that Cambridge was the place one went to do serious science. I went to Clare College in Cambridge; both my uncles had been there. It was a family tradition. These days, this is completely out. Family connections used to be an advantage to get into a Cambridge college; now they give you a worse chance. Today people like to be squeaky clean and not to give anybody any favors; that is, the people who are letting you in. Recently I tried to help the daughter of my very good friend, who was very good, to get accepted by my old college, but I did not succeed.

The King of Sweden handing over the Nobel Prize Diploma and Medal to Tim Hunt, 2001 (photograph by M. Hargittai).

Changing the topic and exaggerating a little, I attended your Nobel lectures in Stockholm in 2001 and my impression was that Hartwell and Nurse found it very natural that they were giving their Nobel lecture whereas you seemed to be somewhat bewildered.

This is very true.

It was unusual because people usually try to project the impression that it was natural that they had been selected.

Let me put it this way: I knew that I'd made a Nobel-Prize-winning discovery, very much on day one when I made the discovery, but I was also aware of the fact that I was terribly lucky to have been catapulted into this distinguished society.

One of the 2001 chemistry laureates, Barry Sharpless declared repeatedly that others had predicted that he would win the Nobel Prize, they had even predicted the year, so he was prepared.

People say that sort of thing. Even one of the members of the Nobel Committee, his name is Anders Zetterberg, asked me some years before

At the Nobel Prize award ceremony, 2001. Tim Hunt is in the center first row, to the left, Hartwell, to the right, partly cut off, Nurse; behind Hartwell, to the left is Greengard, to the right, Stanley Prusiner, behind Greengard and Prusiner is Charles Townes; and behind Nurse is Martin Veltman (photograph by M. Hargittai).

about who should get the Nobel Prize; it was clear that our field was Nobel-worthy. There must be an important field and there must be some pioneers of the field, the Watsons and Cricks as it were, and then you should ask yourself, "Who should actually get it?" So Anders asked me, "How about you?" I told him that I didn't think that people win Nobel Prizes just because they are lucky. I'd known him for a long time beforehand and he was the one who made the presentation speech at the award ceremony in 2001 because he knew the field best.

So you are not a "professional" Nobel laureate.

More than that, I've always felt to be an amateur scientist. I let myself too easily astray, not serious enough. Some of these people are so driven. Growing up in Cambridge, I was tremendously well aware of all these laureates. They came and lectured to us; one met them in seminars; the Cricks, the Brenners, Perutz especially because he played quite a bit part in our teaching, Fred Sanger and his colleagues, to say nothing of the physiologists next door whose sons and daughters I knew. They seemed to me god-like figures.

Which years were these?

Not so much when I was an undergraduate, but more during my graduate studies. I remember when Jacques Monod came to give a talk in Cambridge in 1962 or 1963. Would he be by then a Nobel laureate?

Not yet.

He came to give two lectures; they were remarkable, absolutely fantastic. We knew nothing of what they were doing. It was not presented to us as undergraduates in the lectures; I don't know why. He was just brilliant. So we were confronted with this brilliance. We were only humble students, but we were encouraged. I'm very grateful for this, for my education; we were always encouraged. We were always told to think about these things. "This is what we know. How do we know what we know? What's the evidence of what we know? How we can find out the next thing?" A lot of science education just says, "This is what we know. Just learn it." In Cambridge, that wasn't that at all. I remember a wonderful series of lectures in physiology given by a man who made very distinguished contribution to the study of color vision. He didn't tell us anything about how the eye worked at all; he just interrogated the audience along Socratic lines, trying to elicit from them what the principles of the construction of the eye and what the basic principles and basic design features of the eye must be.

Your undergraduate studies were more than just lectures.

We also had supervision and I remember asking my supervisor about Monod's lectures whether it was true because if it was then it was absolutely remarkable. My supervisor said that it was too early to say, which meant that he hadn't himself read the relevant literature and thus hadn't appreciated the power and elegance of what Monod had done.

Did you have any interaction with the Laboratory of Molecular Biology?

I had a lot of friends there; my friend Steve Martin, for example, went to be a graduate student there. I knew them best much later, in 1974, when our lab burned down in the Biochemistry Department at Tennis Court Road, and we were relocated to across the road from LMB. Max Perutz was very helpful to us. The new lab used to be a teaching facility for analytical chemistry at the hospital site. Our problem was that we didn't have our stores because our real stores were three miles down the road in downtown Cambridge. Also, we didn't have a tea room. We went to ask Max Perutz if we could use his stores for supplies and come and have lunch in their tea room. He very kindly agreed. It was an amazing privilege because we got into interactions with the people of LMB. It took about 18 months before they refurbished another lab for us and sadly we had to move back.

In the year 2000, my wife and I spent three months at LMB, which was wonderful, but in the canteen we didn't find anything of the special atmosphere people talk about.

It was mainly Francis Crick's influence. Sydney [Brenner] never came to the canteen for reasons I have no idea. But Francis was very often there and Max was very often there. You would find them sitting there and you could come and sit with them. This was the time of the nucleosome discovery; and Francis used to like to explain what was going on with the latest results. He was very excited about it and it was very exciting to understand how DNA was packaged. We were just sitting around a table, maybe eight or ten people, just listening to these explanations. Francis was different from Sydney. Sydney liked to show off, to show how clever he was; Francis on the other hand, wanted to make sure that everybody around the table was completely on board, that no salient detail had eluded anybody because on that rests the next argument. He was very generous and a brilliant teacher. He was my great hero in other ways too because he went to an awful lot of seminars, probably a seminar a day. My friend Peter Lawrence and I used to organize a series of after dinner talks in Clare Hall where people would come and talk. We had a huge blackboard and the room was much wider than it was deep. Francis very often used to come and sit in the back of the room and he would very quietly just ask questions for clarification and I learned an awful lot from that. I heard him ask what struck me some rather ignorant questions and it became perfectly clear that he was asking the questions because he didn't know the answer and it was unclear without that. It was purely to elicit information. It was absolutely genuine. I only heard him once to be beastly to somebody and it was a hundred per cent justified. He said something like, "Dr. So-and-so, if what you told us is true, that would contravene the Second Law of Thermodynamics, so I don't think that your hypothesis is very likely to be valid." It was a devastating comment from somebody like Crick, but fully justified. The man was talking nonsense. Francis hated nonsense, above all he hated nonsense. You can see it from his books. He is so anti-clerical, anti-superstition, he writes about the introduction of the so-called Central Dogma and he says that he misunderstood it, but I don't think that he misunderstood it at all.

But he introduced it.

He did, but it's a religious term, a dogma is something for which no evidence exists, and he used it correctly. The term Central Dogma was meant to have this religious overtone. It's a joke, an article of faith. Then people got terribly excited and interestingly pleased when reverse transcriptase was discovered, because it said that the Central Dogma was wrong.

Didn't it make it more difficult to accept it?

No, no, no; it wasn't the contravention of the Central Dogma; the Central Dogma had to be slightly re-defined to say that information from nucleic acids can never get back and it was immaterial whether it was DNA or RNA. It was Crick who resigned from Churchill College, Cambridge, because they wanted to build a chapel, and did build a chapel, and Crick resigned his fellowship over that because he disapproved of it so much. In a curious way that was inspiring, there was a certain integrity.

I talked with him about religion and I sensed a little bit more careful approach about it today...

... than perhaps when he was younger.

He said that before we truly understand how the brain works, it brings only frustration to criticize religion.

Yes.

He didn't change his views on religion, but he almost seemed to have given up the hope that he could easily convince other people about his views.

I think a little humility along those lines doesn't hurt. People like to believe that stuff. I know about this because both my parents were extremely religious and I was brought up that way. As I think about it now, it amuses me. I actually thought that the virgin birth must be possible and that it was not pure faith. It's astonishing, but I did believe that. So it's no surprise to me when people get upset about all that stem cell cloning. It reveals to me dark ignorance and blind superstition and so on and so forth and I don't approve of it and I don't understand why they have a problem with it, but that doesn't mean that you have to go around slamming it in their face.

Jim Watson is also very anti-religious, but because of his position...

... he has to be careful.

Jim thought that Francis would have more freedom in expressing his views and was disappointed that Francis had also moved towards the center. Jim thought that Francis should be more radical.

[Tim Hunt is heartily laughing.] Millions of people believe. If you scratch me, I'm a great rationalist and I do believe that the truth wins over superstitions. If anybody tells me that such and such is the truth, I always want to know why, what's the evidence for that and to take nothing whatsoever on faith, even a scientific protocol. I'm rather slow and sluggish because of that, because I like to find out the truth for myself. This is because I know how easy it is to deceive oneself. All the time we are constantly deceiving ourselves in one way or another and this is not a good thing.

You have mentioned Hugh Pelham and Peter Lawrence. I have met both in Cambridge and was very impressed by them.

This was another thing about growing up in Cambridge. I had a succession of three students. Hugh was the first. The second one was my present boss Richard Treisman and the third was Andrew Murray who is now a professor at Harvard University. These three students came in successive years and they were so brilliant that I thought it was time for me to retire; I really did. They were just miles cleverer than me. They were deep thinkers; brave scientists. There is something that people don't realize what Mark Ptashne spoke of as the psychic risk of being first.

In what sense?

It's actually frightening to be out of the limits, chasing after things that might not exist.

Did you have that feeling?

No, not really, not so much because of the lucky element. I've known many different kinds of science in my life. The most fun is to make a discovery, there's no question about that. To find out something that nobody knew before. I've had a bit of luck in that before I stumbled on the cyclins. I had a very good project. I was trying to understand how heme controlled globin synthesis and I made a discovery that double-stranded RNA was a very powerful inhibitor of protein synthesis, and found it completely by accident. That was very pleasing and very exciting. Although I stumbled into it, it came about in a completely rational way.

Was it your first discovery?

My very first discovery was again by accident. It was a much more trivial thing. I discovered that there are fewer ribosomes on the messenger for RNA on the alpha chains of globin than were on the beta chains of globin. The reason that we discovered that was that we let the centrifuge run too far and the analysis showed this and when we realized it then we went after it. Then, interestingly, we made a silly mistake, we omitted an absolutely

Tim Hunt's portrait by Gabriele Seethaler.

essential control, and we came up with the wrong explanation for the phenomenon, which was later put right by someone else. That was a great lesson for me.

You published the wrong explanation?

We did; the right facts with the wrong explanation. It was irritating because it was a really simple control; I see this in my students all the time; you omit the really simple controls. Since then I've been very fond of controls. This is an interesting difference between biology and physics. Physicists have no controls; the concept is almost wholly alien to them. Biology is more complicated. If you don't check every inch of the way that you are not making a silly mistake — what would've happened if you hadn't added this to the tube or something like that, and so on — then you do make silly mistakes. Ours was a perfect example. Then we also missed something tremendously important in doing those experiments. We could've easily found out that methionine was the first amino acid in all proteins and we didn't. Actually, we were looking for it, but we didn't take our data seriously. You have to notice even little things. We had a peptide on a piece of paper, a two-dimensional chromatogram, and the radioactivity in this peptide was always lower than it should've been. The reason was that the peptide often had methionine stuck on the front of it so it ran somewhere else and we didn't know where it was and we had no way of finding out where it was. We put this low specific activity down to the fact that it was contaminated by a nearby peptide which we knew came from further down in the chain and therefore we thought we had a sufficient explanation, that is, contamination. If it had been a well separated spot then its low specific activity would have immediately alerted us to the more interesting possibility of there being something different.

Didn't the reviewers notice that you'd missed something just by looking at your data?

They didn't notice, and it just worried us. When I was an undergraduate, I was always better with my hands in a practical sense than in writing essays. I wasn't a terribly clever person. I do know how to trust actual results. That became very important later on. If I have a talent, I am good at actually spotting things that don't quite make sense and at realizing that there might be something behind that. It's the unexpected that makes the really great discoveries.

Did you have other misses?

We could have done much more when I was a graduate student. We could've discovered messenger RNA, for example, not the bacterial one, but the mammalian messenger RNA whose existence had been denied. Later on — with Hugh Pelham — we developed a good assay for messenger RNA, a reliable one that is still sold commercially to this day. Very naively, we didn't have the concept of patenting although it was a very good idea. This was in the mid-1970s. Patenting at that time was a very alien concept. The idea of making a lot of money just by selling kits didn't occur to us. Later on we became involved in developing a kit for Amersham International. They were making radiochemicals for biological research mostly, but their most profitable and largest-selling item was ^{35}S-methionine. Their American competitor had the brilliant idea of selling ^{35}S-methionine together with a reticulocyte lysate kit on the side as an added extra. They were making serious inroads into Amersham sales. So the Amersham people came to us for help and Hugh and I consulted their brilliant chemists, who, however, didn't know anything about biology. We had to teach them how to make assays for protein synthesis. It was very interesting to work in this commercial environment because the methodology was exactly the same as what we normally did except we weren't finding out anything more significant than how important the water color was and what the best shape for the tube would be. After a while we got very fed up with doing that kind of routine work because it was so repetitive. It was also helpful because it made me realize why I didn't want to work in industry. It was fun and challenging in the beginning, but ultimately it was purely a matter of engineering and money-making.

Then came your work with the heme.

It was very good also because it introduced me to the concept of scientific competition. In 1966 I'd originally gone to work with a guy called Irving London, who is still alive. I spent three months in his lab in the Bronx, at the Albert Einstein College of Medicine. He was the head of the Department of Medicine, so he was a very big figure in the whole institution, but he and I got along very well. We shared this common interest of how heme controlled globin synthesis. I went back to his lab in 1968 for about two years as a postdoc. Then he moved to MIT, but he didn't have a lab there so I didn't want to go with him and I returned to Cambridge (England) where I worked with my friends, Tony Hunter and Richard Jackson. We began studying this heme business and I devised a very nice assay. It was more by accident than by design, we solved the problem and found that a protein kinase was involved and that relied importantly on the discoveries I'd made as a postdoc. As I mentioned earlier, the discovery that the double-stranded RNA was a powerful inhibitor led to other discoveries. We were seriously baffled at various stages and then the lab burned down, which was a very purifying thing. It felt like making a perfect confession; we had had a lot of confusing data and after that, very quickly, the findings just fell out.

You tend to use religious expressions. What was your background?

I grew up in a very high Anglican tradition.

So there were confessions.

Oh, yes. We used to go to the stations of the cross, and our church looked exactly like a Catholic church. It was within a hair's breadth of being Catholic. My mother towards the end of her life became very religious and she was Catholic in all but name. I don't quite know why.

When you made the reference to your knowing from day one that it was an important discovery, I suppose that was the cyclin discovery.

Yes. It came in a period of my life when I was getting very depressed. The word depressed may not be the right word; it's just a feeling that you'll never make another important discovery in your life. You're barking up the wrong tree and nothing quite works and the experiments you're doing are not particularly important. I was in such a mood. I was trying to explain how sea urchin eggs turned on protein synthesis after fertilization. This is a difficult problem and it hasn't been solved yet and I'm talking about something that happened 25 years ago. I was doing this simple experiment, which showed that this protein suddenly disappeared. I knew enough to be aware that this was a real result; the protein did disappear, it must have been broken into pieces, it just vanished. But we had no idea how it was possible. At that time the ubiquitin system, although it had been discovered, it was not yet interpreted in the right way; it was thought to be a garbage can in the cell. The idea that it was used for identifying particular proteins to get rid of them on a biological signal was very far in the future. I had a conversation with John Gerhart who knew about the magical stuff of the maturation promoting factor (MPF) in frog eggs and he told me — and it was the day after I had observed the protein go away — that you needed new protein synthesis for the reappearance of MPF at mitosis II in frog egg development. It sounds like a totally technical thing but it was actually exactly the kind of behavior I was looking for. It suddenly said that you need protein synthesis to enter mitosis and you need then to degrade a protein to get out of mitosis before you enter the next interface. I couldn't believe my luck because it was very clear that the result was true and I knew it was profound and meaningful. I knew all this, but I had to explain it to people and even my own graduate students didn't understand. I'm probably very bad at explaining myself, but for me it was clear that I'd made by far the most important discovery of my life. It was precisely because of its unexpectedness and its improbability; it was so improbable that even the experts of cell cycle control not even playfully had suggested such a possibility. People had deeply missed the importance of this; they hadn't anticipated this at all, which is why I said earlier that I'd seemed amused and I was bemused precisely because I was bewildered. Why me, why was I given this particular discovery? Luckily, I have come to terms with it and thought that if I hadn't been Tim Hunt, I would've taken Tim Hunt very seriously as a candidate for the Nobel Prize. But because I am Tim Hunt and I know what it means to be in my own skin, I can think of an awful lot of people who are a lot more respectful, who are much more deserving. There are lots of people who are never going to win the Nobel Prize, who are better scientists, but didn't have the good luck to be in the right place at the right time.

Does the Nobel Prize set you apart from your former colleagues?

[after a long silence] Well, I don't know.

How do you feel about Peter Lawrence?

I'm still terrified of him. There are a lot of scary people around. But — as my wife said — the Nobel Prize has greatly increased my self-confidence. People automatically take you more seriously.

Fortunately, you don't convey the feeling that you are aware of this. There are others whose outlook changes and who seem to take themselves more seriously after the Prize.

Most Cambridge people did not. Francis is a good example. There is a funny story about Fred Sanger who is also an extremely modest and self-effacing man. When Sanger's second prize was announced, people came together in the canteen of the LMB and Fred made a small speech. He said that, "A lot of people say that I'm a very modest man, but, actually, I'm bloody good."

In our correspondence, this topic of modesty came up and he signed his next letter to me as "Fred Modest Sanger".

I once had the wonderful privilege of sitting for two hours together with Bart Barrell who was Fred's faithful technician for many years, who got his Ph.D. on the basis of all the papers they published together, which is a stack two or three feet high. Bart explained to me how they invented DNA sequencing, which was a wonderful exposition of the randomness of biological research. There was no design of that research. I used to have dinner with Fred's postdocs in the hospital canteen and even his own postdocs couldn't explain it clearly what they were doing; everything in the "plus and minus" method seemed to be so complicated. Then he invented the dideoxy method and suddenly everything became very simple. But as they were working their way up, some people didn't take them seriously; Sydney in particular didn't take Fred seriously. Fred was always very kind to me, but if you met him, he didn't strike you as a sort of person of a great intellect; you didn't hang on his every word; he could be the gardener or the janitor, more or less. But he had this fantastic intelligence in his fingertips and he knew what he was looking for, it was the multitude of all those little things that he knew that made his method possible, but it wasn't the things that the Cambridge people really respected.

How about Max Perutz?

My impression was that he always had a little higher opinion of himself than he would've liked to have of himself, if you know what I mean. It's hard to explain what it was, but he was a very great director of the Laboratory.

He was fantastic in carrying through the studies with the isomorphous replacement method in his protein studies, but I also felt that a little more credit might have been given to Robertson and others who had introduced the technique for small molecules and to Crick who first suggested its application for proteins.

People should be very sensitive to priorities in science. I would like to make a remark about that in connection with the cyclin discovery. Unlike the development of the rabbit reticulocyte lycate as an assay system — which was done by Hugh Pelham at my suggestion — and unlike the solution of the heme and the double-stranded RNA inhibition — which was done by my graduate students — what was nice about the cyclin business was that I conceived and did the experiment entirely by my own hands without any input by anybody else. It was absolutely and entirely my own from the very first moment. That was tremendously comforting to me. It's a weird thing, the business of giving credit. But this was an exceptional case and especially in biological research where things are so complicated, interactions among scientists are very important. "One tiny fact can illuminate a corner previously dark," said the great German biologist T. H. Bovari at the end of the nineteenth century. It's exactly that.

Of course, there was the controversy about the information from Rosalind Franklin's X-ray patterns channeled without her knowledge to Watson and Crick.

I am very much for channeling such information by whatever means if it helps science. Once I heard Crick talking about this very nicely, about how simple things were and how you had to keep an eye out for them.

I attended a seminar once where Francis and Sydney spoke about how they'd determined the sequence of the chain-terminating codons. It was their first announcement of this tremendous discovery and it was astonishing how trivial the experiments were. It was a question of the bacterium growing on this medium but not on that. The actual mechanics of the experiments was nothing, but the intellectual force that was involved in the interpretation was grand and brought out this great truth about how the genetic code actually worked. The magnesium concentration was very important and you better be aware of it, otherwise the experiment wouldn't work. It's hard to convey the sense of how research in biology is supported by that sort of silly little random-seeming facts.

You work now for a cancer research institution.

It's called Cancer Research U.K.

How does what you do relate to cancer research?

With great difficulty. We are trying to understand cell division, but just because you understand how cell division is controlled doesn't mean that you can understand how to control cell division, particularly not at cells that are going out of control. My view on this is that despite the great advances that many of my distinguished colleagues have made during the last thirty or forty years, so that we do understand why certain cancer cells go out of control, we are still very far away from understanding why they grow when the normal tissues don't. We don't yet understand many ordinary things. Take our nose. It grows together with our face and then it stops growing at a certain point and then it stays the same, does not shrink either. How is this controlled by DNA, we don't know. How DNA controls growth and division is deeply mysterious. Me and my colleagues have discovered the fundamentals of the control of cell division; it's a fine place to start, but it's only the very beginnings. In a sense it is important for understanding cancer, but it's only a very modest contribution to it; it may even not be a very relevant first step because we don't understand how growth and division are coordinated; the two are very different. We can have growth without division and division without growth. It's not yet understood how the two are coupled together and how that coupling can be abrogated.

The drugs that are around?

Many of the drugs that are around block cell cycle progression. We now understand how they do that a lot better than we did twenty years ago. But why they have any preference of killing cancer cells is in most cases not very clear.

Do they differentiate?

Some of them do, but most don't. The depressing truth of the matter is that they just kill everything. If, for example, you inhibit cyclin-dependent kinases, it's just like standing close enough to an atomic bomb so that you get a lethal dose of ionizing radiation, but not close enough so that you don't get burned up by the infrared. After a very nasty seven days or so you will die. The same would happen if you gave people a drug that would inhibit all these enzymes; all the cells would stop subdividing and you would have no more blood, no more hair, and your guts would no longer have a lining, and that would be the end of you. There is no particular reason to think that the cancer cells have in them the problem. The cancer cells divide all too well, unfortunately, and if there was a problem with their cell division control machine, they wouldn't divide so well. I am all for developing these drugs and to see what they do on cells of mice because you never know when you may be lucky. However, as far as I know, so far nobody has been lucky. Commercial research is funny stuff; a lot of these data are not published; most of it is done behind closed doors in drug companies.

How does this work with intellectual property? There is a murky line between university research supported by public funds and company research for the benefit of the company.

I came across some colleagues in the Biochemistry Department when I was already a faculty member who worked on these toxins that killed insects and that have been made into transgenic plants being very good insecticides. They refused to talk about it because their work was patented and I hated that. I didn't think that belonged in an academic institution. But I'm not against patenting and we should have patented some of our discoveries and I probably would be much richer had we done so.

Did you ever patent anything?

Never, never. I find nothing wrong with it but I don't like the secrecy aspect of it. It's partly a European problem; the disclosure regulations are a little different from what they are in the States. It's easier for people in America to publish stuff at the same time as having a patent. In Europe, once you published it, you can't file for a patent. Secrecy is a terrible enemy of science in my view.

You were born in 1943 and you remember the difficult post-war years and you mention in your Nobel autobiography that you used to "break eggs separately into a cup lest one should be bad and the week's omelette be spoiled". I'm just trying to provoke you, but is this the most negative experience you have ever had?

I count myself a very fortunate man. I grew up in a very loving family. I wasn't pushed particularly hard by my parents. The expectations were high but not explicit. At times in my life I have been intensely lonely and at other times unhappy, usually because of girls, and there was a period of three months when I had no income. It was because of some administrative mix-up.

You've met so many stellar scientists. Did you ever feel intimidated? Did you ever have any hesitation choosing this career?

No, because I must have formed my self-confidence along my way. At the end of my first year in Cambridge I was convinced that I would not pass my exam and that I would have to leave. I realized how little I knew and how little I understood. But I actually got a first-class degree. That was to me quite extraordinary and quite undeserved because I was aware of my limitations. I just kept going. I was very lucky not having to make any decisions. While I was single, I took three months off every summer and followed my nose, but passed it very valuably. There's something else. Today people are trying to measure everything, how many papers you write and how many citations you receive. It's not a very good measure. What actually matters is the opinion of your peers.

What would you do being an administrator?

I would never be one. But I sit on lots of evaluation panels and I realize that it's a difficult thing. When Paul [Nurse] was in charge of ICRF [Imperial Cancer Research Fund], he and I often used to sit on appointment panels and we differ a great deal in attitudes and politics and backgrounds and upbringings, but we always agreed about candidates. We had a very similar taste in scientists, which was very often not shared by our colleagues. So there is a question of taste rather than actual productivity.

Taste in science?

Yes. There are lots of ways of doing good science. There are certain things that you look for, little twists, a certain wit, a measure of unexpectedness about discoveries. Somebody may follow a very winding trail through to the end to great effect, and sometimes along the way you don't have much to show for it. You have to have a lot of trust in people to allow them to follow those trails.

You have made this remark that it is very different to teach crystallography and biology or biochemistry.

Shortly after learning that I had won the Nobel Prize, my daughter Celia asked me this funny question, "Why is the ceiling opaque?" I looked out of the bedroom window and while I didn't have any problem with the

Mitsuhiro Yanagida, Aaron Klug (then President of the Royal Society), Kim Nasmyth, and Tim Hunt after Yanagida receiving foreign membership of the Royal Society (London).

ceiling being opaque because of the absorption of light, but I asked myself another question, "How do the photons get through glass?" I realized that I deeply didn't know the answer to that question. Soon there was an occasion at the British Ambassador's reception for the British Nobel laureates in Stockholm, I was sitting next to Aaron Klug and I asked him this question. He told me that I needed to understand Schrödinger's equation. At that my heart sank. Not only did I not understand Schrödinger's equation, I didn't even know what Schrödinger's equation was. Later on I had to give a talk in Alpbach, which is where Schrödinger and his wife are buried and his equation is engraved on the headstone of his grave. There is a mistake and the physicists passing by have corrected it. They had to black out a dot on both sides of the equation. Subsequently I reread Schrödinger's book, *What Is Life?* and also his very interesting biography. He was constantly falling in love with people and god knows what his marriage was like. I also discovered that there is a huge amount of unexamined papers; I learned about it from talking to his daughter. So I started learning about how light gets through glass and I realized how hopeless it is for a biologist to begin to understand quantum mechanics. The only exam I ever failed, apart from a driving test, was in mathematics. I've realized that there are big differences between the physical sciences and the biological sciences. Basically we have a simple Aristotelian view of the world; when you push things they move and when you stop pushing they stop moving.

So what is the difference in teaching the two subjects?

Teaching biology is a mutual voyage of discovery for the teacher and the taught. That made it interesting to teach because you didn't know all the answers. Teaching crystallography is much better defined. You understand Bragg's law and inverse space.

Reciprocal space.

And I have no idea what reciprocal space is. Even Max Perutz's "Crystallography made easy" course passes me by. I accept that it works, but how it works is completely opaque to me. Now I have some understanding of how photons pass through glass although I still feel a lot of mysticism about the details.

How many children do you have?

After the Nobel Prize award ceremony, the Hunt family with their Swedish hostess: Kristin Forsgren (attaché), Tim Hunt, Agnes B. Collins, Mary K. L. Collins, and Celia Collins.

Two little daughters; second marriage; no children from the first. People ask me what it's like to be an older father and I have no idea because I never did the control. Celia Daisy is now ten years old and Agnes Beatrix is six and a half.

What does your wife do?

She is Mary Collins, Professor of Immunology at University College London. She works on viruses, engineering the AIDS virus to make it practically useful. She was my student originally, then we were friends for many years; she pursued me ruthlessly.

You have been in the forefront of your field for quite some time. There are others in the biomedical field who make a discovery, maybe become rich and disappear in oblivion.

There are different scientists; some make discoveries and others like to clean up the field and I belong more to the latter although I have made discoveries too. But I feel more comfortable letting people go ahead and me moving at my own relaxed pace. Very often the two don't go very well together. Go back to our hero Crick, he only did one famous experiment about the genetic code with Leslie Barnett and he told him, "Leslie, you and I are the only two people on Earth who know that the genetic code is a triplet." I love that.

What is the connection of your work with the ubiquitin discovery?

That was pleasing to me because I always thought that it was probably ubiquitin. One of the reasons for thinking that was because the pioneers, Hershko and Ciechanover worked in reticulocyte lysate, so I knew about their work quite early on. At one time I was teaching in Woods Hole and Ciechanover delivered a lecture, a most beautiful lecture, to the physiology course, some time in 1982 or 1983. Then Ciechanover, Varshavsky and Dan Finley published their two papers very close to the time when our publication on cyclin came out and I wrote to them. I asked them if they thought that the two might possibly be connected. They sent me a rather guarded letter back. I think the letter was signed by Varshavsky and Finley and I didn't pursue the matter further.

Embryology course at Woods Hole, 1979; Tim Hunt is fourth from left (with glasses).

Tim Hunt, John Wilson, Martin Raff, James D. Watson, Gavin Borden, Julian Lewis, Bruce Alberts, and Keith Roberts in San Francisco, 1989, marking the second edition of *Molecular Biology of the Cell* by Watson *et al.* and the first edition of *The Problem Book* by John Wilson and Tim Hunt.

Did you ever come across Gordon Tomkins?

I worked in his lab for two weeks during my stay with Irving London; I went out to San Francisco. He was a very inspirational man. When he was giving a lecture he had the audience helpless with laughter even though he was talking about a very boring enzyme. That's when I met Hershko for the first time because he was a postdoc with Tomkins at the time.

Why was the ubiquitin prize in chemistry rather than in physiology or medicine?

Because the medicine prize would have been too difficult to decide who should be included. It was easier in the chemistry prize.

Do you feel comfortable to talk about this?

I'd rather not talk about this because I know a little bit about what went on behind the scenes so it's better not to go into it.

What do you read?

Most of what I read is scientific memoirs. I've read Jacob's *The Statue Within*, and I enjoyed Abraham Pais's books about physicists.

What do you like to do outside your lab?

I like to meet other laureates and I like to meet astronauts. I have met two American astronauts, one French, and one Japanese. Of course, I had a chance for this because of the Nobel Prize. The astronauts are usually small people to save fuel, but they have wonderful personalities because you can't risk having personal animosities up in space.

Do you have any comment at the conclusion of this conversation?

Not really, but it's nice to be focused on.

Seymour Benzer, 2004 (photograph by I. Hargittai).

6

SEYMOUR BENZER

Seymour Benzer (b. 1921 in New York City) is James Griffin Boswell Professor of Neuroscience, Emeritus (Active) at the California Institute of Technology (Caltech) in Pasadena. He received his B.S. degree from Brooklyn College in 1942 and his M.S. and Ph.D. degrees in 1943 and 1947 from Purdue University, all in Physics. First he was at the Physics Department and then at the Biology Department at Purdue. He has been at Caltech since 1967. He is most famous for his discoveries on genes and behavior. He created Drosophila mutants by chemical intervention and studied their behavior. Prior to that, he was one of the pioneers at the cradle of molecular biology, with his work on the fine structure of the gene, and its relation to the structure of DNA.

Dr. Benzer has had extensive postdoctoral experience with worldrenowned professors, including Max Delbrück at Caltech, A. Lwoff, F. Jacob, and J. Monod at the Pasteur Institute in Paris, Francis Crick and Sydney Brenner in Cambridge, U.K., and Roger Sperry at Caltech.

He has been exceptionally distinguished by various awards and other recognition. He is a member of the National Academy of Sciences of the U.S.A. (1961), the American Academy of Arts and Sciences (1959), the American Philosophical Society (1962), foreign member of the Royal Society (London, 1976), the French Academy of Sciences (2000), the Indian Academy of Sciences, and member of many other learned societies. His numerous awards include two Gairdner Awards (1964, 2004); the Lasker Award (1971); the Prix Charles Leopold Mayer (French Academy of Sciences, 1975); the National Medal of Science (1983); the Wolf Prize (Israel, 1991); the Crafoord Prize (Sweden, 1993) and, more recently, the Bower Award for brain research (2004), and the Gruber Award for neuroscience (2004).

There is a full-length book about him and his work by the Pulitzer Prize-winner Jonathan Weiner, *Time, Love, Memory: A Great Biologist and His Quest for the Origins of Behavior* (Alfred A. Knopf, New York, 1999). We recorded our conversation in Seymour Benzer's office at Caltech on February 5, 2004.*

We started our conversation by remembering some members of the famous group of the great Hungarian scientists.

Paul Erdös was one. It was in the 1950s, and I was at Purdue University doing physics at that time. Erdös was a Visiting Professor in the Mathematics Department. I had many friends in mathematics and used to talk with Erdös; I have a picture of him with my daughter Barbie sitting on his lap. He was very kind to children, but only up to a point. Like Szilard, all of a sudden, like closing a door, it was finished. Once Erdös was at our home for dinner and offered to wash the dishes but we didn't let him because we were afraid that he would smash them. He was deeply offended, and protested that Ulam had allowed him to do his dishes. As you know, the anecdotes about Erdös are endless, and there are full books about him.

I remember seeing Teller and Szilard having a debate on TV, and they were disagreeing on everything. The remark I remember best was when the interviewer asked Teller about a particular situation: "If you were President of the United States, what would you do?" Teller said that his being President would be such a tragedy that he hoped it would never befall him. But, if such a disaster should happen, the first thing he would do would be to consult with Szilard. Obviously, Teller was a very brilliant person, and developed the hydrogen bomb. His interaction with Robert Oppenheimer was a dilemma and he lost all his friends. Concerning his passion for SDI, I'm not impressed. It keeps failing, and doesn't seem very promising.

After Szilard turned to molecular biology, and later to neurobiology, I met him on many occasions. We were both involved in the formative period of the Salk Institute. The decision to build it in La Jolla was in part due to Szilard's claim that there was a good Chinese restaurant in

---

*István Hargittai conducted the interview.

San Diego. I later checked it out — it was not great, but La Jolla was a fine location anyhow. Szilard could sometimes be abrupt. On one occasion at the Salk Institute, after listening to a neurobiologist's explanation of how the brain works, Szilard declared, "Your theory is OK for *your* brain!" and he walked out.

Where did your parents come from?

They came from a small shtetl outside Warsaw called Sochocev; they knew each other there, but did not really get together until after they had emigrated to New York about 1910, and married there. I was born in the Bronx and we moved to Brooklyn when I was 4 years old.

Biology had actually been my first interest in high school. When I came to college, I thought I knew enough biology to take advanced courses, but they told me that I would have first to take all the elementary courses on plants and animals. Being an arrogant young jerk, I said, "The hell with it," and took chemistry instead.

How did you happen to be at Purdue?

I studied chemistry and physics at Brooklyn College and became a physics major, and was offered a graduate student scholarship by Purdue. My physics professors told me that it was a rapidly developing department under Lark-Horovitz, which was true. It was very exciting for me. I was 20 years old, married Dorothy Vlosky and, on the same day, we left New York. It was a Saturday so our wedding had to be after sundown, and we had to rush to the train, leaving the others to celebrate.

For religious reasons?

It was a Jewish wedding, performed by the same rabbi who had circumcised me! I was not, and am not religious. My parents were nominally observant; my mother used to bless the candles on Friday nights and maintained a kosher kitchen. But on Sundays, when we often went to a Chinese restaurant, she would pull the bits of pork out of the chow mein and put them aside. I went to Hebrew school and did the Bar Mitzvah thing. My father would go to the synagogue only on Yom Kippur and Rosh Hashana, and I went with him out of respect. I used to bring a physics book to read on top of the prayer book. I remember, specifically, Stern and Gerlach, *The Principles of Atomic Physics*.

You started out as a physicist.

As a graduate student at Purdue, I worked on germanium semiconductors, having joined the department in January 1942, just as the United States had entered the war. I worked in a group headed by Karl Lark-Horovitz, who was head of the Physics Department. We had a project with the National Defense Research Council to develop detectors for radar. The problem was that, at high microwave frequencies, they couldn't use vacuum tubes. They were using just a metal wire in contact with a semiconductor, the old-fashioned crystal rectifier. These were rather delicate in two ways. One was mechanical breakage, and the other was burnout when a large signal would come in; they had frequently to replace the detectors. We had to make crystal detectors with greater stability.

Did you succeed in developing something that was eventually used?

The first crystal diode that came into use indeed was our germanium diode. In the course of the project, we discovered how to prepare the germanium. First we had to make it very pure, then dope it with impurities to provide surplus electrons or holes to produce the *n*-type or *p*-type materials. One of the things we developed was the right combination so that the contact would withstand a high inverse voltage. We also developed information about the properties of the junction between *p*- and *n*-type germanium and photoelectric effects, including a version of a "solar battery" using germanium, although silicon, being much more abundant, later took over. We were persuaded that our techniques would be better exploited by Bell Labs and in the interest of the war effort, we should give them all the information, so we did. I had been trying to make my own version of a transistor by a very sophisticated method, based on tunnel effect. I had also been playing around with having two wires contacting the semiconductor, one passing current and the other detecting what was happening around it, in response to a magnetic field. But I never had the idea of using that configuration as an amplifier, and that's what the people at Bell Labs succeeded in doing. At the time the transistor was announced, I was attending the bacteriophage course at Cold Spring Harbor. Lark-Horovitz called me to go to the demonstration at Bell Labs, a grand PR event. As I came in, Walter Brattain, one of the inventors, told me, "This should have been you!" But I was already into biology, so didn't care very much.

When did you switch to biology?

While at Purdue, I was captivated by learning that the genes can be mapped on the chromosomes. A key event was reading Schrödinger's book, *What Is Life?* You must have had several customers in your *Candid Science* books who made the same switch about the same time. I felt that one might be able to define the parameters in biology, as one does in physics. Francis Crick was also among those who switched to biology because of *What Is Life?*

He read the book but as far as I know it was only one of several books he read when he was trying to find his place. By the time he read these books he had already made up his mind to move to biology and was only hesitating between molecular biology and neurobiology. He was attracted by the mysteries of life and consciousness.

There were some wonderful ideas in that book, for example, that the gene is an aperiodic crystal, which was an incredible insight.

It was remarkably prescient because aperiodic crystals were discovered in 1982 and are known today as quasicrystals.

The thing that got me in that book was a chapter called Delbrück's Model in which he describes a mutation as a transition from one metastable state to another. That was close to what I was familiar with in semiconductors. I was also interested in what happens across the nerve membrane and discussed that with a friend, Lou Boyarsky, who was involved in neurophysiology. The difference in potential across the nerve membrane reminded me of the metal semiconductor contact. I was looking at neurophysiology also as a possible place where I might go.

The real turning point for me was a trip to Bloomington, Indiana, to a Physical Society meeting. I went with Ralph Bray, an old classmate from Brooklyn College and a fellow graduate student in the same semiconductor lab. Ralph had been invited to dinner at the house of his friend Zella and took me along, and there I met Zella's husband, Salvador Luria. I asked him if he knew anyone at Indiana University who worked on viruses, because I had become interested in viruses from Schrödinger's book. And Luria said, "Yes, *I* work on viruses." When I asked him whether he had ever heard of Max Delbrück, he laughed, and took out a picture showing them together. Luria told me about their work on bacteriophages, and recommended that I go to Cold Spring Harbor to take the phage course, which I did the following summer. The course had been founded by Delbrück but, by that time, was being taught by Mark Adams. After three weeks of total immersion in bacteriophage experiments, I was completely hooked.

I understand that Max Delbrück was an important influence on you as on many others. He had switched from physics to biology because he had expected that new laws of physics might be found there.

Delbrück was captivated by Bohr's idea of complementarity and the uncertainty principle, which Bohr had tried to apply to everything: history, economics, etc. I don't know whether it was Bohr who first articulated the idea, but when I came in, Delbrück was expressing the notion that, in order to predict the future of a cell, you would have to make so many measurements that the cell would necessarily be killed. In other words, life and determinism were two components of complementarity. To me, that was a very captivating idea, and I wondered how one might define these components.

Would you care to try to formulate Delbrück's scientific contribution? He was a leader with a large following undoubtedly, but the people whom I have asked and who had been close to him find it difficult to formulate his research contribution. He received the Nobel Prize in 1969 together with Alfred Hershey and Salvador Luria with an umbrella citation for their discoveries in the replication mechanism and genetic structure of viruses. Did he make a big discovery?

His influence went beyond that. He created a whole school of thought. In my case, Luria suggested that I ask for a copy of three lectures Delbrück had given at Vanderbilt, where he was then in the physics department. One lecture was precisely about the complementarity principle, and that was quite fascinating to me. The others were about the methods he had designed for quantitative work on bacteriophage. I had naively thought that a bacteriophage was the closest thing to a naked gene, and if you wanted to duplicate the gene, that was the best system to work with. What Delbrück taught us was how to do the experiments in a stepwise, quantitative way, which was very appealing to a physicist, in contrast to typical biology, which was descriptive. This was a system that you could analyze. It was also a mystery; you start with a bacteriophage particle that attacks the bacterium, and 20 minutes later a hundred descendents come out. What was happening in between? It was a tremendous challenge. That was what I tried to work on when I started in biology, using ultraviolet radiation to penetrate the cell, to analyze what was happening during this latent period.

Although Delbrück had tremendous stature, it was legendary how many times he could be wrong. But, as Jim Watson once told me, it was always for an interesting reason — in contrast to some other people. Max was always a challenge. Gunther Stent summed it up very well when he said that Delbrück was "a conscience and a goad"; he set a standard for honesty, openness, and objectivity, but at the same time he stimulated you by being skeptical. Even after you discovered something, he would say, "No, I don't believe it." One famous example was when my student Ron Konopka and I discovered mutations in flies that change their sense of time, the rhythm of their daily lives. I told Max about it and he said, "I don't believe it, that's impossible." Even when I told him that we had already done it and we had found the gene, he only repeated, never mind, that's impossible. I must tell you that that was not an unusual event. Yet, somehow, he was a higher-level person, an inspiration to the people around him, and I was very fond of him and his spirited wife, Manny. Working in his lab at Caltech brought me in touch with wonderful colleagues like Renato Dulbecco, Jean Weigle, Gunther Stent, Elie Wollman, and others.

When the double helix was discovered, did you realize its importance?

I was there when Watson announced it at a Cold Spring Harbor meeting in 1953. That's when I met Sydney Brenner, and he explained its ramifications to me. Sydney and I started to cook up schemes to work out the correspondence between gene structure and protein structure. Soon afterwards, I discovered a system where I could detect genetic recombination on a very fine scale, sensitive enough to detect recombination even between two nucleotides. That made it possible to map the internal structure of a gene and relate it to the DNA sequence. I was able to show that the gene has a linear structure, the same as for DNA, and that there were many sites of mutation within the same functional unit. In that way, one

Edward Lewis and Seymour Benzer imitating a fight in Benzer's office at Caltech in 1999 (photograph by I. Hargittai).

could define the boundaries of a gene, and we showed that one could make mutations by putting in base analogs or other chemicals that change the bases. We also showed that there were different hotspots of spontaneous mutability within the gene. Brenner and I wanted to have a gene for which the protein was known, so we could measure the sequence of amino acids in the protein corresponding to the sequence of nucleotides within the gene.

When was this?

In 1954–1955, Sanger had determined the amino acid sequence of insulin. But that was before the double helix. I remember a lecture by John Edsall in Urbana in which he showed the insulin sequence and asked, "Where does that come from — one can only gaze and wonder!" When the DNA structure came out, the question was, how to understand the correspondence between gene and protein. I had already started to work out the sequence of the *rII* gene in great detail, but did not succeed in finding the corresponding protein. Then I spent a year in Cambridge with Brenner and Crick, trying, among other things, to find a bacteriophage protein for which the gene's fine structure could be mapped. In the end, Charles Yanofsky was the first to succeed, using bacterial tryptophan synthetase.

Who thought of it first that there should be what we call today the genetic code?

I believe that Watson and Crick thought of it first, as a direct consequence of the DNA structure. Gamow had a conjecture about what an actual code could be, but that was quickly ruled out. Watson gives a historical description of all that in his book *Genes, Girls, and Gamow*.

Crick felt — on structural grounds — that you could not go directly from the nucleic acid to the protein, but that an amino acid had to have a transfer RNA adaptor that would read the appropriate nucleotide sequence, which Brenner named the codon. Benard Weisblum at Purdue and I were working on which amino acids are attached to which transfer RNAs. Our idea to test the adaptor hypothesis was to take the adaptor for one amino acid, then chemically change the amino acid to a different one, and see whether it goes into protein according to the specificity of the adaptor or the amino acid. Bill Ray, one of our colleagues at Purdue, suggested that cysteine could be changed to alanine by a chemical reaction, using Raney nickel. Before we had completed the experiment, Chapeville in Fritz Lipmann's lab did it, and we all published a paper together. The alanine went into protein in the place of cysteine. That was the proof of the adaptor hypothesis. I have always liked that kind of experiment, where you keep one thing constant and twist the other one.

We then started fractionating the different transfer RNAs. Robert Holley had developed a method for fractionating the tRNA adaptors for the different amino acids by a method called countercurrent distribution, using repeated transfers involving positioning between two solvents. Eventually, two molecules, even with slightly different affinities, become separated. We were able to show that, for a given amino acid, there were multiple adaptors corresponding to different codons, thus providing a physical basis for degeneracy in the genetic code. By that time, the coding problem was hot. After Nirenberg's breakthrough discovery in 1961, he, Ochoa and others rapidly found out which triplets corresponded to which amino acid. Later on, the triplet nature of the genetic code was elegantly demonstrated by Crick and Brenner, using my *rII* mutant system.

Before I started my interviews about ten years ago, I was not interested in biology, so my ignorance could be excused. But Marshall Nirenberg, Max Perutz, and Paul Berg told me about you and I remember Nirenberg saying that Benzer is a biologist's biologist, a scientist's scientist and he told me in no uncertain terms that I had to visit you. What does it mean to you being a scientist's scientist?

I suppose, like Erdös was a mathematician's mathematician. He didn't serve on presidential committees for mathematics, or try to operate a big enterprise, get in the headlines, or go on TV. One sees some scientists who, like people in many other fields, are enterprising, looking to make money from their science, or develop power of one kind or another. I am personally not interested.

They wrote a book about you, still in your life, which is not very usual.

Jonathan Weiner did a nice job of describing the *Drosophila* behavior research in popular terms. It won a National Book Critics' Award.

Did it impact your life, your work?

Not particularly. I got a lot of e-mail. The one I liked best was from someone in Brooklyn who said, "I've read the book, I'm about the same age as you, I grew up in the same neighborhood in Brooklyn. Maybe I knew you, but my memory is not very good. Will you please tell me whether I knew you?" I get a lot of inquiries from students who want help with their school essays.

There's another book that will be coming out, of quite a different character. Larry Holmes, a noted historian of science at Yale University, managed to finish it just before he died. It covers an earlier period, up to my working on the fine structure of the gene, and is much more of a scholarly book. Holmes actually went through my old research notes, page by page, making hundreds of photographs of them, and tried to trace, day by day, what I was doing, and what I thought. Weiner's book was on behavior, my third career after the physics and molecular biology periods.

Are you still in this third period?

Yes. I have found it interesting working out the roles of genes in behavior and how the nervous system develops, how it works, and what can make it fall apart. Recently, a good portion of my work is directed at aging and how the genes enter into that process.

How do you go about it?

One way of going about it is to find gene mutations that extend lifespan. For instance, we've found the *methuselah* mutant that extends the lifespan of flies by 30 per cent.

What does aging mean in terms of the gene? Does the gene change in time?

The expression patterns of various genes change. Also, the genes may themselves be damaged during aging. One of the things that can happen is oxidative damage to the DNA itself, causing mutations, which accumulate and cause dysfunction. There are dozens of theories about aging, and it is possible that all are correct in some degree, because so many things are happening.

But if we have the same DNA in all our cells, how does this damaging happen? They can't be all damaged simultaneously.

They can become damaged at different rates in different organs. Even in different regions of the brain, there are great differences in deterioration of the mitochondria with age. In addition to finding genes which can extend lifespan, a converse approach is to find mutations which shorten lifespan. Those define the genes that are necessary to maintain normal life. When we take those genes and over-express them, in some cases that actually extend lifespan. In this sense, there are good genes and bad genes. Although the same genes are all over the organism, each tissue expresses different combinations of them. We have methods of causing expression in specific tissues. Some genes, when we over-express them in the muscle, extend lifespan. Other genes are best expressed in the nervous system. We can determine for each gene which is the key tissue.

You are doing this for Drosophila.

Yes.

Are there any attempts to extend these studies to humans?

People do similar work with mammalian models like mice. I picked *Drosophila* in the first place because it's so much easier and rapid for doing experiments.

Are there companies that are trying to apply these findings?

Several are using *Drosophila* as a model. When the *Drosophila* genome was sequenced, it was found to be about 70 per cent homologous to the human genome. We can work on the fly to analyze the function of a gene that is homologous to a human gene. We can also take the human gene and put it in the fly to study it. This approach allows for a much quicker study of the effects of drugs, for example.

What happens when you put the human gene into the fly? Don't you expect to produce a monster?

Some genes may produce monstrosity, some genes may be lethal. We screen them and work on the ones that function normally.

Were you involved with the Asilomar meetings?

I was not.

Is there any chance that you could produce some human-like organism with the wings of a fly?

We're not putting the fly genes into a human, just human genes into the fly.

Others may deliberately try to produce a flying man.

Some people might try to produce super humans of one kind or another. But that's an extremely distant possibility.

When you say extremely distant, do you think in terms of 5, 50 or 500 years?

It depends on what you're talking about. What is appealing is the possibility of overcoming genetic defects in humans, using the knowledge obtained from fly models. A genetic defect in the embryo is already there, and the question arises, should you abort the fetus, or is it possible to put in the correct gene, or an appropriate drug, to fix it so that the baby develops normally. It's a wonderful possibility.

This is the approach that Jim Watson is advocating very forcefully. But this would be in conflict with some religions.

There are so many religions that some are bound to be against it. They will say that God is in charge and we should leave things as they are. That doesn't make any sense to me.

Wouldn't the legislation try to regulate this and thus introduce limitations?

Of course.

As of today, are there any regulations in effect that would prevent you from doing anything that you would like to do?

Not for fruit flies, fortunately. But as you know, in the U.S. new embryonic human stem cells cannot be developed, although it is allowed to work with ones that are already in existence. But the stem cells already in existence are in the hands of companies, who are not anxious to release them, and many of them are defective in one way or another. Progress of science can be seriously stifled by this kind of attitude. What happened in Asilomar was the development of a sense of awareness and concern on the part of the scientists about the dangers, in order to keep a watch on them, not to be completely irresponsible about it, so that the science can go ahead. But you can't control everybody, especially out of the country.

Let's return to your history.

After working on the fine structure of the gene and molecular biology, there was another transition. My current career was triggered by interest in behavioral differences between people. When my second daughter was born, she behaved completely differently from the first. It occurred to me that this must have something to do with different genes. A book that was influential, that I read at that time, was by Dean Wooldridge, called *The Machinery of the Brain*. It told about nervous system development and function. It had the same kind of influence on me as Schrödinger's book, *What Is Life?* When I was working on the gene, the question became how the information from the gene gets into protein. Now, the extended question was how the information from the gene gets into the nervous system and how does it control the behavior as a result. For that reason, I came to Caltech on sabbatical to work in Roger Sperry's lab, because Sperry had developed ideas about the chemospecificity of neurons.

You may know the experiment. The fish optic nerve connects the eye to the brain with a precise, corresponding mapping. When Sperry cut the optic nerve and turned the eye around 180 degrees, the nerve would regenerate and project back precisely to the right places in the brain, even though the eye was upside down. His notion was that, via chemical tags, the neurons knew exactly where they should connect. It seemed to me that the genetics of that process would be a key to understanding how the nervous system developed.

Alexander Varshavsky, Edward Lewis, and Seymour Benzer at Caltech, Pasadena, 1999 (photograph by I. Hargittai).

I came to Caltech in 1965 on leave of absence from my faculty position at Purdue. Incidentally, when I first got involved with biology, I was on leave of absence one year after another until I had pushed it to four years. I almost got fired at that point. Fortunately, Lark-Horovitz was very supportive of my interest in biology, even though I was in the Physics Department. I came to Caltech with the idea of working with Sperry on the chemospecificity problem. That was fine, except that, by the time I got here, he had lost interest in that problem, which he considered solved to his satisfaction. He had moved on to his famous work on the left and right brain in humans.

I recently talked with Arvid Carlsson who told me that he and a few others had a hard time convincing most people in neurobiology that chemical interactions were of great importance for example in degenerative processes in the brain.

Sounds stupid to me that anybody might have thought that chemistry was not involved. What are brains made of?

Looking back it is puzzling that people were so ignorant but today we may be ignorant of other important things.

That is always true — it's what science is all about. Many people thought that genes were not involved in behavior; it was all environmentally determined. There was the dichotomy between the communists and the fascists; according to the Stalinist dogma, everything depended on the environment; Lysenko managed to completely ruin Soviet genetics by convincing the politicians of that. At the other extreme, you had the Nazi ideology that everything is in the genes, so you had to breed the super race.

The old American school of psychology believed that everything is environmental. Skinner could put a pigeon in a cage and teach it to play the piano, by properly timed food reward. The opposite was the European school, Konrad Lorenz emphasizing that it was all instinctive, genetically controlled. It's ridiculous to say that it's only one or the other because there are so many examples where you can show it's both. When you acquire a dog, you first choose a breed that is most likely to have the behavioral genetic background you desire, but then you have to train it.

You are stressing the importance of both.

Both are essentially important, yes, and so is their interaction. People used to say maybe 80 per cent of this and 20 per cent of that, but that's ridiculous. The only way you can separate these two factors is to keep one constant and change the other. Come to think of it, that may be a possible formulation for a principle of complementarity!

Is it possible?

With flies, you can keep the environment constant and change the genes, and see what happens. When I changed the genes of the fly, I got every kind of behavioral difference you can imagine. At the same time, you can also change behavior of flies of a specific genotype, by changing the food, the temperature, and the fly's previous experience.

Did you do that too?

Yes. For instance, we showed that flies can learn to avoid an odor that is paired with electric shock, and found mutants that could not remember. One of our studies was on the circadian rhythm; flies have a daily rhythm just as we do. They run around during the day and they go to "sleep" at night; they just stand still. You can put a fly in complete darkness and measure its movement with an infrared detector. Day after day, it'll run around for about 12 hours, then stand still for 12 hours. It's an internal clock. At that time, there was still a debate about the origin of the 24-hour rhythm. In fact, NASA spent several million dollars to send flies up in a satellite to go around the earth in 90 minutes and see if it changed the rhythm.

Did it?

I never saw the results of their experiments. But we knew that in the laboratory we could change the period by manipulating the genes. The *per* gene (for period) that my student Ron Konopka and I discovered has a close counterpart in humans, which is also a component of our biological clock. The difference in behavior between late people like me (the "owls") and the early birds (the "larks") has been related to changes in that particular gene. Once found, that gene was the key to finding other genes that interacted with it. People have been able to work out an entire complicated genetic network that controls our circadian rhythm. The strategy is that we discover genes in the fly, then look for their counterparts in the human, and that has worked for several different kinds of important genes. The flies provide keys to human physiology.

Does the knowledge of the Human Genome facilitate your work?

It has been an enormous help. The moment we find a gene of some interest in the fly, we can immediately look at the Human Genome to find if a related one is there. In many cases, there is already some evidence of the gene's function in humans, which provides a clue to what the function may be in the fly, and vice versa. The human has become, in effect, a model system for the fly! It has developed into a symbiotic relationship.

Do you interact directly with people who are involved with human genes?

Mostly, we follow each other's publications, and, of course, exchange notes at meetings.

Is there a complete genome yet?

Yes, for fly as well as human. The methods of sequencing now are so rapid that the genomes of more and more species are becoming available. It's very interesting to compare sequences in two different species of flies to see which genes are conserved and which changes may be related to differences in behavior. That has become tremendously powerful. The genome collection is becoming analogous in a sense, to the *Handbook of Chemistry and Physics*. It is an incredible resource for us, and progress in this field is exponential.

There is now talk and expectations that very soon Americans will live 20 years longer.

It's probably true.

Will just the old age be getting extended?

The idea is not to live longer as an old person, but to extend healthy lifespan. We have mutants that live longer and they're staying healthy longer. It's not just that they're hanging on; they are vigorous for a much longer time. It will involve understanding which genes we should be paying attention to, and what metabolic systems they regulate in the body. That's the ambition. We can already do it with a fly. To paraphrase what Eric Kandel said about his improvement of memory in the mouse, "If you are a fly, I can help you."

We met five years ago. My impression is that if anything you look younger. Do you do anything to yourself?

Nothing special, but it's nice to hear!

How do you feel?

I feel about the same. It's not that I've discovered the secret of longevity. I suspect that a good night's sleep may be an important element. One thing I try to respect is my natural rhythm. I'm an "owl" type; I often work to 2 or 3 o'clock in the morning, then sleep till I wake up. I may miss telephone calls during the day, but there is a great advantage because at night it's quiet and I can get things done. If forced to work on a normal schedule, that takes a toll, a palpably stressful perturbation of my natural rhythm. Perhaps an unconstrained schedule, with spontaneity, may be a secret to staying young, if you can get away with it. Being a scientist's scientist, rather than engaging in the power struggles, may also help.

Pasadena is a peaceful place.

Caltech is as close to an ideal place to work as one can hope to find. The nearby Huntington Gardens is a calm refuge. And Los Angeles is close by for cultural attractions, not to mention the flowering of the San Gabriel Valley as a paradise for Chinese restaurants! Far exceeding San Diego, by the way. Too bad that Leo Szilard could not enjoy this.

Who sponsors your research?

I have had the good luck of continuous funding from the National Science Foundation for almost half a century. Originally, they funded my molecular biology research, which worked out well. When I got interested in behavior, there were still some funds left for the phage work. Remarkably, they let me switch the funds to *Drosophila*, which was a wonderful boost at the very beginning of the behavioral genetics work. I have always felt deeply indebted to enlightened administrators such as Herman Lewis and the recently deceased De Lill Nasser. More recently, I have support from the National Institutes of Health, for the work on neurodegeneration and aging. Some support has also come from the Boswell, Ellison, and McKnight Foundations. At present, I have eight postdocs. Some bring their own fellowships.

How long do they stay?

They typically stay for four or five years, until they find a job. Only three of the current group are American.

Biology has become very important at Caltech.

Millikan had the insight of starting biology very early, bringing Thomas Hunt Morgan here in 1928. I don't know how it came about that Millikan, a physicist, thought biology was promising. At first, as far as student majors were concerned, biology was an outlier, but in recent years it has boomed, as biology has become an important field, both intellectually and industrially.

You are completely absorbed in your work.

Much absorbed, yes, but not exclusively. Even at home, my wife Carol Miller and I discuss research; she is Chief of Neuropathology at the University of Southern California and works on the human brain. We constantly talk about the correspondence between the fly brain and the human brain, and have published a couple of papers together. Carol's specialty is the molecular genetics of Alzheimer's disease. But we have other interests, including music, art, movies, and unending exploration of the abundant local spectrum of ethnic foods.

Any parting thoughts?

I enjoyed this conversation, but I'm curious whether you will be able to extract anything from it.

Christiane Nüsslein-Volhard, 2001 (photograph by M. Hargittai).

7

CHRISTIANE NÜSSLEIN-VOLHARD

Christiane Nüsslein-Volhard (b. 1942, Magdeburg, Germany) is Professor and Director at the Max Planck Institute of Developmental Biology in Tübingen, Germany. She received her diploma in 1968 and her Dr. rer. Nat. degree in 1973 from the University of Tübingen and her Ph.D. from the Max Planck Institute for Virology, Tübingen, in 1974. She received the Nobel Prize in Physiology or Medicine in 1995, together with Edward B. Lewis and Eric F. Wieschaus "for their discoveries concerning the genetic control of early embryonic development". She has been honored with numerous awards, among them the Leibniz Prize (1986), the Carus-Medal of the German Leopoldina Academy (1989), the Rosenstiel Medal (1990), and the Albert Lasker Award for Basic Medical Science (1991). She is a foreign associate of the National Academy of Sciences of the U.S.A. (1990) and foreign member of the Royal Society (1990). She has been a member of the Ordens pour le mérite (Germany), among many other distinctions.

We recorded our conversation in Dr. Nüsslein-Volhard's office at the University of Tübingen in July 2, 2001.*

---

*Magdolna Hargittai conducted the interview.

What made you interested in science in general and in biology in particular?

Nature, at a very early age. I liked flowers and birds and generally I was curious.

Would you care to tell us something about your family background?

My father was an architect and my mother had no profession but she took care of the children; we were five: four girls and one boy.

Were your parents supportive of your interest?

Certainly but not in particular. They were not scientists themselves. But my mother supported very much all her children's interests. One of my sisters was interested in painting and drawing; she got crayons. I got books on animals and plants. I also had a lot of music and she was supportive in that as well. In our family it was normal that we all went to high school and it was also sort of accepted that we get good grades. I think that my father realized early on that I was scientifically oriented and he liked to talk with me about these things a lot. That, probably, also helped. I often told him about my math exams, about biology and others — he was a person with very general interests.

My interest was a general biological interest, not a general scientific interest. I also liked literature and music, so in a way had a rather broad general interest. Also, in high school we had very good teachers, especially in the last year in biology. Our teacher was exceptional and told us about many interesting things such as genetics, evolution and animal behavior. I also read a lot; I was in the birds' club, and did many interesting things. At the end of high school I gave a talk on the language of animals, and when I was preparing that speech I read a lot from Konrad Lorenz and other German biologists.

Then I went to Frankfurt University to study biology. This was not very nice, I liked it much less than high school because I had the misfortune that I knew all the exciting things already. The courses in biology at Frankfurt University were rather boring that time. I studied some physics and mathematics, that I enjoyed but eventually realized that they were too difficult for me. Finally, I realized that my true interest was biology and since just about that time they started to teach a new biochemistry class in Tübingen, I decided to go there. Later I realized that I didn't like that course too much because there was too much chemistry and too little biology but on the long run it was useful since it gave me a solid background in many basic areas. Also I enjoyed being a student at this little old town. During the last two years two new professors came who taught genetics and microbiology, that I liked very much. There were also good seminars by visitors and thus I had the chance to learn about many modern things, such as DNA replication and protein biosynthesis.

I got my diploma in biochemistry and then I had to find a place to do lab work for a year. I worked at the Max-Planck Institute in Tübingen and this was interesting, I could learn a lot. I was also in contact and good friends with other people in the institute — some did work on Hydra, others on DNA replication, again others on the genetics of DNA replication, and so on. I was advised by these people to not continue to work on bacteria and molecular biology but rather explore some more novel fields. That time developmental biology was particularly interesting. It was novel and not much studied yet, and practically had been abandoned by modern biologists. There was a group in Tübingen, which worked on Hydra, a small invertebrae organism that can be cut into pieces and then they regenerate and this is quite fascinating. They were looking for morphogenetic components. These are substances, molecules that induce the development of particular structures at particular sites. This work had been rather frustrating because they could show by transplantation experiments that they are organizing factors in the embryos but they could not analyze the molecules. With my experience in molecular biology and in particular with DNA replication using genetics as a tool to dissect a complex process, I tried to find a system where I could combine the question of morphogenesis with genetics and use genetics as a means to get at the molecules as the products of genes that you could mutate and identify. This has already been done in bacteria by Jacob and Monod earlier, but I thought that maybe we should try to use the same approach for higher organisms and the only organism it was feasible to do it with was *Drosophila* flies. There were also mutants available in *Drosophila*, I read a review on that, they were not very good but they were there. So I decided to work on that.

Drosophila *flies had been used before, for quite a while, for genetics research. What was the advantage of using them?*

Flies were the animals on which genetics was developed. Of course, Mendel used peas but they are not so easy to breed. Among animals, why *Drosophila*:

Thomas Morgan must have had already *Drosophila* around and then they found the first mutant, which was white-eyed, and then they crossed it and found that it was sex-related. All this was very quick to see because there was a linkage to the sex chromosome. Then they found more mutants and so on. The fly was very easily mutable and they collected a very large number of mutants and then they discovered the linkage of the mutations of the chromosomes, discovered recombination, discovered the physical order of the genes in the chromosomes; all this was done in flies. During the course of this time people also picked developmental mutants, which were not characterized by the alteration of a particular trait visible in the adult fly but characterized by different morphologies. There was the big question whether development would also be dependent on genes or was it only the features that you could see in the adults. Soon they realized that the genes had the essential effect.

The mutation that gave the Prize to Ed Lewis the same time as Eric and I got it, caused a change in the body plan.

I read that your results on Drosophila are relevant to higher organisms as well. Did you know that this might be possible when you started your research on embryonic mutants?

No, not really. We did not expect it to the degree that it now appears to be the case. We thought that we would study flies because they were easy to study and we wanted to focus on a particular question. It was a difficult question to answer so we wanted to try it on a simple organism on which it could be answered, rather than something that other people would be interested in.

When did it become clear that these results could have wider relevance?

This is not our work. This has been found by other people, for example, Ed Lewis. They cloned the first gene for which they found that it had homologies in other organisms. It was not restricted to flies but very similar ones were also found in mice and in humans. It is a very curious result and we thought first that it was an exception. We could not believe that the ground plan of vertebrates and invertebrates would be based on the same principles. Then more examples came, also from our laboratory. We found genes, which had very curious similarities in vertebrates and flies. Eventually more and more data accumulated and it became clear that humans and flies have a common ancestor.

Is it possible that your result could eventually be utilized to prevent defects in humans that are caused by such or similar mutations?

No, I do not think so; that is not a relevance to our study and results. It is only that from the study of simpler organisms, such as invertebrates, as the fly, we can learn something of the biology of higher organisms such as humans. But it has nothing to do with curing diseases. It is just a general biology that we learn something about.

Please, tell me something about your current research.

There is a group here that is working on flies and following up on the molecular biology of some of the mechanisms that we have discovered. This operates in the fly embryo to organize the body pattern. We also work on fish and try to understand particular aspects of the neurobiology in fishes, using genetics again. We are isolating genes, identifying genes, which have particular functions and study their properties. We use zebrafish for these studies.

Why did you choose them as your object of study?

Just as with flies, they are easy to breed. Vertebrates are generally bigger than insects, and more demanding with respect to care, space requirement and facilities. Of course when you do genetic research it is very important that you be able to grow a very large number of these animals and the fish is easier and less space-consuming than, say, the mice or other vertebrates.

What is your opinion of genetic engineering?

What do you mean "opinion"?

Just that. Do you like it; do you find it important, etc.?

It is not a question of opinion. It is a method, which is very useful. It is a very elegant method that allows many very sophisticated studies to be done that otherwise could not be done.

Should it be used to treat sick people?

Again, this is a too general question. I mean you use genetic engineering to analyze diseases, analyze infectious agents, and analyze, for example, how cancer is developing. It was incredibly important in understanding the mechanism of cancer and other frequent diseases. It can also be used to develop treatment for these diseases. So in my opinion it is a very powerful method.

I was mostly referring to its ethical implications. Should, for example, genetic engineering be used in medicine to improve the quality of a person's life? Can it be used to improve human intelligence?

No, it certainly cannot be used to improve human intelligence. It can be used to develop therapeutics to treat human diseases, yes.

Jim Watson quoted Francis Crick saying some time ago that a child should be considered to be alive only, say, after two days of birth by which time it can be seen that he/she does not have any serious defects.

That is a very extreme view and I would not adopt that view.

What do you think, if it becomes known that a fetus or an unborn child has a serious defect, should there be an abortion performed?

I think that this is up to the mother, she ought to have the right to abort the child if she feels so and if the child has awful problems.

Many Nobel Prize winners stop doing science after receiving the Prize. How could you manage to continue your research?

They are usually much older and many often had already stopped doing science before even receiving the Prize. I got the Prize at a much younger age, I've got a job, and I can't run away just saying that, "Now I am important, I stop working, I will not do my job anymore." I have a job, I have a contract until the age of 68 and I love my work. So this is one reason why I have not stopped research. This is my profession. My profession is not "to be famous", that is not a profession.

There must be much demand on your time from outside science.

Yes. I am more and more involved with politics, science politics, and also in more general things, like giving talks about ethics and such topics. I am also serving on many committees.

Has the Nobel Prize changed your life?

I became busier. I have a lot of obligations and I am very visible. This is something that sometimes is rather painful because everybody quotes what you say and what you think and you have to be careful that you do not make mistakes. You have to be more on your guard.

Being a Nobel Laureate in a relatively small town, does it make you a celebrity in Tübingen?

No, not particularly. People are modest here, they are polite and sort of ignore this.

Are you a tough woman?

No, I do not think so.

I would like to bring up the idea of symmetry. Not because I myself am very much interested in it but because it has a lot of connection with pattern formation that you also are interested in. What do you think of this concept?

I don't understand the question.

Two of your mentors, Hans Meinhardt from Tübingen and Klaus Sander from Freiburg, ...

I would not call them my mentors.

Both of them wrote a chapter about the importance of the symmetry concept in development and early embryogenesis. They appeared in a book on Symmetry[1] that my husband edited years ago.

Sorry, no, it is not a concept. If you ask if I am interested in symmetry, then, no, not particularly. I think that polarity is more important than symmetry. I do not see symmetry as a concept; it is the result of polarity. If you have polarity in one direction then symmetry, more or less, automatically follows. Only in primitive organisms, the symmetry is very important. But in higher organisms, you just have bilateral symmetry as the ground principle of development of these organisms, and I do not think that you need an extra concept for that. Maybe I am not a philosopher.

Chirality is a special kind of symmetry. There are several simpler organisms, shells, for example, which can appear in both left-handed and right-handed outer forms, still both these forms are built from the same types of molecules, L-amino acids and D-nucleic acids. How then does the different chirality of the outer shape come about?

Probably it is not dependent on the chirality of the individual building blocks but on some higher order phenomena and not the molecules themselves. I don't see that there is much of a problem here. In the early development you already see the chirality in the embryos, in the cleavage patterns and it could actually be random, but it usually is not random. Why it is not random, it actually is a good question but I do not know the answer. Of course, it usually is determined by the chirality of the mother but sometimes it is not. It is the argument of Hans Meinhardt, once you have symmetry breaking in one direction then it is enhanced, and perpetuated, but if it goes in the other direction, that will be enhanced and perpetuated. So probably there is not much mystery to it, just the process of symmetry breaking in the higher order organization of molecules.

You write somewhere that you were very shy as a child. Are you still?

Yes, sure.

Christiane Nüsslein-Volhard around 1992 (courtesy of C. Nüsslein-Volhard).

It is interesting to hear about a Nobel laureate woman calling herself shy.

They don't ask you about this before that, you know.

Your career ran somewhat parallel with your fellow Nobel laureate, Eric Wieschaus. Can you comment on this? Was there any importance of you being a woman and him a man?

I don't see that actually. Neither is there any parallel in our careers. He grew up in the U.S. and went to school at Yale; I grew up in Germany and went to school in Tübingen. We met in Basel and realized that we have common interest and we had the chance to share a lab for about three years. Here, because we had this lab together, we decided to collaborate. We found projects, which were interesting for both of us, this was common. Then we separated again. Eric did what every American postdoc did at the time, he tried to find a job, he found one at Princeton, and went there and is still there. I tried to find a job in Germany, tried to find

Christiane Nüsslein-Volhard with her co-recipient of the Nobel Prize Eric Wieschaus, around 1978 at the European Molecular Biology Laboratory in Heidelberg, Germany (courtesy of C. Nüsslein-Volhard).

an independent position, found this and later I got promoted. So there is nothing special about that.

You said that you have always been interested in animal behavior. Do you do anything along this line?

I've got interested in animal behavior when I was in school and studied Konrad Lorentz and other behavior biologists. I liked to observe animals, always. But I do not do anything particularly in that area. I have cats at home and I know them very well.

What does your double name stand for?

I was married once to a guy called Nüsslein. When we married I changed my name to his and dropped my maiden name. This is the name under which I published. Then when we got a divorce, I added my maiden name to it, this is how I have a double name. As a scientist I did not want to get rid of my previous life.

How long were you married?

Seven years. This was long time ago, before I moved to Tübingen.

Have you remarried?

No.

Do you have children?

No.

Was it your personal choice not to have children?

Of course, it is a choice. But nothing particular; I never said "I do not want children" — it just did not happen.

Was your husband also a scientist?

Yes, he was.

Is it possible that the break up of your marriage had anything to do with the fact that you may have been more successful than he was?

It could have been, but usually it is not just one thing, usually there are more than just one reason.

Have you experienced any negative discrimination due to the fact that you are a woman?

Oh, yes, plenty! But probably it is better not to dwell on it too much. First we can go back to the time when I grew up. The general problem then was that women simply were not considered to be professionals; professional enough to have big important jobs. Therefore, they were just often overlooked and no one entrusted them with important things. I have to say that my science was never discriminated against. So I had no problem whatsoever in getting my science recognized. But in the practical aspects, to get jobs, to get money, to get lab space, women have not been treated equally, I think.

At the time when I was a young scientist, men often had family and children and they got the better positions automatically. The professors always said, "But this is a man, he has to support and nourish a family, so this is why he is going to get the job and you will not." This happened repeatedly to me. The same thing with promotions. They often said, you are a woman and there is a man and he deserves it more. Actually, my worst experience happened with my Ph.D.; I collaborated with a man, I did most of the work, I wrote the paper and then he got first authorship because my boss said, he has a family and for him this is important for his career. You are a woman and married, so it does not matter. This was particularly unjust because he gave up science right after his thesis and went to teach where he did not need the publication at all. Whereas, I did suffer in trying to find jobs later because I did not have this publication with my name as a first author. So discrimination started right away at the very beginning.

When I did my diploma thesis in this institute, it was customary that the diploma students got some stipends, but the big boss said, "She does not need it, she is married." This kind of thing is over by now. But the old professors still think that it is totally legitimate to pay men more if they have families to support. This still happens. If there is a single woman who is good and does a perfect job and there is a man who is mediocre but has a wife and family, he will get the promotion.

This is surprising because, at least in the developed countries, nowadays special attention is paid to this question, and often there is a "positive discrimination" about women in that even if she is not that qualified as a male candidate, the woman may get the job.

Yes. Some women say that it happened so long and so often the other way around, why not have some advantage of this, for a change? I do not agree with that. However, it turned out that when you push the women a little bit more and you get more women in positions, they all turn out to be better than you have thought.

In the Max Planck Society they created some jobs specifically for women and they raised the percentage of women, also in the independent group-leader level, dramatically in all their institutes, and filled these jobs in a short period of time. It turned out that all these women are very good in their jobs. This tells that these women were simply overlooked previously. It seems that you can do a little push without the danger that these women will not do well. Generally they had been so underrated that it is good to give them better status.

Christiane Nüsslein-Volhard in her office during the interview in 2001 (photograph by M. Hargittai).

Do you think that a woman has to work harder to get the same recognition as a man?

At least she can't afford to make mistakes, at least not as much as men. When a young men does that, they say, oh, he is ambitious and it can happen to anyone. But when a young woman does that, it is usually overrated. In particular when women are aggressive, people don't like that at all. When men are aggressive, it is perfectly normal; it belongs to their normal habit. When a woman is aggressive and behaves in the same situation as a man would, they always charge that against her. This is bad because, aggression, in a positive sense is absolutely necessary to succeed in a job. You have to push your point and not that of your neighbor. Unfortunately it is part of the profession.

What do you think: if a woman has a family, is it a disadvantage for her in science?

It could also be an advantage, I don't really know. I think it does not really matter very much; at any rate, matters less than people think. When you have a husband and you have two salaries, you can afford to hire a nurse or a house-help, you can afford a daycare, and then life is much easier than living alone and doing all the work in your household. I don't know.

Do you have any advice to young women who want to have science careers and also a family?

Just go for it and try to find professional help for your household and for daycare as soon as possible and don't be stingy on that. Get the best help, pay a lot of money to the best servants and let them do the job for you. Many people do not do that, they just try to do both jobs and, of course, then they crash.

The few high-positioned women are often looked for in committees, public appearances, and so on. Does this make it hard to keep your eyes on the ball?

Oh, yes. It is very hard. There is also this women's issue on top of it; it is hard to deal with all of them. I am on many committees. I am on committees for prizes, for founding institutions, for hiring people, for European commissions on ethics, and so on. This is a lot of work. I also sometimes appear on TV about science.

What was the greatest challenge of your life?

I can't answer that question.

What are your present ambitions?

I have scientific ambitions. I just want to get more things done in my lab, and my ambition is to have my head a little free for that. I also would like to do my own experiment again, but that probably will not happen. These public duties, as you said, are just too overwhelming.

Who are your heroes?

Ghöte and Darwin.

Can you mention any particular mentors who helped you during your scientific career?

No.

Is there a key to your success?

Genetics. I do not know if there is a secret. It is a combination of some advantages that just happened. But I also have some difficult drawbacks. I am very un-orderly and not very organized. I also tend to be moody. Good and bad moods, varying, and to depression and it changes very quickly. It is sometimes difficult to control.

You grew up in the Federal Republic of Germany, which was already free of Nazism. How much was it around nevertheless when you were a child?

When I was a child I don't know. But I do know that we discussed it a lot; Nazism was discussed in school a lot. We also talked about this with our parents many times. We, the children, also asked what the family did during that time.

What did your family do during the war?

They tried to survive, I guess. They were not in the resistance. Neither of my grandparents was a party member and that created some problems for them. It was very difficult to do anything against Hitler. One of my grandfathers died from depression right after the war, he could not face it. He lost his only son. My other grandfather had to retire early; he had a chair in medicine in Frankfurt. Then his successor, who was a Nazi, was pushed out of this job after the war, and then my grandfather was reinstalled in his previous position, which pleased him very much. I do not want to boast but I think that my family behaved as correct as they could considering the circumstances. One of my aunts was almost killed in a concentration camp because she said something rash about Hitler. It was very easy to get killed that time.

Looking back, it seems today that German academia was in a state of deep amnesia after WWII. There has been an incredible history of German human genetics and the question has been pushed mostly under the rug.

No, it just has been discussed. A few weeks ago there was a big celebration in Berlin with survivors and there was a ceremony and the President of the Max-Planck Gesellschaft gave a wonderful speech. He apologized in public and now all this is documented and investigated.

Do you know Benno Müller-Hill and his activities, his book, Murderous Science? *He found documents in the Bundesarchives, in Koblenz, showing that there was a famous professor first in Berlin, then in Münster, von Verschuer, and that Joseph Mengele was his postdoc during the time when he did his infamous experiments in Auswitz and that Mengele was actually paid by the Deutsche Forschungsgesellschaft for the work done in Auschwitz. Nobody knew about this, or at least, this was never discussed.*

Yes, this is a very bad thing. And I think that such things happened at many universities because they did not have people really. I do not know about von Verschuer particularly, and one should have wondered because they did kick out Nazis from their positions after the war. I do not know how he escaped, you ask Benno. But I bet you that there were lots of ex-Nazis in university positions because there were no others. If you are in a country where 90% of the people are members of the party and try to be in important positions, you can't kill them all, it is very hard. There was also the tendency to try to pretend that everything was normal. Just let sleeping dogs lie and sleep and just get on with your jobs. I think that in practice there was a big problem of having enough people who were totally clean for positions in educating people.

Two images during a discussion at the meeting of Nobel laureates with students in Lindau, Germany, 2005 (photographs by M. Hargittai).

Do you think this was right?

No, of course not. It was wrong not to prosecute them and they did prosecute many and some escaped. They probably could have done better but in general I do not think that they have done it so badly. I think that especially foreigners cannot understand what a traumatic situation Germany was in at the time. People just tried to forget all because it was so horrible. It was not just the injuries done to the other people, like the Jews or the Romas, it was also the terrible things that happened to the German families. All these sons that were killed, all these people who lost their homes, their cities were bombed, so one desperately tried to get something normal again and did not waste too much energy on fighting against old people who also, of course, were not unused to being powerful. Of course, I do not know exactly, this is just how I imagine it may have happened.

What do young people think about the German past?

Well, you saw the reaction to the Third Reich that came in the movement of 1968, when all these young people just could not bear it anymore that their parents would not talk about these things. They could not bear that their parents were involved with these things; they did not know what to do. Just imagine if you lived in such a time, who are we to judge.

Is German science climbing back to its pre-1930s preeminence?

German science is quite good now. It probably will not get much better, it is quite good. Preeminence? Oh, no. Why? We just have to keep up a good standard in international comparison.

Are there any questions you would have asked in my place, but I did not?

No, I think you asked a lot of questions. You even asked the question what I would advise to young women to ask.

Reference

1. Hargittai, I., Ed. *Symmetry: Unifying Human Understanding*. Pergamon Press, London, 1986.

Werner Arber, 2005 (photograph by I. Hargittai).

8

WERNER ARBER

Werner Arber (b. 1929 in Gränichen in the Canton Aargau, Switzerland) is Professor Emeritus at the University of Basel. He studied at the Eidgenössische Technische Hochschule (ETH) Zurich and received his Diploma in natural sciences (nuclear physics) in 1953. His earned his Ph.D. in Biology in 1958 at the University of Geneva where, following a postdoctoral stint at the University of Southern California, he began his research career in 1960. After another research year in the United States at the University of California, Berkeley, he moved to the University of Basel in 1971, and has been there since.

Professor Arber shared the Nobel Prize in Physiology or Medicine in 1978 with Daniel Nathans[1] and Hamilton O. Smith, "for the discovery of restriction enzymes and their application to problems of molecular genetics". Numerous learned societies have elected him to be their members, including the European Molecular Biology Organization (1963); the Pontifical Academy of Sciences (1981); the National Academy of Sciences of the U.S.A. (foreign associate, 1984); the Swiss Academy of Engineering Sciences (1988), the Academia Europaea (1989); the Swiss Academy of Natural Sciences (honorary member, 1995); and others. He has been active in various scientific societies, organizations of science politics, and others, both in Switzerland and internationally. We recorded our conversation in Lindau, Germany, on June 29, 2005.*

---

*István Hargittai conducted the interview.

I would like to ask you about your family background.

I grew up in a village about half way between Bern and Zurich, in the German-speaking part of Switzerland. My grandparents and parents had been farmers. As a boy I worked in the fields and observed the plants but did not think of becoming, say, a botanist.

What turned you to science?

I was interested in nature and I liked the science subjects in school. I considered becoming a teacher. I was also interested in religious education. Our family is Protestant. For some time I wondered whether I should go into theology. Finally, I opted for natural sciences.

Are you still religious?

Yes. It is important for me. I am not religious like strongly religious people are, but as a scientist, I am religious in an integrated way. In recent years I have tried rather extensively to stimulate a good debate aimed to reconcile the worldviews based on traditional wisdom as it is reflected in the Old Testament and the knowledge that has accumulated from scientific investigations. There was, for example, a big confrontation between Darwinism and traditional wisdom. As I have been working on understanding Darwinism on the molecular basis, I find it important to abandon such confrontation. We should find consistency between the two big fields, human wisdom and scientific knowledge. In order to reconcile the two, we scientists should be flexible to interpret the data obtained for updated knowledge, and the religious side should also be willing to reinterpret some of the old, classical texts, which had sometimes been interpreted too literally. An example is when the origin of the Universe and the origin of life is said to have happened in just a few days of creation a few thousand years ago. That doesn't hold in the light of our scientific knowledge.

You have been very concerned about sharing scientific knowledge with the public. Your scientific area is very much involved in the public concern, just to mention the controversy about genetically modified food. There is a lot of misunderstanding about it.

That's true. I am one of those who from the very early times said that one should care about this problem and make some risk assessment, and devise even political guidance for meaningful and useful applications. I'm not saying that there is no problem whatsoever, but I think we should face these problems on a solid scientific and ethical basis. Some of the biologists who had been educated in the classical way, without understanding the molecular processes that go on in living organisms, they make merely claims on the big danger of genetic engineering. They say it is a big danger because it is not natural. This view is sometimes taken as the expert opinion by politicians in search for adherence by the general public. Then comes the influence of the religious belief in the formation of the people's opinion, and this is why religions could not just be ignored. If you believe that God created the world and that God is in love with the world, which is his creation, then, of course, you assume that the world was made the best that can be. So it should be an optimized situation. Therefore, if genetic engineering changes something in the genome, the result can only be worse. If God made the optimum of the human beings as well as the plants that human beings eat and so on, then you don't have the right to change something because you would counteract the will of God.

What is the way out of this situation?

The only way out is to have a debate. We should try to reconcile different views rather than to enhance confrontation. In addition, some people are against genetic engineering because they do not want to eat genes in their food. They don't believe that they eat genes in their daily food and that's the way they understand nature.

I would've thought that the Asilomar meetings and what followed demonstrated that scientists recognize their responsibilities so that should have at least alleviated some of the fears of genetic engineering.

I was in the Asilomar conference in February 1975 and I was very pleased by the process and the outcome. One of the results was the recommendation to introduce stringent control and regulations. It was proposed, for example, that if you don't know for sure what kind of effect can be expected to be exerted from the genetically engineered organism, you should assume that it would be highly pathogenic and harmful. Once you have shown under appropriate laboratory conditions that it isn't harmful, then you can relax the precautions and go to less stringent conditions for further explorations and eventually for its lasting application.

Nature herself has been a genetic engineer.

All the time, yes. I have used this argument for several decades. I have indeed come to the conclusion that genetic engineering uses precisely the same strategies to alter genetic information as nature does to produce genetic variants for biological evolution. In the natural reality there are three different strategies to generate genetic variants. One strategy is changing single nucleotides or just a few; these are local changes in the DNA sequences. Another strategy is to rearrange some segments of DNA within the genome of the organism. The third strategy is to acquire a segment of DNA from another organism by horizontal gene transfer. Precisely these same three strategies serve in genetic engineering. The problem that I still face today is to convince my colleagues in biology, even in evolutionary biology, because they cannot understand these molecular processes, mainly since these are and must be quite rare events. Therefore it will take quite some time until the scientific community of biologists will take it as a granted knowledge and can go to politicians and to the general public to explain to them how natural reality functions. That's the way how scientific progress goes. It does not suffice that a scientist has his own conviction and interpretation of experimental data. It is only upon wide acceptance by the peers, the community of experts in the specific fields that novel interpretations can become part of the scientific knowledge base.

István Hargittai and Werner Arber in Stockholm, 2001 (photograph by M. Hargittai).

Often we agree about the data. However, since genetic variation is a very rare event and since it is not and should not be reproducible from event to event, it's as a matter of fact only statistically reproducible, therefore the interpretation of the natural strategies is not straightforward for people. If you read textbooks on genetics and on evolution, they give the impression that genetic changes are always due to accidents and errors in the genetic material. I feel that this is a wrong understanding of evolution. Evolution is rather something active, and it is relatively careful with the genetic information, all the species that live on our planet have a certain genetic stability. However, in the large populations, some individuals may suffer some kind of genetic change, each of a different type, at a different location in their genome. If the living conditions accept such a change the new genetic variant will be maintained and sometimes favored as compared with its parents, and it may eventually overgrow the parental population. That's the natural selection that's working there.

Turning to another question, the relatively large proportion of Jewish Nobel laureates has been observed. Without going into any definition of who is counted as Jewish, I wonder if you would be willing to comment on this. This question is often asked of Jewish Nobel laureates, but it may be of more general interest.

My understanding is that perhaps there is a higher proportion of entrepreneurs among the Jewish community than among the rest of us. To me entrepreneur means to have the drive, being flexible, perceptive to innovation, and to have openness to new knowledge and ideas. I'm not sure, but this may be a component of success in science. Forced emigration may have also contributed to this phenomenon, but, again, I'm not sure. The driving force in response to emigration is perhaps a more general trait.

Your own research has much contributed to the explosion in genetic engineering. Looking back how would you assess the importance of the discovery of restriction enzymes?

The discovery of restriction enzymes contributed to bringing in new directions in the field of research in genetics. These new research strategies eventually helped to locate genes, to sequence the genes, to analyze their functions, and all that by having means for reproducible fragmentation of DNA molecules.

Did you realize this importance right after the discovery?

It was the basic experimental investigations done in 1960, which convinced me that DNA restriction is something new and would be worth following up in more detail. The first publications in peer reviewed journals appeared in 1962. Then I was asked to write a review, which was published in 1965 in *Annual Review of Microbiology* (*19*, 365–378). I didn't quite remember what I wrote there, but some time ago one of my colleagues made me aware that I made some extrapolation in the last chapter about the possible use of the knowledge on restriction enzymes. At that time there were no purified restriction enzymes yet available, but I said that these were enzymes, which, once purified, might provide a tool for sequence-specific cleavage of DNA. As a possible application I did mention the sequencing of DNA molecules.

Your path seems to have been smooth and without obstacles. Have you had any challenges in your career?

I realize how lucky I have been. I have never applied for a job; jobs have always been offered to me. Having completed my basic studies at the ETH in Zurich to qualify to teach natural sciences at the high school level, the gymnasium, I felt that I might do something else. I had decided towards the end of my studies to go into physics. So my Diploma Work was in experimental physics. When I had completed the requirements for the final exam in the fall of 1953, I went to see my physics professor, Paul Scherrer, and I asked for his advice for continuation. He told me that he had received a request that might be of special interest for me because he knew that my interests included not only physics but biology as well. He suggested to me to move to the University of Geneva where there was an electron microscope laboratory looking for a physicist to run the microscope and to do research in biology. In response to my interest in this position Professor Scherrer phoned immediately to fix an appointment for me in Geneva.

You have mentioned that you made your first literature report at the University of Geneva about the double helix structure of DNA. Did you realize its importance at that time?

I did indeed. We were a small group of people, but there was a good discussion. The discovery of the double helix structure of DNA also contributed to a wider recognition of the discovery made some years before by Oswald Avery and his co-workers that DNA was the genetic material.

Werner Arber after his thesis examination, 1958, together with his two advisors, Jean Weigle (on the right) and Eduard Kellenberger (in the middle). Courtesy of W. Arber.

There were some big names among your teachers and peers, starting with Paul Scherrer, an important contributor to X-ray crystallography, then Max Delbrück, Gunther Stent, Joshua Lederberg, Salvador Luria, and others. Whom do you consider your mentors?

My major mentors were Jean Weigle and Eduard Kellenberger, both in Geneva. Even after Weigle had left for Caltech to work with Max Delbrück, he used to return to Geneva in the summers and came to the laboratory daily and he talked to us and he also did experimental work. It was mainly through him that we had also contacts with the people at the Pasteur Institute in Paris and with Delbrück and the whole phage group. You have mentioned Gunther Stent. I was on three occasions in his laboratory in Berkeley for extended periods of time. He actually suggested to me to question whether DNA methylation could be involved in DNA modification. He introduced me to a specialist in methylation of enzymes and nucleic acids in Berkeley, Rabinowitz. As a follow-up of our discussions, I started to do some experiments and everything looked good; DNA methylation turned out to explain modifications as an epigenetic phenomenon.

Did you have interactions with Daniel Nathans and Hamilton Smith who would then become your fellow laureates in 1978?

Prior to the Nobel Prize we did not have direct contacts, but Hamilton Smith's mentor at the University of Michigan in Ann Arbor, Myron Levine, had spent a sabbatical year with me in Geneva in 1966–1967. I knew about Ham Smith because Levine told me that in his absence, Smith was in charge of his laboratory. For this reason, Ham was aware of our research. When he isolated nucleases from *Haemophilus influenza*, and saw that these nucleases cut DNA into specific fragments, he came to the idea that those nucleases might belong to restriction systems; he did a genetic test, and they were. So he was aware of what the function of those nucleases could be, from our early publications. By then he was at Johns Hopkins University and Daniel Nathans was there as well. I did not know Nathans personally before we got the Nobel Prize.

Matthew Meselson had also worked in your field and he had achieved much fame before with what is called the most beautiful experiment in biology. Why didn't he become a Nobel laureate?

I have known Matt Meselson for a long time. He did this very nice experiment on the semiconservative replication of DNA with Frank Stahl at Caltech in the late 1950s. I knew about it because Jean Weigle told us about this experiment. I used the same technique of labeling with heavy isotopes in the early 1960s to show the effect of modification on DNA.

In 1968, Matt Meselson together with Bob Yuan isolated for the first time a type I restriction enzyme from *E. coli K*. That was before Ham Smith, but Ham was lucky to isolate a type II enzyme. Type II cuts the DNA reproducibly while type I cuts relatively randomly. For this reason type I enzymes are not as useful for genetic engineering as the type II enzymes.

You mentioned earlier that my career was smooth and without difficulties, and this is true although I have had my share of difficulties, and I would like to mention an example. At one time we were working with a few restriction modification systems, all in *E. coli*. We also made some hybrids. I was convinced that one could make combinations of different systems and thus arrive at new restriction enzyme systems. I tried to accomplish this by recombination and by mutagenesis. I worked on this project for about a full year without any success. Two years later some other scientists working with two different restriction systems of Salmonella found by chance a recombinant displaying a novel restriction specificity. This showed that my idea was right, but I worked with an inappropriate system.

Did you describe the failure?

No, I didn't follow it up.

I would like to ask you about science in Switzerland. Switzerland is not in the European Union and another Swiss Nobel laureate had told me that it would not be worthwhile for Switzerland to participate in the scientific programs, because it would cost Switzerland more than what could be gained from it.

Such an approach does not show much solidarity. A relatively healthy country like Switzerland should also try to help some less healthy countries for their scientific development.

Direct interaction might be more efficient.

That's possible, but is it done?

You have two daughters.

One is in neurobiology and the other is an M.D.

I have used in some of my lectures the story one of them told about your research when you received the Nobel Prize.

Of course, our daughters were much younger then. Silvia was ten years old and Caroline was four. It was Silvia's story. She came to my laboratory and saw some plates on the tables. They contained colonies of bacteria, which she imagined as a city with many inhabitants. In every bacterium the DNA represented the king, who was long and skinny and who had many thick and short servants, the enzymes. The king had all the secret instructions about the work his servants had to perform. Silvia described my discovery in such a way that I found a servant who had a pair of scissors. If invaders came, this servant would cut them into pieces and by doing so this servant protected the king from the invaders. Servants with the scissors could also be used by scientists in their investigations of the secret instructions held by the kings.

Has your wife been involved with science?

She is not a scientist but worked as a secretary in the Institute for Medical Microbiology. She was often more familiar with pathogens than I was. She was always interested in my work and we often talked about it.

Werner Arber and his family, 1978, just when he received the news about his Nobel Prize (courtesy of W. Arber).

Has the Nobel Prize changed your life?

This is difficult to assess. Usually I pretend that it has not changed and maybe it did not. Yet I think that it has given me opportunities that I had not had before. Of course, there is no control to know it for sure. In any case, I continued my research and teaching. However, I spent an increasing amount of time on science policy, for example, as a member of the Swiss Science Council, and for three years I was President of the International Council for Science, ICSU. Without the Nobel Prize this might have not happened.

Have you done any research of comparable importance to the one for which the prize was awarded, since the Nobel Prize?

This is a difficult question and, also, I have worked jointly with others. We have been involved with what we call Molecular Darwinism, which is an exciting development to understand Darwinian evolution at the level of the biological macromolecules. It should deserve broader publicity and discussion among biologists, especially among evolutionary biologists. It has not happened intensively so far. I have also been interested in the question of how biological evolution is perceived by the people. I have been involved in debates with different representatives of the civil society, including religious organizations. This I have mentioned before.

Do you have heroes?

No.

Do you have interests outside science?

Yes, I have rather wide interests in cultural evolution, including the fine arts and music.

Would you have a message?

No. But maybe I can say something. From my point of view, some people are afraid of the development of science. They fear that applications of science will ultimately lead to the destruction of life on Earth. Maybe they are right to some degree. For me, biological evolution is an absolutely wonderful thing on this planet. There may be biological evolution elsewhere in the Universe as well, but we can so far only evaluate what happens on our planet. I'm convinced that if living conditions would change so drastically that human life and the lives of some other highly developed organisms would no longer be feasible, life would still continue for a very long time because other organisms could exist under very different conditions. That makes me an optimist. I should add that I am not so much anthropocentric to say that for me only the human life form is what counts. For me, life as such is more important than the specific existence of human life. Therefore I have a big hope for the continuation of life on our planet as long as some living conditions continue to exist. We are just one life form among many and if some other organisms would survive, even if they would be very simple organisms, they could have a chance to develop into higher complexity again by the given means of biological evolution.

Reference

1. An interview with Daniel Nathans appeared in Hargittai, I. *Candid Science II: Conversations with Famous Biomedical Scientists*, edited by Hargittai, M. Imperial College Press, London, 2002, pp. 142–153.

David Baltimore, 2004 (photograph by I. Hargittai).

9

DAVID BALTIMORE

David Baltimore (b. 1938 in New York City) is President of the California Institute of Technology. He received his B.S. degree from Swarthmore College in 1960 and his Ph.D. degree from Rockefeller University in 1964. He did postdoctoral work at MIT, the Albert Einstein College of Medicine, and at the Salk Institute in La Jolla. He was at MIT from 1968 to 1990, at Rockefeller University from 1990 to 1994, when he returned to MIT before accepting the Caltech presidency. He shared the 1975 Nobel Prize in Physiology or Medicine with Renato Dulbecco (b. 1914) and Howard M. Temin (1934–1994) "for their discoveries concerning the interaction between tumour viruses and the genetic material of the cell". He is a member of the National Academy of Sciences of the U.S.A. (1974), a fellow of the American Academy of Arts and Sciences (1974), and foreign member of the Royal Society (London, 1987). He received many awards and distinctions, among them the Gairdner Award (1974) and the National Medal of Science (1999). We recorded our conversation in the office of the President of Caltech on February 5, 2004.*

As I was preparing for this conversation, I read the recent book about you, Ahead of the Curve. *Is this an authorized biography?*

It is not exactly authorized, but I did give him a bunch of interviews and I did read through the text, so it's pretty accurate.

---

*István Hargittai conducted the interview.

To me your family background is conspicuously missing.

The kid who wrote it, he was a kid. He was an undergraduate when he started the book and he finished it when he was a graduate student. He was interested in science and was not interested in me as an individual. He was not very probing.

I've read what I could find about you yet I find it difficult to form an idea about you. You have been tremendously successful, have had an impressive number of high-profile appointments, and you seem to be always ready to move on to a new position. But I have found very little about you as a human being. So, could you please, complete my picture about you?

OK. My father's family was from Lithuania and my mother's family was from Odessa. All my grandparents came to the United States. My parents were both born in New York City. They were quintessential New Yorkers, very intellectual, lovers of the theater, of all New York life. They were not particularly wealthy, but during the Depression, my father was a coat and suit manufacturer in Manhattan and my mother also worked in the business. They were quite successful during the Depression, which is unusual. But they never were successful again, because after World War II, the whole structure of people's life changed and they stopped buying as many clothes. What was de rigueur, which was having an Easter outfit and a fall outfit, most people stopped buying an Easter outfit and half of the business went away. They spent their money on washing machines and housing and real estate. So the business that my father was devoted to, it rode down over that period. We were never poor, but we were never particularly wealthy, though we had moved to a wealthy neighborhood, Great Neck from Queens. We moved because my parents knew that the schools in Great Neck were the best public schools in the country. I remember that my parents were always drumming into me that we were not as wealthy as our neighbors. First of all they did not want me demanding the kind of lifestyle that the kids around us had, particularly when I went to high school. They came from King's Point which was the fancy part of Great Neck, and which gave a different viewpoint of what life could be like.

Were you a religious family?

I personally am not religious, but I did have my Bar Mitzvah, my father was very religious in the sense that he believed although he did not spend a whole lot of time on ritual. My mother was from an atheist family and had absolutely no interest in religion. I went to a conservative schul and it was serious. My mother came from a family of long liberal tradition and they were very proud of the fact that they never joined the communist party. They understood the totalitarian nature of communism even back then, even at the time of the Depression. They were socialist in their leaning and good supporters of Roosevelt.

If I may make a big jump, you are president of Caltech and Linus Pauling resigned from Caltech because his leftist views were so incompatible with what would have been expected of him by the trustees of Caltech. This is an oversimplification, of course.

The world is different and the trustees are different. Today we have a Board chairman who is Jewish, the head of the campaign that we are running is Jewish, many of the Board members are either Asian or Jewish, something other than WASPs that dominated the Board in the early days of Caltech. Caltech drew most of its trustees from San Marino, which is a very wealthy town nearby here, and that was a very WASP community and notably anti-Semitic. The views of some of the early leaders of Caltech were thought to be anti-Semitic. There is some debate about that. We had Jews on the Faculty, but clearly there was an effort to minimize the number of Jews on the Faculty. On the other hand, you could not have a great academic institution without Jews. The aerodynamicist Theodore von Kármán was a notable example, Jessy Greenstein was a major figure in astronomy. You had to be that much better to be admitted to graduate school at Harvard, Columbia, Yale, but if you were that much better, you were admitted. Columbia had its Jews and so did Harvard and Yale. Millikan, who was the president here, was trying to minimize the number of Jews on the Faculty, but still Caltech had Jews. There is a story, and it may be just a story, that Einstein might have come to Caltech instead of Princeton, but Millikan did not want to offer Einstein as much money as it took, which was a thousand dollars more. He was a little leery of getting such an outspoken Jew here.

Speaking about Caltech, don't you feel a little intimidated by the history of this school, by what the expectations might be, having the responsibility to live up to its tradition?

I don't think I can be intimidated; I am more honored and feel a sense of responsibility to the history of this place, to the quality that it has always represented.

What do you do to live up to it?

I try to insist on a very high standard of excellence in everything what we do.

I can imagine your doing this in hiring, but how does this work otherwise?

We have a big decision to make in offering someone tenure. An institution like this must be tough about that. And it can be very unpleasant because it means sometimes turning down people who you know are perfectly good scholars and who may be very nice people. But unless you're willing to be rigorous about that, you're simply not going to live up to the history of excellence that this place has.

Considering your science, would you identify your work for which you received your Nobel Prize as your most important piece of work?

Yes, it is the most important one.

The discovery of the enzyme, reverse trancriptase.

Yes.

What would you single out from your production since?

I have produced some things that are important. The discovery of tyrosine phosphorylation was one, which was made jointly with Tony Hunter. I discovered it in the context of the Abel gene and that has ended up to be the target for the most successful new anti-cancer drug, Glivec, which is a miracle drug for people who are suffering from chronic malign leukemia. It was created by rational drug design at Novartis against the target that we discovered twenty years ago.

By rational drug design?

It was not particularly rational drug design but it was targeted drug design.

How do you feel about combinatorial chemistry?

It has not lived up to its promise.

It still may…

And it will, but the way most people did it was fairly simple substituents on the same chemical background and they did not produce enough chemical diversity. They weren't concentrating on producing diversity that was drug-like. They have produced lots of compounds but virtually no drugs. Much more well-thought-out schemes of chemical design need to be used to screen for drugs.

Do you patent your discoveries?

I have patented discoveries although I did not in the early days, so I did not patent the reverse transcriptase and I did not patent tyrosine phosphorylation. It was 20 years ago anyway. I have an interest in a potentially very exciting thing. We discovered a transcription factor called Fκb. This was more than 15 years ago, but the patents are recent. It is now a major target for drug development.

You grew up in and around New York City and you live now in Pasadena, Southern California. Do you live a sizzling intellectual life?

I also lived in Boston, which is probably the most intellectual community in America because the universities are such a high fraction of the overall activities around Boston. As university president, it's hard to live a sizzling intellectual life because you've got so much to do.

Socially?

David Baltimore and Edward Lewis, Pasadena, 2004 (photograph by I. Hargittai).

Socially and for fundraising proposes and for organizational purposes, political purposes.

You socialize out of obligation rather than personal preferences.

That's right. It's a sacrifice that I make consciously because that is what it takes to be a university president or to be a head of any large organization. As head of the organization, you have both a managerial role and a symbolic role and to maintain the symbolic role is as important as to maintain the managerial role. If you are a great manager but you don't handle the political and social side of it, then you don't have the visibility as a leader that an organization deserves. It's a very demanding job, a very absorbing job.

A few years ago when I was here just before Ahmed Zewail received the Nobel Prize, I could sense the dedication of Caltech to make this happen.

Caltech, as an institution does not lobby and I don't personally either. I make nominations. There are two things that can help somebody get a Nobel Prize, but only a little bit. One is to be visible to the Swedes, so he should go to Sweden when he has the opportunity, and make sure that people know about him. The second thing is that his friends are supportive. The most important thing is to get a series of other prizes so he is branded as prize-worthy. It's generally a long process and there is no guarantee that it would work. In the end, I do believe that the Swedish Academy is extremely independent and I know better the Karolinska Institute which handles the physiology or medicine prize and there is very little one can do to influence its representatives. Of course, a lot can be done, but anybody who thinks it can be influenced is kidding himself.

Whatever it is, Caltech seems to be doing it successfully.

I don't think Caltech is doing it successfully; I think we just have the right people. It was never a question whether Ahmed Zewail would win the Nobel Prize. It was only a matter of when and not of if. The same is true of most of the people who have won the prize in the field I know. Jim Watson didn't have to go out and campaign for the Nobel Prize; it came to him. I have been involved in the Lasker Award and I see how complicated it may become once you have selected the right discovery for the award.

Would you care to predict a Nobel Prize for Alex Varshavsky for his ubiquitin research?

He will receive it within five years.

Did you ever get the Lasker Award?

No. They gave it to Howard Temin alone or with another group. They made a judgment and it was wrong.

And you may never get it because they don't seem to give it to Nobel laureates unless they expect the laureate to get another Nobel Prize.

Not likely.

You shared the Nobel Prize with Temin and Dulbecco. Temin was close to you by age and you made the same discovery independent of each other. I just read the story of Temin's paper in Nature[1] *that the editors changed the order of the two authors and put Temin first in spite of the original submission in which Temin was the second author. The editors did this without consulting the authors.*

It was a very great surprise to me because I had the preprint from Howard, which was Mizutani and Temin.

Was the order of authors really important?

No. Howard would've got the Nobel Prize anyway and Mizutani wouldn't. It didn't matter. But for some reason they did it. Howard has told me or written somewhere that he didn't know that they'd changed the order. I don't think the editors have the right to do that. But I didn't have to worry about that because I was the sole author.[2]

Reading about you my impression is that you are more conservative today than you used to be.

I am more conservative today. I no longer believe that government is the right solution for everything

Did you use to believe that?

That's basically the socialist notion and taking it to the extreme, it's the communist notion. I give you an example where I am absolutely convinced that it is not the right answer. That is when comparing the educational systems in the United States and Europe. I mean higher education. In Europe, all higher education is a state function. All research is a state function. The notion of a private university is very rare. It's coming, but it's still very rare. It's clear that the strength of the American higher educational system comes from the fact that we have a lot of private universities. Even the public universities are largely privately funded. There is no state university in the United States, which gets more than 50 per cent of its budget from the state. Most of them are in the range of 20 or 30 per cent. There is a tremendous power and energy in private activities.

When recombinant DNA was first discovered, I said to a friend of mine that this was a clear case when the government should take over the technology and to develop it in the interest of people. It should not be allowed to be a source of profits to individual companies. It was a great opportunity to do that because it was new, the government funded the research, and they could have easily stepped in and take it over. My friends, who were in the real world — I was a professor at MIT at that time — they patted me on the head and told me that the only way to develop it was through private enterprise. I learned a huge lesson through that because that was correct, the government couldn't have done that effectively, and the market place drove it and it drove us to enormous successes in the use of modern biology in the interest of making new pharmaceutical agents.

The lack here is anybody working in the public interest, and that's a huge lack. I would love to see a government agency and an international agency try to use modern biology in the interest particularly of the less developed world. But I don't think that the leading edge of this business should be managed by government agencies.

So, yes, my views about how society should be organized have changed, very much in parallel with the general changes in this country. Whereas in the days of Roosevelt, socialist notions had a lot of cachet, they have very little cachet today anywhere in the United States.

Could we come back to your discovery of reverse transcriptase? To me your discovery seems to have had importance for basic science in that it overturned what used to be the Central Dogma of biology. For the arena outside science, it served as stimulus for President Nixon's announcing a war on cancer, almost like President Kennedy having announced to send a man to the Moon a decade before.

What Howard and I did was the opening edge in modern cancer research and that's why I consider it the most important thing that I did. It determined mostly what happened in the rest of the 20th century in cancer research. We knew this in 1971, we knew that it was the opening wedge, but we also knew that it would be very dangerous to have a crash program. It seemed though that there was a scientific basis for a crash program in cancer research, but there wasn't. There was only a scientific basis for seeing that we had a new wedge into the problem of cancer. We had no idea where that wedge would lead us. We did not know that cancer was a genetic problem and without that knowledge we couldn't go anywhere. So I was very much opposed to this way of doing research. What happened was — and I give Benno Schmid, who was on the President's cancer panel a lot of credit for this — that the money was steered away from traditional cancer research into molecular biology. The money was much better used than I was fearful it might be. When I look back on the War on Cancer, I realize that there were sources that were made available to the scientific community that had in fact accelerated our understanding of cancer. That's why I changed my view, but it came from my skepticism about the ability of the government to run a targeted program.

Weren't you afraid that given such a crash program, people might tend to find a solution even if there was none?

I was. That's what I implied when I was afraid of the government to run such a program. However, today, I would make a very different argument. We have now 34 years of research since we discovered the reverse transcriptase. In that 34 years, we have gone from not knowing what cancer was to having a good idea about what cancer is. We know a lot about the genes involved, we know about the cellular disruption, and we know the nature of those genes and how they fit into the overall economy of the cell. So if today you have a targeted program, it can make a lot more sense. So I have gone from being an opponent of targeted programs to encouraging the government to think about targeted programs. The government has completely won over the RO1 mentality.

What's that?

It's the code name of the program of individual grants to individual investigators. They are generally of small size and most of us run our labs on such grants. The government officials had it beaten into them by desire of the community to fund research through RO1 grants. So they are all supporters of RO1 grants. What's happened is that we've done a lot of what we can do and we should continue with strong RO1 support, but we also need to think about how we take the basic knowledge that we've got and make it applicable to the disease. In the end, our compassion is the reason why we were funded and why the public supported us.

Of course, you cannot be driven by compassion in any individual piece of research or in an experiment because that may color your judgment of your findings.

Yes. You've got to be objective and rigorous and you've got to be innovative. But we have to also understand that there are people dying of this disease and if we have within our ability to make a difference in the number of people dying, it's our responsibility to go there and try to do that. But we have to do that with the rigor that we were talking about, as opposed to continuing investigation of increasingly arcane issues. This is a very different kind of research; it takes a much more organized kind of research, dollar-eating kind of research, which is very different from the standard RO1 kind of research.

Do you think that SDI – Star Wars may have also had such a positive side effect on research?

No. The Star Wars program is built around particular technologies designed for particular functions and has nothing to do with basic research. At least not that I am aware of. A lot of the Star Wars work is classified, so I wouldn't be aware of it. I merely listen to the people who say that it is worthless, people I admire, lots of them. They just laugh at it.

You don't think that it may have helped to bring down the Soviet Union?

What we did not understand about the Soviet Union and what we did not understand about Iraq, was how totally corrupt and how totally fragile the whole structure was and that how easily it would come down if something happened. History is moving in the direction of that something happening.

Your family originated from the Russian Empire.

I know and I am not insensitive to these questions.

Where does your name come from?

Probably from Ellis Island.

Any idea what the original name was?

I met an elderly relative at a party once and I asked him about it. He said he had looked into it and he thought that he knew the answer, which was Butromovich. I went to the Holocaust Museum where they have a list of all the shtetls in Lithuania and there was a shtetl named Butro. So that would make sense. That would have been my father's name. My mother's name was Lifschitz. That was clearly taken from the old country.

Again, coming back to the discovery of the reverse transcriptase, would you comment on its fundamental significance in bringing down the Central Dogma according to which DNA makes RNA makes protein?

It captured everybody's imagination. It also gave us a richer appreciation for information flow in biology.

Asilomar was a big issue at one time. It was about the dangers of recombinant DNA, about the potentials of letting the genie out of the bottle. Has it completely disappeared?

No, no, it comes up all the time. Every time there is a sticky issue, like stem cells or cloning, somebody will say, "We need an Asilomar meeting on this subject." They've sought us out, all the organizers of the Asilomar meeting and tried to get us to convene the next Asilomar meeting. However, the world is different, the questions are different, Asilomar was a unique moment.

Was it right to organize it?

I believe it was.

In retrospect?

In retrospect.

Why?

Because we truly did not know where we were going and for the scientific community to take charge of the process of investigating the potential dangers of this research was the right thing to do. It staved off legislation;

we managed to get no legislation of this issue in spite of the fact that it was in front of Congress and Congress always feels that if there is an issue, it has got to do something about it. That was because we demonstrated to Congress that we could handle it. We asked them to give us the time to develop the structures that would make it possible for us to responsibly handle the issues.

When the Cambridge, Massachusetts, City Council decided that there might be monsters coming out of the laboratories at Harvard and MIT, and used that to political ends, they appointed finally a citizen's commission to look at the dangers of recombinant DNA. They really asked the question whether the NIH guidelines were sufficient to protect the people of Cambridge against Frankensteins coming out of the labs. This group of people, who were literally the people of Cambridge and not experts and not even intellectuals, they came out with a report that said, "Yes, the NIH guidelines are the right thing to follow." According to their report; what the Cambridge City Council should do is to make sure that the laboratories are following the NIH guidelines; to get an independent report on that; but that was the only thing they had to do. That was a tremendous validation of the Asilomar process. I would even call it a surprising validation; we did not expect such a support. But the members of that panel made themselves knowledgeable, they had long meetings, they listened to testimonies, before they came to their decisions and recommendations.

I think that what we did was to facilitate the development of this science because we staved off political interference. A lot of other people disagreed. Jim Watson disagreed.

First he was for it.

First he was for it, but very quickly turned around saying that there was no danger, and I could agree with him. If I had to make a judgment about whether there was danger or not, I would've made the judgment that there wasn't any danger. That didn't mean that I could convince anybody of that. There was no data around, only intuition about the situation. Nobody had ever done experiments that had that kind of goal in mind. The idea that we should look at this question and make some decisions seemed to me a very good approach. I still think that we did good for the world.

You have been a lot of places during your career so far.

But most of it was at MIT.

Cold Spring Harbor Laboratory, MIT, Albert Einstein Medical College, Rockefeller Institute and later University, Salk Institute, and Caltech, and at places repeatedly. Perhaps more than most people.

I guess that may be true.

How would you formulate your ambitions today?

For the future?

Some people say they have no more ambitions, others may have a whole list.

I will not list any entries on such a list. I don't have any well-defined ambitions. I might say that this experience, that is, being president of Caltech has been a very fulfilling experience as well as a very involving experience. I am now 65 years old.

Some people get their ambitions at this age.

I became ambitious a long time ago. I was generally viewed as premature in my activities. I was head of an institute when I was forty. I got my Nobel Prize when I was 37. It did not stop me at all because it was not the thing that I was focusing my life on. It was a very nice validation of what I had focused my life on, which was doing research. You asked a question before with the implication that I never did anything after the reverse transcriptase that was as important as the reverse transcriptase. And I don't mind that because I do science to do science and the importance of the science is something, which is decided by history. It's not decided by me.

It turned out that that experiment was the right experiment to do at the right time, but I hadn't focused my life on doing an important experiment. I have focused my life on doing science and learning about the things that I cared about. I cared about viruses; there was a chance to learn about different viruses and I took that chance, particularly because it was very easy. I did those experiments and they turned out to open up a whole new world. But I never had the temptation to stop doing research and I still don't have such a temptation. One of the things I am going to do when I retire is to continue laboratory science.

I do have an itch to do more focused research on disease than I've had. I have a whole plan. I would like to do research in particular on HIV and cancer. The point is to use gene therapy protocols to change the genetic inheritance in the stem cells so that they would do things that I think would have therapeutic effects. I call it an Engineering Immunity Initiative.

Any ambitions in administration?

I think I have done administration and I don't particularly want to do more.

I have to ask you about the Imanishi-Kari affair, though not too much because it has been written about so much and I would not like it to dominate our conversation. On the other hand, I would not like omitting to ask you about it. Was there a lesson to learn from it?

The problem was that Theresa was a non-linear thinker and non-linear speaker — and I suppose she still is — and not notably neat in the way she maintained her laboratory and her life. This made her vulnerable to forces that no scientist ever imagined would have to deal with, namely, the Secret Service and national committees. But when her work was put up against those standards and those people, they found lots of reasons to question whether she had been entirely honest. Everything that they brought up had another explanation, but because she had been so chaotic, she did not respond to the questions in a nice, crisp, simple way. They thought they had a case of scientific irresponsibility and there were political forces that wanted to use her case, and my association with it, to demonstrate that scientists needed political oversight. It was a clear power play by the political forces to gain control over science. I knew that and I saw that because I knew that there was no substance to what they were saying. I knew Theresa and her experiments to know that this was not fraud, it may have been sloppiness but not fraud, and fundamentally there was nothing wrong with that paper and nothing has ever been shown to be wrong with that paper.

Was this why you did not want to have the experiments repeated?

I would have been perfectly happy if the experiments had been repeated; there was nobody who wanted to repeat the experiments. It would have been an incredible amount of work to repeat the experiments, all of which would have led you nowhere.

You were in charge of the laboratory.

I said then and I believe now that if in the scientific community you cannot have faith in your colleagues then you can't do science, certainly not collaborative science. We must live with this kind of vulnerability. We must make judgments: whom we trust and whom we don't trust. But there has to be trust. If there is no trust, then there can't be collaboration. I felt vulnerability, but I felt my vulnerability was no different from the vulnerability of everybody around me. I just had to find myself in one of those horrible moments of history and all I could do was to try to extricate us from it.

You finally succeeded.

But it took 10 years.

Mad cow disease?

That is a change.

I wonder if it has any relevance to your research and to your interest. It may become a big problem and science may have not done enough to figure out the mechanism of spreading the disease.

Probably not. We know a lot and the big thing we know is that there was no huge epidemic in England where this thing was spreading like wildfire. If there was no epidemic in England, I'm not worried about an epidemic in the United States. Today we know much better how to protect ourselves than we did before.

I just find it intriguing that it may be a structural problem of predominantly alpha-helical conformations turning into predominantly beta-sheet arrangements.

That is a very good question. A great question. There are biologists working on related problems, but this is a chemistry question that you raised.

Major textbook, national committees, presidency of Caltech, a long list of achievements. What else?

I'm a lousy skier.

Good point because Harold Varmus stands on the top when there is a bicycle competition with his participation (even when he does not win it).

I never wanted to be a physical superman.

Your Caltech home page is introduced with a statement that you are "perhaps the most influential biologist of your generation".

People say that.

How do you feel about Rockefeller University?

I will tell you something very surprising about Rockefeller University. They're giving me an honorary degree this spring.

Why is it surprising?

I got them in all sorts of trouble, but they are finally realizing that I did more good than bad. I love Rockefeller University, I always have.

I interviewed their new president back in Budapest when he was in transition.

Paul Nurse is a wonderful guy and he may have a lot to do with this. I was on his Advisory Board in Lincoln's Inn Field in London for five years and had a lot of meetings with him. If you meet him, he seems like an intelligent London cabdriver. In Britain you come up with an accent and it tells where your background is and he is clearly not from the cultured side of British society. But he loves that about himself. He encourages that view. But he also has a side of him that is able to administer something, to be an effective, incisive, and thoughtful manager. Unless you actually work with him on issues, you don't see that side. But he has this whole other dimension of capability.

Do you have a message?

For whom?

For the readers of Candid Science.

They are very nice books. The message that I take from what I have seen of the world over the last 50 years is that the actual doings of the world are extremely messy, largely because they are so mired in politics. But if you take a very long baseline and you look at the way the world is going, rational considerations, humane thinking wins out. I watched the country which I was born into, which was viciously segregated, which was totally opposed to many different groups, blacks, Jews, anything foreign to basic WASP society, to evolve, over my lifetime, into a country that if not totally accepting the differences, is much, much more accepting them. It is much, much more judging people as individuals than as members of groups, to the point when Massachusetts yesterday said that same sex marriage is the only thing that fulfills the requirement of the Supreme Court, which comes out of Brown versus the Board of Education. It said that separate facilities are intrinsically unequal. They have carried that as far as to say that a separate code of marriage for homosexuals is not equal, it does not provide equal protection under the law. That's an enormous distance for a country to go. With all the ups and downs of democracy, to see that we've actually run to a point like that is incredible. It's not over yet and it may be a dominant topic for the upcoming elections, but it indicates which way the country is going.

References

1. Temin, H. M.; Mizutani, S. "RNA-dependent DNA polymerase in virions of Rous sarcoma virus", *Nature* **1970**, *226*, 1211.
2. Baltimore, D. "RNA-dependent DNA polymerase in virions of RNA tumor viruses", *Nature* **1970**, *226*, 1209.

J. Michael Bishop, 2004 (photograph by I. Hargittai).

10

J. MICHAEL BISHOP

J. Michael Bishop (b. 1936 in York, Pennsylvania) is Chancellor of the University of California at San Francisco. He graduated from Gettysburg College and then received his M.D. degree from Harvard Medical School. Dr. Bishop first worked at the Massachusetts General Hospital as house physician, then participated in a postdoctoral training program at the National Institutes of Health in Bethesda, Maryland. Following its completion, he has been at the University of California at San Francisco where he became Chancellor in 1998.

Michael Bishop shared the 1989 Nobel Prize in Physiology or Medicine with Harold E. Varmus — at that time they were both at the University of California School of Medicine in San Francisco — "for their discovery of the cellular origin of retroviral oncogenes". He has many distinctions of which only a few are mentioned here. He is a member of the National Academy of Sciences of the U.S.A. (1980); the American Academy of Arts and Sciences (1984); the American Philosophical Society (1995); he received the Albert Lasker Award for Basic Medical Research (1982); the Alfred P. Sloan Prize from the General Motors Cancer Foundation; the Gairdner Foundation Award (1984); and many more, most jointly with Harold Varmus. We recorded our conversation on February 13, 2004 in his office in San Francisco.*

---

*István Hargittai conducted the interview.

I would like to ask you to summarize your Nobel Prize-winning discovery.

Simply put, Harold Varmus and I directed research that uncovered the first cellular genes that had the capability of becoming cancer genes. Although we had known the cause of some cancers, like cigarette smoke, sunlight, chemicals, some viruses, when I started working in the field thirty some years ago, there really was no hint of how a cause of a cancer acted on the cell, to turn it into a cancer cell. There were various theories, some of them rather strange. One of those theories was that the causes of cancer acted on genes, thereby changing the genetic program that controls the activity of the cell and converting the cell to cancer. One of the problems of thinking about this is the fact that cancer seems to arise slowly. It's mainly a disease late in life, middle age and later, and it is assumed that it is a product of accumulation of several different adverse effects or events. But no one could quite imagine what those adverse effects or events might be. The only hint that genes might be involved was the fact that perhaps ten per cent of all of the cancers that occurred in developed nations are inherited in the family and appears in the same form of cancer. That's clearly a genetic phenomenon in which there is a mutation in one or more genes, typically in one gene that creates a predisposition to cancer. For reasons we really don't understand, the predisposition is usually tissue-specific. Some inherited cancer genes cause tumors in the eye, some others cause tumors in the gastrointestinal track or the breast. There was a clear indication that genes could be involved in cancer. But for the vast majority of cancers, there was no evidence for that whatsoever. The work that Harold and I and our colleagues did started with a cancer gene in viruses that had been recently discovered. This was a virus that was first found in chickens around 1906 by Peyton Rous. It causes a particular kind of tumor in chickens; it can also cause tumors in rodents, at least in the laboratory, not necessarily naturally, but not in humans because this virus can't affect human cells. That was one of the reasons we were working with it, because it wasn't dangerous.

Was it infectious for chickens?

It was but it is not an epidemic in chickens. It was actually a bizarre anomaly that Peyton Rous happened to pick up this virus. It has perpetuated in laboratories but it is not a problem in the field. This virus had arisen only once and the cancer gene in question is readily lost from the virus. It really was an accident of nature that had no importance to chickens. But the ability of the virus to cause cancer, and very rapidly actually, made it a very valuable experimental tool. A colleague, Steven Martin, at the University of California at Berkeley, across the Bay, had reported around 1970 that it was possible to make mutations in the genes of this virus that affected the ability of this virus to cause cancerous growth. That work was published and we were wondering why that gene would be there. It didn't seem to serve any purpose for the virus; it was readily lost by the virus; when grown in the laboratory, the virus easily lost its ability to cause cancer unless you continuously select it for that capability. It occurred to us that perhaps that gene was acquired from the cells in which the virus arose.

Viruses are a unique form of self-reproducing material because they rely on another organism, the cell, for their reproduction; they can't reproduce themselves outside the cell. We knew that the virus reproduced inside the cell and that part of that reproduction involved the insertion of viral genes into the cellular genetic material, and then being copied back out again. It's a complicated replicating cycle, but knowing that in this reproductive cycle the virus interacted directly with cellular genes, we wondered whether there might have been an accidental transfer of a cell gene to a viral gene. If that were the case, the cell gene was a potential cancer gene because in the virus it had become a cancer gene. So we were looking for a gene analogous to the viral cancer gene in the cells and found it.

It was not easy because this was before the era of recombinant DNA and the technologies we had to develop to look for this gene in the cellular DNA were difficult and time-consuming, but they enabled us to come to definitive conclusions. It was clear what was going on, but there were people who doubted it until recombinant DNA came along and we were able to isolate the gene from the cell. We not just detected it, but isolated and showed that it was a cellular gene, very similar to viral gene, however, there was a difference.

In the virus, the gene has been mutated and that's what made it into an actual cancer gene. That directly supported our operating hypothesis that genes might be involved in cancer. All the causes of cancer like cigarette smoke, which we knew to be a mutagen, sunlight, which we knew to be a mutagen, and environmental chemicals, which we also knew to be mutagens, etc., might also work by damaging genes. We had found a case study of this where an accidental interaction between the virus and the cellular genome had led to a mutation that had converted a normal cellular gene into a cancer gene. This cancer gene was then incidentally transplanted from the cell into the virus where it became readily accessible for study. It quickly became apparent that there were many such genes in the cell. The viruses could acquire them and mutate them in a particular way into cancer genes. By now, we know dozens of them.

We coined the term protooncogene. Oncogene is tumor gene. The gene in the virus was called an oncogene because it is a tumor gene, it creates tumors. The parent or predecessor of that gene in the cell, we called protooncogene. These terms are now widely used, you can find them even in high school biology texts. We now know that there are hundreds of protooncogenes in the human genome, hundreds of genes that, if mutated in a certain way, they can contribute to the genesis of cancer. We know this not only from studying the viral acquisitions of these genes, but from our ability to go into cancer cells and look directly for genetic anomalies.

The first example of this came from the discovery that one of the protooncogenes first found by using viruses is affected by a chromosomal translocation in certain tumors of the human lymph system. In the translocation, chromosomes exchange parts and that exchange actually damages a previously identified protooncogene in such a way that the protooncogene now works continuously rather than being regulated by the cell. That continuous activity initiates the genesis of the lymphoma as these tumors are called. That sort of effect on the gene is now known to be widespread in tumors. This was the first example of finding a protooncogene in a human tumor that was converted to an oncogene by genetic damage. Other examples followed and we now know hundreds.

The paradigm that has emerged from this is that most if not all causes of cancer work by affecting normal genes in such a way that they become tumor-producing genes. It might be direct chemical damage to the gene or it might be switching the gene on so that it runs relentlessly and has adverse effects as a result of that. That is now the standing paradigm under which all studies of cancer are performed. I would say that the most practical outcome of that is the fact that some of the newest therapies of cancer are actually targeted at specific execution of the damaged gene. We can't fix the gene directly, but we know how the gene works as it encodes a protein, and the protein is the handmaiden of the gene, it implements the function of the gene. We can design inhibitors of that protein and that has been done successfully in several instances now. The most dramatic is the drug Gleevec, which is spelled differently in the United States and Europe. That drug was developed to selectively inhibit the protein that carries out the function of a tumor gene in the human leukemia. The protein is the bad actor and if you can block its action, which distracts us, the tumor cells regress.

Couldn't the genes themselves be blocked?

There have been efforts to do that but they have not been uniformly successful. Right now the most promising approaches target the protein product of the gene.

Side effects?

I know; they are very specific, and that's the beauty of this technology. You can tailor them, in principle, and it's true in the case of Gleevec, you can tailor the drug that acts on the abnormal protein only and not on the normal version of the protein.

Is this for a specific kind of cancer?

Yes. Gleevec, for example, acts on the early stages of some specific kind of leukemia. It has made headlines. It's extremely specific; the specificity means that there are no or very little side effects. It also means that it is not a cure, it's not a panacea. This kind of treatment is highly targeted and there are vast amounts of money to develop this kind of targeted therapy for each of the major forms of human cancer. But you first have to understand the nature of the genetic damage in the cancer and you have to understand the nature of the protein that the gene encodes, and you have to know what that protein does. Then and only then can you intelligently design a screen for drugs that would inhibit the function of the protein.

How can we delineate your contribution in this from the contributions by others?

First we discovered the fact that there are such genes and this is why we were awarded the Nobel Prize. We discovered that there are such genes in normal cells and it was a quite controversial discovery at first.

Was it difficult to publish your results?

We didn't have difficulties in publishing, but there was a lot of skepticism about the generality of our finding. We proposed that this was probably a general way of how cancer arises. That was controversial for a while, until this kind of damage was found in human cancers. We went on to study the function of the protein to find other examples and eventually to authenticate the ability of these genes to cause cancer by putting them into mice and reproducing the same human tumor that the genetic damage is normally associated with. You take the damaged gene out of a human leukemia and put it into the genetic material in the mouse, you create an inherited disease, just like the human cancer. This is essentially the ultimate authentication that the gene is doing what we had all suspected it was doing. So is the fact that if you treat the function of the gene, you inhibit the function of the gene, you cause a remission of cancer and also an authentication of the idea. The overwhelming majority of mechanistic research on cancer and the overwhelming majority of searches for therapies now are guided by this view of cancer that the central problem is damage to genes that has been elicited by one or another of the causes of cancer. Alternatively, it may have been damage in the past, which is now being inherited. In any case, at some point, something must damage the gene.

What is the greatest remaining puzzle in cancer?

In my mind, the greatest remaining puzzle is what are all the causes in different cancers. We really don't know that. We only know the causes of a limited number of cancers. We don't know the cause of breast cancer, we don't know the cause of prostate cancer, we don't know the cause of colon cancer, and we could go on and on.

How much does the knowledge of genome help in predicting the occurrence of cancer?

We can't do that yet with a few exceptions. We can do that with certain inherited cancers. Breast cancer is the best example. There are two genes, BRSA1 and 2 that are known that if they are damaged in certain ways then they predispose the inheritance of breast and ovarian cancer from one generation to the next. So there is genetic testing for them. But for the overwhelming majority of cancers that we call sporadic, that are not inherited, there is no sure test. We know some of the genes that are typically damaged in these cancers, but the damage does not pre-exist the cancer. It occurs when the cancer begins. You could not do testing in the uterus or in the neonate or at age 1 and know that the person has this damage and develop colon cancer, for example. It is thought that each of us has a personalized predisposition to certain cancers by virtue of a particular combination of genes that we have inherited from our parents. This is much different than single gene transmission of disease; this is a combinatorial effect of multiple genes creating a predisposition. That's a hypothesis and that hypothesis can now be pursued by genomic technologies as we have the human genome completely mapped and sequenced. That's being done and it's a huge project and it involves looking at very large populations and analyzing vast amounts of data. In the end, if it turns out that there are a large number of genes combining to give this predisposition, it may be useless clinically; we don't know yet. We don't know how large an effect each gene might have, but there is a strong suspicion that each gene is going to have only a moderate effect on our predisposition to cancer. Even if such a predisposition could be uncovered, for the moment, we don't know how we would be able to deal with such information. All what we could do would be to be more vigilant. So we know one or two genetic lesions that if inherited predispose to colon cancer, but these are rarities because ninety per cent of human cancer is sporadic. It develops during the course of our lives and we have no way to predict that at the moment.

That does not sound very optimistic.

But look at the change in a decade. More optimistic is the idea that we will be able to develop specific therapies with very few side effects, cancer by cancer, gene by gene. That's been worked on very intensively.

In your discovery, what was the decisive benefit from your partnership with Harold Varmus?

I don't think that either of us would have started to work on this problem separately. It was the combination of our interests and conversations and our work together with the younger people. I rather doubt that either of us would have pursued the problem the way it was pursued. Just knowing our individual interests before we came together, I doubt that it would've happened. Someone else would've made the discovery; it's inevitable in science.

Can you delineate your individual contributions? Can you compare your partnership to the partnership of Watson and Crick?

[Bishop is heartily laughing.] I wouldn't compare it to Watson and Crick. The fundamental difference between Harold and me is that I am impatient with detail. I pay attention to detail, but I don't enjoy it. If left to my own devices, I am more likely to deal with sweeping generalities and concepts. Harold was the person always immersed in the details, always driving for more details. This is not to say that he was not a master of concepts because he was, but this was a fundamental difference between us. Now I do pay attention to details in the way I conduct my professional life, but I don't get the sheer pleasure out of it that Harold was getting in those days. He may have also changed, I don't know. Apart from this difference, we were very simpatico, we shared many interests outside of science and we shared many interests inside of science; we had the same values in science.

Did you have very different backgrounds?

Yes and no. We're both products of the American culture; we're both products of professional families; we both value having a purpose in life; we're both products of families in which culture was valued. In my family it was mainly performance culture, music, for example. In Harold's family, there was probably a broader purview in that. Although one is of Jewish extraction and the other of Christian, we had very similar values about personal lives and about science. The combination of us had under control our anxiety about competition from other scientists and also set an example for our younger colleagues so that they were not inordinately obsessed with just getting something done before another person, but rather took pleasure in exploration and discovery. We shared those values and we brought them to the collaboration of the group of scientists we directed.

In your book How to Win the Nobel Prize? *you discuss the case when you missed an opportunity to make a great discovery, that of the reverse transcriptase.*

Yes, I did. When I took my position here, which was my first independent academic position, I was working on polio virus, how polio virus reproduces. Next door was a man who was working on the Rous virus. He was biologically inclined and I was molecularly inclined. It seemed a good idea to team up and to see whether we could bring the pair of modern techniques to the study of this virus. The way it reproduced and the way it converted cells to cancerous growth, they were both utter mysteries at the time. The first time I watched this virus convert a cell from normal to cancerous within hours, I was hooked. The first question was how the virus reproduces and I was well positioned to think about that because the polio virus has a single stranded RNA genome and the Rous sarcoma virus has a single stranded RNA genome. Little did I know that the two viruses replicate in very different ways. At least I was well positioned to think about it. So we started doing some fairly conventional experiments, which were not very productive. At that point I was called upon to teach a graduate course and one of my assignments was to teach RNA tumor viruses. I had to read thoroughly in the literature where I came upon the fact that a scientist named Howard Temin had done a series of experiments that were extraordinarily controversial, almost universally denigrated. Those experiments led him to the conclusion that this virus replicated by virtue of having its RNA genome copied into DNA, which was then assimilated by the host cell. Then its DNA was copied back to its RNA. Copying RNA into DNA in those days was considered heresy.

It was against the Central Dogma; that the order was from DNA to RNA to protein.

Right. Subsequently, Francis Crick would say that the Central Dogma really said that you could not go back from protein to nucleic acid.

That still holds.

Absolutely. But it was widely believed that you couldn't possibly go back from RNA to DNA. Coming with a completely fresh eye, knowing nothing about the topic, having never met Temin, nor the people who criticized him, but looking at his experiments, I thought that some of them were quite persuasive. There seemed to be a DNA intermediate in the replication of this virus. Then I went to a meeting and heard him talk and came away with the notion that the virus must have an enzyme that copies RNA into DNA and that Temin must be right. I decided to look for this enzyme. Upon my return, we did a number of experiments, but I didn't know enough about enzymology to design the experiments very well.

Did you tell Temin that you would be looking for the enzyme?

No, I didn't. He was a more senior figure than I was and I didn't even want to talk to him at the time. I hadn't met him, I still hadn't met him. But I took the advice of somebody who knew more about enzymology. I also had been running several enzyme reactions for polio and they were just fine. If I had followed the same approach that I used with polio, it would have been very easy to find the reverse transcriptase. But I changed the experiment according to someone who was an expert in DNA synthesis and it completely misled me. After some failures, I couldn't continue, because I had a grant to study polio and I had to return to it. But it was quite a lesson. The first I heard about reverse transcriptase was in a telephone call from Cold Spring Harbor where Temin and Baltimore both reported this. I went to the lab immediately, did the reaction that I had originally conceived before I got the advice, and it worked in ten minutes. I was pathologically depressed for six months.

Did you do anything in addition to repeating the experiment?

We did, we published early in the field and became recognized for our work on reverse transcriptase. However, the thought that I had figured it out on my own, but was stupid enough not to change things a little bit, it was just unbearable. Of course, the history of science is full of such examples.

But not everybody likes to talk about it.

You are reluctant to talk about it not because it's embarrassing but because it sounds as if you were making some sort of priority claim. Which, of course, I'm not. I'm only pointing out what an idiot I was. It's a good lesson for young people.

Did you think at that time that you would have a second chance to make a big discovery?

It never occurred to me that what I was looking for was so important. It seemed so farfetched that I didn't give it any thought of what the implications might be. It just seemed to me that it was an amusing idea and that I would test it in my spare time and this is what I did. And we blew it.

Did you ever talk with Temin about it afterwards?

Yes, sure. We became very close friends. He had been sitting on his own writings for quite a while. There is a nuance here that we should explore. The enzyme was inside the virus particle and that was the trick in looking for it. There were a few precedents in virology that I was very familiar with. That fueled my own suspicion. There were two possibilities: either the enzyme was made after the virus entered the cell, or else, the virus carried the enzyme into the cell with it. There were precedents for the second idea that I was quite familiar with and in that case it would be relatively easy to find it because you could purify the virus and get away all the cell components as we were working with purified virus and testing it for the enzyme. Temin had that idea, but he waited three or four years because he had never set up such an enzyme experiment, he was strictly a biologist. It wasn't until he got a postdoctoral fellow from Japan, Mizutani, who would actually set up an enzyme reaction, and it worked very quickly. David Baltimore had run enzyme reactions and knew of the precedents that I knew about and that Howard knew about. Once he got interested in this idea, he got it to work very quickly too. If you came at it from the right direction with the right precedents in mind, and had the right, simple technology, you could make it work very easily. The hardest part was to purify the virus. It was hard work in those days. Baltimore actually borrowed the virus from somebody else; he was not growing the virus at the time.

Do you think that Mizutani might have been included in the Nobel Prize?

I don't think so. He came to Temin's lab; Temin had already had this idea, he had the drive to do it, and told Mizutani exactly what he had to do and he did it. It could've been a skilled technician. But I do not know the details, of how much originality Mizutani brought to setting up the enzyme reaction in Temin's lab. If you would include everyone who executes an experiment, the Nobel Prize might as well go out of business because it would never be able to keep it at three.

Temin and Mizutani submitted their report as Mizutani and Temin as authors and the editors of Nature *changed the order to Temin and Mizutani. Do you know about it?*

No, I don't, but it would have been the conventional way to do it with the postdoc being the first author.

But Nature *changed it.*

Nature *used to do strange things. Maybe they still do.*

You worked on the polio virus and you wrote a book about how to win a Nobel prize. Do you think that Sabin and Salk should have received the Nobel Prize?

No. I think that given the premise of the Nobel Prize, which is very much directed towards fundamental discoveries that make other things possible, it went to the right people.

They created the technique of producing large quantities of the polio virus on tissue culture that others had failed to accomplish. But I did not mean instead, I meant, in addition.

You would be hard pressed to find many examples in the history of the Nobel Prize when multiple prizes would be given out along the same line of discoveries. Theiler got the award for developing the yellow fever vaccine. But he was the first after Pasteur who died before the Nobel Prize was established, when someone got the prize for a laboratory-prepared vaccine. If you start with the premise on which the Nobel Prize is typically awarded, it's not unreasonable that Sabin and Salk were excluded. I would also say that they like to limit the award for a particular line of work to one. People have often asked why Cohen and Boyer did not get the Nobel Prize for recombinant DNA, for example. The explanation is that the earlier ground work had been done by the people who were given the award.

You wrote this book, How to Win the Nobel Prize?

It's a product of my irreverent wit. It started out as the title of a lecture series in Jerusalem where the humor was recognized. In the lecture I could at the very beginning exploit the title and make a joke at my own expense. I was going to change the title for the book, but the marketing people thought, this was a great title. My original title was *Notes from an Unexpected Life in Science.* Part of it then became the subtitle of the book. I liked my original title because it stressed that it was not an autobiography. Many people now think it is and criticize me for the title not corresponding to the contents.

Which was the moment when it first occurred to you that you might win the Nobel Prize?

I was at a meeting in Cold Spring Harbor sharing a room with a younger colleague. We had a double-decker bed and I was in the upper deck, we turned the light off and I was dozing off and we were talking about scientific colleagues and suddenly he said, "You realize that you are going to Stockholm." It was at least five years before we got the prize. It had literally never occurred to me. We never talked about it until it became widely rumored that we might win it.

How many years elapsed between your discovery and your trip to Stockholm?

Our first experiment that worked was in October 1974; we didn't publish the work until 1976 because it was very difficult technically and it took us that long to get it in a credible state. The prize was in 1989 and the interim represents the time for the generality of our discovery to become apparent and the applicability to human cancer to become apparent. Once the applicability to human cancer became apparent, then it became a serious contender, but I was not thinking about it.

How did you feel when Baltimore and Temin got the prize?

By that time I had become accommodated to the fact that I had blown it and they hadn't. I admired Temin's courage and Baltimore's bravado. By that time I had already earned the recognition of our colleagues for our work; irrespective of prizes, the respect of his colleagues is the most important recognition for a scientist, nationally and internationally. Your colleagues make an astute, perceptive and diverse group of people, which brings to bear a great variety of judgments on the importance of your finding. It's the best jury you can possibly conceive of for judging your achievement.

How do you sense the judgment of this jury?

Word of mouth, invitations to participate in scientific exchange, recognition of your expertise by being asked to do peer review.

Memberships in science academies?

The National Academy of Sciences is much too narrow of a purview. The world is full of supremely successful scientists who are not in the National Academy of Sciences.

In your Nobel lecture you quoted Karl Kraus, not a typical ingredient of American culture.

I discovered him when I was forty years old. One of the things Nature blessed me with was an endless curiosity. It's a handicap in a way because it distracts me easily. My interest in music came from my family background, but I have developed a broader interest in the arts, which came from my education in liberal arts. I had never been in an art museum until a college professor took me to one. After I finished a two-year postdoctoral fellowship where I was given a permanent appointment, I spent a year in Germany, in Hamburg, my salary being paid by the Federal Government. I discovered German expressionist art and I developed an interest in *fin de siècle* art in Europe. The next step came when I first visited Vienna where I discovered a particular and very distinctive subset of expressionism. I was intrigued by the multidisciplinary salons they had where scientists and writers and others were talking to each other. It was through my reading about Vienna that I encountered Kraus. I am a bit of irreverent myself and I admire iconoclasts and Kraus was a quintessential iconoclast. I read whatever I could find of Kraus in English translation.

What did you bring from home?

First and foremost a sense that you should try to do something worthwhile in life, not necessarily material, something that in some way served the community.

Was it a religious family?

My father was a Lutheran minister. It was a very religious family. My father was very liberal but we lived in rural Pennsylvania and it was a conservative society. I grew skeptical of religion very early so I have not been religious since I was 15.

How did your father take it?

Very well, he never attempted to engage me about it although he was probably disappointed, but he was not a proselytizer. My mother was very stricken by it. The point is that the spirit of the community and of my family was that one had to have a purpose and the purpose was to do something worthwhile, rather than making a pile of money or acquire power.

You have been successful in both in a way.

I could use more money, although I may be overpaid, but I don't have a lot of power. Some think that the Chancellor of the University of California at San Francisco is a powerful position but it is not. Anyway, that's what I brought from home.

Would you single out one or a few mentors in your career who had a decisive impact on you?

First of all, Elmer Pfefferkorn, who was an instructor at Harvard Medical School when I decided to take on my first serious research project during my third year in medical school. I was a complete naïve about research. I had been interested in science since college but I had no clue about research. Elmer took me into his lab and let me work there. I got very excited about doing research. His example of intellectual hospitality and the quality of his intellect of how he went about science probably had as much as anything to do with convincing me that I really wanted to pursue research. Then, I also learned a lot from my classmates at Harvard Medical School who were very sophisticated and had prior research experience as undergraduates; I just gravitated to them naturally and instinctively. They were not trying to talk me into doing science; at the time I thought that they considered me as someone who was not up to it because I was so naïve, but they set an example for me that I wanted to emulate. I liked the way they talked and I liked the attitude to learning they brought with them.

Do you have heroes?

I have to think about that a little. It's so easy to say the obvious things. Adlai Stevenson was a hero of mine because he was a bright, articulate, straight-forward political leader who proved to be ultimately not quite effective enough to go the whole way.

Wasn't he too intellectual for politics?

He wouldn't have been too intellectual for European politics. But as Governor of Illinois, he had a touch that people liked.

And in science?

Certainly, Francis Crick. He is a hero for a variety of reasons. I particularly admire how throughout his career he has maintained this remarkable ability to think about biological problems productively. He has also been visibly removed from the kind of ambitious drive that sometimes makes scientists less attractive than they might otherwise be. I have never seen Francis driven by any ambition except inquiry and discovery.

How did the Nobel Prize change your life?

I don't think it changed my life very much. It increased the demands on my time. My wife also thinks that my life did not change and she is an observer of my life.

Did her life change?

I never asked her.

Would you have become Chancellor if you had not received the Nobel Prize?

There might have been a good chance of it still given the trajectory of my career. I had begun to emerge as someone who could organize. I am only the second Nobel laureate who has been a chancellor in the UCSF system so this is clearly not a requisite for that job.

May it have helped prompting you in this direction?

I would not say that. No. When you are dealing with the general public, being known as a Nobel laureate does give you a certain cachet, which I have found to be pleasant not because I want to pontificate but because it helps me connect to people. They are not intimidated but they are curious.

Would you have a message for budding scientists who read these interviews?

First of all, endure. I was very lucky. I took a year off after the second year in medical school because I felt confused. I was offered a one year fellowship by the Professor of Pathology at the Massachusetts General Hospital to come study pathology and do a little research project. I took that year off and that's also when I got married. My wife was working as a schoolteacher. I had a lot of time to read during that year. When I went back to my third year medical school I took an elective course on viruses because I became interested in viruses in my own reading, how virus research had been instrumental in creating the field of molecular biology. There was one large class of viruses that was open to study that had not been studied yet from the molecular vantage point. They were the viruses that attack mammals. I took a course in animal viruses and that is when I met Elmer Pfefferkorn. The third year was casual but the fourth year was tough and I wanted to continue with my part-time research. I went to the Dean of Students at Harvard and I told him that I wanted to do research as part of my fourth year rather than taking courses because I wanted to become a scientist, not a doctor. The outcome was that he said, OK. He helped me draw a plan how to complete my studies but concentrate on my research. He said that I was committing a professional suicide, but he let me do what I wanted. There were still some requirements that I had to comply with. For example, I had to participate in three deliveries. So I went to the hospital for a weekend and personally, with my own hands, delivered ten babies. No other medical student did that. I also spent some time on the surgical service. Completing my M.D. and spending two years on postdoctoral studies was still faster than how long it takes for Ph.D. students today to become a scientist.

My second message is take a chance when you have the opportunity. I failed my chance with the reverse transcriptase, but luckily, I got a second chance. You have to have courage and convictions and have to be willing to take chances. You have to be judicious about it because if you spend all your time taking chances that would not do either. In science, there is the expectation that you be productive. If you look at the truly successful and creative careers in science, you almost always find that a path breaking risk was taken. If it works, it's an exhilaration beyond description. But you should find pleasure in every little peek into how Nature works. If you don't get pleasure from that, probably you shouldn't be doing science. If you are only happy with the grand slam home run, you are in the wrong business.

Listen to your seniors, but not slavishly so because they have biases and preconceptions that are unwarranted, outdated, mistaken. Be collegial. Collegiality in science is very important. It's been one of the main sources of pleasure for me to know that in every major city in the United States and in every developed nation in the world I have people whom I consider as friends. We share pleasure in science and we share purpose in science. Collegiality is the great privilege of science, to be a part of that huge international community. It's probably the most cohesive and enlightened international community that exists.

Harold E. Varmus, 2002 (photograph by I. Hargittai).

11

HAROLD E. VARMUS

Harold E. Varmus (b. 1939 in Oceanside on Long Island, New York) is President of Memorial Sloan-Kettering Cancer Center (MSKCC) in New York City. He grew up in Freeport, New York, received his undergraduate degree in English literature from Amherst College, and his M.D. degree from Columbia University in 1966. He joined the University of California San Francisco in 1970 where he became Professor at the Department of Microbiology and Immunology in 1979.

In 1989, Harold Varmus and Michael Bishop (of the same department) shared the Nobel Prize in Physiology or Medicine "for their discovery of the cellular origin of retroviral oncogenes". Dr. Varmus headed the National Institutes of Health (NIH) of the United States between 1993 and 1999 and joined MSKCC in 2000. He is a Member of the National Academy of Sciences of the U.S.A. (1984); of the American Academy of Arts and Sciences (1988); and of the American Philosophical Society (1994). He has received, among other distinctions, the Albert Lasker Basic Medical Research Award (1982) and the Gairdner Foundation International Award (1984). We recorded our conversation in his office at the MSKCC on October 28, 2002.*

---

*István Hargittai conducted the interview.

Reading about you I formed the impression as if your tremendous career in science happened almost against your original intentions.

There's some truth to that. I had a very late start, which in some sense I regret. Because just by watching my own students, I've seen there's a lot to gain by starting work in the laboratory long before I did. I was put off by science early in my career and I thought that I was not going to be a scientist.

I usually ask my interviewees, "What turned you to science?"

Your question still pertains, but it only pertains to the age of 28. Up until that time, I was always interested in an academic life and always felt an urge to do something scholarly, but I didn't have an experience in college and even in graduate school that made me feel that science was especially interesting or fun to do. I wasn't exposed to teachers who excited me.

Harold Varmus during the interview (photograph by I. Hargittai).

Our Long Island high school did not have distinguished teachers; some of our science teachers were completely uneducated in science; others knew some science, but had a strain of eccentricity or even sarcasm that wasn't very attractive. Then, in college, I got very quickly seduced by other things. I was much more interested in literature and didn't take many science courses. At that time at Amherst College we had the so-called "New Curriculum" for the first two years. Everybody in the class, even future businessmen and English teachers, took calculus and modern physics. Teaching was good, but because everybody was taking the same course, it didn't have a personalized flavor of some science courses that I was exposed to later. It wasn't until medical school that I really got excited about scientific thinking and the nature of exploration in science.

So for the education in grade school and high school many may be lost for science.

Sure, a tremendous number. We do a very bad job of teaching people what science is in grade school and high school. I'm not very active in this area, but I've helped Leon Lederman who is a hero in the efforts to try to change secondary education, even primary education in this country. There is no doubt that we don't use science sufficiently as a vehicle for teaching reading and writing. This is the first thing. Secondly, we teach science as knowledge instead of as inquiry. There is a big difference because knowledge is boring and inquiry is exciting.

Is it correct to say that up to the Nobel Prize you had primarily a research career and then, eventually, you went into administration?

I wouldn't say it that way. I would say that I was very productive until about the late 1980s and I became less productive just after the Nobel Prize. But even when I went to NIH, I always maintained a laboratory. We've done a lot of interesting things. My focus has shifted quite a bit, but I'm still running a laboratory here. We do some things that are good although there're only a few really novel things. I certainly look upon what my lab has done during the last few years with pride, considering all the other responsibilities I've had. I train good people and they get good jobs; we publish good papers.

Nonetheless, after the Nobel Prize you went into administration.

I did, but I don't like to think of it as administration. I've never been a particularly good administrator. What I've been good at is looking at policy issues from the perspective of a scientist, being reasonable about the way I've pursued things. I've been fairly effective politically, and I don't mean it as an administrator, but politically in both building institutions and getting them supported. I always knew I had the ability to do some of these things, but as a faculty member at the University of California San Francisco, I never took any interest in the kinds of things I would've been asked to do. I've never been interested in being a department chairman and being sure that everybody was getting the right salary and that everyone was having the right amount of space. These are questions that don't interest me. When I look back on what happened to me, I see that I've always been able to run my life the way I wanted to. When I was a faculty member, I was teaching students and did a lot of research, being able to be as productive as I could be. When I had this crown put on my head, I was able to have more influence than I would've had before, and I was able to take on the issues that I really think are important. They had to do with funding of research, training of investigators, support young investigators, reasonable cost policies between government and academia, issues that concern publishing, issues that concern regulations, some of the issues that influence the sharing of research facilities among people, relationships between academia and industry, and misconduct in science. All these are issues that have been much more contentious and much more interesting and reflect on how science is done. These are things I was able to take on once I had this platform for speaking.

Had it not been for the Nobel Prize you would've stayed then a Professor.

Sure. I was happy; I was in paradise for a scientist — teaching, doing research, living in a beautiful place, and having a lot of smart friends. It was only the prospect of actually going in one jump that attracted me. If they'd offered me to be the deputy director of some institute, I would've never done that, but to be the head, that was tempting.

How much did the budget of NIH jump during your tenure?

I was director for six years. When I went it was a little under 11 billion and when I left it was a little over 18. The more important thing was that I got to Washington when people were talking about cutting back the federal budget and I kept NIH ahead of inflation for a few years. Then, as the national budget became stronger, I was able to work, with a lot of help from others, to mount a campaign to double the NIH budget in five years. When I left, we hadn't finished yet, but we were on the trajectory. My relationships with the Clinton Administration and with Congress were important in getting us on that track. At the end of this year the budget will be 27 billion. I see the increase from 11 to 27 between 1993 and 2003 as being part of my legacy.

Are you a Democrat?

Yes, I'm a registered Democrat, but most of the time I was in Washington, Congress was run by Republicans. My party affiliation never really mattered much. I suspect that the Clintons were not unhappy to find that I had been a co-chairman of the planning committee for Clinton–Gore in 1992, that I was a registered Democrat and an outspoken liberal, but when I got to Washington and I began to deal with Congress, my party affiliation didn't matter very much. Most of my best friends in Congress were Republicans.

You used the label liberal, that is more than being a Democrat. It is much more to the left, at least in the U.S.

It is now. Three days ago Paul Wellstone, the Senator from Minnesota died in a plane crash with some of his family and staff. He was a great guy, very interested in medical research although mainly because he was interested in Parkinson's disease, which afflicted his parents. He was unabashedly liberal. I used to belong to the old school of liberals, as a college student marching against the Vietnam war and interested in civil rights, and he was from the same general political camp. He often would be the one vote against a proposition that was approved 99 to 1. He was one of the few people in the Senate who was willing to say about himself that he was a liberal. Thirty or forty years ago he would've been joined by a lot of other people; maybe half of the Democratic Party would've called itself unabashedly liberal. Liberalism has taken some hits because of such notions that are incorrect. For example, the things that were wrong with the welfare system were ignored by the liberals or that the liberals are too soft on our enemies. But we're getting beyond the bounds of this conversation.

In 2000, I talked with George Radda, the head of the Medical Research Council of the United Kingdom. MRC is a similar though much smaller organization in the U.K. as NIH in the U.S.

I don't think that MRC is exactly the U.K. counterpart of the NIH. There is a difference in magnitude, and MRC is not the biggest funding agency for medical research in England. The biggest in England is the Wellcome Trust.

But it's not a government organization.

But it's not commercial. We don't have anything of the size of the Wellcome Trust in proportion of the NIH. We have Howard Hughes, the Heart Association, the Cancer Society, but go ahead.

George Radda said that while the top laboratories of NIH and MRC are not different in quality, the average quality of science for the 18 billion dollar budget of NIH is not as high as that for the 350 million pound (that is, much less than 1 billion dollar) budget of MRC.

I have no way to answer that. I would say that that's true of the Howard Hughes Medical Institute, but Howard Hughes has a big advantage. They decide what areas they are going to fund, whereas NIH has to do everything. Secondly, everybody wants to be a Hughes investigator because it's so well supported and they're able to cream off the top. Everyone who is in Howard Hughes as an Investigator has spent a lot of NIH money on grants and training. They're the products of NIH. Then they had this extra level of support. Howard Hughes has done a very good job in selecting people; it's very carefully reviewing everybody, trying to spread the resources appropriately. To me it's not a surprise that they have a very high return.

How do you measure the success of an organization like NIH?

It's a very good question. I can try to give you some answers. When I was at NIH a new government policy came into effect, the Government Performance and Results Act. We were asked to evaluate our performance just the way the CIA and other budget organizations do. We decided that there were a couple of ways to do that. One was to look at things that could be enumerated. NIH as an organization has to do things like get the grants reviewed, get the grant money out, provide oversight, be sure that the numbers add up at the end of the year. We had thousands of administrative functions that we could use as a measurement of how well we succeeded as an administrative organization, and we did very well. The hard part of it is, and that is what you are asking about, how we evaluate science, how good it is. It's like a stock portfolio, it depends on lots of things. At the end of the year we point especially to publications that describe significant advances that either move knowledge forward or improve the Nation's health. We do that every year and do that very effectively. We have to wrestle with Congress to continue to increase our budget, so we have to be able to justify the idea that NIH is producing more as a result of this enlarged budget than it did before. That's not very easy to do. We can do it because of the sheer volume of the investigators, but it is more difficult to show the enhanced productivity, more discoveries, more advancement of health.

You are now at a cancer research institute and hospital; actually to reach your office, you have to walk through the hospital part. Your mother died of breast cancer. Did it play any role in your choosing your profession, your field of research, and, ultimately, this work place?

I assume it played some role in my initially deciding what I was going to work on. She was sick at the time I was making career choices. But there is always the temptation to personalize these choices; there may be some role for it but you also have to see the scientific opportunity. There is no doubt that I learned in my studies that there were things that had not been very well pursued and that could be developed more deeply to better understand cancer. In particular, cancer viruses were sitting there not fully explored, very simple genetic units that made normal cells behave like cancer cells. It was clear that there was a revolution about to occur in these studies as a consequence of molecular biology. Even though it was still not possible to clone DNA or sequence DNA, it was possible to try to understand how single genes of the kind that we found in tumor viruses worked, where those genes came from, how they behaved in various circumstances. So the temptation to enter this field was very strong, even without a familial connection. We all know that cancer is a major health problem in our society, and everybody knows people among their friends and relatives who had cancer.

May I ask you about mammography?

I'll give you a very simple answer here. It's an imperfect tool; we can do better than that.

There's a debate.

I know and I want to avoid it. It offers some benefit; it's imperfect; eventually, we'll do better.

Eventually.

Soon.

How about today?

Today, I would recommend mammography. But there will be other methods in the future to detect breast cancer early and more reliably.

When we talk about cancer viruses, does it mean the possibility of infection?

The cancer viruses that I work with have been, for the most part, laboratory tools. These viruses have caused cancer in animal cells, transformed the behavior of normal animal cells in culture into oncogenic cells. Those viruses infect the cell; they may and may not grow in that cell, but they infect the cell; they introduce new genetic information that has the capacity to change the behavior of the cell. At the time of the work that Mike [Bishop] and I were involved in 30 years ago, when the cell was a black box, we knew very little about the composition of animal cells. We knew that there were proteins like globin and immunoglobulin and that there must be genes that encode them, but we had a poor understanding of what those genes looked like in a physical sense. However, we knew that viruses were quite small with very few genes. Those genes had to be sufficient for the virus to carry out the replication cycle, and some of the genes, whether those that were involved in replication or others, had the capacity to change the behavior of a normal cell. Since some of the viruses I was drawn to, retroviruses, had very few genes, it seemed very likely that one or only a few genes were sufficient for transformation. It seemed extraordinary that one gene in a virus could make such a dramatic change in the behavior of a cell, especially one as complicated as a mammalian cell.

Let me add just one thing about cancer viruses; there are viruses that we know play a significant role in human cancer. Hepatitis B is one of those viruses and I spent a lot of time working with it. There also are a couple of retroviruses, the human T cell leukemia virus, for example; the human papilloma viruses; and a few others. Those are infectious agents, they often grow in human hosts, and they cause cancer.

You had a partnership with a fellow professor at UC San Francisco, Michael Bishop. How did it work?

It's interesting. I went to California in 1969 to look for a postdoctoral position. I was drawn to Mike because he was smart, personable, and thought the way I did. He wasn't entirely alone. He was drawn there by his previous mentor at NIH, Leon Levintow, and Leon had recruited another scientist, Warren Levinson. Warren, Mike, and Leon formed a little lab, sharing resources and sharing technicians, so it was already a group endeavor when I joined them. I came as a postdoc, with Leon as my official mentor, although we never worked closely together. Quite soon after I arrived it was quite obvious that I was going to work with Mike. We became colleagues quickly. Later on, others joined this group. People in the group had distinct identities, but we ran things together. We often shared students and postdocs. However, we were sufficiently strong as personalities that there was always the possibility for us to change the relationship. When Mike was offered an institute within UCSF, in 1979, and decided to move to a different floor, I didn't move with him. For the next five years we did things separately as well as continuing our interactions. The frequency of doing things together gradually diminished. We have remained very good friends and, ironically, we're doing similar things right now, although we're working entirely apart. Both of us work now on mouse models of cancer — somewhat different cancers and we emphasize somewhat different genes — but our approaches are similar. Both of us are interested in using gene regulation and virus delivery to try to identify the genes that are required to maintain the viability of cancers. We're asking similar questions with more or less similar tools.

Has cell biology recently become more in the focus of inquiry in biomedical research than before?

Cell biology and molecular biology are closely allied disciplines, and they have been such important tools for making discoveries in biology that it doesn't surprise me that recent Nobel Prizes went to these fields. The important thing is to ask, what kinds of things have we learned? It's quite striking that during the past few years simpler organisms than mammals have featured prominently in some of the Nobel Prizes. A few years ago there was a prize in developmental biology to Ed Lewis, Christiane Nüsslein-Volhard, and Eric Wieschaus for defining the genes that are required for early development in the fruit fly. This year's prize is not about development *per se*, but about cell death, which is a major developmental process. Those studies were done on the worm. Last year's prize was given for studies on the cell cycle that were done mainly with two kinds of yeast. The majority of prizes during the past 30 years used model organisms with cell biology and molecular biology as tools, as opposed to physiology or experimental therapeutics.

What are your ambitions today?

I have ambitions in several areas. In my own scientific work, one is to maintain a reasonable amount of productive scientific work despite the fact that I have other responsibilities. I enjoy the daily encounters with students and postdocs and with the scientific community. This is one part. The other part is more ambitious and has to do with a desire to define elements of the cell control system that are vulnerabilities in a cancer cell, so that drug development can be a more rational process. All of us who do cancer research are very impressed with the fact that at least one new drug has been incredibly potent in cancer treatment because it targets the product of a certain gene that is crucial in the formation of a cancer. In this case the gene is called the Abl gene and the drug, called Glivec works incredibly well in the treatment of chronic myeloid leukemia and a rare sarcoma, gastrointestinal sarcoma tumor. This gene-specific inhibitor has validated the claims that all of us have made for 30 years, namely, that knowing about the molecular basis of cancer would be a way to improve the way we treat it and possibly even prevent it. I'd like to see that happen for other tumors, more common tumors, and my lab's activities are focused on developing models for lung cancer, breast cancer, ovarian cancer, and others that aim to identify the vulnerabilities of cancer cells so that better approaches to therapy can be sought.

Then I have other ambitions too because I have other activities. One, of course, has to do with our institution here, which is primed for expansion. We're building new buildings, recruiting new people, and expanding the disciplines in which we work, trying to get our clinical scientists to work more efficiently with our laboratory scientists, and we're making a lot of progress. I'm feeling very good about how things are going. One obstacle is the downturn of the economy, which could affect our philanthropy.

What's your annual budget?

A little over a billion dollars. About 85 per cent of that is hospital budget. We have a big hospital and I take an interest in it, but not being a licensed physician anymore, and never trained as an oncologist, I can't pretend to be giving specific advice about how people should be taking care of patients. I oversee the hospital but I have a strong physician-in-chief who is a major source of guidance about the hospital. Expanding, improving, and encouraging this cancer center is a major thing in my life.

I have several outside interests as well that have taken a great deal of my time. One is in publishing. I'm a strong believer in the idea that the Internet has not been fully exploited to promulgate the results of scientific investigations. We should move from the traditional mode of private and societal publishers and subscription fees to a free access system in which the authors ultimately pay for publication, after review. I have been working on this from a political point of view and as a prospective publisher. The authors already pay for page charges, photo charges, reprints, and subscriptions. We're all authors writing for each other. The whole system could be made less expensive, and there is no need for the reader to pay for access to publications. The peer-review system would continue as is.

Did you ever have any difficulties in publishing your results?

I've never had difficulties in publishing papers that I thought truly important. The paper for which we won the Nobel Prize was accepted by *Nature*, but the reviewers criticized one experiment, and they were right. That experiment was left out of that paper, and I'm grateful to the reviewers.

Returning to my ambitions, there is one more and that is international science. Science is not used sufficiently around the world to advance health and other aspects of life. I've been working on several fronts to improve the way in which science is used, especially in developing countries and in middle-class countries that are doing OK economically but have very poor science. Science is restricted to about half a dozen countries, and yet there are a lot of countries that could afford to have science and don't. Internet publishing will help poor countries by providing access to science reports; rich countries should spend more on the poor countries in the area of health care and health-related research. I've been giving speeches on this subject ever since I worked on a recent report for the World Health Organization. Thirdly, I've tried to develop a concept called the Global Science Corps that would enable scientists at transition points of their careers — for example, going from training to full-time employment or from full-time employment to retirement — to serve for a few years in less affluent countries provided that there were centers of scientific quality.

I would like to ask you about your family background.

I come from a Jewish family, but I don't practice the Jewish religion. I married an Irish Catholic. My kids have no religion. I never go to temple, I'm anti-clerical, but I feel Jewish culturally.

What does it mean?

I'm not sure what it means except that I know that it has something to do with persecution, the fact that my grandparents left Europe around the turn of the last century, my father's side from Poland, my mother's side from Austria; this was well before Nazism, but they left because of anti-Semitism. This is a slightly negative way to view your cultural heritage, but there's no doubt that it has some binding influence. I respect the traditions that have served Judaism well and served those people well who are brought up as Jews — that is, respect for the Book, an interest in doing good, and a sense of pride in scholarly accomplishment that has always been associated with Judaism. That being said, I don't have any religious feelings, I don't have any interest in going to synagogue, as I did in my youth.

Do you talk about being non-religious openly?

Sure.

Aren't you afraid that fundraising for your Center would suffer from it?

I don't go around displaying in public my distaste for religion; I recognize that other people have different feelings. We certainly provide financial support for our clergy at the hospital, our patients need that. But I myself don't get involved in those activities; it bothers me sometimes that the clergy take advantage of misfortune. If someone asks me if I am religious, I would answer no. But I would never give a speech about my non-religious or even anti-religious feelings because I don't think I have any very interesting thing to say about them.

Do you consider your Nobel Prize-winning work to be your most interesting research?

They may not have been the most exciting experiments. There are other things that I did, having to do with the mechanism of retroviral DNA synthesis and integration, our work on the process called ribosomal frame shifting, our work on the hepatitis B virus lifecycle, the discovery of receptors for avian retroviruses, some work we've done recently on mouse models that I've personally gotten more immediate pleasure from. But there's no doubt that what was done to discover the cellular oncogenes was the most important, had the biggest implications for health, and was an obvious thing for the Nobel Committee to award. I may even have had more hands-on experience with our other discoveries. Most of the experiments in our prize-winning work was done by postdoctoral fellows. The first experiments were done by Ram Guntaka, the crucial experiments by Dominique Stehelin, the follow-up experiments by people like Deborah Spector, Richard Parker, and others. I did a lot of experimental work until I was in my early forties.

Would you single out one project for mention from among your current research?

Most of our current activities focus on building animal models of human cancer. The biggest innovation we've made is to devise a method for delivering genes to animal tissues using a gene that we cloned years ago that encodes a receptor for a virus that we use as a gene delivery mechanism. That's been very helpful in building these models. In addition, we and other groups, including Mike Bishop's, have taken advantage of the ability of molecular biologists to regulate the expression of genes in experimental models. That's allowed us to define a way in which a member of the Ras-gene family [an oncogene] plays a crucial role in not just the initiation but also in the maintenance of cancers of the lung. Of all the things going on in my lab at the moment, the most interesting to me is the observation we made two years ago that we can use a mutant of a Ras-gene to induce cancers in the lung that resemble one of the most common human lung cancers. Most importantly, when we reduce the amount of this abnormal protein, the tumors disappear rapidly; the cells undergo programmed cell death! That phenomenon encourages a lot more work because we might be able to figure out how that cell death process is initiated. If we could simulate that process with drugs, we'd have a way of thinking about how Ras-genes, which are commonly mutated in many human cancers, become required for the continued vitality of a cancer cell.

Do you have heroes? Are there people whom you would like to emulate?

I draw a distinction between my heroes and the people I would like to emulate. I do a lot of bike riding, so Lance Armstrong is one of my heroes, but I don't aspire to win the Tour de France.

I saw your picture, in the magazine of MSKCC, standing on the top step as winner of a fundraising bicycle competition.

I was on the winning team, but I shouldn't have been on the top step; I did not win the competition. Coming back to emulating others, I don't want to have cancer so that I can overcome cancer, which my hero cyclist did. There are also people for whom I have great admiration as scientists, and that's a long list. Then I have cultural heroes.

Do you live a sizzling intellectual life?

It's a sizzling cultural life with social overtones. I have a lot of friends and often go out with board members and other donors. One of the nice things about being in this job in New York City is that you are well connected to the rest of the City. I've come to know many of the most influential people in town and that means attending a lot of nice parties. But it's not the same as intellectual life. For that, you need not be in New York City.

This is partly a research institute and you have a research lab in it. Can it happen that someone bursts into your office with some new finding? My impression is that you are well protected by receptionists and secretaries.

That could happen; but it's more likely that they would tell me about it when they submit the paper about it.

Which is not the same.

It's not the same. It's more likely that the person would go to close colleagues.

I meant more your own research group.

I go there every day. My lab is right across the street and I have another office there. I don't think that any member of my group has been in this office. I eat lunch often with my group. There people don't call me "Doctor", they call me "Harold". And they would be likely to announce an interesting result at lunch.

What will be the dominating science in the coming years?

It will still be molecular biology. Molecular biology has changed somewhat, in large part by sequencing genomes. It will also be changed by the impact of computational science and databases. Imaging is also changing biology, the ability to see molecules, to see how cells operate, and to follow their fate. There's an increasing tendency for biologists to think visually. The three-dimensional picture of how cells are working is important. There are physical tools. Then there is an enhanced role for chemistry. Chemical biology is one of our slogans. Chemistry may have a bad image in the eyes of the general public, but in our science, exploring chemical space, building new substances with combinatorial chemistry are of great interest and therapeutical promise. The merger of structural biology — that is, the three-dimensional structure of proteins — with potential drugs, is among the most important themes in medical science.

Peter Mansfield, 2005 (photograph by M. Hargittai).

12

PETER MANSFIELD

Peter Mansfield (b. 1933 in London) is Emeritus Professor of Physics at the Sir Peter Mansfield Magnetic Resonance Centre, University of Nottingham. He received his B.Sc. degree in 1959 and his Ph.D. degree in 1962 in physics, both from Queen Mary College, University of London. In 1962–1964, he was Research Associate at the Department of Physics, University of Illinois in Urbana. He has been with the University of Nottingham since 1964 where he was appointed professor in 1979 and from which he retired in 1994. In 1972–1973, he was Senior Visitor at the Max Planck Institute for Medical Research in Heidelberg, Germany. Dr. Mansfield was co-recipient of the Nobel Prize in Physiology or Medicine for 2003 "for their discoveries concerning magnetic resonance imaging" together with Paul Lauterbur.[1] He was elected to the Royal Society (London) in 1987 and was knighted in 1993. He has numerous other distinctions and honorary doctorates. We recorded our conversation with Professor Mansfield in his office at the University of Nottingham on January 12, 2005.*

I would like to ask you to delineate Paul Lauterbur's contribution and your contribution in the development of NMR imaging.

Our original work in imaging was independent of Lauterbur's work. I did not know what he was doing. He did publish two or three months ahead of us. By then we had already been working on the idea of imaging

---

*Magdolna Hargittai and István Hargittai conducted the interview.

for about a year. I was, in particular, interested at that time in whether one could produce images of solids. I started the work in 1972. I was just about to spend a sabbatical year in Germany at the Max Planck Institute for Medical Research in Heidelberg. I left a student here in Nottingham who was just coming to the end of his Ph.D. studies, Peter Grannell. Peter took up the ideas which I suggested to him before I left for Heidelberg and we corresponded heavily on various aspects of imaging in solids. He carried on doing the work and made some progress while I was away. This was all in 1972.

What did you mean by imaging at that time?

The original idea that we had was a question, "Could you say anything about the atomic structure of materials?" The very first idea that we had was to investigate whether one could actually say something about the structure, having the right conditions and a large enough magnetic field gradient.

Did you have in mind something comparable to X-ray crystallography?

Yes. That was the original thought. After some calculations that I did, it became obvious that it was possible but rather difficult. Rather than try to image actual materials, that is, rather than doing crystallography of real materials, we started out with a model lattice for which plates of a material, camphor, could serve well. Camphor is a solid at room temperature, but it has a very narrow line width because there is a lot of rotation at the local molecular sites, so the line width one obtains with camphor is much narrower than one would expect generally from solids. We made very thin plates of camphor, something like half a millimeter thick and made a structure of three plates with gaps in between. We made an image of this structure, which was a one-dimensional image. We had three peaks in this image corresponding to the three plates. This we published in 1973, just a few months after Paul Lauterbur published his paper. In this description it may sound that it is unrelated to imaging but it was imaging because we needed a magnetic field gradient in order to resolve these three-plate structures. What we did not do at the time was to rotate the specimen so that we could look at the three plates at different angles. That was a minor omission. We were thinking that if we could get the atomic spacing, we would not need to rotate the specimen, we could get the structure from the diffraction pattern from NMR.

In terms of a diffraction pattern, this would be a one-dimensional pattern so it would be difficult to deduce more complicated structures from such patterns.

At this early stage we were only thinking of cubic structures.

What gave you the idea originally?

In pre-1972 times I was involved in the development of multiple-pulse techniques in solids by NMR. We were trying to remove the dipole-dipole interactions in the solid so that the very broad NMR lines one gets in solids could be reduced to reveal the chemical shift structure in solids. This work was started originally with my Ph.D. studies on NMR in London. I discovered the effect which ultimately led us to think about chemical shift structures in solids. A few years afterwards in America, a fellow called John Waugh ...

We have interviewed him.[2]

He was our major competitor at the time because he independently discovered solid echoes. If you pulse a solid, you produce a free induction decay. If you put a second pulse in, one after the other, very close to each other, the signal, instead of continuing to decay, builds up again to form an echo. It is as though you are introducing reversibility into what was thought to be a completely irreversible process. Once you initiate a free induction decay in a solid, it was considered — I am talking about 1962 and earlier — to be over and done with in a completely irreversible process. During my Ph.D. studies, I happened to stumble upon the fact that if one applies two pulses very close together, the signal after the second pulse starts to grow again into an echo. These effects, the spin echoes, were well known in liquids, but everyone in the business at the time thought and actually said in various publications that spin echoes in solids were impossible. The reason that people had not studied this in any detail was because the free induction decays in many solids disappear very quickly and the problem was the response time. If a very large pulse is applied, the receiver is paralyzed and the signal cannot be observed as close to the origin as one would wish. A very short pulse is applied and immediately after that everything is obliterated. A clear signal is not observed until maybe twenty microseconds afterwards. We talk about microseconds here. The problem with the solids that I was considering was that the free induction decay was over and done with in between four and ten microseconds. For most people with the equipment that was available at the time, they could not really study solids. The signal was lost. Part of my Ph.D. work was to build an apparatus which would recover very quickly so after about one or two microseconds, one could start to observe the signal. In the course of doing these experiments I happened to apply two pulses very close to each other. What I found was that as I got closer to the origin — and we are talking about a gap of two or three microseconds — an echo-like signal appeared after the second pulse. If the spacing between the pulses were increased, the echo would move further out in time. Of course, being a solid, it was of very low intensity. To get decent signals, the pulses had to be brought close together, within one or two microseconds. All this was new at the time and there was no commercial apparatus around to do this experiment. I had to design and build it myself.

You said that you stumbled upon the echo phenomenon. Was it an accident that you put in the second pulse immediately after the first one?

It was an accident because I was not supposed to do multiple pulse work at all at the time. My supervisor, Professor Jack Powles, had asked me to build a pulse spectrometer. All he was concerned with was producing a

Peter Mansfield with Count Björn Bernadotte at the 2005 meeting of Nobel laureates (photograph by I. Hargittai).

single short 90° pulse. But having produced this pulse, it was relatively easy for me to produce a second and a third pulse. I say relatively easy, but it was not completely straightforward because the system that I built was not a coherent system. If two pulses were produced close together, there was no guarantee that the phase of the first pulse and the phase of the second pulse were the same. In fact they were tettering about all over the place. There was a random phase shift between the two pulses and as a result, the spin echoes that I observed after the second pulse were sometimes there and sometimes they were not there. This was a mystery to me at the time. It was a mystery to my supervisor as well. We did not understand why we could put two pulses very close together and sometimes we could get an echo and sometimes we did not get anything. There was nothing in the literature that we could have consulted about this. It was something completely new. It was investigating why these things were happening that we eventually realized that there was a spin echo and furthermore that the spin echo was only there if the phase of the two pulses was correct. The phase tetter was what really led to the discovery. Eventually I managed to put these two pulses together and hold the delay between the two pulses very accurately so that the phase of the second pulse could be guaranteed within a fraction of the wavelength of the signal. Anyway, that was the original work and that was done in 1961. I continued to think about these experiments while I was working in America where I spent two years with Professor Charlie Slichter in Illinois. Slichter is a physicist and I was working on the NMR of metals. There was no opportunity to continue the work that I started in London, but I did think about it and I wrote a paper on pulsed echo effects while I was in America. When I returned to England, I came to Nottingham and set about designing and building a piece of equipment which would allow me to apply not just a few pulses but any number of pulses after this first pulse, a multiple-pulse apparatus. It was during the building of this equipment, which I had started in 1964, that I had a Canadian research student, called Donald Ware, join my group. We had been working on this equipment for two years and towards the end of that period I was informed that John Waugh had discovered this multiple-pulse effect. He says that he didn't know about the work we'd already done at a very early stage. He came into it from a completely different angle. He discovered the effect while they had been evaluating a commercial machine. Such spectrometer equipment became available in the early 1960s. Most chemistry departments either had or were expecting to get such equipment to study NMR in liquids. But there were very few departments that had equipment to look at solids. John Waugh either bought or was thinking of buying such equipment and, during the process of evaluating the equipment, he stumbled upon the solid echo effect. This was in about 1966, quite a few years after we had started the work. But this was a digression.

It is very interesting because it shows that your imaging studies and Paul's imaging studies originated from very different roots. He started with studying substituent effects in a series of compounds to investigate their chemical consequences by NMR. Then he heard about Damadian's experiments and that set him onto the course for imaging.

My *bête noire* was John Waugh.

When did you first think of imaging?

In 1972, the following idea occurred to me. If one can remove the dipolar broadening from the signal, then effectively the free induction decay signal could be turned from a solid into something that looked like a liquid. If one could do that — and all these were ifs — then maybe with the application of a small magnetic field gradient, one could observe the structure of the solid itself. That was the reasoning. One of our best results at the time was on calcium fluoride, which is a crystalline solid. Normally the fluorine line width is about two kilohertz. We were able to reduce it to one hertz. It struck me that if we could achieve that with a regular solid, there was a good chance that if we did it together with a magnetic field gradient, we might see atomic structure in the solid. But it was harder to do than we thought. Here we return to the description I gave you at the beginning of our conversation, when I told you that instead of a real solid, I proposed to use a model lattice and sheets of solid camphor served for that purpose. Without the field gradient we just got a broad line. But when we did the line narrowing experiment, we obtained a single line but with a greatly reduced width. Adding the field gradient instead of the single line, we got three lines. The idea, according to which we would remove the dipole-dipole interaction, apply a magnetic gradient, thereby resolving the structure, worked. This is what we had set out to do. But this was only in a model solid. We are now in about 1973. The next step was to do a similar experiment with liquids. That, we reasoned, should be much easier, rather than starting with a solid, then effectively turning it into a liquid before proceeding with the experiment that we wanted to do in the first place. So my student and I decided to do the imaging experiment with liquids.

Did you still consider the atomic arrangement at this point or were you already thinking of biologically interesting objects?

I was thinking increasingly of biological applications. I was first thinking of layered structures, like the skin where we could direct the magnetic field gradient orthogonal to the layers. We did not do any experiments because it seemed to be a very narrow area, so I am talking about thought experiments at this time.

Was this a turning point in your research?

Switching from solids to liquids was a turning point. The year I spent in Germany was at the Max Planck Institute for Medical Research in Heidelberg. There I had been working in an ostensibly medical environment. It is not particularly surprising that I was thinking about medical applications. Once we decided to do imaging in liquid-like systems, in 1975–1976 I went into the garden and took some flower stems, twigs from trees, and anything that I could lay my hands on that was small enough to put into the NMR sample coil. We were detecting the signals from the liquid components. We were determining the water distribution in these systems in the presence of a magnetic gradient. The coil diameter was one and a half centimeters. We also used chicken legs. The NMR apparatus was home made, originally built for solids, but we could change it easily for the study of the items I just told you about. Because of the limited sample size, in early 1976 I suggested to another of my students, Andrew Maudsley, to put his finger into the coil. He was my second student in MRI and this experiment was a major turning point because this was the first image of a live human finger. This experiment created a lot of excitement at the Medical Research Council (MRC) in London. A conference was held there to decide what MRC should do about NMR imaging. At that time we called it NMR imaging. There were other developments in NMR imaging in the Physics Department of the University of Nottingham. By 1976, there was a second group operating and run by the head of Department, Professor Raymond Andrew, which made things very difficult. I was a lecturer at the time. We had only one professor of experimental physics in the Department at that time. Professor Andrew wanted to get much more involved in imaging and decided to have his own independent group. Fortunately, both groups managed to get funding, but the situation created all kinds of difficulties. I don't really want to get into that in this interview. I hope to write about it myself. These problems continued from 1976 to about 1984 when Andrew left the Department and went to Florida, USA. It's a long and complicated story, but what I wanted to mention was that the two imaging groups at Nottingham eventually ended up as three groups because Andrew's group split into two and they were at each other's throats.

That must have relieved you a little.

It did. It was a blessing in disguise. In 1984, everyone in these other two groups left the department. They all went to America. That was the second blessing. I was left here on my own. I was promoted to professorship in 1979 while Raymond Andrew was still here. Whatever was between us it was not so bad that he was able to hold me back. In that sense he was completely fair. It was more to do with the idea that he felt left out of imaging. It was natural enough; he saw the opportunity and he saw the huge applications, and he wanted to be part of it. I must add that the period of 1976 to 1984 was the most productive period that we had. The number of papers that came out of Nottingham was unbelievable, from myself, from Raymond Andrew, and from his colleagues. We were at the top internationally.

Did competition play a role in this?

It did, absolutely.

Did you have joint seminars?

Oh, yes. We had regular colloquia, they gave papers, and we gave papers. Besides, Raymond Andrew was a great organizer, so there were conferences and there was the very first MRI conference here, before the international society of magnetic resonance in medicine was formed.

When was the N dropped from the name?

That was due to the Americans; it was dropped probably in the mid-1980s. There are a couple of explanations for why it was dropped. The official reason was that there was squabbling in America with the nuclear medicine people. Because NMR had the word nuclear in it, they wanted it to be in their department. The radiologists in the States, who have an even more powerful lobby, did not want the word nuclear in the title because they wanted imaging to be in radiology. That's what then happened. The Americans decided between themselves that there should not be NMR imaging, it should be MR imaging. We played no role in it although much of the development happened here in Nottingham and not in America up until the early 1980s. Then in the early 1980s, it started taking off in America and, like most things, if the Americans get involved, it is great and very soon it will be considered having been invented in America. They did it first, never mind what happened in Europe. That's what happened with MRI. It happened with the manufacturing as well because the companies got interested in the 1980s. One of the first companies was a subsidiary of Johnson & Johnson, which was involved for five years before it decided to pull out. In Europe, we had Siemens, Phillips, EMI, and a whole range of smaller companies. The very last company to come into the MRI business was General Electric. They came in during the mid-1980s; they waited and waited before they came in. When they did come in, they dominated completely and they are now making well over half the total number of MRI systems in the world. They got the biggest share in the market. It should have been Europe, but that was not to be.

Does your company produce equipment?

How do you know about my company?

You wrote about it.

OK [Peter Mansfield is heartily laughing.] My company is very small; it has two employees only. We do experimental work and try out ideas. It is not a company that is ever going to build anything. It is also too late to get into the MRI business because there is huge competition; no one is able to compete with GE.

If we can get back to the development of MRI, last time you mentioned that your student put his finger into the NMR machine. That is still far from putting a whole human body into it.

All the interested parties attended the 1976 meeting in London that was organized by the Medical Research Council. All involved in NMR imaging were invited to give presentations at this meeting. It was at this meeting that I presented our results on imaging the fingers. Virtually all the imaging that had been done by that time had been done with dead targets, mice, rats, twigs, plants. The only live imaging presented at this meeting was ours and it was part of a living human being. So it created a lot of excitement, particularly among the medical people. As a result of that we were invited to submit an application for a grant and I quickly put together an application. It was to build an MRI machine, but not for fingers, rather, for the whole body. We got the grant and we built the machine. At the same time, Raymond Andrew submitted an application to the MRC to build an imaging machine for the hand and wrist. He got his grant a little earlier than we got ours. So, again, we were starting off in some sort of competition, Raymond Andrew working on his wrist imaging system and we working on our whole body system. I'm glad we went our route because we were able within a year or so to produce our first whole body images, in 1978.

Were you the first?

No, because we were pipped to the post by a few months by Raymond Damadian in America. I didn't know a great deal about him; I'd seen him at a conference in Heidelberg in 1976. The best they could produce at the time were very crude images of animals. The next time I heard about Raymond Damadian, he had managed to build a whole body imaging machine. However, the techniques that he used for imaging were not terribly good. He was proceeding on a completely wrong track. He was working on a point scanning technique, which he called FONAR for Field Focused Nuclear Magnetic Resonance or something like that. There was a little bit of competition between these two groups. Have you interviewed him?

No, although I would have liked to. I wrote to him but never received an answer.

He's got a big, big chip on his shoulder. He feels he should have shared the Nobel Prize but the Nobel Committee did not think so. If you look up the images that he produced at the time, and the way he produced them, I think he didn't have a viable imaging system. The machines that he now makes in his own company use our ideas, not his own.

Did he use your patents?

Yes, but the patents have now expired. He also patented and I looked at his patents; he used rather general descriptions in his patents; it is a little wishy-washy.

Whose patent is General Electric using?

General Electric is using our patent for slice selection and they are using other patents of ours as well.

Paul Lauterbur never patented anything because at that time his university discouraged him.

I know he says that. Of course, we did not just file one patent; we obtained maybe ten or twelve patents at rather different points in time. Even if Lauterbur missed the first opportunity to patent, my question to Paul would be why didn't he file patents later?

He says that he preferred to be completely open about his research. Did you feel in any way confined in communicating your research?

No, because once we filed our patents, we immediately published our results. You can see all our ideas in publications. We never held anything back.

Did you feel that Damadian should have been included in the Nobel Prize?

In terms of what he in the end achieved, probably not. In terms of the actual steps he took to do imaging, he should've been included, perhaps, for that. He claims that what he did prior to imaging, that is to say, the work on tumors, and the fact that he showed unequivocally in a paper published in 1970 or 1971, that he could see elevated relaxation times in tumors that would delineate the tumor rather than normal tissue, should have allowed him to get a fraction of the Nobel Prize. I don't know. I think the Committee that deals with this looked long and hard at all these things and decided that he should be excluded.

Do others come to mind who might have been included?

There were other people involved; companies and also people in Aberdeen, Scotland; one of their patents is still being used. There are only three people allowed in the Nobel Prize, as you know.

There was one slot unused, of course.

There was one slot unused, but sometimes there are two slots unused. If you look at Richard Ernst's Nobel Prize, for example, one could argue that there were lots of other people that should have been included. Two other chemists could have been included easily. John Waugh could have been included, why not? We are not privy to the Nobel Committee's thinking. My thinking is that if three people were allowed, then I would have had three people. The question then would have been which three? Everyone seems to feel that number one should be Paul Lauterbur. Then the question is, who should be number two and who should be number three? We were in the business early on and I would say that we were neck and neck with Paul. But maybe there are people around who don't feel that my contribution merited the Nobel Prize.

Would you have a candidate for the Nobel Prize for solid-state NMR?

John Waugh would be a good candidate. He did some very nice things. I know that we were at loggerheads on certain things, but overall he made enormous contributions both theoretically and experimentally. Of course, he had a big team as well who helped him achieve all these things. I think it would be appropriate for him to get a prize.

You spent some time at the University of Illinois and Paul Lauterbur is now at Illinois. Was there any connection?

They were looking around in Illinois for someone to set up imaging. I was asked if I wanted the job. I went over there for interview, when everyone in the other groups was leaving Nottingham University in 1984. I did consider the possibility of leaving myself. Things got very difficult here also because it became hard to get funds for continuing research. At the time I am talking about, we were still working with an electromagnet rather than a superconductive magnet. We did not have the money. We were very much held back by research funding. I went over for three interviews; one in Illinois, for the job that Paul Lauterbur now has; another in Cleveland to work at the University of Cleveland (but the work would be supported by General Electric); and the third interview was down in Alabama at the University of Alabama. I got offers resulting from all three interviews. I could've gone to Illinois if I had decided to. All the offers were very attractive financially. For the trip, my wife and the family, that is our two girls, were given very good treatment over there; we were almost persuaded very much in terms of the way of life and so on that we should go over. I didn't make my mind up while I was there at all. I said we'd let people know. When I got back to England, there was a letter waiting

Peter Mansfield in front of the Sir Peter Mansfield Magnetic Resonance Center at the University of Nottingham, 2005 (photograph by M. Hargittai).

for me here from the Department of Health; they had suddenly found that they had money for a superconductive magnet. Was I interested? Whether that was pure chance, I just don't know, but I grabbed the offer with both hands. The rest is history.

We are recording this conversation at the Sir Peter Mansfield Magnetic Resonance Centre of the University of Nottingham. What is the primary goal of the research of this laboratory?

The functional imaging of the brain. This is the major interest of the guy who has taken over from me, Professor Peter Morris, who is a former student of mine.

In your two-minute Nobel banquet speech you mentioned that you receive many letters of gratitude from patients, but also that a few mention the claustrophobic effects of the tight confinement in the machine, and some mention the rather high level of acoustic noise during the scanning process. Then you added that these problems are not being ignored. Are you involved in looking for easing the situation in this respect?

My small company is involved in research to reduce the level of the acoustic noise in the MRI machines. Have you ever been inside such a machine? They are very noisy. Every time the gradient switches on and off, considerable noise is generated. First, we are concerned with the question: why is that noise associated with the gradient change? Secondly, can we do anything to reduce it? Then another thing that the company is particularly interested in is the reduction of the electric field, which is automatically created as described by Maxwell's laws of electromagnetism. An electric field is associated with the magnetic field. When that electric field is created, it is strong enough to cause currents to flow in the body; it can cause muscular twitch; so every time we switch the gradient, the body twitches. That is not good for patients. There are ways of reducing the electric field associated with the magnetic field gradient, by not switching the gradient too fast. If the gradient is switched on and off more slowly, the electric field component can be reduced. The problem is if the magnetic field gradient is switched more slowly then the time taken to do the imaging is increased. Therefore that is not the approach we wish to take. My company is looking at ways of reducing this electric field. They are the two major problems we are working on and there may be others as time goes on.

Please, tell us about your family.

I have two daughters. The eldest is Sarah, 37, who is married with two daughters. The second daughter is Gillian; she is also married, with two sons, and her husband is one of my students here. Sarah is a qualified nurse, but she is now running a small business selling women's clothing. Gillian is qualified in banking, but is more interested in looking after animals. She is doing a course right now in one of the colleges in Leicester. She hopes to become a veterinarian nurse. My wife's background is secretarial. She has not had a job for the past twenty years or so.

What did your parents do?

My father was a gas fitter. He worked for a company that supplied gas for homes. My mother had part-time jobs. My father died in 1966 and my mother died in 1984.

Your schooling was somewhat unusual.

I left school at the age of 15. There was no option; I had to leave. The school was an ordinary secondary school, nothing special. Everyone left at 15.

So you were not a dropout.

I was a dropout in a sense when I went to an ordinary school in the first place. It's a complicated story. When I first went to the secondary school, we had in England three grades of secondary school. There was the ordinary secondary school, there was the central school, and there was the grammar school. In order to get into a grammar school, you had to pass an examination. If you failed, you went to the secondary school. If you failed but not too badly, you had the option of going to a central school. This was just after the war. I was eleven years old when I came back from evacuation to London; I had to take this examination within a week of returning. I was not prepared for it. I did not get the requisite pass level so I went to the middle level school. Though I did not know and I guess many people at that time did not realize, at that particular time considerable changes were happening in the structure of secondary schools. Many grammar schools were converted into secondary schools; they called them secondary modern schools to differentiate. The central school that I went to ceased to exist after one year. I had one year of the central school and after that the central schools vanished and I had to continue in an ordinary school. When I left school at the age of 15, I had no option to continue my studies. I went for an interview with the school career advisor where I was asked what I wanted to do. I was interested in science and in particular I was quite interested in space travel and rocket science. The interviewer burst out laughing. He said, "You are from a secondary school, how can you possibly be a scientist? Ridiculous." He told me that I would never have a chance to become a scientist; he asked me to be sensible and tell him what I wanted to be. I told him that one of my hobbies was printing. He said that was more like it. So I went into printing and ended up initially as a bookbinder. I did this for a period of time but I was always very interested in being a typesetter, or compositor. I had a small printing machine plus one that I made, a flat-bed printing machine, so I could print things up to a limited size at home. Eventually I managed to switch my job from being a bookbinder to a trainee compositor. I did this for three years. But the day I started work, I also joined evening classes. I was working in the City of London and in the evenings I had to rush to the evening class at the Borough Polytechnic. It is now called the University of the South Bank. I enrolled to study for O levels and later for A levels. I did this five nights a week for three years until I passed the O level.

So the educational system almost failed you.

There were alternatives. If you were really persistent, you could improve your education. You could also take examinations, which were the same examinations, incidentally, that people were taking who were still at school. So I did this. Then, at the age of 18, I came across an article in the newspaper, the *Daily Mirror*, and this article described how some young boy, 17 or 18 years old, managed to join a group of professional scientists working at the rocket propulsion department in Westcott, Buckinghamshire. The article said how he had managed to pull himself up and get into this establishment where he was now working with the scientific staff there. I was very interested in this article. I wrote to the editor of the *Daily Mirror* and I got a reply saying that if I wished to follow this up, I should contact people of the Ministry of Supply in central London. So I wrote to the Ministry of Supply and surprisingly I got an invitation to go and talk to these people in London. They gave me an interview and at the end of this interview they said, "We think you should go to the Rocket Propulsion Department to be interviewed there." They arranged for an interview for me; I saw at least one of the scientists there, and they offered me a job without qualifications. I was still studying part-time. The job was offered with the condition that I actually continue my studies and pass the examination to get O level. The O level is like matriculation. That's what I had to do. I took the matriculation examination and passed and so my job was OK. I was only a scientific assistant, which is the lowest grade, but it was a start. Within a few months of matriculating, I got my call-up papers; I was by then eighteen and a half. I got called up to do national service. I served for two years in the army. When I eventually was demobbed from the army, in 1954, I continued studying. This time, for advanced level, for university entrance. I did it in about a year in three subjects: physics, pure mathematics, and applied mathematics. I also did O level chemistry. That got me into university. I started university in 1956 as a full-time student at Queen Mary College, University of London.

Who paid for it?

Initially, it was a county scholarship from the Buckinghamshire County Council because I was living in Aylesbury and had taken these examinations in the county. I applied to the local authority for a grant which I received. Then within a couple of months of getting the grant, I was invited by the Ministry of Supply to attend an interview for a scholarship that they offered. I went along to the interview and they decided to award me a Ministry of Supply bursary to continue at university.

Was your greatest challenge ever to get into university?

Once I was in, it was all fairly straightforward because I was driven. Unlike many of the students in my year, I was the only student who knew exactly what I wanted to do. Most students were younger than I was; many of them came straight from school; they didn't know what they wanted to do. I was in a privileged position. When I look back and weigh up all together, my greatest challenge was to get the research group together in Nottingham as a team to work towards building and improving the MRI systems. There were all sorts of challenges there.

After you had returned from the United States?

Yes. Each step was a little challenge, but the big challenge was trying to get everything together and sorting out all the problems, some of which I touched on earlier, in particular the internecine squabbling that went on. All these things affected and hardened me to go forward to make MRI work. If things come too easily, it does not feel much of an achievement. If it is an uphill struggle, you feel more satisfied if things work in the end.

Is my impression correct that the Damadian controversy did not touch you, it did not concern you?

It was not anything to do with me. It was really a feud between Paul Lauterbur and Raymond Damadian. I knew Raymond Damadian only by virtue of his presence at conferences. There was one instance, which I just mention here briefly because I know that you have to catch your train. I don't quite remember the date, but it must have been after 1979 and before 1984, when I had been invited to a dinner here on campus in Willoughby Hall, which is one of the halls of residence for students.

Every now and again professors get invited to these student functions. We had almost finished our dinner, it was about nine o'clock in the evening and I was sitting at the top table when someone came along and said, "Professor Mansfield, there is a Dr. Damadian outside who would like to see you." I couldn't believe what I was hearing. I just finished my dessert and went out to see him straight away. Outside there was Raymond Damadian, on campus, at nine o'clock at night. He came completely out of the blue. I had no idea that he was here. Anyway, to cut a long story short, I said, "I am in the middle of being entertained here; I'll see if I can make my excuses and leave so that I can spend some time talking to you." I went back, made my excuses to leave, and took Raymond Damadian back to my home. When we got back there, he started to tell me his various problems, mainly with Paul Lauterbur, and how he'd been cheated by Paul. We sat at home until about — and I don't exaggerate because my wife excused herself at eleven thirty — one o'clock. At that point I asked him what he was going to do, and he said he was going back to London. I did not want to let him go to London at that late hour and although we did not have a spare bedroom, we could make some arrangement. However, he would have none of it and insisted on going back to London; he had a car, and off he went at about one thirty. All that we had discussed was how he had been screwed by Paul Lauterbur. He was just pouring his heart out to me. His main concern was that his idea to do MRI had been stolen by Paul and that he was not getting due recognition. He saw me as being a person that he could confide in. I must tell you that after that meeting I did actually feel that he was not as bad as many people were saying. There were people saying that he is a showman and that he is everything but a genuine scientist. I did feel after our long discussion that there was very little that I could actually do, but he probably found it helpful that he could confide in me and tell me about his problems. I could see his point of view, but, of course, Paul Lauterbur denies much of the story, and I find myself in an extremely difficult position. I didn't know quite whom to believe. After that meeting I had considerable sympathy for Raymond Damadian. In fact, some years after that I was elected a Fellow of the Royal Society (in 1987) and we had a Royal Society meeting in London on imaging. I was instrumental in inviting Raymond Damadian to that meeting. He came and gave a paper at the meeting, but unfortunately he blew all possible sympathy that he may have had from the audience.

What did he do?

He just behaved in a ridiculous way. He came to give a scientific paper and started out by showing us a picture of President Reagan and telling us that he had seen people at the very highest level and he had evidence that he had been done down by the various national societies in America. He told us that Lauterbur had connived to make sure that he didn't get a grant, and it just went on and on. He had a twenty-minute talk and he spent most of that time carrying on and on. It was very embarrassing because I was responsible for bringing him there. There were a lot of people there who had heard all these stories about Damadian and some of them felt that he should not be there at the Royal Society giving a talk.

Coming back to your life, if it would be possible to chart your life from the start, what would you have done differently?

I should have been more decisive in doing things. Looking back, everything one does in life is affected by somebody else. You do something and you do it in a particular way because of the circumstances. If I had been bolder, more assertive, I might have achieved what I did at an earlier age. But unfortunately, I wasn't. I was quite happy to keep my head down and not cause problems. It's really about decisiveness; if I had been more decisive, I might have been where I am now, twenty years ago.

Do you feel that you are a very good role model for people who have a hard time taking off?

That may be, but if they were to hear me speaking now they might not think so because what I am saying is that I was not really decisive enough. It also may be that if I had been too assertive, I would not be where I am. It's hard to say. How can one be sure that one wouldn't have upset things? You have to accept things as they are and what you've achieved. I regret that I took early retirement — I retired at 61 — and I spent the last ten years with this small company. We've made some progress but not nearly enough; we have been far too slow.

Wouldn't you need more people?

I know, but it's money.

Were you induced to take early retirement?

No, not really. I took early retirement because I wanted to spend more time on research. I've done this, but I think I've spent too much time on this particular area that I mentioned, reduction of acoustic noise. It's an important problem and it would be nice to solve it. In my case what's happened is that I have made some progress, but I have not solved the problem completely. The question is how much time should one spend on a problem before you give up? You either solve it or you don't solve it. Unfortunately, my character is such that I find it very difficult to give up on a problem. I keep thinking, maybe tomorrow I'll solve it.

Is this stubbornness or perseverance?

I'm stubborn, that's it; that's the fault. If you're too stubborn, you end up wasting time, and that's what I've done. I've identified another problem, which is the E-field problem, but the acoustic problem still remains to be solved. We have some ideas how one might do it, but I've spent fifteen years on this problem. That's almost more time than I'd spent on the whole of NMR and MRI together. The time it took to come up with the original ideas for MRI took me four or five years. It wasn't one idea either; it was a whole world of ideas. At the end of my life I've spent fifteen years on one problem and I still haven't solved it.

Could you afford it?

I couldn't.

I didn't mean only time-wise but also financially.

There were patents taken out on imaging. There was a period between 1987 and about 1995 when there was considerable royalty income, and I made a lot of money. That was another reason why I took early retirement. I didn't feel that I needed the job here. I ran the company and I'm still running the company, but it's costing me approximately a hundred thousand pounds per annum to keep the company going. That's why I feel that I've wasted a lot of time. We've taken up a very difficult problem trying to reduce the noise. If one could solve the problem, it might lead first of all to new designs of scanner and also it could be very lucrative.

Everybody would have to change the machine.

But it's not going to happen.

Do you sleep well?

Reasonably. I sleep better now than I used to, when I was still in post here. I used to spend hours at night half-asleep and half dreaming about solving particular problems.

Are you religious?

Yes, up to a point. When I was young, I was a Salvationist. Are you familiar with the Salvation Army?

We know about their charity work.

It was started by William Booth in the 1800s in Britain, in Nottingham actually. In Germany, it's known as *die Heilsarmee*. I always counted myself a member of the Salvation Army except when I was in the army. I found my old army documents recently and I discovered that I was registered in the army as an atheist. I don't think it was ever true because I was never an atheist.

References

1. An interview with Paul Lauterbur appeared in Hargittai, B.; Hargittai, I. *Candid Science V: Conversations with Famous Scientists.* Imperial College Press, London, 2005, pp. 454–479.
2. Hargittai, I. "John Waugh" *Chemical Heritage* **2001** (Winter), *19*(4), 26–29.

Avram Hershko, 2004 (photograph by M. Hargittai).

13

AVRAM HERSHKO

Avram Hershko (b. 1937 in Karcag, Hungary) is Distinguished Professor of the Technion – Israel Institute of Technology in Haifa, at the Unit of Biochemistry, Faculty of Medicine, The B. Rappaport Faculty of Medicine. He received his M.D. degree in 1965 and his Ph.D. in 1969, both from The Hebrew University–Hadassah Medical School in Jerusalem, Israel. He served as a physician in the Israel Defense Forces in 1965–1967, was lecturer at the Department of Biochemistry of The Hebrew University in 1967–1969, and spent a postdoctoral stint at the Department of Biochemistry and Biophysics of the University of California Medical Center in San Francisco in 1969–1971. He has been at the Technion since 1972. He is best known for his pioneering research of the ubiquitin system.

Dr. Hershko shared the Nobel Prize in Chemistry in 2004 with Aaron Ciechanover (b. 1947), also of the Technion, and Irwin Rose (b. 1926) currently of California, but previously of the Fox Chase Cancer Center. The citation for their Nobel Prize read, "for the discovery of ubiquitin-mediated protein degradation". Prior to the Nobel Prize, he had received the most prestigious awards in biomedical sciences, among them the Israel Prize in Biochemistry (1994), the Gairdner Foundation International Award (Toronto, 1999, with A. Varshavsky), the Wachter Foundation Award (Innsbruck, Austria, 1999, with A. Ciechanover), the Alfred P. Sloan Prize of General Motors Cancer Research Foundation (2000, with A. Varshavsky), the Albert Lasker Basic Medical Research Award (2000, with A. Ciechanover and A. Varshavsky), the Wolf Prize for Medicine (Israel, 2001, with A. Varshavsky), and the Louisa Gross Horwitz Prize (Columbia University, New York City, 2001, with A. Varshavsky). He is a member of the Israel Academy of Sciences (2000) and a foreign associate of the National Academy of Sciences of the U.S.A. (2003).

Dr. Hershko usually spends the summer months at the Marine Biology Laboratory in Woods Hole, Massachusetts, and that is where we recorded our conversation on August 8, 2004, just two months before the Nobel announcement.*

Would you please introduce us to ubiquitin?

Ubiquitin is a small (76 amino acid residue) protein, which is highly conserved in evolution. It was discovered in 1975 by an immunologist, Gideon Goldstein, in a search for immunopoietic polypeptides from the thymus. Goldstein mistakenly concluded that it stimulates the differentiation of thymocytes, but was surprised to find that it is present in all eukaryotic cells examined. He therefore called the protein "ubiquitous immunopoietic polypeptide", or in short, "ubiquitin". Later it was found by others that the immunopoietic activity was due to a contamination in the preparation. Thus, the name ubiquitin was originally a misnomer! So when we joined in into the ubiquitin field, the protein was known, even the sequence was known, but the function was not known. Well, the sequence was almost well established, because they missed the last two amino acids, two glycines, which are important because they link to other proteins. This created some confusion for a while, but all the rest of the sequence was correct.

What have been your most important findings in this research?

I got interested in the problem of how proteins degraded in cells when I was a postdoctoral fellow in the laboratory of Gordon Tomkins in San Francisco in 1969–1971. I found then that the degradation of a protein, tyrosine aminotranferase, requires metabolic energy, and got interested in the problem of how proteins are degraded in cells and why energy is required for this process. Following my return to Israel and setting up my laboratory at the Technion, I continued to pursue this problem, For this purpose, I have used approaches of biochemical fractionation-reconstitution. The breakthrough took place in 1978–1980, when we fractionated an ATP-dependent proteolytic system from reticulocyte extracts, and found that a small heat-stable polypeptide is required for its activity. Aaron Ciechanover was my graduate student at that time and he was the first author on the publication

---

*István Hargittai conducted the interview.

and I was the last author.[1] Our paper appeared in a not very high-profile journal called *Biochemical and Biophysical Research Communications*.

It was furthermore found in my laboratory that this protein, which we called then APF-1 (ATP-dependent proteolytic factor 1) is covalently ligated to protein substrates in an ATP-requiring reaction. Based on these findings, we proposed in 1980 that proteins are targeted for degradation by linkage to APF-1.[2] The similarity of APF-1 with ubiquitin was subsequently noted by others. Our discovery was unexpected and it happened during a summer when I was working in Fox Chase Cancer Center in Philadelphia. There was a postdoc who had a friend, who worked on a certain histone, which was known to be a branched protein. This is really the third component of the story. This branched protein was discovered by Harris Busch in Texas. So it was a histone, which had another protein attached to it.

Then Margaret Dayhoff, who edited an atlas of protein sequences and who had collected all the known protein sequences — just by looking at the sequences — noticed that one branch of this branch protein was ubiquitin. So what Busch discovered was a ubiquitin linked to a histone. The two postdocs at Fox Chase, who were friends, talked about it at Fox Chase. After we had discovered the conjugation, and we thought that there was no precedent of having one protein ligated covalently to another protein, these two friends noted that there had been a precedent. The discovery came from a combination of biochemical work, immunological work, and Harris Busch's work on liver regeneration. Our discovery was made independently from these other two pieces of information, but these independent pieces eventually came together.

Where was most of the work done?

Most of the work was done at the Technion, but the actual breakthrough occurred during the summer of 1979, when I was at Fox Chase Cancer Center in the laboratory of Irwin Rose and with his great help. We discovered the protein at the Technion and we purified it at the Technion, but we thought initially that it was a complex with another protein. The breakthrough of recognizing that it was a covalently bound ligation happened at Fox Chase. It was published in 1980.

What techniques have you been using in your work?

I have been using mainly techniques of "classical" biochemistry, which has become somewhat of a "vanishing art" in the times of molecular biology.

The research group at the Fox Chase Cancer Research Center in Philadelphia in 1979. Avram Hersho is sitting on the left; behind him on the left is Aaron Ciechanover; on the right, Irwin Rose (courtesy of Avram Hershko).

I used this approach not only for the initial discovery of the role of ubiquitin in protein degradation, but also for the subsequent identification of different enzymes of the ubiquitin system.

When you made the breakthrough discovery in 1979, you were 42 years old. What did you do before?

I studied medicine in Israel, then I became interested in biochemistry during my medical studies. I stopped my studies for a year to do research. It was not a formal M.D./Ph.D. program, but there was a possibility of doing some research. I liked research, but I finished my medical studies, though I never practiced medicine except for neighbors and in the Army where I served as a doctor. I finished my Ph.D. after my Army service.

Then I went to do my postdoctoral studies in San Francisco with Gordon Tomkins, who was a very well-known biologist at that time. His main interest was steroid hormone action, but I got involved in how the proteins that he was interested in degraded. They used a protein called tyrosine aminotransferase, which is induced by steroid hormones. The story is that

Gordon Tomkins
(courtesy of Avram Hershko).

I saw that there were 20 postdocs in Tomkins's laboratory and they all worked on the same subject, so I asked him for something else and this is when he suggested to work on the degradation of the same enzyme. Then I noted that the degradation of the enzyme required energy. I was very much impressed that protein degradation needs energy because proteases do not need energy. At that point I knew that we needed biochemistry to understand the process of how proteins are degraded. It was clear to me that protein degradation is important because that is one way to regulate protein levels, which is important, of course, in all the activities of the cell.

When I returned to Israel, I looked for a cell-free system or a classical biochemical method to approach this problem. Eventually, we isolated this small protein, and we found that it is needed for activity, but we did not know how it does it. It is a very small protein indeed, it consists of 76 amino acids. First I thought that it might be an activator of some energy-dependent protease. Activators usually bind to proteins, which they activate, but then I was very surprised that it was linked covalently to other proteins, it was bound not to an enzyme but to substrates.

What was the pivotal point?

Recognizing the importance of protein degradation and going after it. The rest was biochemistry, luck, and perseverance.

Jacob Mager
(courtesy of Avram Hershko).

Did you have enough biochemistry for it?

Yes, I had had a very good biochemical background from Jacob Mager, who was my supervisor at The Hebrew University and I spent my one year off from my medical studies with him. I continued working with him during the rest of my medical studies, and I did my Ph.D. research with him after my army service. He was an introverted man, but a great biochemist. Unfortunately, he did not leave much mark on biochemistry because he was interested in so many things. He worked on about five different subjects all at once. I worked on four different subjects while I was with him. It helped me build a broad-based foundation in biochemistry. For me, it was great. In addition to getting a solid background in biochemistry from him, his example also taught me that I should concentrate on one subject in my research.

What was his background?

He came to Israel from Poland before the Second World War. He got his Ph.D. and also his M.D. degrees, but he never practiced medicine. He was an excellent biochemist, but died very young of lung cancer. He was a heavy smoker.

Any other important teacher?

Of course, Gordon Tomkins, who was a well-known scientist at that time.

Wasn't he Marshall Nirenberg's supervisor at the NIH before?

He was.

Is he alive?

Unfortunately, not. His story is a story that happens only to doctors. He had a benign brain tumor. If it had been removed early enough, nothing would have come of it. But he did not show it to a doctor and, eventually, he went to a friend, who said, it was nothing, but then it got out of hand; they operated on him but it was too late and he died soon after the operation, and he died young. Mager and Tomkins were my two teachers who had important impact on my career and they were each other's opposites. Mager was a solid biochemist who taught me to do all the controls, do everything in duplicate. Gordon never cared about controls, he cared about ideas. His main contribution was to give ideas to other people. I got a lot of stimulation from him during my two years with him.

Anybody else?

I should mention a third person, Irwin Rose at Fox Chase Cancer Center. I owe him a lot, for support and advice, and it was in his laboratory (and with his active participation) that the breakthrough in the ubiquitin story took place. His field is enzyme mechanism, a field, which I never really understood well. I had met him at a meeting and he told me that he was interested in protein degradation. But I had never seen anything published by him in protein degradation, so I asked him why he did not publish anything in this area. He told me that there was nothing worthy of publishing in protein degradation. He was a character and I liked him and I asked him to let me spend my sabbatical with him at Fox Chase. I came to Fox Chase, but I continued there what I was doing in Haifa. He not only let me do it, but also gave me a lot of help and excellent advice. When I found the covalent conjugation of ubiquitin to a protein, the question was about the kind of bond formed between the ubiquitin and the protein, whether it was an ester bond or an amide bond, and so on. Rose knew a lot of chemistry and he helped me find out. He is alive, at 78, retired, living in California, still doing research.

How do you assess your discovery?

I realize that it is a big discovery and has made a huge impact on biomedicine. It is involved in every type of regulation: cell division, differentiation, inflammation, all kinds of diseases, such as cancer, Alzheimer's, Parkinson's; it's involved in almost everything because proteins are involved in almost everything. I was surprised how much it is involved because it is a wasteful mechanism. You make a protein and you destroy it. A lot of energy is being wasted in these processes. But nature does not care about energy, it cares more about regulation. That is the price that nature is willing to pay for regulation. It is also an effective way of regulation. You cross a bridge and burn it, can only go forwards, you cannot go backwards.

Do you have heroes?

In biochemistry, my heroes are Fritz Lipmann and Arthur Kornberg.

How can you measure the impact of your discovery?

When I started, there were ten papers a year on ubiquitin; now there are at least ten a day. I can't follow the literature any more.

What connections are there between your work and other research on the roles and activities of proteins in the cell and on cell cycles?

In the last 10–15 years, the ubiquitin system has been shown to be important by many other laboratories in many different cellular processes.

What is your main current research interest?

Currently I am working on the role of the ubiquitin system in the cell division cycle. This is what brought me to Woods Hole. When we started working on the ubiquitin system, I knew it was important, but I did not know, it was that important. Cell division is driven by oscillators. It's like a clock, and this clock is moving the cells ahead. An important protein was discovered here, cyclin, by Tim Hunt, and he got the Nobel Prize for it. This protein goes up and down in mitosis, when the chromosomes separate. It goes up by synthesis and goes down by degradation. I became interested in how cyclin is degraded, and this can be studied by biochemistry. Here I got a very good biochemical system in the eggs of marine vertebrates in large quantities and I am using clam, which is the same clam that you eat in clam chowder. The clam makes hundreds of millions of eggs because they get dispersed in the ocean and their sperms have to find them. This is why they make so much of it. Here you can take them out of the female clams and fertilize them. All these millions eggs begin to divide and one can make extracts and we ship them to Haifa on dry ice and we isolate the enzymes that are involved in cell division. We have found a very important enzyme that is involved in degrading cyclins and the same protein degrades some other proteins that are important for the separation of the chromosomes. Right now I am working on the same enzyme, which has important control functions. For example, the enzyme is inhibited until all the chromosomes are aligned correctly. We discovered this enzyme in clam but it is a universal mechanism, it is conserved from yeast to man. I am working on how ubiquitin-mediated degradation is important in cell division.

Has there been any practical application of your research yet? Of course, cancer research comes to mind but there may be other areas.

Since the ubiquitin system is involved in many aspects of cell division, it is not surprising that aberrations in this system cause cancer. Last year, the first drug targeted against the ubiquitin-proteasome system, Velcade, has been approved by the FDA for the treatment of multiple myeloma, a bone marrow cancer.

You have received several awards jointly with Alex Varshavsky. Has there been a division of work between the two of you? How would you characterize your interactions, is it more cooperation or competition or neither?

We have never actually collaborated. However, the molecular genetic work of Varshavsky has complemented my biochemical work.

Aaron Ciechanover's name appears on some of your publications and on some of Alex's publications. Is he a bridge between the two of you?

Aaron Ciechanover was my graduate student at the time of the discovery of the ubiquitin system. Later, when he was a postdoctoral fellow with Harvey Lodish at MIT, he collaborated with Alex Varshavsky, who was then also at MIT. So you may regard him as a bridge, at least in conveying information that he had learned in my laboratory.

At the conclusion of your postdoctoral stint in San Francisco, did not you play with the idea of staying in the United States?

Judith and Avram Hershko in Woods Hole in the summer of 2004 (photograph by M. Hargittai).

No, because I wanted to live in Israel.

Is it more difficult to do top science in Israel than in the United States?

Yes, but it is possible.

Here we are in your office/laboratory at Woods Hole, a rather small room housing your office, laboratory, assistant, and student. I suppose you have a much larger complex at the Technion. Why is it important for you to come to Woods Hole, where it is not inexpensive to rent even such lab space?

You are right. This lab is the condensation of the four rooms I have at the Technion. But it is important for me to come over — not for instrumentation and not for anything that I cannot do in Israel — for the peace of mind. Israel is a very compressed country where you listen to the news almost every hour. I find the environment much more creative in Woods Hole. The peaceful atmosphere here gives me a lot. It has become a pattern that I build up the background for my work at the Technion and the breakthroughs come to be during my stays over here. At the Technion, when something breaks down, I have to call for the technician. If it is not the professor who calls, he would not come. I have all kinds of diversions there including the committees although I am rather good at staying away from them. What I like is teaching.

So you accomplish more in Woods Hole. Who calls for the technician when something breaks down?

Everything is very well organized here although the conditions are more primitive. The conditions are relaxing and the atmosphere is creative. For me, it is important to work with my hands; this is something very personal, but I think through my hands. I think much better when I am doing an experiment than when I am telling others to do the same experiment. Also, I don't answer the fax and the e-mail during the day, just once a day, during the evening. So more thoughts and ideas come to my mind when I work here in this relaxed atmosphere. This is also what happened with our discovery of the ubiquitin system. We knew in Israel that a small protein was required, I already knew that it gets attached to proteins, but at the beginning I thought that it was a complex with the protease. But the breakthrough that it was a covalent linkage, that happened during a summer at Fox Chase in Philadelphia. I was much better concentrated than I could be at home.

Do you have interactions with other researchers here in Woods Hole?

Some. I am not an interaction person as opposed to Ciechanover, for example. I like to work by myself. I do talk with people when they come for advice but I rarely go to others.

Isn't there a coffee room where people get together?

There is, but I don't go there. I go to some of the seminars. I like to work by myself. But I always bring one of my students with me and I encourage my student to interact with others; this is also why I bring my students here, not only to help me in my work. There are also excellent courses and my students attend them.

But you never thought of leaving Israel?

No, that's my personal belief that we need a Jewish State and that we should live there.

What is the background of your family?

Karcag was a small town (~ 25,000 inhabitants), which had a small Jewish community of about 1,000. My father was the schoolteacher of the Jewish

At the award ceremony in the Hungarian Parlament, from left to right, Sylvester E. Vizi, the President of the Hungarian Academy of Sciences, Avram Hershko, István Hargittai, and Ferenc Gyurcsány, the Prime Minister of Hungary, 2005 (courtesy of the Office of the Prime Minister).

elementary school, so everybody in the Jewish community knew him. You may know Dr. Gergely from the Hungarian Academy of Sciences; he was one of his students in Karcag. More than two-thirds of the Jewish people from Karcag perished in the Holocaust. My father wrote a book to commemorate the Jewish community in Karcag. He was born in Biharugra. His family as far as I know came from what used to be Northern Hungary in the old days. My maternal grandparents were from Békéscsaba and my mother was born there. Both of my grandfathers were cantors. My maternal grandfather was very musical, he collected Jewish melodies, he directed choirs, and he organized workers' choirs, which was not denominational. My mother was also very musical. My uncle, my mother's twin brother emigrated to Israel (then Palestine) in 1936. He is still alive; he is now 94 years old.

I was born on the last day of 1937. My parents had two sons and my mother wanted to have a girl, but they realized that bad times were approaching and they decided that they would not have more children.

Did you ever talk with your parents about whether they considered emigrating from Hungary before World War II?

We did not talk about it, but I don't think that they considered it. The Jews in Hungary knew about the dangers and they even met some refugees from Poland, but they thought that what those refugees told them could only happen in Poland. They could not imagine that it could happen in Hungary. The Hungarian Jews thought themselves first Hungarian and then Jews. My father might have been a little bit of a Zionist even before the war, but even he did not consider emigration. My uncle who went to Palestine might have left mainly to escape from military service.

The Jews in Hungary (except most in Budapest) were deported to Auschwitz. How did you survive? Were you in the group that was diverted to Austria[3]?

First we were in a ghetto in Karcag and I have some faint recollections of the gendarmes coming and ordering us out of our home. In the ghetto, my mother tried to give us as normal life as was possible under the circumstances. I do not consider myself a Holocaust survivor because others had to endure much harsher conditions than we did. The worst was when after about a month we were transferred from the Karcag ghetto to the Szolnok ghetto where we were put into a sugar factory where they concentrated the people from all around Szolnok. It was an open-air camp; it was raining and there was nothing to sleep on in the night. There was no food. People were crying. Some tried to escape and were severely beaten and we had to watch that. I was six years old and my brother was eight, so he remembers everything better than I. We were put on trains. Some trains with some of my relatives went to Auschwitz and our train went to Austria. First we were brought to a concentration camp called Strasshof where everybody had to strip and then we entered some chambers, which were, however, not gas chambers, because water came out of the faucets. Then trucks took us to a little village near Vienna called Guntramsdorf. There we were put into a stable with straw on the floor. Our group had about thirty or forty people. The grownups worked in the fields and in the winter they worked in a factory. The Russians arrived some time in April of 1945 and we walked back on foot to the North of Hungary. Eventually we could take a train and arrived in Budapest from where we went to Karcag. I remember that our house was looted and my mother went to find our furniture and she found a few pieces here and there. Nothing was returned voluntarily.

My father was not with us, he had been taken to the Russian front in the forced labor service. Most of their guards were tolerable, but there was a drunken major who ordered them to strip and run out into the snow. My father was captured by the Russians before the others had a chance to kill him. My father returned to Hungary at 1947. He wrote in his book that there were hundreds of farmhouses in the vast land around Karcag, called tanya in Hungarian, and it would have been so easy to hide some Jewish families there, but there was not one case in which help would have been offered. My father was quite bitter about it.

What happened to your maternal grandparents?

They perished in Auschwitz. My mother never talked about them. She could never watch any movies about the Holocaust. My little cousin also perished in Auschwitz. Her mother, my aunt, survived and came back. I remembered her, before the war she had black hair and when she came back her hair was all white. Another of my cousins still lives in Szolnok, the only one of my relatives who still lives in Hungary.

Avram Hershko's father, Moshe Hershko, with his pupils around 1947 in the Jewish school in Budapest (courtesy of György Székely, Budapest).

I do not have very good memories of Hungary. I speak the language, but not too well. My parents decided to immigrate to Israel in 1950. My father continued to be a teacher in Israel, and became a very successful writer of math books for elementary school. He passed away in 1998 at the age of 93, and my mother passed away two years ago at the age of 91. I don't have much connection with Hungary, except for my cousin whom I have visited twice. My brother is very different from me; although being older, he remembers more about the Hungary of the Holocaust, and he is not happy about that, but he resents Hungary less than I do, and he has a lot of friends there. He is a well-known hematologist in Jerusalem. I also have some good memories of Hungary from the time when I was a very little child. We had a nice house; my father was an amateur gardener; I inherited from him the love of gardening and we have a little garden around the house where we live in Haifa. But then my father was taken away as everybody else and we did not hear about him until his return in 1947. He wrote a book about his life, but it is in Hebrew, it was meant for his grandchildren. My father was a teacher and soon enough he started teaching refugee children. He was not just a teacher; he was a pedagogue, an educator. Then he was discovered and brought to Jerusalem to a teacher's seminary to teach other teachers. He wrote some books and he made quite a career. I grew up in Jerusalem and had a good life.

How far did your parents witness your success and recognition?

I received the Israel Prize in 1994 and both my parents were there; they were very happy. Then my father passed away in 1998 when he was 93 years old, so he did not see all the other prizes, but my mother was still at the Wolf Prize; we brought her over in a wheel chair and our grandchildren were there too. The ceremony was in the Knesset.

I would like to ask you about your present family.

I am married, more than 40 years now, to Judith (nee Leibowitz). She is a biologist, and has helped me tremendously, both at home and in the laboratory. We have three sons and six grandchildren. Needless to say, the grandchildren are our greatest joy now. They all live near Haifa, so we see them a lot. Our oldest son is a surgeon; he is a Senior Lecturer of Surgery at the Faculty of Medicine of the Technion. The second son is a computer engineer and works in a start-up company near Haifa. Our third

István Hargittai, Judith Hershko, Magdolna Hargittai, and Avram Hershko at the Hungarian Academy of Sciences, Budapest, 2005.

son is studying medicine in Budapest. There is a large group of Israeli students, almost a fifth Israeli medical school in Budapest. The instructions are in English; the level of teaching is good.

Where is your wife from?

She was born in Switzerland and came to Israel to work for a year. We met at the Hadassah Hospital of the Hebrew University; we married and have been already together for 41 years. Her grandparents came to Switzerland from Poland before World War I. Her grandfather was very active before World War II, saving refugees from Germany and Austria by bribing officials.

I would like to ask you about your Jewishness.

I feel very Jewish, which for me is tradition and not religion. I am completely non-religious even though my father was an orthodox especially during his later life. They kept kosher and did not work on Saturday. He was also liberal in the sense that he knew that when we came to visit on Saturday, we drove to his place. He was a modern orthodox, but he never got any of it on me. Maybe it was the Holocaust. I remember when we were in the ghetto in Karcag, I thought if there was a god in heaven, would he permit this to happen? This is how it started for me.

Do you keep kosher?

No, and I eat pork, the good parts. My wife would not touch it. There are all kinds of shades of these approaches. I am definitely not religious, but I strongly believe in Jewish roots and traditions and heritage.

How do your sons feel?

They feel more Israeli than Jewish, I think.

It happens often with discoveries that make a huge impact that after a while it becomes difficult for the leaders of the field. You seem to be easily identified as the initiator of the field.

There is the record of who published those papers in 1978 and 1980 and it is clear.

Does it make you nervous that everybody expects you to be awarded the Nobel Prize sooner or later, probably sooner rather than later?

After all these prizes and my election as a Foreign Associate to the National Academy of Sciences of the U.S.A., I realize, of course, that there is a chance, but I am not waiting for it. I don't want to get the "October Syndrome". I do not think that I am nervous, and I tell people that ask to wait very patiently.

Would it bother you when and if you get the Nobel Prize if Hungary would boast to have yet another Nobel laureate?

It won't bother me; although I don't like Hungary, I don't hate Hungary; I hope that there will not be a street named after me in Karcag [Hershko

Avram Hershko with the bust of Eugene P. Wigner in the garden of the Budapest University of Technology and Economics, 2005 (photograph by I. Hargittai).

is heartily laughing]. When I came to Israel, I started a whole new life and I forgot about Hungary except for talking Hungarian with my parents. But I resent that the Hungarians helped the Germans in the extermination of the Jews of Hungary. It was a minor shock when I went back to Hungary for the first time in 1990 and saw the old, pre-war coat of arms and the soldiers in the uniforms of the Horthy times. I went back to Karcag and found only eight survivors of the once thriving community there. I did not go to Békéscsaba, but I have heard that the old synagogue had been converted into a storage house of furniture. I am not religious, but I resent it.

Is your Hebrew perfect?

It is, with a slight Hungarian accent. They tell me that my Hungarian accent is even more noticeable in my English.

Do you have a message?

My message to young students in biology is to use biochemistry, whenever it is needed, especially in the post-genomic age, when the function of most of our genes is still not known.

References and Notes

1. Ciechanover, A.; Hod, Y.; Hershko, A. "A heat-stable polypeptide component of an ATP-dependent proteolytic system from reticulocytes", *Biochem. Biophys. Res. Commun.* **1978**, *81*, 1100–1105.
2. Hershko, A.; Ciechanover, A.; Heller, H.; Haas, A. L.; Rose, I. A. "Proposed role of ATP in protein breakdown: conjugation of proteins with multiple chains of the polypeptide of ATP-dependent proteolysis", *Proc. Natl. Acad. Sci. USA* **1980**, *77*, 1783–1786.
3. There are close similarities in our lives in the childhood period, and I wrote in detail about my experiences in Hargittai, I. *Our Lives: Encounters of a Scientist*. Akadémiai Kiadó, Budapest, 2004.

Aaron Ciechanover, 2003 (photograph by I. Hargittai).

14

AARON CIECHANOVER

Aaron Ciechanover (b. Haifa, Israel, 1947) is a Distinguished Research Professor at the Vascular and Cancer Biology Research Center, The Rappoport Faculty of Medicine and Research Institute, Technion – Israel Institute of Technology, Haifa. He was co-recipient of the Nobel Prize in Chemistry for 2004 together with Avram Hershko (also of the Technion) and Irwin A. Rose (formerly of the Fox Chase Cancer Center, Philadelphia) "for the discovery of ubiquitin-mediated protein degradation".

He received his M.Sc. degree from the Department of Biochemistry of "Hadassah" and the Hebrew University School of Medicine in 1970 and his M.D. degree from the same school. He received his D.Sc. degree from the Department of Biochemistry at the Faculty of Medicine of the Technion – Israel Institute of Technology, Haifa, in 1981. He has had academic appointments at the Technion since 1977, serving as Full Professor since 1992. He was director of the Rappaport Family Institute for Research in the Medical Sciences between 1993 and 2000. He had numerous visiting appointments, including short-term ones at the Fox Chase Cancer Center in Philadelphia in the years 1978–1981 and at Washington University School of Medicine in St. Louis, Missouri, in the years 1987–2002, and others in Japan, Sweden, and the United States. He did his military service as military physician in the Israeli Navy and the Unit for Research and Development, Surgeon General Headquarters in 1974–1977.

His awards include the Wachter Prize (Innsbruck, Austria, 1999, together with Avram Hershko); the Lasker Award for Basic Medical Research (2000, together with Avram Hershko and Alexander Varshavsky); the Israel Prize for Biology (2003); and others. He is a Member of the European Molecular Biology Organization (EMBO; 1996), the European

Academy of Sciences and Arts (2004); the European Academy of Sciences (2004); the Israeli National Academy of Sciences and Humanities (2004). He is also a Foreign Fellow of the American Philosophical Society (2005), the Royal Society of Chemistry (UK; 2005), and member of other learned societies.

We recorded conversations with Aaron in two portions, one in September 2003 in Sweden and another during his visit in Budapest, May 4–9, 2005. In the 2003 conversation, the emphasis was on the history of events leading to Hershko and Ciechanover's ubiquitin-related discovery. The present account blends the two sets of conversations.*

A year and a half ago we recorded a conversation, but it remained incomplete. On that occasion you gave a detailed historical background of research on protein degradation. At that time, you were not a Nobel laureate yet. We are now embarking on a second recording and I wonder if you could briefly summarize the historical background before we move on to other topics.

Although it started much before 1942, 1942 was a turning point. It started with a belief that proteins are static and they don't exchange. We are born, we grow up, we accumulate them, and then we die with them. This would be like wood or metal. Then the famous Jewish scientist of German origin, Rudolf Schoenheimer (1898–1941), who came to the United States and worked in the Department of Biochemistry at Columbia University, used heavy isotopes for the first time for labeling amino acids. Schoenheimer found that amino acids are going into proteins then they are coming back from the proteins. He summarized his findings in a nice book that came out after he died, in 1942, *The Dynamic State of Body Constituents*. Schoenheimer said that the old hypothesis that proteins were static in our body was wrong. Rather, they are extensively exchanged.

His ideas did not precipitate for quite a while. Even in the mid-1950s there was a paper by two famous scientists, Jacques Monod and David Hogness claiming that proteins were static. Actually, they used the very word "static" in their paper! The field was running again into dormancy, until two things happened. One was that people started to reproduce Schoenheimer's experiments in different ways. Melvin Simpson at Yale University found that not only were the proteins exchanging, but also that proteins needed energy

---

*István Hargittai conducted the interview.

for their degradation. It was strange thermodynamically, because proteins are high-energy compounds, so why invest further energy in order to degrade them into low-energy compounds? Then came Christian de Duve, who discovered the lysosome. The lysosome was an organelle that contained proteases inside, so this provided the machinery. There were also doubts in that the lysosome was the organelle, in which *intracellular* proteins were degraded. At the same time people started to see that different proteins have different half-life times. The phenomenon was there, but people had doubts about the machinery.

Then came Brian Poole, a student of Christian de Duve, and others and they used lysosomal inhibitors. They observed that following inhibition of the lysosome they cannot affect degradation of *intracellular* proteins, but only the degradation of the proteins that are coming into the cell from the outside: proteins that are endocytosed. Poole said in a beautiful statement that the lysosome is involved in the degradation of *exogenous* proteins[1]: *"... In this way we were able to measure in the same cells the digestion of macrophage proteins from two sources. The exogenous proteins will be broken down in the lysosomes, while the endogenous proteins will be broken down wherever it is that endogenous protein are broken down during protein turnover."* He was prophesying the existence of a non-lysosomal proteolytic system in the cell, and he called it *"wherever"*! For me it was the ultimate proposal in one sentence: *wherever* they are degraded. He made a clear distinction between exogenous proteins and endogenous proteins! I think this was the point from where Avram [Hershko] started to collect scattered pieces of information: 1. There is degradation. 2. The degradation requires energy. 3. It is non-lysosomal, so there must be a different machinery. This was the starting point.

When did you join in?

There were two steps. We started the project that led to the discovery in October, 1976. I joined Avram in October, 1972, when Avram returned to Israel. Avram was a postdoctoral fellow with Gordon Tomkins between 1969 and 1971 at the University of California in San Francisco, where he worked on intracellular protein degradation. He decided, that it is of interest for him and he discovered, actually corroborated the finding of Simson that degradation — not of the entire cohort of cellular proteins, but now for a specific cellular protein called TAT (tyrosine aminotransferase) — requires metabolic energy. Then he returned to Israel and decided to work on the mechanism of this degradation.

When I first joined Avram, I was a medical student, but it was before my military service, so I joined in for a short while and worked on something else: on pleiotypic response and metabolism of phosphatydil-inositol. After my discharge from the army, I came to the lab in October of 1976. Avram at that time was still searching for the right system. So it took him several years to go.

How long did you serve in the army?

Three years, from the October war of 1973 until 1976. I served as a medical officer and that was the only time ever that I practiced real medicine. Also, while I was on active military duty, I came to teach biochemistry to medical students in Haifa, in Avram's department. I was part of his department in an "informal" way, just moonlighting.

Why did you choose an advisor before you went to serve? It was for a very short time.

I started my career as a medical student in Hadassah Medical School in 1965. During the years in medical school I got disillusioned in medicine and I decided to take one year off, and do my Master's degree in biochemistry. I joined a very good lab in the Hadassah Medical School. At that time — this was 1969 — Avram had graduated already from the same school and left for his postdoctoral fellowship. I had heard about him. I had to do a small thesis for the medical school and it was at the time — in 1971 — that Avram returned from his postdoctoral fellowship and I decided to join him and just to ask his guidance for my mini-thesis.

When did you return?

I came back from the army in 1976 and decided never to go back to medicine. In October 1976, I joined Avram as a Ph.D. student in Haifa at the Faculty of Medicine of the Technion.

When did you go to the United States?

Avram left for his sabbatical with Ernie Rose in the summer of 1977 and he left me alone in the lab. During the ten months we worked together before his sabbatical, we characterized the system in an extract of reticulocytes, in a maturing red blood cell. We started to fractionate the lysate on an anion exchange column. We found that we broke a paradigm. There was

Aaron Ciechanover in his laboratory at the Technion in Haifa (all photographs courtesy of A. Ciechanover unless indicated otherwise).

a problem as we fractionated the extract that had the proteolytic activity into two complementary fractions. Neither of the fractions had a proteolytic activity, but the reconstitution of the two gave the activity, which was not known until that time for any other protease. If you take trypsin and a protein and you put them together, the protease will digest. The fact, that we had to recombine two fractions, was new.

We didn't know what it was. Despite not knowing what it was, we decided that we were going to purify the active component in one of the fractions. It was the fraction that contained hemoglobin and we thought that it should be easy. That was the point when Avram left. He left me with the mission to purify the active component in one of the fractions. I had a substitute supervisor appointed by the Institute, Mickey Frey — Michael Frey.

I tried to get rid of the hemoglobin in the fraction, but did not succeed. Then one day Mickey came with a crazy idea. He said, boil it! Just boil it! Maybe it's not a protein. We boiled it, and the protein hemoglobin precipitated like mud, like a cooked egg, and all the activity remained in the supernatant. Then we showed that it's a heat-stable protein. I sent Avram all data to the United States, and from there he wrote our first paper, that he also mentioned in his review.

Was Mickey on the paper?

Mickey was not on the paper, he just gave the advice, but it was unbelievable advice. There was another student of Avram that was left behind, that joined me, Yaacov Hod. He was the second author and Avram was the third.

So the real boiling and initial characterization of what later turned out to be ubiquitin was done in Haifa in the absence of Avram. In the summer of 1978, I joined Avram in Fox Chase for the first time at the end of his sabbatical and I spent three months with him in Fox Chase. Then Avram returned to Israel and so did I. We continued to work on the system in Israel.

When did the seminal discovery happen?

In the summer of 1979, but all the preparatory work was done in Haifa, the purification of ubiquitin, all the enzymes, everything was done there. At that time it was not known that it was ubiquitin.

What I consider to be a decisive moment, happened in Haifa. I remember the evening when I discovered it. When I took labeled ubiquitin, (iodinated ubiquitun) and incubated in the presence of ATP with the crude reticulocyte extract, all the radioactivity shifted to the high molecular weight region following separation on gel filtration chromatography.

For months we played around, we didn't understand what was going on. The understanding that it was a covalent conjugation of ubiquitin to the substrate was made in the summer of 1979, with the huge help of Ernie Rose in his lab in Fox Chase. Both of us went to Fox Chase for the summer in 1979, just for the summer, and we did the same the following summers. We spent "mini-summer-sabbaticals" at Ernie's lab.

The findings of this summer of 1979 that an adduct was generated with a covalent linkage, were published in two papers in PNAS [*Proceedings of the National Academy of Sciences of the U.S.A.*]. They were several months apart. The adduct was between ubiquitin and the substrate, but at that time we didn't know it was ubiquitin, we called it APF-1, or ATP-dependent proteolytic fraction 1. The convergence to ubiquitin came also in Fox Chase, a year later.

How did it come?

Ernie had two postdocs, Keith Wilkinson and Arthur Haas. Keith had a friend in another lab, Michael Urban, and they talked about what was going on in research. We knew that our protein was a small protein. We knew the molecular weight, it was like 8000 dalton, a relatively small protein and we knew that it was conjugated to large proteins. Michael came up with an idea that there is precedent to a small protein called ubiquitin being conjugated to histone. The structure of this conjugate between ubiquitin and histone H2A was known. I said, maybe our protein is ubiquitin, the substrate is like histone, or whatever it is and you get this conjugation. *They* took purified ubiquitin, and put it in our system and found that it was doing the same. Then the amino acid composition was confirmed between APF-1 and ubiquitin. The amino acid composition was identical. But our case was different, because we found that multiple molecules of ubiquitin, or APF-1, were attached to the substrate, while in the case of histone, it was a monomodification, a single modification.

But to me — and I'm just trying to provoke you — it seems that these two friends' finding was very important.

The identification of APF-1 as ubiquitin was extremely important in the sense that it led to the convergence of the fields. Let's say that we would've proceeded without them. Then we would have proceeded with our own unknown protein to the same very mechanism. It wouldn't have changed the basic mechanism.

Did you publish your findings jointly?

No. We published our paper with their paper back to back in the *Journal of Biochemistry*. Our paper came first, it was the characterization of a small polypeptide, and we published the amino acid composition. The next paper said that APF-1 from the previous paper *is ubiquitin*. There was a little tension at this point between Avram and the postdocs because Avram said that everything was ours and Ernie didn't want to be involved at all, but his name figured on both papers. In any case, they had access to all our data. In the end, if you look at it in retrospect, it was their idea to combine the two areas, there is no doubt about it. We were not in the histone field. It was the friend of Keith who came and said there is a precedent. They purified ubiquitin, they iodinated it, and it did exactly the same as our APF-1.

It might be easy to downplay their contribution.

But it's not downplayed. Their achievement is the way into deciphering the details. Without them, we could have been walking and swimming in this endless ocean for a long time until resolving the problem. But once we knew that it was ubiquitin and we knew the bond, the road to deciphering the mechanism — that was carried out later by Avram and myself — was open. It accelerated the process tremendously.

Did they become famous?

Yes, they became famous in the ubiquitin field. They came from Ernie's lab. Keith is now doing a lot of work on isopeptidases. He is invited to all the conferences, he wrote a review on the Nobel Prize later on in *Cell*. Arthur [Haas] has worked on kinetics. Yes, they've become very good, established scientists.

But they didn't go to Stockholm.

They didn't go to Stockholm.

I mean, nobody invited them?

Oh, they came to Stockholm, Ernie invited them. I thought you meant it metaphorically. Both Arthur Haas and Keith Wilkinson came. Urban, who was the bridge to the histone lab, was not there.

How long did your cooperation last with Avram?

I wouldn't call it cooperation; let's be precise, I was his graduate student. I worked for him from October 1976 until August 1981, roughly 5 years.

How long did your cooperation with Rose last?

Mine lasted through 1981, but Avram continued to go there for additional summers. I left them and went to MIT in September 1981, and I went on my own.

You continued ubiquitin work.

Very actively. When I purified ubiquitin with Avram, there was a discrepancy between the dry weight of the protein and its spectrophotometric measurement at 280 nanometers as well as according to the determination of the molecular weight by Lowry's method.

The famous Oliver Lowry, the most cited scientist ever?

I knew him very well because later I had a joint appointment at Washington University in St. Louis. Oli Lowry developed an assay for proteins and the reading of ubiquitin was very low and it was the same at 280 nm in the spectrophotometer, but the dry weight was very high. Avram suggested that it is a ribonucleoprotein and the nucleic acid makes the waste. Avram said to me: Digest ubiquitin with DNAse and that kills DNA! But there was no effect on degradation. Then he said: Put ribonuclease! I put ribonuclease and lo and behold there was a complete inhibition of proteolysis of one substrate (BSA, but doesn't matter)! Then Avram said to me: Do the same experiment with another substrate! I did it, and there was no inhibition at all. So, one substrate was inhibited by ribonuclease, and one was not. Initially Avram was encouraged, than he got discouraged. He dismissed it as an artifact.

I didn't believe it was an artifact. When I was alone at MIT, I pulled it out, and I published a paper; it is described in detail in my review. Actually I got advice from Alex [Varshavsky] in this work, just technical advice. I published the paper on my own. That's very interesting: this paper became part of Alex's N-end rule. It's one chapter in Alex's N-end rule that evolved completely independent of Alex. I was dragging the finding from Haifa that Avram dismissed as an artifact and it all evolved independently from him.

What could you tell us about Oliver Lowry? When they prepare statistics, they leave out his paper, because it upsets all statistics.

I know, because everybody uses it. Not now anymore, by the way. Now he is declining. People are using other techniques and there are better reagents. But Oli Lowry was amazing.

He was the Chairman of Pharmacology at Washington University in St. Louis. I came to know him via my good friend Alan Schwarz that I just mentioned. Alan said to me: Come and I want to introduce you an amazing scientist. And we wrapped it to the story that we needed hexokinase. Oli was kind of the king of all the enzymology, and he made his own enzymes by himself. He built his own spectrophotometer. Then we went to his lab. We were sitting there and he pulled out the enzyme. He had his own refrigerator. He was already 80 years old, or 78, but he was still running the department very actively.

We started to talk about discovery of enzymes. Oli was the master. He is known for this reaction but he wrote a book. I don't know if you read the book. It is a very important book on biochemical reactions and how to divert them so at the end you can monitor it. Actually, he was ingenious in the sense that from any enzyme that you wanted to measure along the glycolytic pathway at the end he managed to generate NADPH, which is this reducing agent, nicotinamide adenine dinucleotide-, either with phosphate or without phosphate-, hydrogen. And this reagent absorbed at 340 nm. So Oli defined a series of chain reactions that at the end you generate this 340 nm absorbing material and it was all the time stoichometric. So whatever the first product you had upstream, you could always quantify by measuring the 340 nm absorbance of NADPH or NADH. I was talking to him and came back and back. At the end he died tragically, I think he got Alzheimer's and died. For me, he was not my hero, but he was the ultimate biochemist.

Do you know if he cared about his citations?

We didn't talk about the citation. I must tell you, when I walk into a room of somebody who is famous, I feel tense. Even to these days, when I walk into the room of a Nobel laureate, like Elie Wiesel, I feel tense. Maybe these people are beyond me. Oli was so famous. Everybody said Lowry, Lowry, but when they said it they didn't know what it meant. They said it like some name, like Toshiba, Sony, Panasonic. And I said: you must be Mr. Lowry. For years I have been measuring this reaction of yours. And he said, it is not my most important achievement, it is technical, it is a reaction, it is not a breakthrough in biology, nothing. So, he kind of dismissed it. He was a very interesting character. He had white hair, very impressive.

Did you go to MIT as a postdoc?

I was a postdoctoral fellow with Harvey Lodish; he, like I, was a graduate student with Avram Hershko before. These things must be made extremely clear. We did a lot of work together with Harvey. We published very important studies on the iron uptake by transferrin receptor. That was a series of papers that became classic. They are now in all textbooks of cell biology. I was moonlighting in all kinds of things. I was working very hard, so I managed to work in parallel. Gradually, I left what I did with Harvey and devoted all my work to ubiquitin on my own and with collaboration.

Did he pay you?

No, he never paid me. I had my own fellowship, but nevertheless I was in his lab and he was very generous.

So, this was the end of your interaction with Avram Hershko.

The formal one. Then I came back to his department. I was recruited back to his department in 1984 as an independent Senior Lecturer. Avram was the Department Chair. In that department there were three senior scientists, Avram, Mickey Frey, and myself. When I came back, another lady was recruited, so we were four. Avram was the Chairman of the department and he still is.

During your stay at MIT, you started collaboration with Alex Varshavsky.

I started, and completed. It was an important chapter. Alex noted a mutant cell, isolated by a Japanese group that loses the histone-ubiquitin conjugation at high temperature. Alex approached me at MIT and along with a very talented graduate student of Alex, Daniel Finley, we deciphered a defect of this cell. Daniel Finley is very famous in the field and he is a professor at Harvard, now.

The defect was in the first enzyme of the conjugation cascade. We found that a consequence of the defect was that the cell didn't degrade proteins. This was the first strong and direct evidence that the system also operated in mammalian cells. I must admit though that the first indirect evidence that the system is not limited to the red blood cell had come already from Avram's lab, before I came to MIT. Avram and I developed an antibody to ubiquitin, and using the antibody we found a strong correlation between the levels of degradation and the level of the conjugate. But this was indirect, and the cells had many other advantages. The discovery with Alex was strongly confirmatory and added a lot of information. Alex was in the ubiquitin research in a totally unrelated area from ours. He was in ubiquitin modifications of histone as related to the structure and function of the nucleosome. I mentioned this ubiquitin-conjugated histone to which the field converged already in 1979 and 1980. This structure was known, and Alex was studying the function and the role of this histone-ubiquitin conjugate as related to the function of the nucleosome. Then he bumped into this cell that loses ubiquitin-histone conjugate and I think this was his entry into the proteolytic field.

Do you still work on the ubiquitin system?

All my life. I have been, and will work.

How would you characterize Avram as a mentor?

Superb. Actually, I followed Avram's steps exactly ten years behind, even though I didn't know about him initially. In my student years my mentors were Jacob Bar-Tana and Benjamin Shapiro. With them I worked on fat in liver. That's why I decided to work with Avram on phosphatydil-inositol. I brought the lipids into the lab. I really fell in love with biochemistry. From them I learned that biochemistry is my love.

Avram and you are different personalities. Did it ever cause any conflict?

As a grad student, no. We wrote reviews together and I graduated. Retroactively, I discovered the letters that he wrote to the graduate school. He said that this was an exceptional work and all the reviewers of the work, all the examiners thought that it was. He recommended that the Technion would exceptionally grant this thesis with a summa cum laude. They never do it. So we had excellent relations during my graduate studies although we had scientific discussions. The story that I told you about the discrepancy in the weight of ubiquitin, Avram dismissed it as an artifact and I turned it into my first work and we published several papers, including in *Science* and in *Nature*. This was one of my re-entries back into the ubiquitin field as an independent scientist.

I think, even for a good mentor, you still have to re-evaluate what the mentor tells you. But I think that I followed Avram very carefully, I followed his way of thinking, and I followed his stubbornness. Once he holds a bone, like a dog, he never lets it fall. If there is a problem, you just fight it, don't give up, do controlled experiments, clear experiments. I learned a lot from him, a lot.

I don't know if it was an active mentorship. I cannot tell you whether it is easy to be a good mentor and an excellent biochemist at the same time and it also depends on you, whether one or the other is more important for you. Nobody doubts that he is an excellent biochemist. I was his student and I learned from him.

Do you have heros?

I don't think that I have heroes. I don't know what a hero is. Samson was a biblical hero. A hero maybe Ben-Gurion. I think. Ben-Gurion is my hero, the founder of the Jewish state. He was a huge hero. I didn't know him personally, but because of his courage. You take an idea, you take a few refugees, coming from Europe and you establish with them a state. In a state you immediately have problems: to establish an education system, to build an army to fight the Arabs, religious and secular, and now, 55 years later, or even it was much earlier, you see, that the democracy was established in the Middle East. Certainly, Ben-Gurion was a hero, not Herzl, by the way. Herzl was the prophet, but Ben-Gurion did it.

Ben-Gurion, unlike current leaders in the State of Israel and leaders worldwide, was a multi-task leader. Leaders these days probably work sequentially. They say: there are so many things on the table that we cannot resolve. We take one, and we don't resolve the second one until the first one is resolved. But Gurion had to build an education system and he built a state-founded superb education system. At the same time he had to disintegrate all the underground organizations, the Etzel and the Lechi that were fighting the British, and to build a unified army that was under the control of the civilian government, so that there should be no coup d'etat. He built the army, he built a health system. On top of it, he studied the *Bible* in his house. Every week, there was a course of *Bible* in his house and Ben-Gurion wrote a book himself on studies in the *Bible*. It's called *Lyunim ba Tanakh* in Hebrew.[3]

Ben-Gurion was, in my eyes, in many ways, King David. No doubt, he is my modern biblical hero. He built the State of Israel, the ultimate shelter and place for Jews all over the world. This is a major achievement. I wouldn't pick as hero somebody, who deciphered a pathway or the structure of a molecule, although I may admire him. I don't think that these were heroes, but again, it depends on your definition of a hero. Ben-Gurion is clearly my hero, maybe the only one. I never thought of it, but you caught me and surprised me with your question.

Is there a letdown after the Nobel Prize for yourself or for your students to continue research?

Not for me. There are some distractions now, but I am already back in the lab, writing papers, reviews, grants, but it's hard. It is like a beauty contest and I'm a beauty queen in the first year. I'm now waiting for the next competition to end and in October 2005 the projectors will be shining on somebody else. Then, I'm sure it will level off a little bit, though not completely. This is only a few months after and certainly you are in the middle of the storm. But I am determined now to bring my life back to normal and to continue to do science as strong as I can do it. I am not going to do any administration.

Especially in a small country as Israel, a Nobel laureate takes a much more important social position, than in the United States.

It is both funny and pathetic. I look at it in a humorous point of view, because I take life in the appropriate proportion. Immediately, after the prize announcement, and even now, people come to me as if I was a leader: what do you think about this or that? — And they hand you a microphone. They want to hear what you say about religion and what you think about the engagement in the Gaza strip and mostly about education in Israel, which is deteriorating. It is a major problem in the country. A week after the prize, I said: listen, ten days ago, before the prize, you never came to me; my opinion was not important. Now, some guys in Sweden gave me some stamp on something that I did 25 years ago, which is completely unrelated to the question you are asking me about. And all of a sudden, I became somebody, whose opinion on an unrelated matter becomes important. How do you link these different pieces that do not belong to one another? One thing is factual: from last week, to this week I lost about 10^7 cells in my brain, which is a constant process, we always lose cells. So, if at all, from last week, to today, I became more stupid. Why is it important, what I think about this or that? Because, some Swedish decided that the ubiquitin system that I worked on as a graduate student got now the ultimate recognition? So I take it cynically, that's it.

Of course, this will continue to some extent.

I realize so, but I suppress it. Now, I became a little bit nasty. Now I say no. People invite you to all kinds of funny things: for the graduation of the school of nursing, technicians. Then you learn to use politely the word "no". Actually, I learned something else. Why to say no? You say: "I would love to come to the talk/invitation/conference however, I have an *overlapping commitment.*" The word "overlapping commitment", I don't know from whom I learned it; I stole it from someone for sure. It works magic. I am just occupied, I am busy. So you never say no, say "overlapping commitment".

Did you go back to your old school?

Yes, I went to my old school, and I even gave a talk to the students of my old school. I have a lot of admiration for my old school. Two weeks ago, my classmates made a class reunion in my honor and 36 out of 42 came from my high school. We graduated from high school exactly 40 years ago, in 1965. We had excellent teachers. Now we don't have teachers and this takes us back to the serious part of our conversation: Israel has a deteriorating education system now and we are going to pay heavily for it.

We had wonderful devoted teachers and I really admired them. My teacher of chemistry is still alive and she came to the class reunion. We became very good friends with her many years ago. If you ask me about basic chemistry, everything that I know, I know from her.

I went to the high school, I went to the place where I grew up, I went to the cemetery, visiting. This I did two days after. On the day that the prize was announced was the evening of the Holiday Simchat Torah, on Wednesday. Thursday was the holiday and on Friday I went immediately in the morning to my parents. My brother called me. He tried to reach me after the announcement and then all of a sudden I heard him crying on the other side of the phone. He was on vacation in France. Mostly what we tried to imagine was how our parents are conversing now about the two of us. My brother became also extremely successful in business and in politics. He became one of the most prominent figures in the State of Israel. He was the chairman of El Al, the chairman of the Israeli Discount Bank. Our parents just didn't see anything. I mean nothing. They died when my brother was a young lawyer. He is fourteen years older than I am. Our mother died when I was 10 years old and our father died when I was 15.

Who brought you up?

It was a miracle. The distance between me and delinquency and prison was much shorter than the distance between me and Hebrew University in Jerusalem. I could have easily slid into delinquency. I was alone. I was stealing thongs, these rubber sandals from the Haifa beach from people, who went to swim in the sea. It was a mischief although I don't know if the police would have treated it as such, had they caught me. Then I realized that I am betraying the spirit of home, because at home we studied all the time. It was a spirit of studies. The Talmud, the Mishnah and my brother studied, all of us. Then I said to myself, this is not what

Aaron Ciechanover's grandparents in their family circle, including Aaron's brother standing on the right-hand side of the picture, which was taken probably a couple of years before Aaron's birth.

my father wanted from me, this is something very foreign to the entire spirit of the family. It's like one day you get up in the morning and decide to be different. I don't know whether it happened in one morning, or it took several months. I still remember limping in school and not doing well but after several months I was standing on my feet and starting to run my machine.

So you were brought up by your aunt.

I had an aunt in Haifa, a widowed aunt, the sister of my mother, and I had my brother. My brother was married already, lived in Tel-Aviv, in another city. When my mother died I grew up with my father, so we stayed at home. When my father died I was in the 10th grade of high school, so I had another two and a half years to go. I had two options. My brother wanted to take me to his home, but then I would have had to change school and change my friends and my social environment and he had just married. Then my aunt came with her proposal that I would grow up in

Aaron Ciechanover's parents and his older brother in Haifa.

her home in Haifa. I stayed where I was, went to the same school, it was the same everything. She kind of adopted me; she became my mother, and took me to her home. She died in 1996.

Did you have a role model?

I don't know; my brother was very successful. He was a starting lawyer. I think my real role model was my father and the spirit of studying in our home. I always remember my father holding a book. Whenever I remember him, he is sitting at a small lamp in the night, in the balcony, and studying Talmud, law, something. He never wasted a minute. I didn't think in terms of role model because I was very young, I didn't understand it exactly. But I remember very well, the three, four months that I still failed after my father died and I was 15. And in the summer I made this decision and really entered a different track. So I think it was the major turning point and there was medical school and the rest.

Whom did you take to Stockholm?

This was a very tough problem because they limited the number of people I could take sixteen. I took obviously my wife and my son. I took my

Aaron Ciechanover, Stanley Prusiner, Marion and Elie Wiesel, and Yitzhak Apeloig at the Nobel Prize award ceremony, Stockholm, 2004.

in-laws. Obviously I took my brother and his wife and their children. Then I picked my best friends from different stages of my career. I took my technicians from the lab who are doing everything for me; without them, I am dead. It was an eclectic, but also a rather systematic collection. It was already my allotment, but I begged for a little bit more. We had a negotiation that Elie Wiesel, whom I also invited, is a Nobel laureate, so they didn't count him. It was a tough choice, it was difficult, and I was sitting with my wife thinking deeply on each and every one of them.

Originally, you didn't want to become a scientist, you wanted to become a doctor.

I wouldn't say that I wanted to become a doctor. I would say that I didn't know what science was. Since I came from a Polish family and there was a big pressure at home, Jewish mothers wanted their children to become doctors. Always at home people wanted me to become a doctor so I went to medical school. It was kind of an obvious pathway track. And science, I only discovered the light when I was in medical school. When I studied biochemistry and when I fell in love. It was certainly a gradual transition. I always was involved in biology. In high school I majored in biology, I had a collection of dried plants that I dried in the Talmud and the Mishnah of my brother. I had a collection of bones, skeletons and lizards,

Aaron Ciechanover receiving his M.D. Diploma at The Hebrew University of Jerusalem in 1974.

I was always somewhere around biology. But then when it was time to decide about career I didn't know what to study in biology, so medicine was an obvious choice.

You mentioned your connection with Poland.

You can actually wrap it in a more general question: how much I feel toward Central Europe, World War II, and my relationship to the State of Israel, my Judaism and what I feel is its role. It is good that you asked this question because it goes back to my question about a hero and Ben-Gurion and what is the meaning of a Jewish state for me. It is almost a simple equation. There was a flourishing Jewish community in Europe that contributed a lot in countries like Hungary, Germany, and Poland and they perished, they were murdered cruelly. Without going into any details: they were sent to Auschwitz, to Thereisenstadt, to Westerbruck, from all over

Europe. The State of Israel is the direct historical result of the Holocaust in Europe, there is no doubt about it in my mind.

Rather than talking about Poland, I would go a little bit further with your permission. It is the State of Israel and my Judaism versus Central Europe and what happened in Central Europe. I think we discussed today the book of Amos Elon, *A German Requiem*. It is about the Jewish community in Germany that for two centuries flourished, from Moshe Mendelsohn, the philosopher to his grandchild, the composer Felix Mendelsohn and through Heinrich Heine and Albert Einstein, a huge Jewish community. The book describes them as a very fragile community. All the time they were asking, begging, crawling for their rights. At the end they thought that they were integrated and satisfied their neighbors. They thought that their contribution to the society would protect them. And nothing, nothing helped them at the end. Between 1939 and 1945, they were all cruelly murdered. *Nothing* was taken into consideration by their very next-door neighbors.

So the State of Israel was established and I happened to be born in this country, I happened to be born Jew. These people perished because they were Jews. *Just* because they were Jews, they were not criminals or else. In Poland, most of the Jews perished in Polish ground with the collaboration of Polish people, no doubt about it. I visited Auschwitz and Birkenau. Actually, the extermination was in Birkenau, in the neighboring camp, Auschwitz was only the sorting place.

Now, I am very proud. I am a proud Israeli Jew, living in his own, independent state, holding my rifle and my revolver. Nobody will kill me, because I will fight for my life. I will not be slaughtered; I will not be taken to a truck, like a cow, and taken to a concentration camp and to a gas chamber. Nobody will do it to me, ever. I am a free citizen of this world, on top of being a scientist. This is a side issue, now I am talking about the principles of growing up. Now, when I go back to Europe, I don't hate Europe. What do I have to hate, the ground? Though the ground maybe soaked with Jewish blood, I don't hate it. I have collaborators from all over the world, including Germany. Not in Poland, it just happened so. If there was a good Polish scientist, I would collaborate with him.

Actually, I am going to Poland in two weeks. I am going to Zakopane. I am flying to Krakow. They invited me to a meeting. After I got the prize, the city that my father was born in offered me an honorary citizenship, which I am going take, honoring my father, so I consider it a great honor for me. My father was not persecuted there, he left Poland much earlier. He came to Israel in the early 1920s, nothing happened to him. Other branches of the family perished in Poland during World War II. So I don't hate it, but I can do it from a position of strength. I am now protected by my own country. The State of Israel is the answer to the persecution of Jews in Europe.

I see this event, my visit there for what it is worth. OK, now the Jew is coming back, walking on your streets with an Israeli flag and says, I got the Nobel and now you are coming back to me and asking me for forgiveness. First of all, I cannot forgive on behalf of anybody. I don't have the right to forgive, even on behalf of my wife. I am doing it to honor my father. It is an achievement. He was born in this country. Actually, there were millions of Jews in Poland, and nobody remained. I mean a handful of Jews are living now in Poland. There were four and a half million Jews in one single European country, all perished. Nothing remained, not my neighbors, not hundreds of years of tradition, the contribution to the community, nothing.

But I don't hate Poland, like I don't hate Hungary, I don't hate Germany, like I don't hate France for the current anti-Semitism. It is a general attitude of me that I take people personally. I don't group people as German, Polish etc. People are people. But I think part of this attitude is absolutely due to the fact that I am a proud Israeli Jew, no doubt about it.

You seem to stress Israeli and Jew.

I have to explain this to myself all the time, too. I am a proud Israeli Jew. Because Israel is not a regular state, it was established at a point of time in history after events in Europe as a shelter. It has a Law of Return that every Jew can come and become an automatic citizen, unless, he is a criminal. It has very strong Jewish lines into it, not that I agree with all of them. Let's not misunderstand me about marriage and burial and other things. But the country is certainly a Jewish state. Let's don't make any mistake about it.

A secular state.

It should be a secular state. Here we are sliding into a very complicated question.

I am interested in your views.

My view about Judaism is complicated and simple at the same time. I am not an observant Jew, I don't eat kosher food, I travel on Shabbat, but I do go to the synagogue and cherish the Jewish culture. Jewish cantor music is almost the only music I listen to, because it reflects, symbolically, many things for me. I know the prayer books, I go to the synagogue on the High Holidays, I study Talmud and Mishna, so it is really important to me. I would say that Judaism for me is more of a culture than something that I have to work toward. I don't worship it, I don't serve it, but for me it is a very important culture. It is part of me.

When you are saying a prayer, you don't mean it?

The question what you are asking requires an in-depth discussion of whether I believe in God as is and whether there is a Jewish God. For me, saying a prayer makes me feel belonging to a cultural community. I also highly appreciate its being poetic and that it was written during times in the Diaspora and during the destruction of the Temple in Jerusalem. It's very much contextual, it is not isolated. When I pray for God its meaning is not that I believe in this very God, and what he did. It is much broader for me.

Do you ever ask anything from God?

No, I don't think that God can give me anything.

Einstein said that as long as you pray and ask for something, you are not religious.

You see? Now I became religious in the eyes of Einstein. We slid into the most important controversies in Judaism. It actually goes back to the Bible. Simple Jews would estimate the relations with God as a very simple thing. I work for you and you pay me, you keep me healthy. So, the Holocaust itself became a major question for Judaism, for doubting Judaism. How can there be a God, if six million people perished by Christian Germans? It is very complicated; we cannot go into it. We may have the opportunity in the next five days to discuss it, but it goes back to the Bible. In the Bible, there is a statement. First I tell you in Hebrew, then, I give a translation:

מדוע דרך רשעים
צלחה שלו כל-בגדי
בגד:

Why is the evil successful, and all those who betray are living in peace? And why the righteous people suffer?[4] The Holocaust really accentuated this question in Jews. Why is it that the Jews suffered so much in the Holocaust? There is no answer. There is no direct relationship here. It is not like in a grocery store: you pay 20 pennies and you get a piece of bread, in return. There is no direct relationship between God and his flock.

You said that education is deteriorating in Israel. How can you measure it?

There are many objective criteria. You can measure the achievement of students in mathematics, compared to other countries. You can also observe it longitudinally in the army. The army always tests recruits in the apprehension of mathematics, Hebrew, language skills, and others. This is deteriorating in the army, so on the same test people are deteriorating. Teachers are not given incentives, salaries are going down, teaching has become a second or third class profession. The country just cuts and cuts the budget of the universities. Universities are collapsing in Israel these days. Not only high school education, the kindergarten and elementary school education suffer as well. This is, in my eyes, committing suicide. We don't have uranium, we don't have coal, we are not Saudi Arabia, all what we have are the brains. Everything that is built in the State of Israel is the mirror image of its education: the hospitals, the economy, science, the Weizmann Institute, Hebrew University, music, archeology, the Massada excavations, the Philharmonic Orchestra, it is all education.

Is it being discussed?

It is being discussed. It is the only thing I volunteer to be, to become a mouthpiece for these people a little bit, if it won't take too much of my time. It is very complicated. People see the psychology of the current government to break the elite; it is a battle between the government and the elite. I wouldn't go that far, I wouldn't accuse the government with an intentional destruction of the university and the education system. I wrote now an essay that will probably go to *The New York Times* on the destruction of the education system in Israel.

How would you characterize yourself?

I think I am stubborn. I had a unique life experience, because my parents died when I was very young. I grew up in my aunt's house and my brother kind of adopted me. Then I made medical school on my own. I worked during the night; I moonlighted from Day One. I had to make my living, nobody paid for anything, I made every penny to pay tuition fee. I think I am stubborn by nature. But from Avram I learned that I need to take the stubbornness of life and also to place it into science. Stubbornness for the sake of stubbornness is useless, but in science it has purpose. This is one thing I learned from Avram.

I learned from him to be very controlled, to do several controls for each experiment in order to believe that the result stands out as a real finding rather than an artifact or something that doesn't belong to the main road. I think I learned from him to be very controlled. But I think I grew up independently. I took subjects that he didn't believe in, I focused and developed independently. I learned a lot from Avram and I'm not going to tell you what I didn't learn from him, because I also rejected many things.

I also learned from Avram to be a story teller when describing our research. There should be an introduction, then chapter A and chapter B, and so on, and then there should be an epilogue. You have to be patient; you have to be a marathon runner although when you get too bored, you can make a jump from time to time. I learned from Avram to be a dog that holds onto a bone. I even heard this metaphor from him. While holding onto the bone you should not open your mouth, you should not tell anything unnecessary, and you should not ask unnecessary questions. This I learned in the army.

What is, what you didn't learn from him, but you try to do with your students?

What I try to do with my students, is to try to educate them as broad as possible. I tell them to read, and that they will never understand any context of any problem if they don't read it broadly.

I think that Avram is very authoritarian. I am much less authoritarian, I am much more encouraging the freedom of thinking, of discussing, arguing with your mentor. Saying: let's do this and not do this. I think Avram is less receptive to criticism from his students, to the ideas that wanted to change the direction. I felt that Avram was leading me along a well-defined line. Let's take the ubiquitin discrepancy between the weight and the measurement that he dismissed as an artifact. There was no way that I could continue it in his lab. I had to go to MIT to reopen this drawer.

I couldn't open this in his lab. In my lab students can do whatever they want if the idea is right and it belongs to ubiquitin. My students swim freely in this large swimming pool. I realized that I learn from them as much as they learn from me. There are many examples that the students took me along their own way. I think I am much less authoritarian and more open to the students. This is my personal taste. I cannot tell you, which approach is more successful.

How does Ernie Rose compare with you two?

Ernie is Avram's connection, he is not my connection. Avram met Ernie at a meeting very early. This will probably be shown in Avram's interview. Ernie was working on ATP-requiring processes, mostly in the glycolytic pathway. Avram thought that Ernie may help to decipher the problem of energy in proteolysis. It turned out, in retrospect, that Ernie had also studied protein degradation but never published it.

Ernie is a unique character. He is the very opposite of Avram, absolutely disorganized, very difficult to follow. He may come to your bench and tell you something that you may not understand and come back after an hour and tell you something completely different. Avram is very ordered. Avram tells you what to do in your experiment. Avram wants things to be built like a very ordered structure. Avram is a builder, a constructor and you cannot build the ceiling first and then go to the foundation. Ernie

Aaron Ciechanover with the diploma of his Nobel Prize in Stockholm, 2004.

shoots ideas. He was the extreme mentor for me. He gives you an idea and he forgets about it, comes back and it is really up to you to catch whatever is right. If you are not focused, Ernie can mislead you completely. You start to work on something, then stop to work on it and you never go anywhere. I think it was a very nice triangle, the three of us.

I did an experiment and Avram did an experiment and then the three of us were sitting and discussing. Ernie was pushing in all directions, Avram was pushing it towards the foundation and in between we found what to do in the next experiment. They are very different. Ernie is extremely disorganized, unfocused in many ways. You have to figure out the good nucleus and put it back, but he is a terrific protein chemist and enzymologist. He came at the right time and the right point into our lives.

I would like to return to the story of the identification of APF-1 as ubiquitin. You knew the function of APF-1, but you didn't know what it was. In the NO discovery, this was the essence of the discovery. What is EDRF, what is this material? Then Moncada realized that it was NO. At least you and I think that it was Moncada who first realized it. The Nobel Committee didn't think so. In your case, it was not you, who realized what APF-1 was, but two other people.

I tell you exactly the story. The two other people didn't work in the University in Greenland. They worked in Ernie's laboratory. They were two postdocs in his laboratory and another postdoc in a neighboring laboratory. So, it's not exactly two other people in the geographic sense and in the knowledge sense. We discovered the small protein and purified APF-1 and we found that APF-1 is attached to the substrate and marks it for degradation. Then other groups in the Fox Chase Cancer Center were working on histones and there was a precedent of a protein that modified another protein. A postdoc from a histone lab, Michael Urban came to Ernie's lab, after he heard our finding — Avram's and my finding — and told two postdocs in the lab, Keith Wilkinson and Arthur Haas that there is a precedent to a protein that modifies another protein. There is the molecule called ubiquitin that modifies histone.

Why didn't he tell you or Avram or Ernie?

Actually, he did tell Avram, and Avram wanted to follow it up, but there was a fight between Avram and Ernie, and Avram gave up, and let these postdocs publish it but he made it a condition to have our papers back to back. But we should leave this behind because if it would come out there would be a big fight about it. So let's leave the truth aside. By the way, there was a tiny incident between Avram and Ernie in the Nobel ceremony. Ernie said to Avram something like, you know, it's attesting to my spirit that I let my postdocs publish their findings on their own and I did not put my name onto their paper. At this point Avram said that Ernie did not remember how it happened in reality, because it was not at all that simple and that it was very different from Ernie's kind description. Then they did not pursue this topic any further. Actually, Ernie protested that his name was not on the paper. But the essence of the story is that the postdocs identified APF-1 as ubiquitin. They did the crucial experiment of putting ubiquitin into our system and then APF-1 became ubiquitin. The names of all three postdocs, Keith Wilkinson, Arthur Haas, and Michael Urban are on the paper in JBC. By then we had already known the amino acid composition of APF-1 and it overlapped completely with the amino acid composition of ubiquitin. There was one amino acid deviation that may have been due to some mistake in the analysis.

The identification of APF-1 as ubiquitin was a very important discovery; it was also the convergence of fields. It immediately told us what the connection was, what the bond was. It immediately directed us to decipher the mechanism because it was a peptide bond. It led us to discover the three enzymes that run the conjugation reaction and so on. So in this way it alleviated our road, tremendously. But it was not decisive because at the end it would have been discovered that APF-1 is ubiquitin in one way or another.

But suppose that some other people had identified that APF-1 was ubiquitin. It wouldn't have taken away much from the very importance of the modification that we found for the purpose of degradation. Nevertheless, it was a very important discovery.

It seems that there were two lines of investigations that at one point crossed and then they diverged again, and so did the people.

Exactly.

You carried on ...

We carried on. By the way, the modification of histone is a monomodification, while modification for protein degradation is a polyubiquitination.

Nobel laureates in Stockholm, 2004, from left to right, Linda B. Buck (medicine); Aaron Ciechanover (chemistry); Avram Hershko (chemistry); Irwin Rose (chemistry); David Gross (physics); Edward C. Prescott (economics); Frank Wilczek (physics); and H. David Politzer (physics).

Only now, after 25 years they know what the modification of histone means, physiologically.

It is only modification, not degradation.

It is not degradation, absolutely not degradation. What we learned from them is the chemistry. I mean that the molecule is the same, but the *purpose* of modification is totally and completely independent. At that time it was not known why histones are modified and it became known only in the 2000s, it took another 20 years to decipher the role of this modification. But no doubt, that it paved the road to the mechanistic deciphering of the conjugation reaction.

For you.

For us, yes. No doubt.

You have talked about the contribution of the two other laureates, but not about your own yet.

I left it like that, because again, I know what Avram did, I know what Ernie did and the dynamics of my participation. I cannot tell you that I did something in particular because I cannot repeat the experiment without me. I cannot tell you that I discovered this and this. I can tell you what I discovered independently, like the boiling of the protein, the purification while Avram was away. He was absolutely not aware of what we were doing, that we were boiling the protein, etc. Then I can tell you about the further developments at MIT with Alex and by myself. They were independent steps that I can identify. But when I worked with Avram alone and together with Avram and Ernie, it is like operating on Siamese twins. How can you tell between Watson and Crick, Brown and Goldstein and other known pairs, who exactly did what.

Another question that is impossible not to ask is Alex Varshavsky's contribution to this whole research area, having in mind also this letter that appeared in Science[5] *that congratulated your Nobel Prize but then said that Varshavsky should also at one point receive the Nobel Prize.*

I came to MIT in August 1981, after I completed my Ph.D. studies with Avram to work in the laboratory of Harvey Lodish. I started my work on receptors then I deviated to do some independent work on ubiquitin that I dragged from Avram, this artifact about the discrepancy in the dry weight of ubiquitin. Then Alex Varshavsky, who worked at MIT, approached me. He said to me that he noted in the literature a mutant cell that was isolated by a Japanese group of Yamada. In this cell the histone-ubiquitin conjugate that we talked about, disappears at high temperature. The mutation is only a high-temperature mutation, at low temperature everything is normal. Once you raise the temperature you lose the ubiquitin adduct.

For losing this adduct at high temperature there can be one of two explanations. Either, you accelerate deubiquitination, so some enzyme is accelerating its function on a mutation-dependent manner, or you decrease ubiquitination. The molecule is gradually deubiquitinated but you cannot reubiquitinate it. Because what you see in every moment is the steady state. I think, we believed more in the steady state, in the loss-of-function-type of mutation, because most mutations are loss of function, rather than gain of function. Acceleration of deubiquitination didn't look highly likely. What looked more likely was loss of ubiquitination.

Since, in Avram's laboratory we developed all the techniques during my Ph.D. of how to isolate all the enzymes. I thought that with Alex it would be very easy to identify the mutation in the cell. That's what we did. A very talented graduate student of Alex joined us, Daniel Finley, who is now Professor at Harvard, a very successful one. Together we identified the mutation in the cell to be mutated. It was E1, the first enzyme in the ubiquitin-conjugation cascade. We published it. Consequently to this mutation the cells were also defective in protein degradation. This was strong and direct evidence that the system is operating also in mammalian cells: if you inactivate the enzyme, you don't get conjugation and you don't get degradation. Because the cell was also defective in the cell cycle, we predicted the involvement in the cell cycle. This was not a discovery, it was just a prediction. Alex said subsequently that he discovered it, but that is an exaggeration. To say that he was the discoverer of the ubiquitin involvement in the cell cycle is a blunt lie.

What we did was that we proposed in the discussion that because the cell was defective in the enzyme and because it was also defective in cell cycle, the two phenomena are related. Therefore the ubiquitin system is important in driving the cell cycle, which later became true. It was proved initially by Bart Kirschner and then by Avram. So in our case it was only a proposal. It is important to put things in context. When I was a graduate student with Avram we had already a proof that the system is broader than residing solely in the red blood cell that we were studying. We already raised antibodies to ubiquitin and investigated a liver cancer cell model and showed that the ubiquitin system is there, operating protein degradation. I would say though that the proof was a little bit less elegant than the one we found with Alex. In that sense, I am not sure that the discovery I participated in when I collaborated with Alex was a major breakthrough. It was yet another nice, elegant corroboration of what I call the ubiquitin signaling proteolysis hypothesis.

Then I think that Alex contributed significantly to the development of the field. He took it from the biochemistry tools into the genetic tools. He cloned all the enzymes in yeast. Then the system started to develop into specificity: How different proteins are selected at a certain point to be degraded. Why protein A is degraded and protein B remains intact? Alex defined the first rule, or I would say the first recognition signal that is called the N-terminal signal, the N-end rule. But this also evolved in parallel in my own lab and in Avram's lab using less sophisticated and less elegant biochemical technologies. Alex's technology was very elegant and therefore more systematic, but the principle evolved in parallel. Alex also educated a great generation of students and postdocs that are really the founders of the field.

We used to have a very active interaction with Alex. I entered his room freely all the time and we spent hours together in a row and he was in the lab all the time, from early morning to late night. We discussed every detail of our experiments. To summarize, Alex took the ubiquitin field from biochemistry to genetics. He cloned many enzymes, we defined the first signal, and he did a lot of work on polyubiquitination. Alex's contribution to the field is significant.

Have they tried to lure you to America before the Nobel Prize?

Many times, before and after.

Obviously, they didn't offer you what would have made you stay in America. What would it take to make you move?

Let's make one thing clear, I will never leave Israel completely. Even if I'd go to America one day, it would be part-time only. Israel will always be my base and my home, physically, not just theoretically. Many Israelis

The three 2004 chemistry Nobel laureates in Stockholm on the day of their Nobel lectures, from left to right: Aaron Ciechanover, Irwin Rose, and Avram Hershko.

that live abroad say that Israel is their home, but they are never there. For me it is a physical home. Always.

To answer your question about what would take me there, it would be the opportunity to do much better science than I do in Israel and certainly not my personal benefits. Maybe, the opportunity to do broader, bigger science. So that I would be able in the short time that is still left for me, to be in science, to expand it exponentially with other people, rather than to walk on my own road.

America is the ultimate environment for the survival of those who fit. Fitting means that you compete and competing means that you are able to recruit your own money by convincing the funding authorities that you are better than the others. That includes your salary, your telephone, everything. America is the ultimate in what I call soft money science. Nothing is hardened, nothing is fixed, nothing is guaranteed for life. All the time you are walking on the tip of your toes. This is American philosophy in general. Now, after the Nobel Prize, probably the offers that I have now, without going into details, obviously freeze me for all this and take me for what I am, with my ideas to build and to crystallize. But again, I think America is driven by economy, so maybe they understand the potential of me bringing in money, even without bringing it in directly. By luring in good people, by crystallizing an institute around me would substitute for the money. For the Americans, to think about competition is not necessarily to think about in a very personal way, it maybe in a more general way.

Can we consider your Nobel Prize as a success of Israeli science?

Israel is a miracle in my eyes. Regardless of the government it is a miracle. It is a consequence of the people that live in this country. In 50 years we managed to establish everything — although not exactly from scratch because we brought lots of tradition from Europe. Everything was cast into the desert and the swamp. We started physically from scratch, but a lot of knowledge flew in from Europe. There were scientists and physicians that came instead of going to the United States. Many came to Israel after the Holocaust to build universities and the education system and farming and the agriculture. But in 50 years we achieved, miraculously, hospitals, our health system, music, the Israeli Philharmonic Orchestra, theaters, every aspect of cultural life. Everything is new. When you walk in the street you don't have the sense of the heavy architecture and structure than when you walk in the streets of Budapest or Paris or London or of any other major European city. Everything was established almost from scratch, starting a little bit before the formal establishment of the country, still under the British regime. It is not going further back than 70 years. All the Israeli universities are 70–75 years old.

You asked me about the Nobel and I don't think that it is a success of Israeli science. We have done our research at the Technion and the Technion is mostly a technical institute and we have been its stepchildren. I said to the president right these days and this will go to the interview in these very words: we have been for decades, stepchildren at the Technion. I think that we pushed it against the stream and against the spirit almost without support from the Technion. Israeli science is great, but it is very personal. You can say that the Israelis developed a great education system and I grew out of this education system, but in many ways it is very personal. Obviously, without our universities we couldn't have done it. I couldn't have done it in the basement of my house. So the universities provided me with a physical environment, I wouldn't say with a spiritual environment, certainly not the Technion. They changed now, but they changed as a result, not because of understanding.

You have had a strange situation with awards. Until the Lasker, you hardly had any.

You cannot compare me with Alex and Avram who have been showered with awards.

But regardless of them, on an absolute scale, you didn't get awards. You got an Austrian award before the Lasker.

It was from Innsbruck University, called the Wachter. This I got in 1999. This was my first award.

Yet I didn't have the impression when we had our first conversations in Sweden in 2003 that you would be insecure.

No, I have never been insecure. I am not in the award committees. I don't know how awards are being granted and who works behind them. I saw that Alex and Avram were harvesting many awards along the way. I never had the feeling that I was left alone, I didn't feel any bitterness. We can look at Ernie Rose, he didn't even get the Lasker. I am not an award person and I never sought them, I never worked for getting them. I don't have the connections and I never pushed any cart for them.

In spite of the lack of awards you gave me the impression in 2003 that you might be in for the Nobel Prize.

For sure I knew that the discovery of ubiquitin system is of the caliber of discoveries that are being awarded a Nobel Prize. I am a biochemist. I have been in biology for 30 years. I knew every year of my life in my profession whom the Nobel Prize was awarded both in chemistry and in physiology. So I could put things on a comparative scale. I was a graduate student at the time that the discovery was made. I carried the knowledge to MIT and I continued on my own and then I collaborated. I made another important discovery along with an MIT scientist, with Alex. I played an important part in the story.

On the other hand, I didn't know about the policy of the Nobel, how they relate to graduate students; maybe they are secondary players n their eyes. I didn't know whether former graduate students receive it or not. How the committee can judge whether he played an active role or a passive role. But one thing I did on purpose all along the years. I protected myself from the Nobel Prize. I didn't talk about it, I didn't think about it. Certainly in Israel nobody can help you. I didn't live in a big American institution. I prepared myself for the day — without even thinking of it — that Avram and others may get it or may not and I will not.

I knew, that I had to continue afterwards and in a healthy way and in a productive way. I really removed it from my agenda. Obviously, I thought about it, it flipped, but came and went. I cannot tell you that I didn't think about it for a minute, but not in any active manner. Absolutely not, because otherwise, I think it can be devastating. I wanted to avoid in a planned, cold manner, the devastation. Now, I cannot tell you what would have happened otherwise. I am happy not to be in that position, but I prepared myself, I wouldn't say to the worst, because it is not the worst, it is a only a prize, but I prepared myself for the day that the Prize will be given to others that did discoveries in the ubiquitin field and not to me. I didn't want to commit suicide or to fall into depression the next day. I wanted to continue.

You have met with the students of the Lauder Jewish School in Budapest and one of them asked you about how do you reconcile religion and science.

I am not facing such a problem, because I don't believe that I am a religious person. I am very Jewish, but I am not a religious Jew. There must be a distinction. I don't obey the orders, I don't do anything or I do very few things. For me, Judaism is something completely else. I am very keen on Judaism. Over these days we have had a lot of discussions and it is inevitable that I am going to use some of your words in addition to mine. For me, Judaism is tradition, culture, history, remembrance, it's parents and home, scholarship, it's many things. I don't feel that I have to reconcile science and worshipping something that I do not understand and that is above me.

Judaism for me is a real deep value that appears in a great variety of appearance. Let's take my hobby. The music that I like the most is Jewish cantor music of which I have a huge collection, mostly from classical Jewish cantors, like the Koussevitzkys, Zawel Kwartin, Yosselle Rosenblatt, the Malavskis, etc. It is also the prayers themselves, not just the music. It is a kind of opera music; some of the Jewish cantors were very famous opera singers. Take Richard Tucker or Jan Peerce, who were opera singers, so the question may arise, what's the difference between Jewish cantor music and the Barber of Seville or Carmen or Samson and Delila or whatever? There is a huge difference in the sense that the Jewish cantorial music carries the prayers and the suffering of Jews along centuries, the conversations between them and their God. Some of them are written in an extremely strong and poetic way.

Again, it's not the God in the sense of a Jewish God. It's the God in the sense that places us, human beings, in the right place where we belong, in terms of humbleness in appreciating creation, appreciating nature, the complexity of biological sciences. We need to be very humble because the world is so complex that we don't understand it. You can get a Nobel Prize on the brain, but we don't understand the brain. We can get the Nobel Prize in advancing some of our knowledge on cancer, but we don't have a clue what cancer is, or how to cure it.

I give just one example. It takes me back to my father. There is a very famous prayer before the Musaf prayer. The Musaf prayer is a very important prayer and it takes a real turn of importance during High Holidays, during the Rosh Hashannah (first day of the year) and Yom Kippur (the Day of Atonement). In this prayer, the Shaliah Tsibur, the Chazan (the cantor, the representative of the community) sings the prayer for the community. He stands before his God, before the Ark and says, in my translation, "I am nothing. I am a piece of dust and I am worth nothing. I am just a simple human being. But I have a responsibility. I was sent by this community, thousands of people that are gathering in this synagogue to pray for you that you will forgive them. I am just their messenger. So, please open the gate of heaven for me, because it's not for me, it's for them and for everybody".

I shiver when I read this and it doesn't matter which God, it can be any God. My father told me about this prayer, which he learned in the central synagogue of Warsaw. Before the war there were 4.5 million Jews in Poland, one of the largest Jewish communities in the world. Many of them lived in Warsaw, hundreds of thousands. Many were in the synagogue, but it was nothing compared to what was on the surrounding streets. They were all underneath the tallitot, this white sheet that the Jews cover themselves to dress in kind of holiness, because you are praying to your God. You couldn't see it, because they were covered. They don't expose themselves to God. There are many ceremonial parts in it. And you can see this huge field of heads that are nodding underneath. Then the Chazan in the middle asks God. You see that if he fails, he fails all the many thousands of people. It is something huge — I am talking about the responsibility — because he speaks on behalf of all of them. So, this is one thing. Again, it's a prayer to God but it puts the entire community and the cantor in the right place.

How did the Jews live? They lived a religious life. They wrote their codex of laws, the Talmud and the Mishnah. What is the Talmud with the Mishnah? — it is a civilian law codex. What happens if I rob you, what happens if I rape your wife, what happens if I violate this and that? They assembled it into a way telling everybody how a community should live. This has become a religious script, but it is basically a civilian script. It is one of the oldest codes of laws in the world. So, obviously, I appreciate it, it is Jewish, but it has nothing to do with God. It is how human beings should live together in a community. Obviously, there are some laws of sacrifice in Jerusalem. You go to the Temple on the High Holidays and you sacrifice to your God and bring part of your harvest, agricultural products to God, but you give it also to the poor. On every field you harvest, you always keep the corner of the field not harvested so that the poor people can come and take it, because they don't have it. There are many social rules built into it. This is all in the Talmud and the Mishnah, which are religious scripts, but these are the ways we should live.

Then it obviously has to do with my parents. My parents grew up in Poland and then they made it to Israel before the Holocaust and they grew up orthodox. They left me the tradition, this is the tradition part and now we are coming to the State of Israel. The State of Israel was recognized by the United Nations on November 29, 1947, and was established as Ben-Gurion's independent state on May 15, 1948, as the direct result of the Holocaust. It wasn't established in the previous century, nor was it established 50 years after the Holocaust, it was established right after the Holocaust. And what is the Holocaust? It is the destruction of Jews *because* they are Jews. They were cruelly murdered because of it, persecuted all over Europe, and murdered by the Germans. So, how can I *not* relate to it? I *must* relate to it, I am part of it. I am living in a country that is the direct result of the Holocaust, a religious persecution. It was racial in the eyes of the Germans, but in my eyes after many years of anti-Semitism and persecution, this was just the end of it, the culmination.

Then Jews established their own state and established their own army with two purposes. One that they would be able to protect themselves, doing something that nobody can do it for them. And to bring Jews from all over the world and tell them: if you are a Jew, you are entitled to be here and this is your shelter, you are not asked any questions. This is the only state in the world that gives automatic citizenship to Jews just because they are Jews. This is a shelter from a physical threat, not just from a religious threat; things are linked together. It has nothing to do with whether I get up in the morning and pray according to one book or another. It is a chain of history, the Talmud, the Mishnah, the persecution, the culture, and my parents, my state. It altogether establishes a very unique — in my eyes important for my very existence — structure that I live in. I don't tell you, there is no problem of reconciliation between science and religion, but I never faced it. I never made an excuse for it either and I never felt that I needed to make an excuse for it in order to satisfy people who ask me, "How can you be a scientist?" There are scientists who wear a kipa and are really religious in the full sense of the word. They may find it difficult, and I am sure that they have thought of an explanation.

Today, we met the world-famous writer George Konrad. He asked you this question about the relationship between Israelis and Jews in the Diaspora.

It was a very interesting conversation with Konrad. Israel is a very young country. Most of the Jews that settled in Israel before the war or immediately after the war came from Europe. Those who came much earlier came because they were after Zionist ideas and also for biblical ideas. In general, some of the Jews went to the United States, some stayed in Europe, and some went to Israel, this is the rough division. Let's just simplify and let's talk about our relationship with the most important Jewish community for the State of Israel, most important, because it is the biggest, the richest, and the most influential, the American Jewish community. With the American Jewish community some relations have evolved to my dislike. There is some kind of a division of tasks, very simple. You help us financially, because we are here. So *you* pay for *our* presence there. There were many organizations that were established to transfer the funds from one side to another. It was really an exchange. It worked, but then it started to get complicated and certainly could not continue in this way, because of many reasons. I tell you several reasons.

One of the reasons is, that the members of the Jewish community in the United States are American citizens and they owe loyalty to the American government. They are not Israeli citizens and not everything they do has to comply with the interests of the State of Israel. A certain schism has developed. Traditionalists go with the Democrats in presidential elections, though many Republican presidents, like Nixon and either Bush, are more pro-Israeli than the democrats. The Jews in America grew up in their own culture, most of them are not orthodox; the orthodox Jews are a separate issue. The American Jews have developed their own religious, cultural frames, the conservative and the reform movements. In the State of Israel, only the orthodox stream was adopted by the government. It is gradually recognized more and more that the relationship between the American Jews and the State of Israel cannot be reduced to a simple transfer of funds. We have to understand that they have different religious attitudes and different social attitudes. It is hard, not for us, but for the government, to accept. Actually, I don't remember any time that anybody in Israel had a real serious discussion on how we hug them, how we take them into ourselves and at the same time don't alienate them.

Meanwhile another interesting thing happened. We are now 60 years after the end of the Holocaust in Europe. The first generation in which this financial relationship was found was the generation of the Holocaust. They really felt first hand that they owe us. Two parts of the families split before or immediately after the Holocaust. Some went to Israel some went to the United States. They suffered the same and now their brothers went to Israel, to suffer more. Not the same suffering, but suffering. Now it is different. Now it is the age of the children. For the second generation this argument doesn't work anymore. They are American citizens, second or third generation American citizens, we have to convince them with a very different reasoning that they should support Israel. And the different reasoning is coming. Now we have the Technion, we have universities, we have high technology, and we can now come up and say: hey, guys, support the best side of Israel! I mean, the technology, the export, let's share! We have shared values. But it becomes more and more difficult. We cannot rely any more on this simplistic, cheap approach of ours, saying that we work hard, shed our blood and sweat, and you are the rich, fat living-in-peace Jews, and you owe us. We are now approaching an era with the second generation and I am also the second generation. I am not anymore the generation of the Holocaust. Now, we are approaching an era of understanding, of sharing, and of charity. We are now approaching a point that we are starting a new relationship with the American Jewish community.

We still have a problem with Jewish communities all over the world. There are smaller Jewish communities in the United Kingdom, in France, and elsewhere, and this is not settled yet. There is anti-Semitism in many of those countries and we still don't know what to do with it. Just to mention that a little while ago the Israeli Prime Minister called for all the Jews in France to leave France and immigrate to Israel, because there is anti-Semitism. This is not an attitude for 600,000 Jews to adopt, to leave their workplace, their job, their property, leave everything, and come to Israel. To do what? What solution do we offer them? So, we still need to work out our relationship with the Jewish communities. It is a continuing process. It is a problem. But now, in Budapest, and I was in Athens, I see the joy, the revival of whatever Jewish life is there. We are there, we are here, we live open, we don't have to hide it. I mean, it is still hidden. I am not fooling myself about the openness, but the synagogue was packed with tourists in the morning just to see it as a sightseeing site, not as a praying site. People are interested in Jewish life and in what happened in the Holocaust and there are memorials all over the place. For me it's wonderful. I don't think that all Jews should live in Israel, they should live, wherever they want, where they feel like living, but I realize that this is not a simple issue.

It's little more than half a year that you have been a Nobel laureate how different is it from what you had imagined it would be?

I am a fresh laureate and I am not sure that I have absorbed it in full. I believe that I am still in the process of absorbing its full significance. It's very big you can see it. People are talking to you, who wouldn't have talked to you before. It is not a matter of fans that talk to you, but you get a really different attitude. I was in Japan and it was extreme, the honor and the respect. This attitude doesn't stick to me well. Some of it may stick in the future, but I really want to remain me.

On the other hand, I feel that it gives me some positive powers to say things that will be heard, not that I didn't say them before. Maybe to influence on positive processes, maybe on the educational system in Israel. Now, I had another thought today, following our tour of Budapest that I may be able to help the Jewish communities to make the right connections so they maybe able to preserve the Jewish heritage. There are a lot of problems. You need money for it, you need connections, you need to fight against the municipality that wants to develop the area. So I may use it in a positive way, but obviously, not in a way that will inhibit, or disturb my science.

The main mission is obviously to carry on with my science as I did before. I am very interested, there are very interesting things, and the ubiquitin system is far from being resolved. The basics, the biochemistry are far from being resolved, and I am curious. For me the passion to science has remained the same. So it gives me the power to do some, selective positive things. I start to realize that it is big, it is very big. I have never been involved in it, I have never nominated, I have never been in a ceremony. I have never read a book about it and I have never read material on it and I wasn't aware of the big industry that is going behind it. But I knew that it's big and I suppressed the thinking that I could ever win it. I knew that the achievement, the ubiquitin system is in the magnitude of discoveries that win a Nobel Prize. I am a biologist, I know where things stand in context. There are 20,000 papers published in this field, companies are making drugs, I am cited thousands and thousands of times. It is important, it guides important processes in the cell, cell cycle, differentiation, quality control; the ubiquitin is everywhere. I had won one important prize, so I knew I was in the game. But I didn't get any of the prizes the others got. Now it doesn't matter anymore because this one single prize wipes them off, all of them.

I try to imagine the frustration that you may have gone through while Alex and Avram were receiving prize after prize and you were left out.

Obviously, I was frustrated, but not to a point that affected me a bit. I didn't stop writing, traveling, and participating in meetings. At no point did I build my life around prizes.

Did your wife say anything?

My wife is even more extreme than I am. She is out of the game completely, even now. She is extremely happy and proud, but actually she is my guide and my traffic lights to get back to normal life. She says, yes, it's nice, we got it. It helps us to establish ourselves more and to get obviously the money that we can renovate our home. She is very practical about it. But don't burn our life on this altar. This is not a reason to change life. Let's take it in proportion.

But the proportion is tremendous.

Yes, but the prize was given for a work that I did 26 years ago. For three papers that were extremely important. Obviously, I assume that in considering the prize, it was also for later things. I stayed in the field in a very active manner much beyond that I graduated from Avram. I contributed significantly to the field after as an independent scientist. So I was not a bystander, a hitchhiker of the system that happened to be in Avram's lab at one point. I contributed significantly after with Alex and alone for 25 years. I wrote many reviews, lectured in the most prestigious conferences, and was cited thousands of times in the literature. I wasn't by any means a bystander in this game. But not for a moment I victimized myself by thinking of prizes and what would happen if not.

At the same time, I'm also a human being and we get offended, we get insulted, and we get depressed. I cannot tell you that the other prizes didn't touch me at all and that they flew by like a piece of dust over my head. Obviously, I knew that there is the Wolf Prize, the General Motors Prize, and other prizes that I didn't share. What could I do?

I have to ask some questions that if I wouldn't ask them would seem odd. You must have thought of some explanation why you didn't receive the smaller prizes, and then you got the Lasker and the Nobel Prize.

There are nominations written by certain people and they wrote what they thought was the truth in their eyes. Whether the nominees of those prizes themselves had to do with it or not — I mean with who was included and who was excluded — I don't know. I have my own suspicions whether

I was excluded intentionally by this way or another way. I cannot tell you that I know the truth behind it because I never read the letters of recommendations. But these are nominations, and my name was either there or was not there. Probably it wasn't there, because otherwise I might have received them.

You know, what, enough of this convoluted answer to your question. Let's be a little more honest about it and we can decide later whether we want to have it in the printed interview or not, but I don't want to kid you and I don't want to kid myself either. Some time ago Alex must have decided that I was dangerous for him because I was the link between Avram and himself and considerations for me might diminish his achievements. Of course I know the truth, which is my truth and Alex may know a different truth.

I've read what I could about this discovery and to state it bluntly, I was puzzled. You were part of Avram's discoveries and you were part of Alex's achievements as well, but you were left out of recognition until very recently, until 1999 to be exact. If I was puzzled by this, people in fifty years would be even more puzzled with less opportunity and hope to understand it.

István Hargittai and Aaron Ciechanover in the Hargittais' home in Budapest, 2005 (photograph by M. Hargittai).

It is a puzzle for everybody who don't know the history. But I wonder if we are discussing personal views or whether we are writing history?

My interviews concern personal views and I cannot claim writing the history of discoveries. At best these interviews will become additional sources for future historians. But the main purpose is to introduce the interviewees as scientists and human beings.

That's fine with me. However, if and when somebody will attempt to write the history of the ubiquitin field, it will take an entire Ph.D. dissertation. So what I can tell you is my personal views. It is of interest that for the works the Nobel Prize was given, I was never alone, never an independent researcher. I was Avram's student, and later I was with Alex, a sort of postdoc although not in the formal sense because I was a member of a different lab. I am convinced that the Nobel Committee [of Chemistry] made the right judgment, including Ernie. He got his Nobel Prize for being the co-author of a single paper for which he worked for three months. We brought all the details from Haifa and we didn't understand them. We had laid out the puzzle and Ernie had to make sense of it and he did. He didn't think of the importance of the problem and he didn't think of APF-1, all he did was, and all he had to do was, make sense among our data. He did that in three months. What remained after that, were details. Once you lay the foundation of the house — once we had E1 and E2 and conjugation, it was clear that there was a protease and there must be a conjugating enzyme, all the rest came out in a most direct way. Anybody could have picked it up who happened to be walking by and it was Alex who happened to be walking by. This subtracts nothing from his achievement and he is very intelligent, but the foundation was laid by the three of us. Avram can say that he could have achieved it alone, what we did the three of us together, and he is right, but it would've taken him sixty years. These are all useless arguments. By the way, I doubt that even a historian who might write a thousand pages will ever get to the absolute truth.

This is why I find it important to be honest about it at this point. I know it: Alex moved the prize machine and he excluded me on purpose. Future historians will have to examine the letters of nomination for the various prizes. They will have to examine David Baltimore's letters of nomination on behalf of Alex. The Wolf Prize is a case in point. Incidentally, when I received the Nobel Prize and I received a letter of congratulation from the Wolf Foundation, I returned the letter because I didn't want to have it in my collection and I told them so. My exclusion from the Wolf Prize could not have been an accident; there must have been an active involvement on somebody's part. In addition, the Wolf Prize was given after the Lasker Prize, so there was an international judgment about it available to the Wolf people. So my slighting by the Wolf committee was sandwiched between the Lasker Prize and the Nobel Prize. This failure of the Wolf Prize will stick out in its history.

Originally, what turned you to science?

I don't know. It wasn't home. This is very strange.

In my interviews, it's often a chemistry set or a teacher.

I remember that from Day One I was attracted to biology and to science. In my biography I said that when I was 10 years old, we built a self-propelled rocket that we tried to send up to Mt. Carmel.

That was around or soon after the Sputnik in 1957.

It may have influenced us. I was always fascinated. I always read about Galileo. I remember at home, I had one of the first Hebrew encyclopedias. It is not

Aaron Ciechanover in Budapest, 2005, with the Chain Bridge and Pest in the background (photograph by I. Hargittai).

the *Hebrew Encyclopedia*, that is something else. It was an encyclopedia written in Hebrew. I was always attached to it. Then I was very attracted to chemistry. In this there was no influence at home because my father was a lawyer and my brother studied to become a lawyer. For me it was all self-education and teachers. There were no physicians at home either, but I started in medicine. Then it was a big disillusion that we can't cure anybody. Medicine doesn't cure, it takes measures against symptoms. We don't understand diseases. I have to understand more mechanisms. I cannot just live in a set of instructions of what to do with patients. Like the patient got heart attack, make an ECG. I need much more understanding of the processes. Then I had to settle somewhere. When we have a choice of professions we like to settle in a layer, in which we feel comfortable. You can go to elementary particles physics, to subnuclear forces, you can go to solid-state physics, which is more interactions of atoms, it is not inside one atom. Then you can go to chemistry, which is interaction. Then you can go to biochemistry in which there are processes and then you can go to medicine or to physiology, which is all organs. Medicine is the ultimate in complexity but then it has the gratification of treating human beings. Then there is physiology about how it works. There is blood all over the place. Then there is the cell level and there is the subcellular level. When I got disillusioned in medicine, I said, it was too complex for me. But I liked biology all along and it was very close to medicine in the first place. There was no option for real biology at that time, because I had to go to the army. But then immediately I got disillusioned in medicine. I realized that this was kind of a mistake for me. So I came to scientific research relatively late.

References

1. Quoted from Poole, B.; Ohkuma, S.; Warburton, M. "Some aspects of the intracellular breakdown of exogenous and endogenous proteins".
2. See, in Ciechanover, A. *Nature Medicine* **2000**, 6(10), 1075–1077.
3. The English edition of the book appeared with the title: *Ben Gurion Looks at the Bible*.
4. In the "official" English translation, "Wherefore doth the way of the wicked prosper? Wherefore are all they secure that deal very treacherously?" Jeremiah, 12:1.
5. "Varshavsky's Contributions." *Science* 2004, *306*, 1290–1292.

Irwin Rose (courtesy of Irwin Rose).

15

IRWIN ROSE

Irwin Rose (b. 1926 in New York City) is currently at the Department of Physiology and Biophysics of the University of California at Irvine. He shared the 2004 Nobel Prize in Chemistry with Aaron Ciechanover and Avram Hershko (both of the Technion, Haifa) "for the discovery of ubiquitin-mediated protein degradation".

Both Irwin Rose's parents came from secular Jewish families, on his maternal side, the Greenwalds originated from Hungary and on his paternal side, the Roses originated from the Odessa region of Russia. When Irwin Rose was 13 years old, the family moved to Spokane, Washington, to a high and dry climate because Irwin's brother had rheumatic fever. The father remained with his business in New York though. No one in the family was involved with science, but young Irwin worked summers in the local hospital and imagined himself having a career solving medical problems. After high school he attended Washington State College, then served in the Navy, and went to graduate school at the University of Chicago. Initially he was investigating the diet dependence of DNA for his Ph.D. thesis. When, however, the genetic nature of DNA was revealed, he had to look for a new project, which he found in looking at the construction of DNA from various base components using carbon-14 labeled compounds. This was in the very early 1950s when the American scientists had the advantage of using carbon-14 whose export was at the time prohibited by the Atomic Energy Commission. His next interest was in establishing the absolute stereochemistry of enzymatic reactions and determining their mechanistic significance.

Irwin Rose received his Ph.D. degree in 1952. Then he did postdoctoral work at Western Reserve University in Cleveland and at

New York University (the latter with Severo Ochoa). In 1955, he became Instructor in Biochemistry at Yale University Medical School. His interest in protein breakdown started at this time when he learned about the ATP requirement for it in a liver slice system. It was also in 1955 that he married Zelda Budenstein, a graduate student in the Department of Biochemistry. She had a research career of her own and they have four children. Dr. Rose's main interest in the next decades was in studying the mechanism of enzyme action.

In 1963, he moved to The Institute for Cancer Research of the Fox Chase Cancer Center in Philadelphia. That is then where Avram Hershko and his graduate students joined him for the first time in 1977 and frequently afterwards for the next 18 years for sabbaticals and summer visits.

After I had recorded interviews with Aaron Ciechanover and Avram Hershko, it would have been natural to do an interview with Irwin Rose. I knew, however, that I would not have the opportunity to visit California before the completion of this volume of interviews. This is why in April 2005 I sent Dr. Rose a set of questions of which he kindly responded to some. Interviewing by correspondence is far from what personal encounters can offer. However, I am grateful to Dr. Rose for his kind cooperation. This Introduction was also compiled from the material he sent me.*

Many were surprised that the Nobel Prize for the discovery of ubiquitin-mediated protein degradation was given in Chemistry rather than in Physiology or Medicine. Were you surprised?

I think the prize to Chemistry was appropriate. The work was really the enzymatic basis for the system. As an enzymologist it is clear to me that enzymology owes more to the concepts of chemistry than to biology.

Many were surprised that you were one of the laureates. Were you? (Here I must add that I was not because I had recorded conversations with Avram in August 2004 and with Aaron in September 2003, and both had spoken about your contribution with great appreciation.)

Yes, I was surprised. Because over the past few years special awards for the ubiquitin work have been given to others. I appreciate the good words of Drs. Hershko and Ciechanover, but I hope the Nobel Committee made

---

*István Hargittai conducted the interview.

up their own mind. My contribution to the E1, E2, E3 story and especially the discovery of the isopeptidase and ubiquitin aldehyde are significant contributions. I think I may have been helpful in other discussions.

Avram Hershko told me in August 2004 that your field was enzyme mechanism but you were also interested in protein degradation although you had never published in this particular area. My question, before the ubiquitin discovery why were you interested in protein degradation and if you were, did you plan any work on it?

My interest in the ATP mechanism of protein breakdown began in 1955 when I learned from Dr. Mel Simpson who was a colleague in the Biochemistry Department at Yale about his earlier demonstration of ATP activation of protein breakdown in liver slices. From 1955 until 1977 I did occasional experiments looking for a cell free system. Failure to find one turned out to be due to a lysozomal protease that degraded the Ub of the extract. It was not until the paper of Etlinger and Goldberg in 1977 using a reticulocyte extract that it was possible to go ahead with fractionation. Hershko definitely was the leader in this phase of the work. I think the discovery was nothing like we had seen before. We contributed by analyzing the mechanisms and specificities of the new system, etc.

Did the time Hershko and Ciechanover spent in your laboratory impact your further work?

During their stay with me and when my postdocs Haas and Pickart were working on E1, E2, and the isopeptidase story I continued work on my other problems which had to do with questions about the mechanism of synthetase enzymes, what you could learn from reaction stereochemistry, how enzymes recycle after the product is liberated (fumarase), studying reactions of enol-pyruvate, etc. Also, I felt conflicted about competing with Dr. Hershko's priorities.

How does it feel to be thrown suddenly into the limelight?

How does it feel? Disruptive, highly disruptive. I have never paid much attention to the Nobel Prize as a goal and I knew nothing about the great importance it had for friends, colleagues, or Institution to which I had previously been associated. After the amazement I came to appreciate some serious discussion, say, on public TV on aspects of the work for which the prizes have been given. This would broaden the understanding of how discoveries are made and of scientific progress at a time when religious subjects have become such a major influence on the people. I don't think that the many presentations that occur during the Nobel Week in Stockholm succeed in doing this. It should be up to the National Academies to develop such programs.

What turned you originally to science?

I think I was attracted to science from a feeling that doing research was the only way I could think of to make a worthwhile lasting contribution.

Were there any difficulties in your life and career that you had to overcome?

Difficulties: I think if I had not left New York as a child I would have been better educated.

Did you have a mentor, a decisive period in your early career that determined your future path?

I did not have a mentor, no one who pointed the way or set an example. My first experience with people who were ambitious in applying what they understood was in graduate school at University of Chicago. By this time I was so undisciplined that I was careless in choosing my thesis problem or

Irwin Rose in the library (courtesy of Avram Hershko).

in making my postdoc choices. I was only anxious to get out of my own, I think that was an important motivation.

Did World War II in any way impact your life, career?

World War II gave me the G. I. Bill. That allowed me to go to the University of Chicago.

Do you have heroes?

Heroes? Leonardo da Vinci and Benjamin Franklin.

Alexander Varshavsky, 2004 (photograph by I. Hargittai).

16

ALEXANDER VARSHAVSKY

Alexander Varshavsky (b. 1946, Moscow, Russia) is Smits Professor of Cell Biology at the California Institute of Technology in Pasadena, California. He moved to Caltech in 1992, after 15 years at MIT's Department of Biology in Cambridge, Massachusetts. He was born and educated in Russia, and was 30 at the time of his emigration to the U.S., in 1977. In Russia, and for a while at MIT as well, he studied the structure, replication and segregation of chromosomes. Over the last 27 years, the work of his laboratory focused on the ubiquitin system and related fields.

Dr. Varshavsky is a Member of the National Academy of Sciences of the U.S.A. (1995), the American Academy of Arts and Sciences (1987), the American Philosophical Society (2001), and Foreign Member of the Academia Europaea (2005) and the European Molecular Biology Organization (2001). He received, jointly with A. Hershko (Technion, Haifa, Israel), the Gairdner Award (Canada, 1999), the General Motors Sloan Prize in Cancer Research (2000), the Massry Prize (2001), the Merck Award (2001), the Wolf Prize in Medicine (Israel, 2001), the Horwitz Prize (2001), the Wilson Medal (2002), the Stein and Moore Award (2005), and the Lasker Award in Basic Medical Research (2000) (he shared the latter award with A. Hershko and A. Ciechanover). He also received the Novartis-Drew Award (1998), the Shubitz Prize in Cancer Research (2000), the Hoppe-Seyler Award (Germany, 2000), the Pasarow Award in Cancer Research (2001), the Max Plank Award (Germany, 2001), and the March of Dimes Prize in Developmental

Biology (2006). We had initial conversations in 1999 and in 2004 in Pasadena before we embarked on a more formal interview by e-mail during 2005.*

I would like to ask you about your family background. You were born right after WWII, and for years there was hardship in Soviet society. How did the war impact your family?

My mother Mary Zeitlin and father Jacob (Yakov) Varshavsky were acquainted before WWII, and got married in 1942. My mother's family was fortunate to be evacuated in time from Kharkov, a city in Ukraine that was overrun by the Wehrmacht in the Blitzkrieg offensive of 1941. My father graduated from the Chemistry Department of Moscow University two years before the war. He was drafted in 1941, joining a tank battalion of the retreating Soviet army, and was injured soon afterwards. A piece of shrapnel hit his kidney, and chronic damage ensued. He was never sent to the front again. That injury, bad as it was, was a stroke of luck, since few returned home among those who began as soldiers in 1941. Father's older brother Isaac was a military officer on a submarine based in Sevastopol, on the Black Sea. He was in combat from the beginning of war, and died two years later, in 1943, on a destroyer that was bringing him to Sevastopol's base and

Alexander Varshavsky's parents: Mary Zeitlin and Jacob (Yakov) Varshavsky in Moscow, 1950s (all pictures courtesy of Alex Varshavsky unless indicated otherwise).

---

*István Hargittai conducted the interview.

was sunk by a submarine. (In 1943, German submarines could still enter the Black Sea via the Danube river.)

You went to school when Stalin was still alive. What was life like then? Were you much indoctrinated in ideology?

The repression and physical hardships of life in the former Soviet Union were obvious to visitors from the West, especially during the first decades after WWII. But many denizens of Russia didn't see their lot this way, because they lacked a frame of reference. The Soviet regime's excellence in suppressing dissent relied in part on information blackout and incessant propaganda. As a result, many people didn't know that life in countries on the other side of the Iron Curtain was far better, freer than their own. Not everyone was duped of course. Native intelligence varies in outbred population, and some people perceived the truth of their condition even through the thickest of smokescreens. Russian-language radio broadcasts from the West were jammed but often not well enough, so a determined listener with a short-wave receiver could occasionally hear them. Those bits of truth about the nature of the Soviet regime became common knowledge (and even then largely amongst intelligentsia) only in the 1960s, some years after death of Stalin. The state's propaganda scored its greatest successes with children, whose trusting minds were particularly susceptible to lies. Besides, the adults were afraid to share misgivings, let alone hatreds, with their progeny. A cherubic kid might be innocent enough to chat at school about mom's and dad's conversations at home, with dire consequences for the entire family. I was a fairly typical Soviet youngster, and vaguely remember being enthusiastic about the Communist mythology, until early adolescence, when I started to notice inconsistencies in the official propaganda. Its main idea was that we, the Soviet people, were the luckiest people on earth.

You lived the first 30 years of your life in the Soviet Union. People often posit this question to me as well: how one could survive the Communist system.

The sheer cruelty of Communist regimes, in Russia at first and later elsewhere, had few parallels in history. But dictatorships age. Having killed a lot of its people and scared to death the rest of them, a senescing tyranny can afford a modicum of relaxation. By the time I had begun to understand anything worth discussing, Stalin was long dead, and had been denounced by Khrushchev, a sidekick who clawed his way to the top and began a less bloody rule. I was born into a family of steady professional occupations (my mother a physician, my father a scientist), and was insulated from physical privation. But living in a Communist country was a psychologically difficult affair, easier for some, more trouble-prone for characters like mine, with hopes and reality on different planets. Such people trap themselves by their dreams. Before managing to escape from the Soviet Union in 1977, I had a few brushes with disasters that would have left me unable to become a scientist, had I not been lucky. One near-calamity, recounted below, stemmed from the writing of my first scientific paper. In all misadventures, the fault was mine, not the system's: the latter didn't hide but I still ran into it, daydreaming a pillow ahead, instead of granite.

What turned you on to science?

I grew up in a scientist's family. So my interest, and later love for science were a case of "nature" and "nurture" together. My father, now 87, retired and living in Salt Lake City with my mother (and my sister's family nearby), was a physical chemist in Moscow, devising methods for production of heavy water. That work, a blend of fundamental and applied physical chemistry, was a part of Russia's atomic bomb project. He got interested in DNA in the 1950s, joining other physicists and chemists who were leaving their fields at that time for the nascent world of molecular biology.

My father, mother, I, my sister Marina (born in 1954, when I was 7), and our nanny Maria ("Marusya") lived in a single room of a communal apartment typical for its time, in a city desperately short of decent housing. The apartment was essentially a corridor with many doors, one of which led to a cramped toilet, another to a communal kitchen, and the rest to about fifteen single rooms. Each of them housed a separate family. My earliest realization, at 5 or 6, that my father received special treatment by the world came in the shape of milk bottles. The largesse of the atomic bomb project trickled down even to people twice removed from it. My father was given a "free" bottle of milk every day, on the grounds of his working with "isotopes". They were stable isotopes, but never mind. He was supposed to drink that milk on the premises, but instead brought it home, where I was told that the milk was from mad cow, a frightening but attractive description to a 5-year old, many years before "mad cow" came to signify a less benign proposition.

Friends of my parents often visited them. Becoming a teenager (by that time, the family moved to a small apartment on the outskirts of Moscow), I began to see scientists among my parents' guests as separate, more interesting people. A din of conversations about physics, chemistry and biology surrounded me at dinner parties, with serious talk dwarfed by jokes, laughter, and political commentaries that would have been unthinkable in Stalin's time. The regime didn't lose its fangs but the willingness to use them had diminished, and people were emboldened a bit. By 16, I wanted to do all of science, mathematics included, dreamt of becoming a writer too, and felt, without evidence, that all of this was possible. That mania grandiosa eventually subsided, perhaps not entirely. To the world outside I was a typical "academichesky malchick", a Russian idiom for "professor's son": an alloy of cockiness, nerdiness and insecurity, the latter camouflaged by arrogance. To myself, I was the Eighth Wonder of the World, a secret knowledge to be sure, but it surfaced with regularity that must have made dealings with me a chore.

My initial interest in science soon upgraded itself to love. Memoirs by scientists, their biographies, particularly of physicists and mathematicians, became my favorite reading, and the personages themselves my closest friends, all the more so because I was ill at ease with real people. At about 17, I read Einstein's remark that one cause of his attraction to science was the desire to be shielded from everyday existence, from its unbearable cruelty and inconsolable emptiness, from the prison of one's constantly changing whims. I was astonished to see he felt that way, for I did too, but didn't discuss the subject with anyone, then.

How strong was your father's influence (knowing that he was a substantial scientist)?

It must have been important, in more ways than one. Here's my father's advice, given when I was 16. "So you are interested in biology. Good. But ignore biology for now, kind of. Get the best background in math, physics and chemistry that you can possibly achieve, then worry about biology. Learning biology is much easier than physics and math, so focus on them first." By then, I knew enough to immediately sense he was right.

Was it difficult to get into that most prestigious school, Moscow State University? Was it particularly difficult for a Jew? Why did you choose the Chemistry Department, rather than Biology or Physics?

I became a student in the Chemistry Department of Moscow University in 1964, having decided to do biology but wanting to learn physics, math and chemistry as well. There were five entrance exams, amidst heavy competition. I received an "A" in all of them (math, physics, chemistry, "literature", and history of the Communist Party), and was admitted. There were several other Jewish freshmen as well, in a class of about 300 students. I heard of course, from people whose veracity I trusted, about other outcomes and discrimination, overt or covert, against Jewish applicants at Moscow University and elsewhere, especially in math departments. My own experience was different, possibly because it was problematic not to admit a candidate who did well on exams. I didn't know much about discrimination because I was a space cadet, barely noticing dangers and obstacles in my path. "A sea doesn't rise above drunkard's knees" ("Pianomu more po koleno") is a Russian proverb that describes my emotional makeup then, and later as well. If such immaturity doesn't lead to disaster right away, it can produce an illusion of invulnerability, before a blow.

I began moonlighting in a biochemical lab during the second year at university, loving that work from the first. The subject occasionally reciprocated. At nineteen, in 1967, listening to a lecture on quantum mechanics, I was visited by the first genuinely new scientific idea I ever had. Around that time, the Lac and lambda repressors, predicted by Jacob, Monod and their colleagues in the early 1960s, were demonstrated to actually exist, by Gilbert, Müller-Hill and Ptashne. But the thought I had was different: what regulates repressors? Could it be that repressors regulated themselves, for example by inhibiting their own synthesis? I was so stunned that such a simple idea might be new, let alone correct, that I bolted from the classroom, went to the library and buried myself there, forgetting such trivia as attending lectures and laboratory courses. Skipping lectures was OK, because one could prepare for exams by reading. But playing hooky with practical courses was madness, since one had to complete them to qualify for next semester. Predictably, I came very close to expulsion from university. That would have been a disaster (of my own making), for I was, by then, of sufficient age to be immediately drafted for a 3-year stint in the Soviet Army. The main reason I wasn't expelled was my having been a straight-A student up to the moment of the repressor idea announcing itself.

I read, wrote and rewrote, preparing a theoretical paper and furnishing it with differential equations that described the behavior of circuits in which repressors regulated themselves. The paper was published in the January

1968 issue of Russian "Molecular Biology", then a new journal. Nowadays, the modeling exercises I labored over in that paper would be called "systems biology". My version of it in 1967 was too simplistic to be of any use. But the idea itself was new, and turned out to be correct for some repressors. Models of biological regulators regulating themselves began to be considered in English-language papers around 1971, naturally without reference to the 1968 publication in a relatively obscure Russian journal.

That idea, and my coming close to expulsion from university because of it, were a searing experience. I saw that I was capable of thoughts that were genuinely new, and possibly even correct. This deepened my commitment to the craft. I also saw, a bit later, that my penchant for equations should be postponed for a remote future, since biochemical systems at hand were too dimly understood to allow quantitative modeling realistic enough to be useful.

Your first and only place of work after university was the Institute of Molecular Biology. Even the name of that Institute signified a change in Soviet life. For example, at the Fifth International Congress of Biochemistry that took place in Moscow in 1961, the term "molecular biology" was

James Watson and Jacob (Yakov) Varshavsky at the Fifth International Congress of Biochemistry, Moscow, 1961.

not permitted to be used. Would you comment on the situation of molecular biology in the Soviet Union over the years?

The shadow of Trofim Lysenko was still visible in the 1960s. He was a textbook example of a power-hungry demagogue. A wily but limited man who posed as a scientist (and probably saw himself as such) but never understood science. In the mid-1930s, Lysenko began his campaign against the community of Soviet geneticists. His power derived from his support by Stalin, to whom Lysenko promised great agricultural harvests. Stalin and Lysenko had much in common as sociopathic characters and chieftains, but differed in their demeanor, with Stalin reserved (in public) and Lysenko a histrionic bully. Stalin held the entire country in his fist, while Lysenko's fiefs were biology and agriculture. By the late 1940s, Lysenko became the tsar of whatever remained of Soviet biology. Several leading geneticists were killed or imprisoned. The lucky ones were fired, or dropped their studies early enough and turned to safer occupations. Lysenko's own brand of "Marxist" biology is not worth recounting. (There are books in English about the life and career of that Rasputin-like figure.)

One consequence of the Lysenko's reign in Soviet biology was the near-disappearance of people who understood genetics professionally enough. Entering science in the 1960s, I could see the effect of that gap. One evening in 1968, a leading Russian biochemist was giving advice to an undergraduate: "Ah, Alex, don't waste your time on genetics. It's mostly obsolete. Beautiful in a strange way, but next to useless. They keep tormenting fruit flies, but it's us biochemists who will produce the understanding that really matters." Having spent a day reading genetic papers, I sensed he couldn't be right, that genetics was essential too. But that was just a hunch on my part, not an obvious truth it would have been in a more enlightened setting.

Khrushchev, the successor of Stalin, continued to support Lysenko, but Khrushchev ran a "milder" dictatorship, and people became less afraid. Leading Soviet physicists, whose cachet with apparatchiks derived from physics' importance in military technology, helped Vladimir Engelhardt, a distinguished Russian biochemist, to organize a new research institute in 1959. Its baroque name was the "Institute of Radiation and Physico-Chemical Biology", since "Molecular Biology" was still verboten. The place was renamed the Institute of Molecular Biology (IMB) around 1970.

Can you describe your early experiences of working in a lab, at the university and later at IMB?

The Institute of Molecular Biology (IMB) at 32 Vavilov Street in Moscow, where Alexander Varshavsky worked in 1970–1977.

From left, Georgii Georgiev, the head of Alexander Varshavsky's lab at IMB; Alexander Rich, visiting from Massachusetts Institute of Technology; Jacob (Yakov) Varshavsky, head of another lab at IMB; and David Baltimore, also visiting from MIT, Moscow 1975.

At Moscow University, both I and people around me knew that "real" molecular biology was being done largely in the West. The latter could have been just as easily on the Moon, for it was impossible to visit either. Russian laboratories were strapped for everything — hard currency (rubles were worthless outside Russia), clean reagents, equipment, and contacts with Western scientists. Promotions and funding were based on merit once in a blue moon, while capacity for intrigue and membership in the Communist Party rode far in front. I sensed the provinciality of my scientific milieu. It depressed me for sure, but I was spared, for a while, the full comprehension of a gulf in quality (notable exceptions notwithstanding) between Eastern and Western molecular biology. One reason for slow awakening was the sheer appetite for work: even washing dishes in the lab was not a chore to shirk from if it could accelerate an upcoming experiment. Another reason was my love of physics and math, where Russian scientists were anything but provincial. Leading mathematicians in Russia were second to none, and theoretical physics was world-class as well: Lev Landau (before his dreadful car accident in 1961), Vladimir Fok, Yakov Zel'dovich, Igor Tamm, their former students, and other terrific scientists. But I wasn't a mathematician or theoretical physicist, and the bleak reality of my circumstances began asserting itself soon after graduation from university. I got a job at the Institute of Molecular Biology, a flagship place for doing such biology in Russia. The laboratory I worked in was led by Georgii Georgiev, one of the best scientists at the institute. We studied chromosomes and RNA. Things were looking up, but there also was, in the midst of excitement, an undercurrent of second-ratedness in my larger surroundings. My fear was that I would cease being aware of that, and become one of many who gave up or never aspired in the first place. What sustained me was youth, love of science, and ambition, a heady mix. It kept me working and hoping, against all evidence, that things might improve. The time was mid-1970s, a couple of years away from a chance to escape it all.

"Doing science is like driving a car at night. You can only see as far as your headlights, but you can make the whole trip that way." Long before I encountered this metaphor, by E. L. Doctorow (his actual remark was about writing a novel), I sensed the attraction of scientists' racket: an air of open-ended adventure, a contest of sorts, with the landscape rough, unpredictable, with other cars racing toward Holy Grails out there, and occasionally colliding with yours, by accident or not quite. A rambunctious life to be sure, but quiet on the outside, with maelstroms and lava flows hidden from casual view. A set of qualities for such a life must contain a genuine interest, beyond mere curiosity, in understanding the world's design, but an ambition too, even with thinkers whose visage suggests otherwise. Photos of old Einstein, a serene photogenic sage, free of strife, do not recall the assertive and ambitious young man in Switzerland, on the cusp of initial success, and his later, often overt, competition with rivals, a great German mathematician David Hilbert amongst them. Hilbert produced (and correctly interpreted) the equations of general relativity simultaneously with Einstein, a fact unmentioned in popular accounts of the subject but known to those who are interested in physics' history. *What Mad Pursuit* is the title, from a poem by Keats that Francis Crick chose for his autobiography. And mad it is, propelled not only by desire for knowledge but also, in no small measure, by desire to impress and awe — oneself, others, posterity. That's where one's genetic makeup comes in particularly strongly, I think: a predilection for life of a certain shape and texture.

My mother tells me that her ancestry is traceable to the chief rabbi of Prague in the early 17th century. Had I been born a few centuries earlier in the Jewish ghetto of Prague, I might have ended up a religious scholar, splitting hairs with fellas at a yeshiva about the subtleties of the Talmud and Kabala. It's nice to hope that I would have discerned, unaided, the incorrectness of religious outlook, recognizing its vacuity in the midst of semi-medieval Prague. Wrong hope, most likely, for it is difficult, nowadays, to appreciate the acuteness of insight, in addition to independence of spirit, that would be necessary for such a discovery before the rise of modern science. But some, quite rare, people in both antiquity and Middle Ages must have glimpsed this insight. That the names of early doubters are largely unknown to us is no accident. From times immemorial to roughly the 1700s in Europe, one could safely declare a disbelief in deities and their deeds only in total solitude. Even centuries later, the nonbeliever's view of extant religions is a majority opinion only among scientists, not in the public at large. But the long-term trend, on the scale of centuries, although a turbulent one, with transient local reverses, is clearly toward secular, science-informed outlook. A stance in which things and phenomena we don't have good explanations for are called mysteries or puzzles, without attempts to camouflage the insufficient understanding by theology and fairy tales.

How, when, and why did you decide to emigrate? How much do you feel Jewish, and how much of your environment considered you Jewish, in the Soviet Union and in America?

By the 1970s, the idea of emigration was in the air, especially amongst Russian Jews, whose treatment by the regime was strikingly inconsistent. It was more difficult, though not impossible, for a Jew to become a student at a good college, more difficult to get a good job, to be promoted, and certain professions, particularly those linked to politics and power, were nearly closed to them. But, somehow, the very same Jews were allowed to apply for emigration to Israel and often actually left the country, a privilege denied to other ethnic groups, Russians included. Besides the usual human inertia and fear of the unknown, a major reason that the entire Jewish population of the Soviet Union, about 3 million, didn't simply pack up and leave (a lot of them did) was straightforward: the possibility of being refused permission and the resulting legal limbo, including the loss of one's job, harassment or worse. The main causes of this policy by the Soviet state — discrimination and preferential treatment at the same time — are well understood: several centuries of anti-Semitism in Russia, agreed with and abetted by the authorities, but also their desire to curry favor with the West, in return for economic subsidies, direct or indirect.

I am non-religious, and know neither Hebrew nor Yiddish, if one doesn't count a smattering of Yiddish words, many of them a part of English slang. Reading about the history of Jews, I saw it as a singular one, but my connection to it was untinged by patriotic zeal. My appearance then did not identify me as a Jew right away. And so I was, from time to time, an uneasy listener to people who complained to me about "those Jews", or even vented their hatred of them. Fascinating experiences, for it was clear, especially in "hatred" cases, that one dealt with a person whose inner turmoil or rage, compounded by lack of introspection, has found a traditional, centuries-old outlet. This is not a treatise on the Russian brand of anti-Semitism, a beast very much alive. The example above is just an illustration of the unease that accompanied the lives of Russian Jews even in relatively benign times.

The idea of attempting to leave the country occurred to me rather late, a couple of years after graduation from Moscow University. I was engrossed in loving my profession and trying to succeed in it, kind of forgetting that a better way would be to change the place of one's pursuits. My marriage history didn't help either. It became even more checkered later on, until I got lucky, in 1990, after many years and less than happy marriages. My first wife was my girlfriend at high school. I was 19 when we got married. Our daughter Victoria was born two years later, in 1968. By 1970, we were divorced. It would be flattery to call my first marriage a caricature of the real thing. Both of us were immature, and selfish in ways that children, not adults, tend to be.

I received my Ph.D. in 1973, and continued to work at the Institute of Molecular Biology, where Georgiev, the lab's head, allowed me to supervise a graduate student and an occasional undergraduate. He was supportive in other ways as well. My subject was the structure and organization of chromosomes, then a mystery. This problem is huge, multifaceted, and remains a partially explored territory. In the early 1970s even "elementary" questions, about the path of DNA at the first level of its folding in chromosomes, and about the role that histones (DNA-associated structural proteins) play in this folding, were unanswered. Our work in those years contributed, in a minor way, to the problem's eventual solution by Roger Kornberg, who worked in England then.

My papers were published not only in Russian journals but in Western ones as well. I began to receive invitations to give talks abroad, but my attempts to receive permission to travel to Western countries were a waste of time. I was allowed, though, to visit other barracks of the "socialist camp". One of them was Bulgaria, the other German Democratic Republic (GDR). Three decades of rule by communists damaged but didn't destroy German industriousness. The GDR was unfree and much less prosperous than its Western counterpart, but its standard of living was still higher than Russia's. I could glimpse West Berlin across the Wall, but that was all I could do, being too macroscopic for teleportation.

Then a stroke of luck, at first unrecognized. In 1976, I received a letter from England, an invitation by Aaron Klug, an outstanding structural biologist, to give a talk at a symposium in London he was organizing. Knowing that permission to travel would be refused, I was about, then decided otherwise and requested an appointment with the Institute's director Vladimir Alexandrovich Engelhardt, a man in his eighties then, a member of the Soviet Academy and one of most highly positioned scientists in Russian biology. I didn't hope for much, and was unprepared for what I heard.

How did you manage to leave Russia?

Vladimir Engelhardt, the founder and first director of the Institute of Molecular Biology (IMB) of the then Soviet Academy of Sciences, in the 1960s.

Engelhardt, the founder and director of the IMB, knew me from the time of my joining Georgiev's lab in 1970, in part because my father headed another lab at the same institute. Georgiev spoke highly to Engelhardt about my work, or work habits (I spent days and nights in the lab), and he became interested. A relationship developed, utterly unequal but a real one. Engelhardt occasionally called me to his office and asked about ongoing work. I learned not to bore him with actual studies, for which his attention span was no longer equipped, and confined myself to just-so stories about chromosomes and stuff.

This time, I told Engelhardt about receiving an invitation from Aaron Klug to attend a symposium in London and give a talk there. He didn't let me finish. "I know Klug. Fine scientist, he. Hmm, and I haven't been to England for some time ... Are you the only one he invited from here?" I replied that I had no way of knowing for sure, but presumed that Klug invited just me. "Very well, then," said the old man and became as animated as I ever saw him to be. "Why don't we travel to London together, you and I?" "Vladimir Aleksandrovich, I would be delighted to, but do you think this idea is realistic?" "But of course," was the reply. "Let's begin by sending a cable to Klug, and inquire whether he can send a formal invitation to me as well. Once we receive that invitation, I'll see what can be done. Please wait." I did, expecting refusal. A few months later, in February 1977, I boarded the Soviet IL-62 for a flight from Moscow to London. I was in coach and Engelhardt way upfront, in first class.

I'll never know how he managed to obtain permission, from the directorate of KGB that handled such matters, for me to travel with him. My case was an open-and-shut one. A Jew, for starters. Divorced, i.e., morally unstable: if he left his marriage, he can leave the motherland too. Well known to the said directorate as someone who tells anti-Soviet jokes at Institute's parties. Besides, I didn't have hostages. In the former Soviet Union, and in places like North Korea today, a person trusted enough to travel abroad was expected to leave behind someone he would have difficulty parting with for good. Children, in a stable family, were best hostages. A wife or husband was a so-so hostage, but better than nothing. Parents were no hostages at all. By the lights of KGB, I shouldn't have been allowed to come close to that IL-62, let alone fly in it. It must have been Engelhardt himself, his limited but tangible influence that made the difference. Being the director of a major institute, he served apparatchiks above him. Apropos, Engelhardt was of German, not Jewish, descent, a plus in his dealings with those characters. He kept the furnace of loyalty hot and burning for years, and in return received occasional favors, such as taking his word that a nerd who didn't deserve to travel abroad and was a defection risk to boot, should be allowed to come with him.

A week-long jaunt to England was over fast, a bewildering experience. I flew back to Moscow with Engelhardt, feeling miserable and believing, on good grounds, that I'd lost my first and last chance of escape. The trip itself soon became a distant recollection, a collage of images. My talk at a symposium in London. Meeting scientists whose work I admired from afar. Traveling from London to Cambridge, where colleagues received me most kindly, and were generous with gifts of reagents and gadgets for benchwork. Walking in downtown London at night, marveling at the shop windows and profusion of lights in the streets. But never telling a policeman that I wished to ask for political asylum.

Why did you decide not to stay in England? What happened next?

Without telling a soul, I planned to defect in London, and was sure I would do so right after arrival. Then something happened to my resolve. The sight of an old man who was kind to me and seemed to trust me; his fragility; and my concern that he might just die from stress and disappointment if I defected were the main reason, an attack of altruism if you will. There was also a smidgeon of fright, a reluctance to cut the knot so abruptly and irrevocably. In 1977, the Soviet Union appeared to me, and to everyone I knew, as a Thousand Year Reich. Leaving it would guarantee my never seeing family and friends again, a price I thought was acceptable, then discovered it wasn't.

Returning to Moscow, I regaled acquaintances with tales of a life they never saw, but felt sad, and was sure I missed the only chance of escape. Then, a few weeks later, an utterly unexpected phone call: "Alexander Yakovlevich? Vladimir Leonidovich is my name. I'm a colonel at Komitet Gosudarstvennoi Bezopasnosty." (Hence the acronym KGB.) "How was your trip to London? Good impressions? Lots of science, I gather." He continued before I managed to reply. "My colleague and I would like to meet with you next week, if you don't have objections." I didn't. A few days later, I was going up in the elevator of the hotel Moskva, near Red Square, having been instructed to knock at the door of a room on the seventh floor. Two men stood up to greet me, one of them in his forties ("Vladimir Leonidovich"; he never told me his last name), and a younger one, who introduced himself as "senior lieutenant". Both wore civilian clothes.

Our conversation, details of which I remembered for years but never wrote down, was lurching from one irrelevant subject to another. I played along, knowing they didn't invite me to hear interminable tales, interrupted by insincere laughter, of their catching a huge pike or carp in the Oka river. Roughly a year later, when a guy from the CIA came to Cambridge, Massachusetts to debrief me at MIT, and heard that the room in question was at the Moskva hotel, he shook his head and said, "Yah, Yah. Seventh floor, a room with an oil painting of a woman on the wall near window, right?" "Absolutely right," I replied, dumbfounded by the CIA's colossal erudition.

Meanwhile, in that very room my camaraderie with the KGB grew in leaps and bounds. They finally got tired of describing fishing trips, and Vladimir Leonidovich suggested, insinuatingly, "I betcha you would like to travel abroad often, wouldn't you?" "Sure, who wouldn't," I replied, playing a level-headed fella, honest to a fault. "I enjoyed the visit to London, and my work at the Institute benefited from that trip." After another digression and burst of camaraderie, a proposition was advanced, simple and clear. "Look, Alexander Yakovlevich. We hear good things about you as a scientist. But not so good things about jokes at parties, ga-ga-ga!!!" Having recovered from mirth, he continued. (The meeting was long and not worth recounting in detail. I learned more about pikes and carps than I ever cared to know.)

"The Soviet Union and the entire socialist camp are surrounded by enemies. Counter-intelligence officers must know what the other side is up to. Genetic engineering — you know about that stuff, do you?" I nodded. "It may soon become an instrument in the hands of American military." To say "military", he used a quintessentially Soviet-Russian word, "voienschina": something to despise, but also to fear. "We should be on the lookout for these bastards' plans, to pre-empt them, and if necessary to develop countermeasures."

Countermeasures, my foot. From their remarks about molecular biology, in between guffaws and small talk, I knew that my counter-intelligence chums had a rather vague idea of what DNA was, but they bravely pretended otherwise. Their orders from above were probably clear: to find out what that new imperialist trick — genetic engineering — was. Not that anyone cared, with carps, vodka and fishing gear looming larger in their minds than adenine or cytosine, but reports had to be produced, *"competent scientists"* consulted, gathered intelligence properly recorded. The rusty machine creaked on, and invited me into its craw. It didn't occur to me then, but became clear later, that Paul Berg, David Baltimore and their colleagues who organized the 1975 Asilomar Conference, where concerns about possible dangers of genetic engineering were discussed in a public setting, have discombobulated not only the more impressionable amongst U.S. public officials, but apparently scared the KGB too! Thus, remarkably, my collaboration with the KGB that made the escape possible was helped by events far away, when the Asilomar conferees decided to discuss their pros and contras in public, with reverberations that reached Vladimir Leonidovich, the man without a last name.

Everything went swell that day. I told them, with a straight face, that my foremost duty as a Soviet citizen was to assist the counter-intelligence branch of "organs" (Russian slang for the KGB) in its valiant efforts. To do the job, one would have to travel to the West, naturally. That was fine, I was told, just fine, and we parted. This time, my good behavior on a trip would be vouched for by KGB apparatchiks I didn't care about, the mother of all understatements.

It so happened that before the visit to the Moskva hotel I received an invitation to give a talk at the international symposium on chromosomes in Helsinki, Finland. A few months rolled by. Near the end of August 1977, the senior lieutenant, in a gesture of seeing an agent off to a mission, gave me a lift to the railway station. The destination was Finland, a Western country but not a good place to defect in, I was told by an acquaintance. The gist of his warning was that the Finnish authorities were appeaseniks of the Russian bear at their border, and would eventually return a defector, after giving him another chance to escape, this time from a Finnish police precinct. Nowadays, throngs of people from Russia and other, now independent, states that comprised the former Soviet Union are attempting to resettle themselves in the U.S. and Europe. Unlike me in 1977, these people have no problem leaving their countries. Their difficulty is to be allowed to enter their destinations, in a world where a defector from Russia is an extinct species. Not so in 1977, when the West would welcome a person who managed to flee a Communist dictatorship. Such people were rare birds, and didn't overwhelm the generosity of the receiving country.

Whether or not the warning about Finland was actually correct I don't know to this day, but my taking that advice seriously made it necessary to cross, somehow, from Finland to Sweden, without a visa to the latter. But I didn't worry too much about it, because I counted on meeting a friend in Helsinki.

Who was that friend, and how did you escape? There are stories about your emigration, but there are also puzzles about it. How could you

From left to right: Robert Hoffman, Leonid Margolis, and Israel Gelfand, Moscow, 1977.

turn up in Germany, for example, when you had been let out to attend a meeting in Finland? Did you have the necessary visas?

A year before my trip to Finland, Robert Hoffman, a young American scientist, came to the IMB for a sojourn of several months, and began working there, sharing his time between IMB and a lab at Moscow University. He was the first American I met for more than a few minutes, a remarkable apparition. Gregarious, friendly, free, cracking jokes, learning new language and taking delight in four-letter words that Russian is justly famous for, Bob Hoffman was a breath of fresh air. It became obvious right away that Bob wasn't exactly a fan of the American political system. He described to me, a denizen of full-employment country, the problem of unemployment in the U.S. and other such nightmares. Sensing that he is far more intelligent than his naiveté (about to be cured) suggested, I pulled no punches in telling Bob what I thought about the worker's paradise he decided to explore. Just two weeks of Bob's exposure to realities of Soviet life produced a complete transformation. I was working at the bench when Bob burst into the room and shouted, mercifully in English: "Alik!" ("Alex" is "Alik" in Russian.) "Do you know you're living in a fascist country?!" Making sure we were alone in the room, I replied that I did, that now he knew it too, and if our neighbors on the floor didn't know it already they would have learned it this very instant.

It didn't take long for Bob and me to become close friends. We walked the streets around the IMB, discussing everything we could think of. I came to trust Bob unconditionally, and confided with him about the approach by the KGB, the impending trip to Helsinki, and my decision to escape from there to Sweden. Bob wholeheartedly approved, and suggested, with warmth and generosity of spirit — his particularly endearing trait — that he, too, would fly to Helsinki, but from Boston, to which he was about to return, being on leave from a lab at the Massachusetts General Hospital. Our idea was that Bob would meet me in Helsinki and help with the escape. The two conspirators promised each other to keep mum about their plans. I knew the dates of the symposium, and we decided, a la le Carré, that Bob would wait for me at the Helsinki's railway station every hour on the hour for three days in the row, and that I would try to find him there if I went to Finland, permission being in the hands of my KGB handlers and still uncertain at the time. Two friends in Moscow, Misha Evgenev and Lucya Ulitskaya, then husband and wife, also knew of my plans. Misha was a geneticist at an institute close to IMB. Lucya

Varshavsky with Michael Evgenev, a geneticist and close friend back in Russia, Moscow, 1977.

shepherded two little children then. She is a well known writer today, an outstanding one, actually. I trusted Misha and Lucya entirely, and left with them a few documents, such as my Ph.D. diploma, that I didn't dare to travel with, lest my suitcase be searched.

Bob Hoffman and I met again in Helsinki, exactly as we had planned. My talk at the symposium was scheduled for the first morning. In the afternoon of that day, there was a reception for participants, at a hall right across from the railway station. Just minutes after discussing science with Francis Crick and his colleague Ruth Kavenoff, I went to the railway station. Bob Hoffman, having crossed the ocean to meet me, was there all right, sitting on a bench. He was conspicuous not only because of his height. The collar of his raincoat was straight up despite good weather, courtesy of spy thrillers. We embraced, Bob looked around, checking for agents with machine guns (none showed up, having more interesting things to do), and we proceeded to concoct the escape plan. A large ferry crossed the Baltic sea between Helsinki and Stockholm every day. It turned out that Bob had flown from Boston to Stockholm, then took the ferry to Helsinki, and noticed that most passengers were not asked to show their passports, a blessed Scandinavian attitude. He would buy two tickets to the ferry from Helsinki to Stockholm, departing next morning, and we would travel together, on the assumption that a ticket checker would let me in without a visa.

The actual escape was nothing to write home about. A guard at the ferry, bored and indifferent, glanced at the ticket and waved me through.

Bob and I went to a cubicle in the ship's belly, stayed there and emerged when the ferry arrived at Stockholm. At Bob's suggestion, we took a taxi and drove straight to the U.S. embassy. Its security officer was courteous but not exactly thrilled by our feat. He needed a Russian defector like a hole in the head. "I'll ship you to our Consulate in Frankfurt, Germany," he said. "It's a big place, you know. There must be folks there who would assist you with getting a visa to the U.S." Hours later, we were in Frankfurt. The Consul was too busy to receive me. A Consulate's apparatchik, talking dishearteningly like apparatchiks I knew in the other life, explained that he could do nothing about the visa, but could send me to Rome, Italy, where lucky Jews who left the Soviet Union legally were cooling their heels, waiting (sometimes for months) for a visa to the U.S. Having escaped, I longed to begin scientific work as soon as possible, so Bob and I decided that I would stay for a few days in Frankfurt, trying to get an audience with the Consul. The next day, we had another idea. I called David Baltimore at MIT in Cambridge, Massachusetts. The call stemmed from a shaky hope that Baltimore might remember me from a few encounters at the USA–USSR symposium in Kiev, Ukraine two years before. David was well known, already then, for his co-discovery of reverse transcriptase, a contribution for which he received the 1975 Nobel Prize in Physiology or Medicine.

David remembered me, was friendly and gracious (in stark contrast to the gentleman at the Consulate), and suggested that I remain in Frankfurt for a few days, while he tried to find out whether anything constructive could be done about the visa. In the meantime, Bob Hoffman had to go back to his job in Boston. He gave me money and left. I stayed at a seedy (as I realized only later) hotel in the Frankfurt's red-light district, a memorable experience for a runaway from a country with puritanical sexual mores. An audience with the Consul didn't seem to be in the cards, but my mood couldn't be deflated by such a trifle. There was a Burger King nearby. I went there for breakfast, lunch and dinner (often followed by a second, late dinner), thinking that cheeseburgers and French fries belonged in the antechamber of Paradise, perhaps at the Place itself. One morning, a black limo made its way to the hotel along a narrow street. The Consulate's apparatchik emerged from it, and half an hour later I was in the presence of the Consul. He smiled at me most benignly, and announced that he had in his hands an airline ticket. First class, nonstop, all expenses paid, from Frankfurt to New York, and my U.S. visa too. The next time I flew first class was 15 years later. I wasn't told the cause of such a startling reversal of fortune, and didn't inquire, having assumed (correctly, as it turned out) that our phone call to David Baltimore was involved. Years afterward, I learned that David called the MIT office of Frank Press, a distinguished geophysicist who was, at the time, the science advisor to Jimmy Carter and had an office at the White House. My visa may have been cabled from an office traceable to that House. (I don't actually know.) That would account for the Consulate going overboard in sending me across the ocean immediately and in style, instead of in coach. Having arrived at New York's JFK, I flew to Boston. Bob Hoffman met me at the airport, and in no time at all I saw my first American apartment — Bob's — in Cambridge, near Harvard University. A nice place, with an unoccupied sofa that became my bed.

Our first priority was to send a couple of cables to Moscow's IMB. The joy of having escaped was weighted with concern about repercussions for my parents and sister in Moscow, and for Georgiev and Engelhardt as well. My cables, written together with Bob, were designed to convey an image of space cadet who didn't comprehend the irreversibility and hurtful seriousness of what he has done, and even "*hoped to return one day*". It didn't matter whether or not the self-portrait, in those cables, of a bumbling knucklehead was believed by the apparatchiks in charge of punishing people who remained behind. The incipient tiredness of the Soviet regime was reflected in its (relatively) laid-back attitude to transgressions perceived as unthreatening. The "evidence", in the form of cables, that I was a loony would give officials a formal pretext to be restrained in their penalties, if they preferred so. I was denounced, in due course, at the IMB's public meeting. Engelhardt remained the director. Georgiev received a formal reprimand. My father was at first expelled from membership in the Communist Party, but was reinstated later. That reinstatement saved his job as professor at the IMB. Everything went just about as well as it could, but only on the surface. Feelings were another matter. My mother and father were flabbergasted by my escape. It took years, and efforts on both sides, to overcome a divide that these events engendered. (My parents left Russia in 1991, long after me.)

You were hired by MIT soon after coming to Cambridge/Boston. How did it happen? How did you manage at the beginning?

Right after my arrival in Cambridge, in September 1977, the laboratory of Bob Hoffman at the Massachusetts General Hospital, across the river in Boston, became my second home. Bob and the lab's head Richard Erbe gave me a bench, and my work in the U.S. began, a dream realized. I was struck by the luxury of having disposable supplies, things like capillaries (precursors of tips in modern pipettes), clean napkins, little gadgets of all kinds, chemical reagents that were neither dirty nor difficult to obtain. In Moscow, I would have washed those capillaries with loving care, reusing them until the end of time.

A few days later Bob and I paid a visit to nearby MIT. We wished to thank David Baltimore for his help with the visa to the U.S. I also hoped to meet with Alex Rich, another outstanding MIT scientist who traveled to Russia and visited the IMB. David wasn't at MIT on that day, but Alex Rich was. He made a suggestion I didn't expect. "You traveled to Helsinki to attend a symposium," he said. "Ergo, you must have slides of your talk. Why don't you give us a seminar about your work in Moscow?" A few days later I gave that talk, and was told it went well. The results I described, while not particularly exciting, were genuinely new and spanned a broad range, from the organization of nucleosomes to the folding of SV40 viral minichromosomes. Unbeknownst to me, MIT's Biology Department was initiating a search for a junior faculty member in the area of chromosome structure. Although my field of work was appropriate for the planned appointment, the idea of offering that job to me must have been a difficult proposition, for I had fallen into Cambridge from the Moon just days before, utterly outside of a formal search.

A few days after the MIT seminar, I received a call from Cyrus Levinthal, of the Columbia University in New York. He asked how I was, and offered me a job of assistant professor at Columbia. (I don't know how that decision was made in New York.) As if the offer from Levinthal were not enough, I learned from Gene Brown, the genial chairman of MIT's Biology Department, that Francis Crick, whom I had met for less than an hour in Helsinki (he attended my talk there as well), has called Brown and inquired whether he could be of help in finding a job for me in the U.S. or U.K. A few days later, Gene Brown offered me the assistant professor's position at MIT, less than two months after my leaving Moscow. I said, "*I accept*" before Gene finished describing the offer.

A large room with seven lab benches and a small adjoining office was cleaned up and ready for me in no time at all. I moved there, and began working in my own lab, alone and happy, going to nearby laboratories when I needed equipment. Most "heavy" equipment at Moscow's IMB was imported from the West, so I knew how to use the instruments. Gene Brown signed my requisitions for consumable supplies, and told me that I would be able to buy equipment and hire personnel in a year or so, once I receive my first grant, which I was supposed to write and submit to the NIH as soon as possible. Since I didn't know about "start-up" funds for equipment and personnel that a newly hired faculty member was supposed to be given, I didn't request such funds, and received none. Learning of that omission several months later, I felt no less grateful to Brown and his colleagues, for I knew that hiring me must have been a gamble on the Department's part.

Colleagues at MIT, particularly Alex Rich, David Baltimore and Howard Green, were the source of help and moral support from the beginning. I can't thank them enough. Roger Kornberg, the discoverer of nucleosomes, who was then at the Harvard Medical School (I met him on the 1977 trip to London), contacted me soon after my arrival to Cambridge and was of tremendous, warm-hearted support. Robert Horvitz, a geneticist who studied nematode biology, began his work as an assistant professor at MIT almost simultaneously with me. His laboratory was across the corridor from my lab. Bob helped me with advice in more ways than one. His science went from strength to strength right away, and was a great example for me.

Bruce Alberts, whom I first met in the Soviet Union in 1975, was a professor at Princeton in the 1970s, before his move to the University of California at San Francisco. Bruce and I met again soon after my coming to the U.S. We have kept in touch ever since. My friendship with Bruce, his attitude to life and work had effects on me that I have difficulty putting into words. Looking at the Bruce's oeuvre, his discoveries in the lab and his involvement with the scientific community, I saw that one could be a first-rate scientist without being inward-bound. But my deck of genes wasn't Bruce's, and there was nothing I could do about it.

In the excitement of the initial months at MIT, I managed to forget that Bob Hoffman may need some privacy, as he eventually told me, in a most tactful way, seeing that I didn't grasp, as yet, the advantage of living in my own apartment, rather than his. He also explained the benefits of sleeping on a real sheet, with a real blanket, as distinguished from rags on Bob's sofa. Having heard this, I encountered yet another miracle in my new country, for it took me an hour, from start to finish, to rent a place in Cambridge. My apartment was a walk-up one-bedroom joint, in a creaky two-story house that was so infested with cockroaches (as I found later) that getting rid of them in my cubicles didn't make sense: the fallen were swiftly replaced by cousins from adjoining flats. I was taken aback, a little, by those cockroaches, as they were absent in my parents' Moscow apartment. But the nuisances of everyday life barely registered. Work was all I cared about, with the intensity even greater than in Moscow, where I managed to get married, divorced, went to parties and found other ways to waste my time. None of that in Cambridge, Massachusetts. The monk was determined, and more single-minded than Savonarola ever was.

I must have nursed a childish pride in my work ethics, as the following episode illustrates. One evening, soon after joining MIT, I was in the midst of an experiment that was to last through the night. Near the elevator, I ran into Boris Magasanik, a great yeast geneticist who worked late that day and was going home. Hearing that I was going to work in the lab all night, Boris inquired how often I did that. "As often as possible, twice a week," answered the happy simpleton. Boris' reply was terse: "That's dull." Having learned, later, of Magasanik's wide-ranging interests, I saw his reply, its lack of tact notwithstanding, as a disappointment in the narrowness of my mind. The immediate result of that encounter was my becoming a bit more worldly in what I said to colleagues. But not worldly enough. A day later, I told Gene Brown that I didn't wish to teach students. Having convinced myself that I was in a breathtakingly free country, where one can finally say what one thinks, I carried the license far and wide. Gene listened to my pronouncement, muttered something under his breath, then collected himself and told me, in an even voice, that things were very simple: if I refuse to teach, I will be fired. That information put a stop to refusals, but it took some time before the illusion of "totally free country" dissipated closer to reality's level. One day, in my first year at MIT, I had a conversation with Graham Walker, a fellow assistant professor and a nice man. Gram told me, in his gentle way, that he heard of my mutiny against teaching duties. "We were hired to teach as well, you know," he said. "And there's another thing to consider: tenure. Teaching figures significantly in tenure decisions. Way down the road, but still ..." I appreciated Gram's advice, and am grateful to the MIT folks for their correct perception of my post-escape self as being "drunk on freedom", as a colleague put it.

The dislike of teaching was not about teaching itself, for I understood its importance. I begrudged the time. No science could be done while delivering a lecture, and no learning either, with my exiting the classroom having the same knowledge I had possessed before entering. Worse than that, I grew excited, despite myself, while actually teaching. So the dervish was tired after a lecture, and had to catch his breath before returning to work, yet another delay. Later, having read about the life of Ludwig Wittgenstein, a semi-nutty philosopher (more than semi, actually), I felt great respect for, indeed identification with, the man's intensity. My respect did not extend to philosophy, the subject of Wittgenstein's labors. But the intensity, the utter immersion in one's work, to nearly the complete exclusion of everything else, was a quality for which Wittgenstein was a paragon. I approved of his way of living a life with all my heart, and had a private name for his and my phenotype: the Wittgenstein Syndrome. It can be illustrated by a passage in Bertrand Russell's memoir. Russell was a tutor, of sorts, to young Wittgenstein, a scion of wealthy Jewish family in Austria who renounced his inheritance, left Vienna and came to study with Russell in Cambridge, England. The story is about Russell, in the company of lady friends and Wittgenstein in tow, going to Cambridge river to watch a boat race. After a while, Russell saw an agitated Wittgenstein going back and forth on the embankment and muttering to himself. "What's up, Ludwig," called Russell. The man-child he addressed swung around and near-whispered, the indignation too great for a normal voice: "How can you, Bert, waste time on this meaningless, mind-numbing exercise, when the work remains undone?!" And he strode away. "Good for you, Ludwig," I thought, happy for Wittgenstein and for myself too, as I shared the attitude wholeheartedly.

How well did you adapt yourself to your new life, to the transition that was so abrupt? Were you well prepared for this job? Did you have to catch up? How did you readjust yourself from the Soviet system of science to the American one, with its competition for grants, etc.?

My first year at MIT, when I worked alone, could buy supplies anytime, and had access to equipment I needed, was happiness itself. I usually walked from my apartment to MIT in late morning, having returned home the night before as late as I could. I often sang aloud on the way to MIT, beginning another day of work I loved, with no bosses in sight and no shortages of equipment or reagents. I knew the experiments to do, or thought I knew, and didn't worry a single bit about grants, tenure and things of that sort. The reason was simple. Having escaped from a constricting professional life in Russia, I sensed my total commitment to the craft, which I did not perceive as a profession at all, but rather as the only thing I would ever wish to do, the only thing worth spending life on. Such a feeling, impossible to acquire at will, renders one much less vulnerable to usual career concerns. I knew there would always be a bench for me to work at. If I were good at what I did, a lab and co-workers would be there too. I resided, at last, in a place where excuses didn't apply: I could no longer blame anything or anyone but myself for professional failures, if such were to occur. My attempting to move to the West was propelled, in part, by desire to find such a place, and I did. That alone kept my spirits up.

The actress Barbara Hershey played in Martin Scorcese's movies, and said this a few years ago to Mark Singer, who was writing a piece about Scorcese: "... Who knows what talent is? ... I don't think talent is as rare as the need to express it or the strength to handle the rejection. I don't think Marty can help it; there is nothing else he can do with his life." Irrespective of what I think about Scorcese's films, Hershey's description of the phenotype is dead-on, including resilience in a world where failure is right around the corner. The accompanying cost — an obsessive, often narrow personality — is endemic amongst denizens of science and other competition-heavy fields. Marvin Minsky's remark sums up the downside: "If there's something you like very much, you should regard this not as you feeling good but as a kind of brain cancer, because it means that some small part of your mind has figured out how to turn off the other things."

I sensed, dimly at first, that I will eventually come to see the rest of life, its everyday's array, including entertainment, travel and even human relations, as too predictable, let alone disappointing, and would begin to distance myself, graciously if I could. This outlook was incipient then, for I was still too young for it, and was curious about the world I fell into. But my later selves kept gravitating to that premonition, with two exceptions. One was literature, both fiction and nonfiction. Finding and reading good books (and occasional schlock) gave me pleasure, laughter, understanding, and kept alive the suppressed desire to be a writer. I also had a pipe dream: meeting a woman who would become a wife as a soulmate, a person whose closeness would fulfill one's entire need for human contact. A hermitage of two, with total absence of inattention and selfishness that are the stuff of life and that I knew many marriages to suffer from. Improbably, that hope came true with my fourth marriage, to Vera, whom I met in the 1980s in

Vera and Alexander Varshavsky in 2001.

New York. She was a physician, and loved her profession even more than I did mine, if that was possible. Her patients were lucky people. We got married in 1990. Our closeness, trust and mutual dependence did not arrive in one day, and are a major blessing, for both of us.

The year 1978, my first year away from Russia, had a quality that never recurred, in part because I worked alone. Being perpetually short of time, I ate in a hurry, mostly at self-service MIT joints or at McDonalds and Burger Kings. Fast food was tasty enough. That regimen went on for a year, until late fall of 1978, when I fell ill, most likely from avitaminosis and other nutritional misbalances. My "diet" of cheeseburgers and French fries ignored fruits and vegetables. Save for a bout of infectious mononucleosis at 16, I was never seriously ill before, and was baffled by rapidly worsening health: aches in the joints, wracking cough, difficulty sleeping, and frequent colds. I was rescued by a kind woman named Elena Erez, also an émigré from Russia. She saw the cause of my condition better than I, guided me to healthier food, and cooked for both of us. At the beginning of 1979, she moved to my Cambridge apartment.

I soon recovered, and resumed the usual hard work at MIT. Elena and I were discovering, gradually, that ours wasn't a durable union when she started to feel unwell, near the end of 1979. After several months of uncertainty, an aggressive cancer, leiomyosarcoma, was found. At the time of diagnosis we were unmarried, but got married soon afterwards. Elena was educated (though not medically) and intelligent, but a diagnosis of cancer often diminishes one's rationality, especially if remedies are meager or nonexistent. I tried to hide the truth from Elena, describing spontaneous remissions and other optimistic stuff. But knowing, or at least suspecting, that hers was metastatic cancer with a poor prognosis, she wanted to try her luck with "healers" in the Philippines, who claimed to cure cancer and most other diseases by "operations" done with bare hands, without knives, as booklets that Elena received had described. The healers were also priests, or claimed to be such, and had their base of operations in Baguio, a city not far from Manila, the capital of Philippines.

We went there in early 1982, and stayed at a Baguio hotel owned by the racket that the "priests" ran. I expected charlatanism, but was still unprepared for its brazenness, and for the willingness of patients to be duped by low-grade magician's tricks. The patients were ill adults, like Elena, or parents with sick children, paying for every "operation", of which there were going to be "many". Unable to offer Elena a cure from her illness, I felt obliged to keep my view of the place to myself. Elena enjoyed living at a tropical resort and seemed not to notice that healers were mountebanks. After two weeks in Baguio, I had to go back to MIT, while Elena stayed at the hotel, where she preferred to be, saying that "treatments" were helping her. Three months later, I flew to Philippines again, and brought Elena home. She knew that her condition was getting worse. My mantra about spontaneous remissions, while politely listened to, was probably no longer believed. A course of utterly useless chemotherapy at the Boston's Beth Israel Hospital ensued, recommended to Elena by oncologists who (I bet) knew as well as I did that their witch's brew wasn't any better than "operations" in Philippines. Elena died in October 1982.

I kept working throughout that time, to the extent I could. By then, I had a functioning laboratory. Back in 1978, I had learned how to apply for grants. It wasn't difficult, just time-consuming. The competition for grants, while considerable, wasn't as depressingly cutthroat as it is today. By 1982, I had received two NIH grants, one for studies of eukaryotic chromosomes, the other for work with circular SV40 viral "minichromosomes", which served as a model of vastly larger (and linear) chromosomes of mammalian cells.

How did you manage at the beginning? What studies did you do during your first years at MIT? How successful was your project, or perhaps projects?

In the first year, from late 1977 to summer of 1978, I worked alone and continued studies that began back in Moscow. I used the SV40 minichromosomes isolated from virus-infected mammalian cells as models of cellular chromosomes, and tried to address the problem of nucleosome arrangement. Were nucleosomes distributed in a pattern that was specific vis-á-vis DNA sequence? Or was the arrangement quasi-random, in addition to being dynamic? (Nucleosomes are repeating superhelical turns of DNA wrapped around the oligomeric structural proteins called histones, with the adjacent nucleosomes connected by DNA segments called linkers.) To reduce potential nucleosome "sliding" (nothing was known about it at the time), I "fixed" isolated minichromosomes with the crosslinker formaldehyde, then treated them with restriction endonucleases, which cut SV40 DNA either once or at multiple specific sites. At first I learned little, but later saw that one site in the minichromosome was much more susceptible to cleavage than any other site. Remarkably, that single restriction site resided in the most "interesting" region of SV40, its origin of replication, an area of ~ 400 base pairs (bp) that also housed transcriptional promoters. Soon thereafter, an analogous experiment with the multiply cutting endonuclease *HaeIII* hit the jackpot: the above ~ 400 bp region could be "excised" from formaldehyde-fixed minichromosomes as a single fragment of histone-free DNA, in contrast to the rest of the minichromosome, which was still an intramolecular aggregate, held together by formaldehyde-produced DNA-histone and histone-histone crosslinks.

This and related advances yielded two insights: that the control region of SV40 minichromosomes was strikingly more exposed to endonuclease attack, and also that nucleosomes were either absent from that region or were in a configuration noncanonical enough to preclude histone-DNA crosslinks. These discoveries have become a major part of the modern understanding of chromosome organization, because later work, by us and many other labs, has shown that the exposed (nuclease-hypersensitive) regions, which allow access to DNA in the otherwise tightly coiled chromosomal fibers, are the universal feature of chromosomes at replication origins, transcriptional promoters, and other functionally important sites. The use of formaldehyde in that 1978–1979 work, which stemmed from my Moscow studies with Georgii Georgiev and Yuri Ilyin, was the precursor of later formaldehyde-based studies in my lab that led to the invention in 1988, by Mark Solomon, Pamela Larsen and myself, of a method for detection of the *in vivo* locations of specific chromosome-associated proteins. This technique,[1] called the chromatin immunoprecipitation (CHIP) assay, has become a key method for mapping and dissecting the interactions of chromosomal proteins with DNA *in vivo*. Various incarnations of the CHIP assay,[1] including its most recent, genome-scale applications, are revealing the dynamic organization of multiprotein structures that assemble on transcriptional promoters, replication origins, and other functionally active sites in chromosomes.

In late spring of 1978, when I was completing the work that revealed nuclease-hypersensitive regions in chromosomes, two MIT graduate students, Olof Sundin and Michael Bohn, had joined the lab and the project. We published the first results in *Nucleic Acid Research* in 1978 and a more detailed account in 1979 in *Cell*, then a 5-year old publication founded and edited by Benjamin Lewin that had already become a leading journal in molecular biology and related fields. Two other groups, Carl Wu and Sarah Elgin at Harvard, and Walter Scott at the University of Florida, independently discovered nuclease-hypersensitive regions in chromosomes, using a different approach that involved nonspecific nucleases such as DNase I.

Bohn soon left MIT for a medical school, while Sundin and I continued working with SV40 minichromosomes. One day in late 1978, I was reading a paper on SV40, and noticed a faint "ladder" of bands of electrophoretically fractionated SV40 DNA. The paper's authors didn't comment on the "ladder". I got a wrong idea, at first, of what those bands might be, but even that idea was exciting enough to suggest to Sundin that we should try to establish whether the "ladder" was for real, and if it was, to figure out its nature and significance. Neither of us suspected that we were beginning a 3-year study that would lead, in 1980–1981, to a fundamental discovery: the first and universal pathway of chromosome segregation at the level of DNA.[2,3]

Briefly: when a circular chromosome such as SV40 begins its replication, two replication forks run from the origin of replication in opposite directions, meeting halfway around the circle and leaving behind two daughter minichromosomes. Analogous processes take place during replication of larger and linear chromosomes containing multiple origins of replication, except that a replication fork meets a fork running "toward" it from the adjacent origin. (In a mechanistically distinct but topologically equivalent model, it is the chromosomal fiber that moves, with replication forks being spatially fixed in the nucleus.) These pictures of chromosome replication had a problem that wasn't even recognized as such at the time:

how do the two converging replication forks (large nucleoprotein structures containing polymerases, helicases and other proteins) replicate the last several hundreds of nucleotide pairs that the forks themselves occupy? We discovered, in part through the invention of high-resolution, two-dimensional electrophoretic techniques for analyzing DNA replication intermediates (these methods are still employed in the field), that before the replicated daughter chromosomes segregated to yield individual circles, they went through a remarkable "topological" dance of being, at first, wound around each other as *multiply intertwined catenated DNA*,[2,3] a new form of DNA at the time, since only singly intertwined catenated DNA circles were detected before 1980.

We had shown that the transition from two circles of daughter chromosomes that are still linked through a remaining (short) DNA duplex to two separate circles proceeds through a set of intermediate structures, dimeric multicatenanes in which the two daughter circles intertwine around each more than 30 times. *In vivo*, under normal conditions, these essential intermediates are rapidly processed, one intertwining at a time, through the action of enzyme called topoisomerase II (topo II). We also found that while decatenation was going on, other enzymes, including DNA ligases, were filling in and sealing the initially gapped or nicked daughter DNA duplexes into uninterrupted (covalently closed) circular DNA. *In vivo*, the two processes could be shown to take place at the same time, so that a population of late replication intermediates was a dynamic "matrix" of many structures (distinguishable by our electrophoretic methods), with different (nicked or closed) states of individual circles and different levels of catenations (intertwinings) in the topologically linked replicated chromosomes. All of these structures, previously unseen and not even suspected to exist, rapidly converged *in vivo* to the final state: two covalently closed, separate chromosome circles, or two separate linear chromosomes, as shown later by other groups. Many subsequent studies have demonstrated that the chromosome segregation pathway we discovered in 1980–1981 with the SV40 minichromosomes[2,3] was both essential and universal, operating in all organisms, from eukaryotes to prokaryotes.

At the time of our work on the multicatenane-mediated chromosome segregation, topo II enzymes were a novelty, having been discovered, in the form of DNA "gyrase", by the Martin Gellert's laboratory in late 1970s, and characterized by his and other labs, notably by those of James Wang, Bruce Alberts and Nicholas Cozzarelli. One aspect of our insight was that topo II was now expected to be essential for decatenating *multiply intertwined daughter chromosomal fibers* that formed at the final stages of chromosome replication and functioned as key replication intermediates.[2,3] That topo II was indeed essential for segregating chromosomes through the decatenation pathway was shown around 1985 by Connie Holm and David Botstein, then at MIT, by Rolf Sternglanz and co-workers at the University of New York at Stony Brook, and by Mitsuhiro Yanagida at the Kyoto University, Japan.

The segregation of chromosomes at mitosis involves two fundamental processes, acting together: one is the *decatenation of multiply intertwined daughter chromatids*, including those at the centromere of a mitotic chromosome.[2,3] The other is the *physical separation of sister chromatids*, pulled to the opposite poles by the spindle's microtubules. A critical part of the second segregation pathway was identified many years later, in 1999, by the laboratory of Kim Nasmyth, through the finding that a specific protease, termed separase, cleaved a subunit of oligomeric protein called cohesin (its molecules hold the sister chromatids together), thereby allowing the separation of sister chromatids,[4] *provided that their multiple DNA catenations had been resolved by the first segregation pathway*, discovered by us 25 years ago.[2,3]

Having been gradually swamped by ubiquitin studies in the lab (they began in 1978), I did not continue chromosome segregation work after 1983, and did not expect to return to that field, which grew from our elucidation of the first chromosome segregation pathway[2,3] into a major arena that encompasses both the mechanics of segregation and its regulation. But fate held a surprise. In 1999, Hai Rao (a postdoctoral fellow) and I saw, in a paper by the Nasmyth lab, that a fragment of separase-cleaved cohesin subunit bore N-terminal arginine, which our previous work had shown to be a degradation signal in short-lived proteins, recognized by the ubiquitin-dependent N-end rule pathway. Hai and I wished to determine whether that fragment of cohesin was in fact short-lived *in vivo*, and if so whether its degradation was functionally important. In 2000–2001, our collaboration with Frank Uhlmann (then a postdoc in the Nasmyth lab) and Kim Nasmyth demonstrated that the N-end rule pathway indeed targeted the cohesin's fragment for degradation.[5] Crucially, this degradation was shown to be *essential* for the proper functioning of cohesin machinery and high-fidelity chromosome segregation.[5] Other investigators, particularly Douglas Koshland, had previously found that "upstream" events, including the activation of separase, are also regulated by the ubiquitin system.

Thus, my laboratory's studies of chromosome segregation ended up to underlie the understanding of this fundamental process at three levels: through the discovery, in 1980–1981, of the first (DNA-based) chromosome segregation pathway, which involves the *decatenation of multiply intertwined daughter chromosomes*[2,3]; through the discovery, in 1984–1988, of *the essential role of ubiquitin conjugation in the cell cycle*[5,6,14] (see below); and through the discovery, in 2001, that high-fidelity chromosome segregation requires the *destruction, by the N-end rule pathway, of separase-produced cohesin's fragment.*[5]

Other non-ubiquitin work of the early years included my finding, in 1981, that growth factors, such as hormones or tumor promoters, can strongly increase the *frequency* of gene amplification in mammalian cells under conditions of cytotoxic stress. (The lab was small then, and I could still work at the bench.) The phenomenon of gene amplification was discovered in 1978, by Frederic Alt and Robert Schimke. What I found was that this process could be greatly accelerated under certain conditions, including those mentioned above. Analogous gene amplification events contribute to rapid evolution of cancer cells in a tumor, and to the emergence of drug-resistant cells during anti-cancer therapy. Thea Tlsty and Robert Schimke independently discovered, also in 1981, the same phenomenon of induced (accelerated) gene amplification.

In 1984, Francois Strauss (then a postdoc in the lab) and I demonstrated that the previously-developed (by Donald Crothers and Arnold Revzin) gel shift assay, which until then was used for studies of purified DNA-binding proteins such as Lac repressor, could be employed to detect specific DNA-binding proteins in the presence of other DNA-binding proteins, including nonspecifically binding ones. That 1984 work[4] converted the gel shift assay into an exceptionally powerful method for detecting specific DNA-binding proteins in crude extracts. Since then, this method has become a major tool in studies of gene expression.

Ubiquitin, a small protein universal amongst eukaryotes, entered my life in 1977. Ubiquitin studies in the lab initially competed with other projects, some of which are described above. The situation changed abruptly in 1980–1981, when I saw that there might be a genetic route to discovering the *biological significance and specific functions* of the ubiquitin-dependent proteolysis. This proteolysis had just been demonstrated by Avram Hershko's laboratory in Israel, in experiments with cell-free systems and isolated enzymes.[6,7] My laboratory's ubiquitin and non-ubiquitin studies continued in parallel for several years afterwards, with ubiquitin gradually taking over the entire lab.

Was there any difficulty with receiving tenure at MIT?

I was granted tenure in 1982, largely on the basis of our non-ubiquitin contributions, some of which are described above. By that time, the lab was still small (if I recall correctly, 4 or 5 people), but reasonably well established. The work that led, by 1984, to the first biological breakthrough in the understanding of the ubiquitin system began in 1981, by me and Daniel Finley, then a graduate student. (Other ubiquitin research by the lab started in 1978, but until 1981 it did not involve proteolysis.) Having been interested in chromosome studies, I had difficulty leaving them, but saw that the lab would have to do it if I was committed to follow my hunch in 1980 that the ubiquitin system (then an interesting *in vitro* finding, undefined biologically) was likely to be both complex and uncommonly multifunctional.

How and why did you begin the ubiquitin work? What was key insight or insights? How did it develop in your lab?

There were few similarities between my Moscow milieu and the astonishing new life. The libraries were one of them. They were just as quiet and pleasant in Cambridge as in Moscow, and a library at MIT soon became my second home. Reading there in late 1977, I came across a curious paper, of the same year, by Ira Godknopf and Harris Busch. They found a DNA-associated protein that had one C-terminus but two N-termini, an unprecedented structure. The short arm of that Y-shaped protein was joined, through its C-terminus, to an internal lysine of histone H2A. The short arm was soon identified, by Margaret Dayhoff, as ubiquitin (Ub), a universally present protein of unknown function that was described (as a free protein) by Gideon Goldstein and colleagues in 1975.

I got interested in that first ubiquitin conjugate, Ub-H2A. Back in Russia, I had begun to develop a method for high-resolution analysis of nucleosomes, based on electrophoresis of DNA-protein complexes in a low-ionic-strength polyacrylamide gel, a forerunner of the gel shift assay (see above). At MIT, my first postdoc Louis Levinger and I developed this method further in 1978–1982, by adding the second-dimension electrophoresis of either DNA or proteins, and mapping the spots of fractionated

DNA by southern hybridization. We located Ub-H2A in a subset of the nucleosomes, succeeded in separating these nucleosomes from those lacking Ub-H2A, and showed that ubiquitin-containing nucleosomes were enriched on transcribed genes and absent from transcriptionally inactive regions such as centromeric heterochromatin.

In the meantime, Avram Hershko, his graduate student Aaron Ciechanover and their colleagues in the Hershko laboratory at the Technion (Israel) were studying ATP-dependent protein degradation in extracts from rabbit reticulocytes. In 1978–1980, they found that a small protein, termed APF1 (ATP-dependent proteolytic factor 1), was covalently conjugated to proteins before their degradation in the extract. In 1980, they suggested that a protein-linked APF1 served as a signal for a downstream protease, and began dissecting the enzymology of APF1 conjugation. In 1981–1985, through the elegant use of biochemical fractionations and enzymology, Hershko and co-workers identified a set of three enzymes involved, termed E1 (ubiquitin-activating enzyme), E2 (ubiquitin carrier protein or ubiquitin-conjugating enzyme) and E3 (an accessory component that appeared to confer specificity on E2). Although our studies of ubiquitin in chromosomes began in 1978, I didn't know about the 1978 APF1 paper by Hershko and co-workers, since the identity of APF1 and ubiquitin was unknown, at the time, to them as well. The disposition changed in 1980, when APF1 and ubiquitin were shown to be the same protein, by Keith Wilkinson, Michael Urban and Arthur Haas, who worked in the lab of Irwin Rose, a collaborator of Hershko during his sojourns at Philadelphia's Fox Chase Cancer Center.

When I read the 1980 papers by Hershko *et al.* and Wilkinson *et al.* that described, respectively, the APF1 conjugation and the identity of APF1 and ubiquitin, two previously independent realms, protein degradation and chromatin-associated ubiquitin, came together for me, suggesting a regulatory system of great complexity and broad, still to be discovered biological functions. I decided to find genetic approaches to the entire problem, because a system of such complexity was unlikely to be understood through biochemistry alone. In 1980, reverse genetic techniques were about to become feasible with the yeast *Saccharomyces cerevisiae*, but were still a decade away in mammalian genetics. I kept reading, as widely as I could. Near the end of 1980, I came across a paper by M. Yamada and colleagues that described a conditionally lethal, temperature-sensitive mouse cell line called ts85. The researchers showed that a specific nuclear protein disappeared at elevated temperatures from ts85 cells, and suggested that this protein may be Ub-H2A. Glancing at their data, I had to calm down to continue reading, being virtually certain that the protein was Ub-H2A: in the preceding two years we had learned much about electrophoretic properties of this ubiquitin conjugate. On the hunch that mouse ts85 cells might be a mutant in a component of the ubiquitin system, I wrote to Yamada, and received from him, in 1981, both ts85 and the parental ("wild-type") cell line.

Daniel Finley, then a graduate student, joined my lab at that time, to study regulation of gene expression. He didn't need much convincing to switch to ts85 cells. A few months into the project, Finley and I made the critical observation that ubiquitin conjugation in an extract from ts85 cells was temperature-sensitive, in contrast to an extract from parental cells. While this was going on, I met Ciechanover, who came from the Hershko laboratory in Israel for a postdoctoral stint in the MIT lab of Harvey Lodish, and was studying growth factor receptors. Presuming that Ciechanover was still interested in ubiquitin (very few people were), I told him about our results with ts85 cells, and invited him to join, part-time, with Finley and me to complete the ts85 study. Ciechanover did, the work continued, and in 1984 we submitted two papers that described, primarily, the following discoveries: (i) mouse ts85 cells have a temperature-sensitive ubiquitin-activating (E1) enzyme; and (ii) these cells, in contrast to their wild-type counterpart, stop degrading the bulk of their short-lived proteins at nonpermissive temperature.[5,6]

This was the first evidence that ubiquitin conjugation was required for protein degradation *in vivo*. (The earlier studies by Hershko and co-workers were done with cell-free systems.) The results with ts85 cells also indicated that ubiquitin conjugation was essential for cell viability, the first hint of the enormous, many-sided biological importance of the ubiquitin system. In addition, ts85 cells were preferentially arrested in the G2 phase of the cell cycle, and the synthesis of heat-stress proteins was strongly induced in these cells at the nonpermissive temperature, suggesting that ubiquitin conjugation was involved in the cell cycle progression and stress responses.[5,6] In 1983, Tim Hunt and colleagues discovered unusual proteins in sea urchin and clam embryos. These proteins, which they called cyclins, were degraded during the exit from mitosis. We suggested in 1984[5,6] that cyclins were destroyed by the ubiquitin system, a hypothesis shown to be correct in 1991, by Michael Glotzer, Andrew Murray and Marc Kirschner, and independently by Hershko and co-workers as well.

It may be helpful to place the above advance in historical context. Despite some hints to the contrary, until the 1980s and the two 1984 *Cell* papers,[5,6] the prevailing view was that intracellular protein degradation was a simple and even mundane process, serving largely to dispose of "aged" or otherwise damaged proteins. Cellular regulation was believed to be a separate affair, mediated primarily by repressors and activators of gene expression, which were assumed, often tacitly, to be long-lived. Among the reason for this lopsided perspective was the difficulty of connecting the long-recognized proteolytic system in the lysosomes to specific pathways of intracellular regulation. Thus, most people studying gene expression in the 1960s and 1970s assumed that the regulatory circuits they cared about did not involve short-lived proteins. As we know now, just the opposite proved true, especially in eukaryotes, where most regulators of transcription are conditionally short-lived proteins whose levels in a cell are determined at least as much by the rates of their ubiquitin-dependent destruction as by the rates of their synthesis. Ironically, the first physiological (as distinguished from artificial) substrate of the ubiquitin system was a transcriptional regulator, Matα2, which Mark Hochstrasser (then a postdoc) and I demonstrated in 1990 to be short-lived *in vivo*, and delineated its degradation signal.[7] As mentioned above, a mitotic cyclin was the second such substrate identified, in 1991.

In addition to having been a breakthrough that indicated the importance, indeed the requirement, of the ubiquitin system for intracellular proteolysis, cell viability and cell cycle progression, the ts85 papers were also the first instance of a study that addressed the *in vivo* workings of this system. In 2004, this pair of 1984 papers[5,6] was selected for re-publication by the editors of *Cell* as being amongst the most important papers that have been published in the *Cell*'s 30-year history. In a review accompanying re-publication, Cecile Pickart, one of the early pioneers in the ubiquitin field, summed up the papers' contribution: "*The two papers ... led to a new world-view; not only was the ubiquitin/proteasome pathway a major proteolytic mechanism in the average mammalian cell, but it was also likely to regulate cell cycle progression. These conclusions are so well accepted today that it is difficult to appreciate the magnitude of their impact at the time the two papers appeared.*"[8]

Although the ts85 discoveries left little doubt, among the optimists, about the importance of the ubiquitin system in cellular physiology, it was difficult to extend these findings in the same system, owing to limitations of mammalian somatic cell genetics, which was hampered at that time by the impossibility of altering genes at will. In addition, the advances with ts85 cells produced little more than hints about *specific* physiological functions of the ubiquitin system, and also did not address another fundamental problem: the *source of specificity* of ubiquitin conjugation, i.e., the existence and structure of degradation signals, the features of proteins that make them the targets for ubiquitylation.

Therefore in 1983, even before the completion of ts85 work, Dan Finley and I, together with other colleagues in the lab, began systematic analysis of the ubiquitin system in the genetically tractable yeast *S. cerevisiae*, a project that soon expanded to occupy the entire laboratory. Between 1983 and 1990, this work revealed the first *specific* biological functions of ubiquitin conjugation. (Our ts85 results[5,6] demonstrated the importance of the ubiquitin system in general physiological terms, such as the overall *in vivo* proteolysis and cell viability, but only hinted at more specific functions.) Briefly mentioned below are key advances of those early years that established the physiological fundamentals of the ubiquitin field.

In 1984, Engin Özkaynak, Finley and I cloned the first ubiquitin gene, and found it to encode a polyubiquitin precursor protein.[9] By 1987, we showed that this gene, UBI4, was strongly induced by a variety of stresses, and that a deletion of UBI4 resulted in stress-hypersensitive cells.[10] These genetically based results validated and deepened the earlier indirect evidence with mouse ts85 cells,[5] thereby establishing one broad and essential function of the ubiquitin system.

In 1986, Andreas Bachmair, Finley and I discovered, through the invention of the ubiquitin fusion technique, the first degradation signals (degrons) that target proteins for ubiquitin conjugation and proteolysis.[11] By revealing the *basis of specificity* of intracellular protein degradation, this critical advance has spawned the field of degradation signals, a major arena of current research. One set of degrons discovered in 1986 gives rise to the N-end rule, a relation between the *in vivo* half-life of a protein and the identity of its N-terminal residue.[11] The seemingly simple N-end rule is underlied by the remarkably complex N-end rule pathway, whose functional and mechanistic understanding has gradually become a major project in the lab. The N-end rule pathway is still a focus of our work, surprising us by what it has up its sleeve, including its functions, which continue to emerge.[12]

In 1987, Stefan Jentsch, John McGrath and I discovered that RAD6, a protein known to yeast geneticists as an essential component of DNA repair pathways, was a ubiquitin-conjugating (E2) enzyme, the first such enzyme to mediate a specific physiological function.[13] We noticed that the sequence of RAD6 was weakly similar to that of CDC34, an essential cell cycle regulator (of unknown enzymatic activity) defined genetically by Leland Hartwell. In 1988, a collaboration between Breck Byers's and my laboratories demonstrated that CDC34 was indeed a ubiquitin-conjugating enzyme.[14] This discovery produced the first definitive evidence for a function of the ubiquitin system in cell cycle control, a role suggested but not proved by our earlier ts85 studies.

In 1989, Dan Finley, Bonnie Bartel and I discovered the functions of the other yeast ubiquitin genes, UBI1–UBI3, which were shown to encode fusions of ubiquitin to one protein of the large ribosomal subunit and one protein of the small ribosomal subunit,[15] an arrangement conserved from yeast to mammals. *In vivo* experiments with mutationally altered yeast UBI proteins indicated that the presence of ubiquitin in front of a ribosomal protein moiety, despite being transient *in vivo*, was required for the efficient biogenesis of ribosomes. Remarkably, ubiquitin acts, in these settings, not as a degradation signal but as a cotranslational chaperone. This first nonproteolytic function of ubiquitin, mediated by its fusions to ribosomal proteins,[15] appeared to be an exceptional case until years later, when Linda Hicke and Howard Riezman demonstrated that ubiquitylation of a plasma membrane-embedded receptor signals its endocytosis. Ubiquitin is now recognized to have numerous nonproteolytic functions.

In 1989, Vincent Chau and other colleagues in my laboratory discovered that ubiquitin conjugation results in a *polyubiquitin* chain of unique topology, with links between adjacent ubiquitin moieties through a specific lysine residue of ubiquitin.[16] We also showed that a substrate-linked polyubiquitin chain was essential for the substrate's degradation by the proteasome,[16] yet another beginning of what, nowadays, is a major arena of ubiquitin studies.

In 1990, Bonnie Bartel, Ingrid Wünning and I, through the use of genetic and biochemical approaches, cloned and characterized the first specific E3 ubiquitin ligase, UBR1, the *S. cerevisiae* E3 of the N-end rule pathway.[17] Many more E3 enzymes, whose mechanistic functions include the recognition of specific degradation signals in targeted proteins, have been identified in 1990s and later, a process of discovery that continues as I write, in part because the number of distinct E3 ubiquitin ligases in a mammal is estimated, at present, to exceed a thousand.

A key feature of the ubiquitin-dependent protein degradation is subunit selectivity, i.e., the ability of the ubiquitin system to eliminate one subunit of an oligomeric protein or a multiprotein complex, leaving intact the rest of it and thereby making possible *protein remodeling*. This fundamental capability was discovered in 1990 by Erica Johnson, David Gonda, and myself, in the context of the N-end rule pathway.[18] Also in 1990, Mark Hochstrasser and I detected subunit selectivity in the degradation of Matα2 (see above), the first physiological substrate of the ubiquitin system.[7] Subunit-selective proteolysis is one of the most fundamental capabilities of the ubiquitin system, a feature both powerful and flexible, in that it enables protein degradation to be wielded as an instrument of protein remodeling for either positive or negative control. Among many examples are activation of a major transcription factor NF-κB via degradation of its inhibitory subunit IκB, and inactivation of cyclin-dependent kinases (which drive the cell cycle oscillator) via degradation of their regulatory cyclin subunits.

In summary, the complementary discoveries in the 1980s by Avram Hershko and co-workers, and by my laboratory, then at MIT, revealed three sets of previously unknown facts:

(1) That the ATP-dependent protein degradation involves a new protein modification, ubiquitin conjugation, which is mediated by specific enzymes, termed E1, E2 and E3.
(2) That the selectivity of ubiquitin conjugation is determined by specific degradation signals (degrons) in short-lived proteins, including the degrons that give rise to the N-end rule.[11]
(3) That ubiquitin-dependent processes play a strikingly broad, previously unsuspected part in cellular physiology, primarily by controlling the *in vivo* levels of specific proteins. Ubiquitin conjugation was demonstrated to be required for the protein degradation *in vivo*,[5,6] for cell viability, and also — more specifically — for DNA repair,[13] the cell cycle,[14] protein synthesis,[15] transcriptional regulation,[7] and stress responses.[9,10] In addition, ubiquitin-dependent proteolysis was discovered to involve a substrate-linked polyubiquitin chain of unique topology that is required for protein degradation.[16] The ubiquitin system was also discovered to possess the critically important property of subunit selectivity, i.e., the ability to destroy a specific subunit of oligomeric protein, leaving intact the rest of it and thereby making possible *protein remodeling*.[18]

The Hershko laboratory produced the first of these fundamental advances (item 1), and my laboratory produced the other two (items 2 and 3). Over

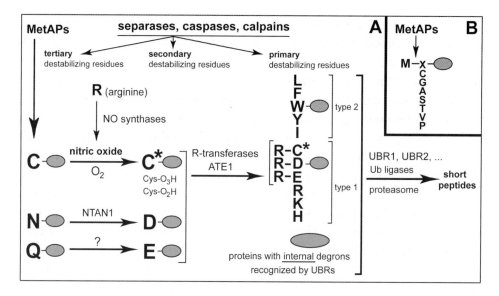

(A) The N-end rule pathway in mammals. This proteolytic pathway was the first specific pathway of the ubiquitin system to be discovered, initially in yeast.[11] It is present in all eukaryotes examined, from fungi to animals and plants. Although prokaryotes lack ubiquitin conjugation and ubiquitin itself, they, too, contain the N-end rule pathway, a ubiquitin-independent version of it.[19] Studies of this pathway, its mechanisms and functions, have gradually become a major focus of my laboratory. N-terminal residues are indicated by single-letter abbreviations for amino acids. The ovals denote the rest of a protein substrate. MetAPs, methionine aminopeptidases. The "cysteine" (Cys) sector, in the upper left corner, describes the recent discovery of a nitric oxide (NO)-mediated oxidation of N-terminal Cys, with subsequent arginylation of oxidized Cys by ATE1-encoded isoforms of Arg-tRNA-protein transferase (R-transferase).[12] This advance identified the N-end rule pathway as a new kind of NO sensor. C* denotes oxidized Cys, either Cys-sulfinic acid ($CysO_2(H)$) or Cys-sulfonic acid ($CysO_3(H)$). The type 1 and type 2 primary destabilizing N-terminal residues are recognized by multiple E3 ubiquitin ligases of the N-end rule pathway, including UBR1 and UBR2. Through their other substrate-binding sites, these E3 enzymes also recognize internal (non-N-terminal) degradation signals (degrons) in other substrates of the N-end rule pathway, denoted by a larger oval.
(B) MetAPs remove Met from the N-terminus of a polypeptide if the residue at position 2 belongs to the set of residues shown.

the last 15 years, these complementary "chemical" and "biological" discoveries in the 1980s caused the enormous expansion of the ubiquitin field, which became one of the largest arenas in biomedical science, the point of convergence of many disparate disciplines. Our biological discoveries,[5-18] together with later studies by many excellent laboratories that entered the field in the 1990s, have yielded the modern paradigm of the central importance of regulated proteolysis for the control of the levels of specific proteins *in vivo*, as distinguished from their control by transcription and protein synthesis. In other words, these advances revealed that the *control through regulated protein degradation rivals, and often surpasses in significance, classical regulation through transcription and translation*. This radically changed understanding of the logic of biological circuits will have (in fact, is already having) a major impact on medicine, given the astounding functional range of the ubiquitin system and the multitude of ways in which ubiquitin-dependent processes can malfunction in disease or in the course of aging, from cancer and neurodegenerative syndromes to perturbations of immunity and many other illnesses, including birth defects. A number of pharmaceutical companies are developing compounds that target specific components of the ubiquitin system. The fruits of their labors have already become, or will soon become, clinically useful drugs. Efforts in this area may yield not only "conventional" inhibitors or activators of enzymes but also more sophisticated drugs that will direct the ubiquitin system to target, destroy, and thereby inhibit functionally any specific protein. I feel privileged having been able to contribute to the birth of this field, and to partake in its later development. The dynamism and surprises of this endeavor remain undiminished even today, two decades after the 1980s.

You have received, jointly with Avram Hershko, just about every major award in biology, including the Lasker Award, which you shared with Hershko and Ciechanover. Were you surprised not to have been included in the 2004 Nobel Prize in Chemistry?

I did not expect the Nobel Prize for ubiquitin work to be in Chemistry, rather than in Physiology/Medicine. The juries of scientific awards I received, jointly with Avram Hershko or with Hershko and Ciechanover, have done their homework, making clear the issues of credit. An easy task, given the unambiguous record of publications. In sum, the answer is yes: I couldn't help being surprised, at first, by the news in October 2004. Things became clear a bit later, when I saw the actual citation of the 2004 Nobel Prize in Chemistry: "*For the discovery of ubiquitin-mediated protein degradation.*" This citation lacks the second, biological (function-based) part, in contrast to all citations of the (earlier) joint awards to Hershko and me. In other words, the Chemistry Nobel Committee paid attention to the citation's accuracy, demarcating the Chemistry award as the one for the mechanistic

Avram Hershko and Alexander Varshavsky at the Horwitz Prize reception, Columbia University, New York, 2001.

("chemical") contribution. In doing so, the Chemistry Committee separated the initial mechanistic discovery by Hershko, his student Ciechanover and his collaborator Rose from the function-based discoveries in the 1980s by my laboratory, and by other groups afterwards. I presume that the careful wording of citation by the Chemistry Committee was intended to avoid interference with a recognition, at a later time, of the complementary biological (physiological) discoveries. I cannot be sure of this interpretation. It seems reasonable. A nonpolitical, courteous letter about the subject above, entitled "Varshavsky's Contributions" and signed by numerous colleagues in the ubiquitin field, including most of its leaders, was published in *Science* in November 2004 (*306*, 1290–1293, 2004). One should like, if one can, to behold prizes through the armor of irony and common sense. But the irony, too, has limits. I was touched by the *Science* letter.

You had a meteoric rise at MIT. Why did you leave it? Looking back, was it worth it to leave MIT for Caltech? How do you compare the two places?

My "career" at MIT was a standard one. Transitions, at a usual pace, from assistant to associate to full professor. I was happy at MIT, worked there for 15 years and liked living in Cambridge/Boston. There are several U.S. universities whose overall quality is comparable to that of MIT (Caltech is one such place), but none of them is "better" than MIT, certainly not in biology. The decision to move to Caltech was prompted by an unexpected event. A letter arrived in 1990 from a colleague there, describing a new "endowed chair", a fancy version of full professor position. The colleague wrote that I could be considered for that position, alongside other candidates, if I was interested. I received, occasionally, such suggestions before, but not from universities as distinguished as Caltech. I showed the invitation to my wife Vera, knowing that she loved Southern California. We visited Caltech in February 1991. The charms of Pasadena's subtropical climate, the scientific quality of Caltech, a warm reception by colleagues there, and the (later) offer of position were compelling to both of us. The lab moved to Caltech in 1992, thirteen years ago. I'll always miss MIT. The colleagues

Varshavsky's laboratory at Caltech in 2004. Back row, from left, Christopher Brower, Jack Xu, Jianmin Zhou, Cory Hu, Jun Sheng; front row, from left, Emmanuelle Graciet, Cheol-Sang Hwang, Zanxian Xia, Janet Dyste, Konstantin Piatkov, and Alexander Varshavsky.

there were supportive and kind. Having begun to work at Caltech, I could see firsthand that it's a great place, unsurprisingly so, for it's akin to MIT in all respects but size (Caltech is smaller). Vera and I live in La Canada, a hamlet on a mountainside, close to Pasadena. Vera's grown-up son Roman, who became my son too, lives nearby. Bob Hoffman, my and Vera's dear friend and my comrade in the 1977 escape, lives in San Diego, California. He is professor at the University of California at San Diego, and heads a biomedical company as well.

What are your current ambitions? What drives you nowadays when you get up in the morning and start a day in the lab? You have had a tremendous career and you are far from retirement, so how do you chart your next years?

Percy Bridgman, a great experimental physicist, worked at Harvard in the first half of the 20th century, achieving previously impossible static pressures and using them to study materials under such conditions. To him belongs a definition of scientific method that I find delightful: "*The scientific method is doing your damnedest, no holds barred.*" Since the time I realized that nothing in life would ever interest me more than science, more than doing it, a description like Bridgman's would convey just about everything there is in my connection to the craft. I worked in several scientific fields, and continue to be interested not only in ubiquitin and proteolysis but in god knows what as well. I read widely, for it's a pleasure, and also because new ideas or directions for the lab's ongoing work might pop up in fields utterly away from it. I don't anticipate leaving the N-end rule pathway, a lovely, many-sided creature. Its new functions, some of them of medical relevance, are emerging left and right, yielding surprises[12] even two decades after the pathway's discovery.[11] There may be other adventures too. I have ideas for some, but the brevity of a day, let alone of life, puts a lid on one's flirtation with playing Leonardo in the 21st century. Even Leonardo wasn't all that good at being the Leonardo of our admiring, reality-distorting perception of him.

Seymour Benzer, a geneticist at Caltech and one of greatest biologists ever, is now 84. He runs a lab and does first-rate science. I hope to follow his example if I can, and if longevity cooperates. Never retiring, unless I'm asked to, or become unfit to continue. Playing this exacting, unpredictable, and deeply meaningful game to a hilt.

Alexander Varshavsky and Seymour Benzer at Caltech, 2004.

Any thoughts about science in general?

"He digs deep, but not where it's buried," averred the poet Anna Akhmatova, in a conversation cited by her biographer. She was speaking, naturally, of another poet. (Compared to writers, scientists positively love each other.) Akhmatova's unkind aphorism captures the predicament of anyone who aspires to innovation, be it a momentous way of stringing words together or major discoveries in science. There is a distinction, here, between a poet and scientist. A poet may despair of finding a way to connect an insight and its form of expression. He may never find that form, or he may find it the next minute. If he does, the result, a verse, is truly his own. While poetry is occasionally about content, it is primarily an alloy of content and form, and the form is poet-specific. (Hence the Robert Frost's definition of poetry as the part that "is lost in translation".) So the poet is, in a way, safe from being scooped. Other dangers, aplenty. But not that one. By contrast, in science a truth is a stickler for accuracy but cares little about the form. One must arrive at a truth first and mark the arrival by a published paper. The form, that unique identifier of individual, barely counts in science, and certainly does not in a long run. Hence the extreme competitiveness of scientists throughout the ages. Their genuine curiosity about the world is warped and smothered by the haste to acquire narrow expertise and tools, lest that buried fruit is unearthed by another digger who is simply lucky, or better prepared, or (most often) both. The only thing I dislike about working in my beloved profession is my inability to enjoy a study by learning at a leisurely, pleasant pace, as broadly and gradually as I care, instead of being focused and intense. There must be scientists of sunny, relaxed, unhurried dispositions, but I never met such people amongst the peers whose discoveries are both first-rate and more numerous than one. Perhaps they are tranquil when they retire or become administrators. But they are not mellow in their prime.

References

1. Solomon, M. J.; Larsen, P. L.; Varshavsky, A. "Mapping protein-DNA interactions *in vivo* with formaldehyde", *Cell* **1988**, *53*, 937–947.
2. Sundin, O.; Varshavsky, A. "Terminal stages of SV40 DNA replication proceed via multiply intertwined catenated dimers", *Cell* **1980**, *21*, 103–114.
3. Sundin, O.; Varshavsky, A. "Arrest of segregation leads to accumulation of highly intertwined catenated dimers: dissection of the final stages of SV40 DNA replication", *Cell* **1981**, *25*, 659–669.
4. Strauss, F.; Varshavsky, A. "A protein binds to a satellite DNA repeat at three sites which would be brought into proximity by DNA folding in the nucleosome", *Cell* **1984**, *37*, 889–901.
5. Finley, D.; Ciechanover, A.; Varshavsky, A. "Thermolability of ubiquitin-activating enzyme from the mammalian cell cycle mutant ts85" *Cell* **1984**, *37*, 43–55.
6. Ciechanover, A.; Finley, D.; Varshavsky, A. "Ubiquitin dependence of selective protein degradation demonstrated in the mammalian cell cycle mutant ts85" *Cell* **1984**, *37*, 57–66.
7. Hochstrasser, M.; Varshavsky, A. "*In vivo* degradation of a transcriptional regulator: the yeast $\alpha 2$ repressor", *Cell* **1990**, *61*, 697–708.
8. Pickart, C. M. "Back to the future with ubiquitin", *Cell* **2004**, *116*, 181–190.
9. Özkaynak, E.; Finley, D.; Varshavsky, A. "The yeast ubiquitin gene: head-to-tail repeats encoding a polyubiquitin precursor protein", *Nature* **1984**, *312*, 663–666.
10. Finley, D.; Özkaynak, E.; Varshavsky, A. "The yeast polyubiquitin gene is essential for resistance to high temperatures, starvation, and other stresses", *Cell* **1987**, *48*, 1035–1046.
11. Bachmair, A.; Finley, D.; Varshavsky, A. "*In vivo* half-life of a protein is a function of its N-terminal residue", *Science* **1986**, *234*, 179–186.
12. Hu, R.G.; Sheng, J.; Qi, X.; Xu, Z.; Takahashi, T. T.; Varshavsky, A. "The N-end rule pathway as a nitric oxide sensor controlling the levels of multiple regulators", *Nature* **2005**, *437*, 981–986.

13. Jentsch, S.; McGrath, J. P.; Varshavsky, A. "The yeast DNA repair gene RAD6 encodes a ubiquitin-conjugating enzyme", *Nature* **1987**, *329*, 131–134.
14. Goebl, M. G.; Yochem, J.; Jentsch, S.; McGrath, J. P.; Varshavsky, A.; Byers, B. "The yeast cell cycle gene *CDC34* encodes a ubiquitin-conjugating enzyme", *Science* **1988**, *241*, 1331–1335.
15. Finley, D.; Bartel, B.; Varshavsky, A. "The tails of ubiquitin precursors are ribosomal proteins whose fusion to ubiquitin facilitates ribosome biogenesis", *Nature* **1989**, *338*, 394–401.
16. Chau, V.; Tobias, J. W.; Bachmair, A.; Mariott, D.; Ecker, D.; Gonda, D. K.; Varshavsky, A. "A multiubiquitin chain is confined to specific lysine in a targeted short-lived protein", *Science* **1989**, *243*, 1576–1583.
17. Bartel, B.; Wünning, I.; Varshavsky, A. "The recognition component of the N-end rule pathway", *EMBO J.* **1990**, *9*, 3179–3189.
18. Johnson, E. S.; Gonda, D. K.; Varshavsky, A. "Cis-trans recognition and subunit-specific degradation of short-lived proteins", *Nature* **1990**, *346*, 287–291.
19. Tobias, J. W.; Shrader, T. E.; Rocap, G.; Varshavsky, A. "The N-end rule in bacteria", *Science* **1991**, *254*, 1374–1377.

Osamu Hayaishi, 2005 (photograph by M. Hargittai).

17

OSAMU HAYAISHI

Osamu Hayaishi (b. in Stockton, California, in 1920) is Chairman of the Board of Trustees of Osaka Bioscience Institute in Osaka, Japan. He received his M.D. and Ph.D. degrees from Osaka University in 1942 and 1949, respectively. He served as a medical doctor in the Japanese Navy in World War II. In 1949–1958, he held appointments at the University of Wisconsin, Madison, the University of California, Berkeley, the National Institutes of Health, Bethesda, Maryland, and at Washington University School of Medicine in St. Louis, Missouri. He was Professor at the Department of Medical Chemistry of Kyoto University in 1958–1983 with shorter periods at the Department of Molecular Biology of the same university and as Professor and Chairman of the Department of Biochemistry at Osaka University. He served as the first Director of the Osaka Bioscience Institute in 1987–1998, and held many other positions.

Of his numerous awards and recognition we mention only a few. Professor Hayaishi was elected Foreign Associate of the National Academy of Sciences of the U.S.A. in 1972, Member of the Japan Academy in 1974, received the Order of Culture of Japan in 1972, the Wolf Prize (Israel) in 1986, and became Honorary Doctor of Medicine of the Karolinska Institute (Sweden) in 1985. We recorded our conversation at the Osaka Bioscience Institute on April 18, 2005.*

---

*István Hargittai conducted the interview.

How did it happen that you were born in the United States although you did not stay there then for a long time?

My father was a very ambitious and brave person. He was born in 1882 in Japan and died when he was 95 years old. His family was not very wealthy and he did not go to a prestigious university. He became a medical doctor and passed the National Board examination and practiced medicine in a small town in Kyoto Prefecture. Then one day he decided that he would go to the United States. He saved enough money and although he did not know anybody there, he went. There were many immigrants from Japan in California and the people he met advised him that there was a very strong anti-Japanese sentiment in California. He wanted to go to medical school and he went to the East Coast, to Baltimore, because he had heard that Johns Hopkins University had one of the best medical schools in the country. However, the tuition there was prohibitively high, whereas he could get exception to tuition at the University of Maryland if he stayed and lived in the State of Maryland. He even won a fellowship. When his professors learned about his medical license in Japan, he was exempted from basic training, and entered the third year studies immediately. In two years he graduated, then passed the National Board examination, and started practicing medicine in California. I still find this unbelievable because it was at the time when Japan was still a backward country and he did not have a good medical training in Japan. He became well known among Japanese immigrants in California and even in other places, so many came to see him with their problems from all over the country. He was happy and made good money. He met a young lady from Japan, who became my mother.

My father wanted to improve his qualifications and at that time most Japanese doctors considered Germany for further education. My father wanted to go to the Robert Koch Institute in Berlin as a postdoctoral fellow. We went there when I was eight months old and we stayed there for two years. During this time my father wrote two articles on immunology and when I visited the Robert Koch Institute some years ago, we found his papers in the library of the Institute. I was proud of my father because he produced research results without any prior training in basic science.

We returned to Japan in 1923 because his mother, my grandmother, was still living there in a small town near Kyoto. At that point my father wanted to become a professor in a Japanese university, but he was not accepted because the Japanese system of medical education was rigid and

Three-year-old Osamu Hayaishi with his mother, Michiko Hayaishi in front of the Victory Column (*Siegessäule*) near the Berlin Zoo (*Tiergarten*), 1923 (all photographs courtesy of O. Hayaishi unless indicated otherwise).

bureaucratic and they did not want to recognize my father's American qualifications. My father was disappointed and decided to practice medicine in Osaka. He opened a hospital and worked there almost until he died.

Did you choose medicine because of your father?

It was mostly his influence. I grew up with my parents; our family was a typical average Japanese family; a very sincere one, and he was a very conscientious doctor, and he was well respected by his patients. We always had dinner together and almost every time my father talked about recent development in medicine, especially in the United States.

Wouldn't he speak more about developments in Germany? This was in the 1930s.

His philosophy and life style was much more American than German. He liked democracy. He did not like the rigid and bureaucratic medical education in spite of his having taken his postdoctoral training in Germany.

Where did you study?

I went to the medical school in Osaka University and even I had some problems with adjusting to their German way of teaching. This was in the late 1930s and early 1940s and most of the professors had been trained in Germany and most textbooks were German. The reason I chose Osaka University was that there was a famous biochemist, Yashiro Kotake, and his lectures were more than half in German. They were given in the typical German "Geheimrat" [privy councillor] style. Most other professors also followed such a style because of their training. The structure of the Japanese medical schools followed the German hierarchic system with one professor at the top. After World War II, this was changed completely. The system of medical education followed the American style. Back in my student time, however, it was the very rigid German system. There was, however, one exception, Tenji Taniguchi, the Professor of Microbiology, who used a textbook in English. He was friendly and democratic and treated his students and young co-workers as friends. He had been trained in Scotland. Eventually we became very good friends and that's how I decided to study microbiology. It was a very naive idea.

Why was it a naive idea?

I did not particularly like microbiology. I could have had some other choice. My father was a physician, my elder brother was a surgeon, and there were many famous clinicians at Osaka University. I graduated and received my medical degree in 1942, during World War II. I had a choice of going into the Army or the Navy. The Japanese Army followed the German system and the Russian system. The Navy was strictly Anglo-Saxon. When I was being trained to become a Navy officer, I was encouraged not to drink sake, but drink only Scotch whisky.

Osamu Hayaishi as a young medical officer in the Japanese Navy during World War II (1943).

Why was it important?

Because everything followed British tradition in the Japanese Navy. The Army and the Navy were just like cats and dogs.

If this was the case, how did your colleagues in the Navy take the task of fighting against the British?

There were many people, although still in the minority, who were more fond of America and England than Germany. However, the Japanese government and the Army were friends of the German Nazis and the Italians and they went into this miserable war.

Where did you serve?

When I became an officer, there were very few ships left, so I was stationed in the Kuril Islands, which are north of Hokkaido. They are now part of Russia.

Did you participate in combat?

American bombers came and bombed our area, and I had to do some operations of wounded soldiers, but did not participate in combat.

Was it the only time when you practiced "real" medicine?

Right. After three years in the Japanese Navy, the war ended. I was in Hokkaido at that time and almost immediately I came back to Osaka.

How did you learn about the end of the war?

It was radio broadcast.

Did you listen to the Emperor's speech on August 15?

I did.

Did you know about the atomic bombs?

Yes.

From the speech or before?

The news came that there was a new big type of bomb, but I didn't realize that it was an atom bomb. We didn't know any scientific details; it was just a new type of bomb that exploded in Hiroshima and Nagasaki. Only many months later did we learn that it was a new type of bomb, but we still didn't understand what atom bomb meant

Was there anything that became imprinted in your mind from those times?

I was still a Navy officer even though I was a medical officer. All my fellow officers and those civilians who helped us felt a great relief when we heard the news. We only hoped that some right-wing people in the Japanese Army would not object the Emperor's decision. We hoped, for example, that they would not try to kill the Emperor in order to continue fighting. The majority of the people realized, even without knowing that the new type of bomb was an atom bomb, that fighting was hopeless and we in the Navy knew this months, maybe a year before others had realized it. The Japanese newspapers did not give us correct information about the lost battles and they concealed the difficulties of the Japanese Navy, but we knew that many of our submarines and ships had been sunk and the whole Navy had almost disappeared.

Weren't you preparing for a final fight?

No, except maybe some radical ones.

There is a debate still going on about the usefulness of the atomic bombs. Some argue that the Japanese were ready to fight for every square inch on the homeland islands.

I only have a feeling that there might have been some people, especially in the Army Headquarters, who were willing to continue, but if I am not mistaken, even they realized that there was no use to keep fighting against this enormous amount of modern weapons, not only the atom bombs, but all the other war machinery.

The sixtieth anniversary is coming up later this year. Is there an anti-American feeling specifically because of the atomic bombs?

There is still a sizeable fraction of Japanese people who have felt that there was no need to use atom bombs in Hiroshima and Nagasaki to kill many thousands of civilians in such a cruel way. They could have dropped an atom bomb on Iwojima or a small inhabited island to show that they have such a powerful new type of bomb, which can destroy almost everything and if the Japanese would not surrender, the next thing would be to drop this bomb on a populated area, over Tokyo or Hiroshima or Nagasaki, to kill thousands or millions of people. They could have at least shown how dreadful and powerful an atom bomb was in a demonstration on a small island, to give a previous warning to convince people and the Emperor. I visited Dresden the other day and Dresden was completely demolished by bombs and several ten thousands of people were killed, and there are still some German people who think that that was a useless murder. People criticize Auschwitz and the Nazis for killing innocent Jews, but the Allies too slaughtered not only soldiers but also innocent civilians in all these wars everywhere. When I came back to Osaka after the war, I saw the entire city completely demolished. Our house where my family had been living for almost thirty years disappeared without a trace. Fortunately, my father and mother had left for the countryside and my father was practicing medicine in a small village. But many of my friends and their families were killed.

You brought up the comparison with Germany. The Germans have done a lot to face their war crimes. Do you think that the Japanese have done enough in this respect in order to re-establish normal relationships with their former enemies? I realize that this is a very sensitive issue, but we have gotten into this topic and now I feel I cannot avoid asking you this question, especially in the light of present-day discussions of this topic.

It is not only sensitive, but also a very difficult issue. I don't think that anyone can discuss this problem within a few minutes. I was recently reading some of the writings of the Pope, John Paul II. Some people say in Japan that it is impossible for other human beings to be like the Pope. He spread peace not by violence but forgiveness and understanding. If all the people on the Earth would be like John Paul II, there would be no more war, no more unnecessary killing. Unfortunately, there are so many different kinds of people on Earth.

I would like to quote from two other conversations I had with Japanese scientists and I would like to ask you for your comment. A famous Japanese scientist who lives in America told me that in a way Pearl Harbor was a response to the suffocating American actions by which America would not let Japan get the necessary oil for its economy. Do you think this to be a right assessment?

There is some truth in what he said. The same thing goes to the Japanese invasion of China, Korea, or Russia. The Japanese said that the Chinese initiated the war; they disturbed and invaded our territory; so the Japanese Army had to fight back. It's very hard to judge which side is correct.

In the other conversation, a prominent Japanese scientist living in Japan told me that in the Japanese schools history is taught in great detail through the Meiji era, but the rest is hurried through at the end of the school year. The Japanese children do not get much information about Japan's modern history. Nobody would say this explicitly, but it appears that this is done on purpose in order to avoid facing what really happened in the 1930s and during World War II. There is then this newly approved Japanese textbook, which, according to the Chinese, falsifies the history of World War II.

To be honest with you, Sir, I have not read through all the history books in China, Japan, Korea, and elsewhere. But isn't it generally true that in most countries, their own history is glorified and emphasized but the bad things are minimized or ignored. If I had read them I would not know whom to believe. I'm an experimentalist and even in my own experiments sometimes I don't know if I am looking at things that are true or false. It's very difficult to judge. Some things, which took place in a faraway country a long time ago, are very hard to assess and understand. These days there are some big anti-Japanese demonstrations in China and they criticize everything what has been reportedly done by the Japanese in China during the war. I'm sure there must be some truth in it, but how much truth is there, a hundred per cent, or only ten per cent, or only one per cent, I don't know. I cannot judge. The same is true with the Japanese history book. I don't know what is written there, I have never seen it, and I cannot say. I do know, however, that our government has strong control of what is there in the Japanese textbooks. However, I have a feeling that Chinese government has a far more strict censorship.

My last question in this respect is related to my personal experience. About eight years ago I visited the Kamikaze Peace Museum in the southern part of Kyushu Island. I did not know about such a museum, but my hosts at Kagoshima University took me there.

I did not know about that museum but might have guessed that there would be such a museum in that area.

It is a nicely constructed memorial place; there is a statue of a kamikaze pilot in front of the building and his airplane is standing there, which must have been recovered or reconstructed. Such a memorial could be interpreted in two ways. One is as a protest against using young people for such a purpose, but the museum could also be interpreted as hero worshipping. My impression was the latter. I saw there a group of children with their teacher; he was obviously explaining to his pupils what happened to the kamikaze and at one point the children started weeping and so did their teacher. The displays, the memorabilia exhibited there, all gave me the impression of paying tribute to heroes rather than condemnation of a political and military establishment that misused these young lives. I sensed pride rather than sorrow.

I have never heard of this museum, but hearing this story from you, I am not surprised that there is such a museum in the Kagoshima region. In that area the traditional Japanese samurai spirit has been very popular, and there is a strong right-wing sentiment there. Secondly, many kamikaze took off from Kagoshima. There are still people there who remember these poor kids, how they were served their last supper, and how they flew off to their deaths. Most people had very sympathetic feelings towards these young people. I think there is good reason to have such a museum in that area. As for the purpose of such a museum, it was probably established to pay tribute to these young pilots and to remind people how foolish and useless the "war" was and to keep peace.

Of course, I have very few opportunities to have such a discussion and I appreciate your patience with me.

I am happy to answer your questions but these are based upon very limited experience and are my own personal views. I am sorry that my English is not sufficient to express my feelings adequately in such delicate matters. I sincerely hope that my intentions are not misinterpreted.

I have read about your scientific research and I would like to single out two topics for asking you about if you would agree with my choice. One is the discovery of the oxygenase enzyme for biological oxidation and the other is your sleep research. I wonder if you would care to tell us about these two areas of your research.

I entirely agree with you. These two are my most favorable subjects. Firstly, on the occasion of the fiftieth anniversary of the discovery of the oxygenase enzyme, the editor of the *Journal of Biological Inorganic Chemistry* asked me to give a concise summary of the story [**2005**, *10*, 1–2] as an introduction to their New Year's issue. This research goes back to the time immediately after the war. Upon my return from military service, I went to see my university and my professor in Osaka, Dr. Taniguchi, even before I went to see my parents.

My professor told me that Japan was defeated completely, but it did not vanish. He told me that it was our obligation to rebuild it and since I was a medical doctor, I should rebuild medicine from its very foundations, starting with basic research. However, I told him that there was no food to eat, no shelter to live in, and there was no water, no gas, and no electricity to provide the conditions for work. I told him that my intention was to go to the countryside where there was fish and rice and I would engage in practical medicine. With this I bid him goodbye, but he stopped me. He asked me whether I knew the old Japanese saying, "The seed of persimmon is better than a bowl of rice." He wanted to tell me that we should work for the future rather than be bogged down with immediate satisfaction of our needs. However, I didn't quite understand what he wanted to tell me at that time.

In any case, I wanted to see my parents and in the train I thought about my professor's message. I came to the conclusion that he might be right. If I went back to my father's practice, I might be able to eat good food, but if I went back to the university, I might be able to contribute to the revival of Japanese medical science. I discussed my options with my father, including the possibility that I stay with him and help him with his practice. He was by then almost sixty-five years old. But he told me that we should go to Osaka together. He closed down his practice and we moved to Osaka with all our belongings. Osaka was a big city and although it was almost completely demolished, people were living there and there were no doctors. He opened his practice, which soon became thriving.

I went back to Dr. Taniguchi and asked him to train me to become a microbiologist. He appointed me to be his assistant and gave me a small salary. Unfortunately, we had no research money, because the government did not yet support basic medicine at that time. We had no conditions for work and no animals to experiment with. But we were receiving journals from our friends from the United States. We read those journals and organized seminars and discussed what we read. At the same time we started to teach medical students.

Then one day, unexpectedly, an older gentleman came to my office, Yashiro Kotake, a very famous professor emeritus biochemist of Osaka University. He brought me a small bottle of tryptophan — an amino acid. He had spent his career on working out the mechanism of tryptophan metabolism. He had retired and had no use for the sample and he wanted me to work with this valuable substance. I realized that it was a precious gift but was at loss what to do with it. I had no experience yet in research. Then one day I read an article by G. S. Mirick in a 1943 issue of the *Journal of Experimental Medicine* in which he described soil microorganisms, which can metabolize almost anything. I went out into the backyard of our institute and took a spoonful of soil and mixed it with a little tryptophan and water in a small beaker. This was the most inexpensive experiment with the most inexpensive "animals". I did not see any change the next day; at the bottom there was the soil, and clear water above it. The next day again, there was no change. But in a few days, a white cloudiness appeared in the supernatant.

Did you run a parallel experiment?

Of course, and what I observed did not happen without tryptophan. The difference was obvious and visible. This was an experiment by an ignorant, young, untrained scientist and yet it opened up my whole career. To make a long story short, I was able to fish out the bacteria and study the metabolism of tryptophan, and I found a new pathway of tryptophan metabolism in this microorganism. It was different from the tryptophan metabolism in rats, mice, monkeys, cows, and humans which had been studied by Kotake and many other scientists in the United States and Europe.

I was delighted and I told myself and my friends that I made a big discovery. In order to publish it I turned for help to my professor. He knew about the new English-language journal of the Japanese Biochemical Society called the *Journal of Biochemistry*. So I published my findings there and it came out quickly because they hardly had any manuscripts waiting for publication. In a month or two after my paper appeared, I received a telegram from America, saying that my paper was of great interest and inviting me to the United States. It was from David E. Green who offered me air

Osamu Hayaishi with David E. Green (then Director of the Enzyme Institute, University of Wisconsin), 1949.

ticket and a stipend. He had been in Cambridge, England, for many years and was then the head of the newly created Enzyme Institute in Madison, Wisconsin. Some of my friends warned me that it would be very dangerous for me to go because some Americans would consider me still an enemy especially as I had served in the Japanese military. But my professor and my father told me that I should accept the invitation and that the American people would be fair and would treat me well.

I left Japan in 1949 and this move opened my research career as an enzymologist/biochemist/microbiologist.

By then, you had received your Ph.D.

Yes, I received it from Osaka University prior to my departure for the United States. However, I must tell you honestly that I was not yet a well qualified and well trained scientist. But I was very curious and ambitious.

Still in Japan, what was the essence of your discovery?

I discovered a novel enzyme, which I named pyrocatechase. It catalyzed the conversion of catechol to muconic acid. I isolated this enzyme from the soil microorganism and obtained it in a soluble form. This is a simple enzyme and it consumes one oxygen molecule per molecule of catechol. The oxygen opens up the aromatic ring of catechol and muconic acid forms in a hitherto unknown type of reaction:

$$\text{catechol} + O_2 \Rightarrow \text{cis,cis-muconic acid}.$$

It is probably the first instance of the oxidative cleavage of the aromatic ring to produce an aliphatic compound. This had not been known before and this was what Dr. Green especially liked about my work.

Did you use a tracer?

I did not because it was not available and mass spectrometry was not available either.

So it was a guess.

It was. My friends in Japan and America all told me that there is no biological oxygenation and that addition to oxygen to oxidize something occurs only in chemical reactions. Such oxygenation does not happen in living systems where oxidation means the removal of hydrogen atoms or the removal of electrons. For example, the oxidation of glucose is its dehydrogenation but not the addition of oxygen. The role of oxygen in the oxidative processes in living systems is that it accepts hydrogen or electrons and is reduced to water. This was the famous "dehydrogenation" theory of Heinrich Wieland, who received a Nobel Prize in Chemistry in 1927. His book appeared in 1932, *On the Mechanism of Oxidation*. He claimed that molecular oxygen is merely accepting hydrogen or electrons and it can be replaced by co-enzymes and even by dyes like methylene blue. His claims have been in all textbooks for decades, and it has been a central dogma of the biological oxidation processes.

So Dr. Green invited you even though he did not believe that you interpreted your experiments correctly.

I believe that he was attracted by the discoveries of a new metabolic pathway and the novel enzyme, which catalyzed this unique reaction. I emphasized more the chemical side rather than the mechanistic side of my findings in my paper. I worked with Dr. Green, then I worked in California, and then Arthur Kornberg invited me to join his laboratory at the National Institutes of Health. I was a postdoc with him for two years and he taught me enzymology.

If I may interrupt you, when you submitted your initial paper to the Japanese journal, which probably did not have a very rigorous refereeing system yet, did you realize that you were going against a basic dogma of the field?

I had a feeling that it was against the main dogma but I was not sure if the famous Wieland theory was very clearly established. I didn't know whether there was unambiguous evidence supporting his theory or whether it was accepted because he was a famous Nobel laureate. When I was in the United States, I did many experiments to further study the mechanisms of biological oxidation, but there was no convincing experiment either way, until I had a chance to use the stable isotope of oxygen, ^{18}O.

Could we say that you were lucky to have submitted your initial paper to this then obscure journal?

That's correct. They didn't have critical referees and they didn't have many papers, so they were eager to publish what they got.

Did you follow Arthur Kornberg when he moved from the NIH to St. Louis?

I was the first member (assistant professor) of his Department there.

It was a great place at that time. Did you meet the Coris?

I did. They were not only great scientists but great human beings as well. They were very helpful to me, perhaps because of their own experience having immigrated from Europe when they were young. I had all kinds of difficulties. I was lecturing in my poor English and suffered sometimes from difficult questions asked by very good students. The Coris had a broad vision and he was a great leader but he could not have succeeded without her as she was not only a good scientist but also an excellent experimentalist. Because of his fame, Carl was always very busy, but Gerty carried on in the laboratory. She was never too busy to talk with me when I had problems. If I had to see Carl, Gerty always arranged the meeting for me. They were excellent partners. They deserved sharing the Nobel Prize.

At which point was your discovery finally accepted?

After two years in St. Louis, I thought that it was time for me to return to Japan. But one day I received a call from the Director of the NIH offering me to come back to the NIH as Head of the Toxicology Department. I have done many things but had never been a toxicologist. But he explained to me that such specifications didn't mean much in the United States. I talked with Arthur and he told me that it was a great opportunity and encouraged me to accept it. My wife was also delighted because St. Louis was too hot for her and air-conditioning was not very efficient in our apartment. Thus I accepted the appointment, we returned to Washington, DC, I hired some new toxicologists for my section, and initiated several new projects. One of them was to further investigate pyrocatechase, which maybe related to the metabolism of toxic substances. It seemed interesting to investigate its metabolism from a toxicological point of view. By then, tracing was a feasible approach, as you have pointed out. I called many places looking for oxygen-18, but people in the United States could only suggest getting in touch with the Weizmann Institute in Rehovot, Israel. They were processing salt water of the Dead Sea and separated oxygen-18. I got in touch with Dr. David Samuel of the Weizmann Institute.

Members of the Section on Toxicology at the National Institutes of Health, Bethesda, Maryland, 1955.

He found my project to be of great interest and he gave me a sample of oxygen-18 as a free gift. Again, I was very lucky. We also needed mass spectrometry, but there were mass spectrometers around at the NIH. The experiments unambiguously proved my hypothesis correct. This was in 1955, exactly fifty years ago.

It was lucky that you could prove it yourself.

Ever since I made the original discovery of pyrocatechase, it has stayed with me as my first love. I thought about it all the time even in my dreams. But when we published our proof, the reaction to it was not very warm. At about the same time Howard Mason in Oregon reported a similar kind of oxygen incorporation in which, however, only one atom of oxygen was incorporated per molecule. He discovered an enzyme in mushrooms and it incorporated oxygen in phenols. He called his enzyme "mixed function oxidase or oxygenase". There were then other reports of similar phenomena using exactly the method I had used. Then, Konrad Bloch at Harvard

From left to right, Shozo Yamamoto, Konrad Bloch, and Osamu Hayaishi at the Oxygenase Symposium, Lake Hakone, Japan, 1981.

University, who later (in 1964) received the Nobel Prize, also reported similar experiments. All this contributed to the gradual acceptance of my original conclusions about the new type of oxidative enzymes, which I proposed to name "oxygenases".

There was an International Union of Biochemistry Congress in Vienna in 1958 where I was asked to organize a symposium on oxygenases. I invited six speakers and it was a well attended meeting although there were still some people who preferred to believe the Wieland theory to the new findings. They thought that our conclusions were applicable to the very primitive organisms only and not to the higher living organisms. Other people, however, joined into the work on oxygenases and showed them not only in tryptophan metabolism, but also in steroid metabolism and in prostaglandin metabolism, and so forth. Oxygenases have now become recognized not as an exception in oxygenation in the living organisms and present not only in primitive organisms but also in mammals, plants and so forth. This reaction is now recognized as an important one in the mammalian organisms playing important roles in the metabolism of essential substances like hormones, vitamins, neurotransmitters and so forth. Moreover, there is now P-450, which metabolizes drugs, carcinogens, pollutants, and many synthetic compounds, a typical oxygenase. In general, the so-called oxidative enzymes can be grouped into two major categories. One is dehydrogenases in which hydrogen atoms and electrons are transferred and Wieland was correct in saying that this system is mainly involved in energy production. The other category of oxidative enzymes is the oxygenases. They are involved more or less in metabolism, transforming one compound to another, producing hormones, vitamins, neurotransmitters, and synthesizing such essential metabolites as well as degrading foreign compounds. This is the essence of the fifty-year old oxygenase story.

In 1964, the International Union of Biochemistry Congress was held in New York. I gave a plenary lecture on oxygenases, which was my best lecture ever. After that I received many offers from the United States and Europe and from Japan. Finally, I decided to go back to Kyoto.

Were you an American citizen?

I was by birth, but I lost my American citizenship because I served in the Japanese Navy. I later tried to get my American citizenship back. However, Japan does not recognize dual citizenship and I might risk my Japanese citizenship in some extreme situations. So I finally remained Japanese citizen only.

I would like to suggest moving to the other major area of your research, sleep.

Prostaglandins are lipid mediators and local hormones. There are about thirty different kinds of prostaglandins in nature. Almost all tissues and cells have at least one or two prostaglandins. They are produced in situ, in the cells. They are involved in numerous functions such as in muscle contraction, kidney function, cardiovascular function and so forth. When you get cold, you have headache, pain, cough, as symptoms, and you take aspirin. Then the fever is gone, the pain is relieved, and the other symptoms also disappear. Aspirin inhibits the enzyme called cyclo-oxygenase. Almost all prostaglandins are produced by prostaglandin cyclo-oxygenase — a very unique oxygenase. It is the most important rate-limiting enzyme. This is why I became interested in prostaglandins. Prostaglandins were then found in almost every type of cells except in the brain. Back in the 1970s, when I was still a professor at Kyoto University, very little was known about prostaglandins in the brain. This made me curious. With the help of my young colleagues, I opened up the brains of cows, dogs, rats, and humans, crashed and grounded brain samples, made an extract, and tried to study prostaglandins. But brains are full of lipids, complex lipids like cholesterol, glycolipids, and so forth. For this reason it was very difficult to purify and study prostaglandins in the brain.

Where did you get the brains from?

We tried many different animals, from slaughterhouses.

The human brains?

From hospitals after the patient died. Many organs are donated by hospitals for medical studies. There are also brain banks.

Were you looking for any particular illness or any brain would do?

At first, we were interested in the prostaglandins in the normal brain. There is a human brain bank where many hospitals send brains that are donated. They are classified under the names of diseases and they are used for medical research. I had not realized that it would be so difficult to study prostaglandins from the brain and I understood why there had been so few studies before we had started. After working for some years, we developed our methodology, and analyzed the prostaglandins of the brain. We found out that prostaglandin D_2, PGD_2, are the most abundant prostaglandins in the brains of rats and other mammals including humans. This prostaglandin was usually not found in most other tissues except in the blood platelets and maybe in a few other tissues, but only in very small quantities. Furthermore, this prostaglandin D_2 used to be considered biologically almost inactive.

But being a unique constituent in the brain, I thought that there must be some special function for it in the brain. I was ready to try some simple experiments. We injected PGD_2 into the brain of the rat in small quantities. The animal went to sleep. Up to then, sleep had been considered as one of the most important and yet least understood physiological functions of the brain. Almost one hundred years ago, a Japanese Professor of Physiology at Nagoya University, Koniomi Ishimori subjected dogs to sleep deprivation. Then he took samples of the cerebrospinal fluid of these sleep-deprived dogs and infused these samples into the brains of normal dogs. The recipient dogs went to sleep. This experiment showed that sleep is controlled by a hormone-like chemical substance rather than by the neural network. This is called the humoral theory of sleep induction. Ishimori did some important, pioneering work and discovery for which he should have received the Nobel Prize. However, at that time no one believed his results. Furthermore, the organic chemistry of the time was very primitive.

Did he identify the substance?

No, he did not. He was a physiologist and not a chemist. He had some thirty co-workers on his publication, which appeared in Japanese and remained unknown to the outside world for a long time. Concurrently with Ishimori's work, and entirely independently, a neuroscientist in Paris, Henri Piéron did almost exactly the same kind of experiments. He published similar results in French claiming that he had found a hypnotoxin or sleep toxin. He claimed that when animals were subjected to sleep deprivation, there is some toxin accumulating in their brains. In any case, these studies were carried out about one hundred years ago, but the nature of their sleep substance(s) remained entirely unknown.

During these one hundred years since then about thirty different so-called sleep substances have been isolated from brains, blood, cerebrospinal fluids, and other organs and tissues of many animals by many scientists in Europe and in America. There were even attempts to treat insomnia with such substances. However, although all these thirty compounds became famous and reported in newspapers, most of them have disappeared in oblivion, because their physiological relevance remained unclear.

In the meantime, in the 1920s, EEG, electroencephalography was invented, and the German psychiatrist Hans Berger took electroencephalograms of brain waves. He could determine quantitatively the amount of sleep. In the 1950s, Jouvet in Lyon and Kleitman in Chicago described another type of sleep, which is accompanied by rapid eye movement. The muscles relax completely. This is different from slow wave sleep, which is now called non-REM sleep. So sleep is not a uniform function at all. It consists of two entirely different stages.

In short, sleep research has a long history and yet even now sleep is one of the most important and least understood physiological functions of the brain. It was about twenty years ago that we found and identified PGD_2, as a sleep substance and PGE_2 as a wake substance. For the past twenty years we have been doing biochemical and molecular biological research on sleep and experiments involving molecular genetics. We are trying to elucidate the mechanism underlying sleep regulation.

This century has been declared the century of brain research. Many brain researchers, including ourselves have been trying to figure out how information is transmitted in the brain. We would like to find out what is the mechanism of memory, thinking, dreaming, emotion, and so on. Neural pathways and hormonal pathways are being elucidated. But one of the most important questions is understanding sleep. What does it mean for the brain cells? Do the brain cells rest during sleep? Why we need to sleep? How are sleep and arousal controlled? No one can answer these simple questions. There are now classified 88 different sleep disorders. We would like to know the origins of these sleep disorders in order to diagnose, treat, cure and/or prevent these sleep disorders on the basis of evidence-based medicine. The problem is that the basic understanding of the sleep-wake regulation is still missing. Understanding sleep disorders might possibly bring us closer to understanding the mechanism of Alzheimer's disease, Parkinson's disease, and so on. We still don't know much about the disorders related to memory in old age.

Twenty years ago I was very ignorant about sleep and brain functions, but hoped to gain an understanding of it and I also hoped that I might be able to find some treatment for other brain diseases, such as Alzheimer's, Parkinson's, and others, and find the way to cure them. I was especially interested in dementia of old people and its relation to sleep.

Osamu Hayaishi receiving the 1986 Wolf Prize in Medicine from the President of Israel, Chaim Herzog, at the Knesset.

You used two different words, "treatment" and "cure" and they have very different meanings. There are already some treatments available, but no cure.

You are quite right. Eventually, we should be able to find not only treatment but cure as well. The first task would be to stop progression of the disease in an early stage. Up until very recently many people thought that brain tissues cannot regenerate. This is possible with liver, for example. Muscles and bones can regenerate to some extent, but nerves and especially brain cells never regenerate. We now know from experience and experimental work that some nerve cells can regenerate to some extent. For example, we can delete genes for PGD_2 synthesizing enzyme or PGD_2 receptors to produce so-called knock out animals. These animals appear quite healthy and sleep well in spite of the fact that there is no rebound and no increase in PGD_2 content in the brain after sleep deprivation. This gives me hope to think that it might not be impossible to cure even brain diseases. I maybe too optimistic but future experiments are necessary to prove it.

How would you summarize what you have reached during these twenty years?

I've opened up the gate to clarify the basic molecular and genetic mechanism of sleep-wake regulation. I've opened up the first stage, which is PGD_2 and PGE_2 and we have found adenosine, which is another compound providing a link between the hormonal PG system and the neural system. Almost all biological processes are controlled by a combination of two mechanisms. One is the hormonal system and the other is the nervous system. Take blood pressure, for example. It is controlled by adrenaline, which is a hormone and by the nerve system in the hypothalamic area. There is a center for blood pressure control, which extends its influence to the blood vessels and the heart. Both the chemical substances and the nerves are important. In sleep, we can definitely say that these two systems are linked by adenosine. It is a metabolite of ATP, which is one of the most important chemical compounds, the energy currency of our body. From ATP, ADP forms, then AMP, and eventually, adenosine. This substance regulates the collaboration of the hormonal system and the nerve system and this function signifies its importance. This means that this end product of the energy metabolism tells you either to go to sleep or to wake up. I think that this is a very reasonable hypothesis. We will need a lot of work to prove this. We know that sleep neurons and wake neurons are present. That much we know. But the final and most important question about the function of individual neurons, that question we are not yet able to answer. We find it even difficult to propose reasonable hypotheses to answer this question.

If we move now to more general questions, I would like to ask you about how people judge Japanese science. Some time before, a frequently-used label was copying, but more recently, the label innovative is being increasingly applied. How do you feel about it?

We have to look at this question from a historical perspective. Japanese culture, including science, has been concerned with making a good imitation. Japan opened itself to western civilization about 100–200 years ago in the Meiji era. Many ambitious young scholars went abroad to study. Most of them came back to Japan and became good imitations of their professors in Germany, England, and other places. Very few people started their own ideas, their own projects. They were top students in Japanese universities who were not necessarily very innovative to begin with. They may have been more of the encyclopedic scholars. Our entrance examinations of the major universities are primarily based on memorization. They are not aimed at testing abilities for creativity. These people whose primary virtue was memorization went out to study under great scientists in foreign laboratories. What they took back home was good imitation rather than creativity. This was also true when famous professors came to visit and stayed in Japan as visiting professors in the Meiji era.

There was a famous German professor, Erwin von Baelz who started the School of Medicine at Tokyo University, which is the most prestigious medical school in Japan. He stayed for twenty years or so in Japan. He married a Japanese lady. When he became quite old, he returned to Germany and gave a famous farewell address. He said that he spent many years in Japan, taught many students, and some of them performed very well. But if he was to give an important advice to the Japanese students and medical people, he would suggest caring more for their own ideas and following their own creativity rather than imitating others.

The Japanese society itself, not only in science, but business, for example, may also be characterized in the same way. It is usually family business, at least it used to be, and the great-great-grandfather gives order to his son and his son to his son, and so on, and everyone says, Oh, yes. This hierarchy has been in effect not only in companies, but even at the universities. If a young person would like to try something new, the elders may not like it. This system of attitude used to be generally characteristic for the entire Japanese society through World War II. It may have been very similar to the German military system. The young students even at the top-notch universities did not have much freedom to try new things. Up until recently the big research grants used to go to big professors in huge amounts of money. The professors would then distribute the money among their young collaborators who themselves could not apply or if they did, they hardly had a chance. The Japanese social system, including the school system and the grant system did not favor the young people to gain their independence early.

My family, my father and myself, we were not typical Japanese. If I had gone to study at Tokyo University, I could have become a good doctor, but I don't think I would have been able to do any independent research. As it happened, I was already carrying out independent research as early as when I was 25 years old. Many people helped me to be sure, but I was working on my own ideas. This was certainly atypical of the era. It was an exception that I could publish my own papers under my own name, without a senior supervisor. As a rule, it was almost impossible at that time. By now, things have changed thanks to mostly American influence. Many Japanese scientists went to Europe and America, and the whole society, even the companies, have undergone some changes, slowly but definitely.

Osamu Hayaishi and Susumu Tonegawa in 1993, on the occasion of the fifth anniversary celebration meeting of the Osaka Bioscience Institute.

The impression you just quoted has been heard from many foreigners and not only from foreigners but also from people like me. When I returned to Japan and started my work at Kyoto University, I completely reorganized the Department of Biochemistry. I served Kyoto University for twenty five years and produced six hundred fifty Ph.Ds. and a hundred and fifty professors, so they further have spread this new approach scattered all over Japan. Of course, I had many young assistant professors and associate professors, who participated in this revolution. My department became a Mecca of biochemistry in Japan.

After my retirement from Kyoto University at the age of 63, I became the Head of this Institute. It may seem to be an early age for retirement, but it is also a good screening point. I was very fortunate that the City of Osaka offered me a huge grant to build and organize this Institute.

Does it belong to Osaka?

It doesn't belong to Osaka but Osaka City provides money for its operations. This is a private institute.

Who owns it?

It is owned by an independent Foundation called "Osaka Bioscience Institute".

Professor Fukui also had his institute after his retirement from Kyoto University.

Yes, he did. We knew each other very well. We went to the same high school and our careers ran in parallel in many aspects and we were competitors in a very friendly way.

May I ask you about religion?

I must confess that I have no religion and never had. In a way Science is my religion. I go to temples, I go to shrines, I pay respect. It's a good place to meditate. But I won't have a funeral and I won't have my grave. I've asked my wife to place my ashes into the Pacific Ocean. I was born in the United States, came to Japan, went to the States, came back to Japan, and spent almost all my happy life in these two countries. The Pacific Ocean connects them, where I would like to rest in peace.

Your family?

Osamu Hayaishi with Takiko Hayaishi, 1972.

Osamu Hayaishi in his office during the interview, 2005 (photograph by M. Hargittai).

My wife and I met in Japan, but her father was a graduate of Johns Hopkins University. Our fathers met in Baltimore. We have one daughter and two grandsons.

Do you have heroes?

Arthur Kornberg. Without him I would not have been here today to discuss things with you. He was my hero as well as my mentor and good friend. I have been very lucky to have so many good friends. Wherever I went, I have always been fortunate to meet good people.

Ada Yonath, 2002 (photograph by Dan Porges, courtesy of A. Yonath).

18

ADA YONATH

Ada Yonath (b. 1939, in Jerusalem) is Director of the Kimmelman Center for Biomolecular Assemblies and Professor of the Department of Structural Biology at the Weizmann Institute in Israel. She received her Ph.D. at the Weizmann Institute in 1968 and was postdoctoral fellow at the Mellon Institute in Pittsburgh (1969) and at the Massachusetts Institute of Technology (1969). In 1970, she returned to the Weizmann Institute and established the first protein crystallography laboratory in Israel, and became a professor and a director in 1988. In parallel, she was lecturing at Tel Aviv University and at Ben-Gurion University in Beer-Sheba from 1971 till 1978, and was visiting professor at the Max-Planck Institute for Molecular Genetics in Berlin (1979–1983). Between 1986 and 2004 she headed a Max Planck Society research unit in Hamburg.

She is a member of the European Molecular Biology Organization, EMBO (1988), the International Academy of Astronautics (1991), the Israeli Academy of Sciences and the Humanities (2002), the European Academy of Sciences and Art (2004), an associate member of the National Academy of Sciences of the U.S.A. (2003) and the American Academy of Art and Sciences (2005). Her distinctions include the Kolthoff Award of the Technion (1990), the First European Crystallography Prize (2000), the Kilby International Award (2000), the F. A. Cotton Medal of the American Chemical Society (2002), the Israel Prize for Chemical Research (2002), the Anfinsen Prize of the Protein Society of Boston (2003), the Paul Karrer Gold Medal (2004), the Massry Foundation International Award and Medal for Ribosome Research (2004), the Datta Award of the International Biochemistry Society (2005), the Fritz Lipmann

Lectureship of the German Biochemical Society (2005), the Louisa Gross Horwitz Prize of Columbia University (2005), and the Rothschild Prize for Life Sciences (2006).

We recorded our conversation during the 22nd European Crystallographic Meeting in Budapest in August 2004.*

First I would like to ask you about your family background.

My parents were born in Poland and immigrated to Palestine in 1933, just after Hitler came to power in Germany. Eventually, they settled in Jerusalem. My father was a rabbi, so were his father and his grandfather. After my parents came to Palestine, the climate change contributed to the deterioration of my father's health and he died when I was 11 years old. I had a little sister and it was a very difficult time for my mother. I had to help her emotionally and also in practical terms. We were very poor.

In spite of all these circumstances, I remember my childhood as a wonderful period of my life. My parents were very supportive. I started earning money at the age of 11, teaching younger children, and I was also caring for my sister. After about three years we moved to the north

Her parents, Hillel and Esther Livshitz, in about 1946 in Jerusalem (all photographs courtesy of A. Yonath).

---

*Magdolna Hargittai conducted the interview.

of Tel Aviv where my mother's sister lived. Eventually I went to high school, but I kept working all the time. My memory is that I never had time for anything because I had chores before school and after school. When high school time came, I was admitted to the most expensive high school in the country, without paying for it, but I had duties.

As long as I remember back, I always felt that I wanted to learn more. I was never satisfied with what the curriculum would give me. In the high school I always went to the school library and I read a lot and I enjoyed reading and learning enormously. I liked literature, but disliked the classes where we had to interpret the authors' aims. This doesn't mean that I didn't have my opinion about what I read; it only meant that I didn't necessarily agree with the teacher's opinion. Nonetheless, I was a good student. I also liked writing.

How did you decide to go into science?

In Jerusalem, mathematics and physics were open to almost everybody who had good grades from high school, but they were very restrictive in chemistry because they needed lab space. The number of chemistry students was limited to fifty whereas they could take more than two hundred for mathematics. Nonetheless, I applied for chemistry and thought that I would move to something else if I did not like it, or if I would not be admitted, but I was. This made me very happy. During the first two years we had plenty of organic chemistry, inorganic chemistry, and physical chemistry, but hardly any biochemistry. By then I had decided that my interest was in biochemistry and biophysics. I always reached my goals even though not always through the most straightforward ways. I ended up with a M.Sc. in Biophysics and continued for a Ph.D. at the Weizmann Institute in Chemistry. I was considering doing protein crystallography, but it was not yet well developed. This was in the mid-1960s. Professor W. Traub suggested to me to do fiber crystallography, and I worked on the structure of collagen. In this work, I collaborated with Efraim Katzir and his group, who prepared our samples.

After my doctorate, I went to Pittsburgh for a postdoctoral year at the Mellon Institute, and did muscle research. However, protein crystallography was still in my mind as my principal interest. So, finally, I started doing protein crystallography in the group of F. Albert Cotton at the Massachusetts Institute of Technology in Cambridge. This meant a major turn in my professional career. It helped also that I had extremely good relations with William Lipscomb at Harvard University. I had contacted him during my collagen work when I was considering coming to work with him, but at that time it did not happen. During my year at MIT, Lipscomb gave a course there, which I attended.

Following the two years in America, I returned to Israel, and started my own group in protein crystallography. I was alone in the entire country. I had an instrument and some limited lab space, and it took almost half a decade before things started working. During this period we did publish papers on simple structures. We solved the structure of triclinic lysozyme, after David Phillips determined the structure of the tetragonal form. One of his postdocs, Dr. John Moult, came to Israel and worked with us in solving triclinic lysozyme, which we then used as a matrix to investigate unfolding and refolding in cross-linked crystals. The lysozyme itself we bought from a company; the cross-linking I did myself. I had to learn a lot of the procedures from books and at meetings.

From the beginning of the 1970s, protein crystallography in Israel was slowly moving to the frontier of science. It was also at that time that I developed a collaboration with Professor Michel Ravel at the Weizmann Institute. He had a technique to prepare what he called large amounts of the initiation factors, which initiate ribosome function in protein formation. The work involved an intensive collaboration with the late Professor Paul Sigler, who came to Israel for a year and a half. We had good interactions with the Chicago group; later I had spent a sabbatical year there. We were trying to grow the necessary crystals, but failed. During this year, at a meeting in Canada, Professor H. G. Wittmann of the Max Planck Institute in Berlin talked about the ribosome and reported the sequence of the initiation factors. We talked and he was interested in collaboration, and eventually I went to Berlin.

It was just a few months before I was supposed to go to Berlin with one of my students that I was riding my bicycle to the beach — it was in February, and the weather can be beautiful in February in Israel — and I fell down in the middle of the street. I had a brain concussion. I was brought to the hospital. My brain concussion was more or less over within two weeks, but there were side effects, which prevented me from flying. Also, I needed an operation. After everything was done, I went to Berlin about five months after the original plan.

When I arrived in Berlin, in November 1979, the initiation factors were almost ready, and in my "free time" I discovered that they had very active pure ribosomes in huge amounts from several bacteria and I suggested

Mounting a crystal on the camera, in the X-ray crystallography laboratory of the Weizmann Institute, 2001 (photograph by Samuel Engelstein).

using them for crystallization. The people were very supportive. I knew that many prominent scientists had failed before in attempts to crystallize ribosomes, and if I would fail too, I would be joining a distinguished group of luminaries, including Francis Crick, Jim Watson, Aaron Klug, Alex Rich, and others. I knew though that this was my big chance. I went about the project very carefully, since I assumed that the difficulties with ribosome crystallization stem from their heterogeneity as well as their tendency to deteriorate. First of all, I went back to the old literature and studied everything that was written about the ribosome, especially techniques developed for maintaining their integrity for relatively long periods, required for crystallization. I took advantage of procedures developed in the sixties by A. Zamir and D. Elson. I spent only two months in Berlin, but after I had returned to Israel, they kept sending me almost every week pictures taken by a light microscope (neither fax nor internet was available in those days). In three or four months, we had micro-crystals, which were much too small to be studied as single crystals, but gave a promising weak powder pattern. Then it took about four years to get the first diffraction patterns that didn't look like garbage. Our paper about the micro-crystals came out in 1980, and for the past quarter century I have been involved in elucidating the structures of ribosome.

Would you give a brief introduction to the ribosome?

The ribosome is an organelle within the cell, which appears in thousands of copies in each living cell. It is a huge and dynamic assembly of proteins and RNA. This is where the genetic code is being translated from the nucleic acids to the proteins. In other words, it is the site for the synthesis of proteins resulting from translation of messenger RNA (mRNA). Transfer RNA (tRNA) that decodes the genetic code carries to the ribosome the cognate amino acids, to be incorporated into the growing chain. The biosynthesis of proteins happens with fantastic speed in the ribosome. When a chemist wants to create a peptide bond, it may take days and high temperature and other extreme experimental conditions, whereas the ribosome can do this fast, within microseconds, and under mild conditions within the living cell. Also, scientists often make mistakes; the ribosome hardly ever makes mistakes.

In principle, any ribosome can read any genetic code. A ribosome in the human body can translate the genetic code of bacteria and vice versa. The ribosome is a factory for making proteins and can follow any genetic instructions. However, the ribosome of a higher organism — mammals and eukaryotes — is more complex than the ribosome of bacteria. The higher complexity is a consequence of additional tasks concerning regulations and selectivity, and it has to do with more interactions with the cell. The differences between bacterial and mammalian ribosomes are subtle. Even the active sites or areas near the active sites contain some differences. This is why ribosomal antibiotics can work. The antibiotics should impact the pathogenic bacteria only and not the patient, not even cause side effects. Sometimes replacement of one single nucleotide can make the difference in the effects on bacterial and mammalian, that is human ribosomes.

Are there direct physiological consequences of how ribosomes work in different individuals?

There maybe all kinds of differences, but not in the ribosomes. The internal signaling system in some people responds very fast to external signals, in others they respond more slowly. This is why the consumption of the same amount of alcohol might induce some persons to feel sleepy while others become more animated. The proteins that degrade the incoming alcohol are not present all the time in the organism, only when they are needed, unless, perhaps, one is an alcoholic, and consumes alcohol continuously. So it depends on the speed in which the signal alerting it to the presence of alcohol reach the ribosome, what the individual response to alcohol consumption will be. The ribosome does not make the decision, it must be notified by a signal, and associate with the corresponding messenger RNA.

More serious may be the consequences of the speed of the signals reaching the ribosomes during some heart problems. Many heart problems can be taken care of by the body itself. In such a case the ribosome produces enzymes that can tackle the problem. Again, those enzymes are not being produced continuously, but only when there is a problem, when an event takes place. This is also why higher organisms have larger ribosomes than those of bacteria, because they have to be able to respond to various events not existing in prokaryotes.

I would like to ask you about the role of symmetry in the ribosome structure.

The ribosome is not symmetrical. Although there have been larger systems whose structures have been elucidated, some viruses, for example, but the ribosome has been the largest asymmetrical structure that has been determined. However, when we looked at the ribosome structure very carefully, we found two-fold symmetry around the active site. This two-fold symmetry is present in all ribosome structures determined to date. The symmetry relates to the backbone around the active site between the position where the tRNA comes in with the amino acid and the position of the tRNA carrying the nascent chain.

Why should there be this two-fold symmetry?

Initially, this was a puzzle, but by now, we know the reason. Once we found this symmetry, we soon understood its functional significance. The stereochemistry of the peptide bond requires this symmetry. The frame is the ribosome. The tRNA substrates follow it, so that there is a certain complementarity in the incoming tRNAs and the outcome structures. Hence, this fascinating structural element provides the architecture required for peptide bond formation, for the elongation of the newly-born protein, and for directing it into its exit tunnel. Not only did we determine this symmetry in our analysis, but we also built models that demonstrated the requirement of such a symmetry.

In the double helix of the DNA structure, there is also two-fold symmetry.

In the ribosome structure, the two-fold symmetry has a defined function. When I talk about the dynamic character of this symmetry, I mean the dynamics of the catalytic event. It took more than a year for us to convince ourselves that the observation of two-fold symmetry was real. It makes the structure even more beautiful. In fact, initially, we were not very popular with this observation. The Yale Group (T. Steitz and P. Moore, who determined the structure of another ribosomal subunit) had previously come out with the idea of chemical catalysis before we had completed our studies. They suggested that the ribosome, acting as a ribozyme, is directly involved in the chemical catalysis of the biosynthesis of the proteins, and this idea found its way into textbooks. There was a lot of controversy, papers and verbal discussions about the mechanism of protein production, but it all converged to our 2-fold symmetry positioning mechanism. In addition, it turned out that the Yale Group worked under conditions that were far from physiological environment, hence ended up with significant disorder in the functionally relevant features.

We worked under physiological conditions. Our findings of the internal symmetry became firm in 2002, but it took us nine months before we could convince the referees of the correctness of our results. Even then we took a very low profile. We did not insist on having found the only mechanism; we did not exclude the possibility of other mechanisms. Even the people who found out biochemically that the mechanism suggested by the Yale Group is wrong, were puzzled by our discovery of two-fold symmetry, which, apparently, nobody else had noticed. Other scientists, who are not crystallographers, did not quite understand the significance of the symmetry, and this also made it more difficult to get our point across. More recently, during the past months, however, there have been publications showing that our results of structure analysis are being increasingly accepted.

Who were involved in these latest studies?

Most of the work was done by one of my students in Israel and in Hamburg. Two Israeli woman scientists are involved. Dr. Ilana Agmon discovered the two-fold symmetry of the ribosome system, based on the interpretation made by Dr. Anat Bashan, the chief crystallographer in my group. During the most intensive periods of the structure analysis, we were about ten people. However, in the past, when we were pioneering the procedures, the number of participants went up to about twenty-five people. In most periods, even when the group was very large, people worked on specific problems individually and publications came out specifically by those who carried out the work. However, in the pioneering stages the whole group contributed equally and appeared on all publications.

Several people have mentioned to me that you should receive the Nobel Prize, people who do not talk about such matters lightly. Have you thought about that?

May I not answer this question? It is embarrassing. When I got the first micro-crystal, I met with a Swedish professor, who is not alive anymore, who was one of the founding fathers of structural biology. At that time I was working very hard, it was an exciting period of my life, I hardly ever slept. He noticed that I looked pale and haggard and asked me why? I told him that I might have crystals of the ribosome. He looked at me and said that this was a Nobel Prize project. This was right at the beginning of the work, in the middle of the 1980s. I never talked about it, but it has stayed with me. When we got the first high-resolution results, there was a scientific advisory committee meeting at the Weizmann Institute, and there, again, some people expressed the same opinion. Such impression tends to leak out, and people very often ask me about the Nobel Prize, a question which I do not like. But I know that this is a project that is very much in the center of attention, so it would be useless to deny that I am aware of the possibility.

Do you feel nervous when October is approaching?

I hardly think about it except when people like you are asking me. It was never the prizes; it was always the intellectual stimulus that drove me. With this I am not saying that I am not happy by getting recognition.

You have moved around a lot between different places. What is your permanent affiliation?

I have always had the Weizmann Institute in Israel as my base and permanent affiliation. Berlin was collaboration, and Hamburg was a temporary position (for 18 years), which ended in the middle of 2004. The people in Hamburg have become well established, and, besides, I have reached retirement age.

How did you manage when you were commuting between Rehovot and Hamburg?

It was difficult and time-consuming, but it was necessary. I could do measurements in Hamburg that I couldn't have done at home. The Max Planck Society established a unit for me in Hamburg. It was very helpful even though for collecting high-resolution data we had to go away from Hamburg to more advanced synchrotron.

With her granddaughter, Noa, at the party given at the Weizmann Institute after receiving the Israel Prize in 2002 (photograph by Shalom Nidam).

You have not retired in Israel yet?

We also have a retirement age, but professors can get an extension, so I have not retired yet.

Are you married?

I was. I have a daughter, who is an M.D. and a lecturer, and I have an eight-year-old granddaughter. She knows all about ribosomes. Years ago, she was watching when I was receiving a big Israeli prize and had to be standing in front of the whole nation, all dressed up. After that her kindergarten teacher asked me to give a lecture about ribosomes to the kindergarten. I consider it my greatest success that I kept them spellbound for an hour.

How did you manage having a child and doing science at the same time?

She was little when I was at MIT and there was a kindergarten. We also took turns with my husband, trying to share responsibilities. It was not easy, but I always found some arrangements. I had to solve this problem for the holidays as well, because I could not stop my research during the holidays. I also did a large part of my work at night as well. Sometimes I couldn't stay for the whole day in the lab, but I always put in a full day's work, mostly more, using the nights.

Receiving the Israel Prize, in 2002. Yonath shaking hands with the President of Israel, Moshe Katzav. Standing, left to right: Limor Livnat, Minister of Education; Ariel Sharon, Prime minister; Moshe Katzav, President; and Aaron Barak, Head of the Supreme Court.

Did your daughter ever complain that you were not there when she needed you?

A little bit, but she also developed to be very independent. She considered it my merit that I was not there all the time. She was already 16 years old when I started my collaboration with the Max Planck Institute, and had to go away for longer periods of time. I asked my mother to be with her, but my daughter decided that she could manage on her own. As we lived within the confines of the Weizmann Institute, it was a safe place and I was not afraid to leave her alone. More than that, eventually she started taking care of my bank account, my car, she paid our bills, the telephone, everything. She learned early how to be responsible. I could completely rely on her.

Did you ever have any experience with discrimination against women?

No. The only thing I can think of is that sometimes I had the impression that they expected more from me than they would from a man in my position. Even this was not a very strong feeling. I was making very slow progress in the beginning because I was inexperienced in crystallography, but this was not because I was a woman. There was one occasion, when we were still together with my husband, the Weizmann Institute did not promote me, and one of the professors told me that they were not worried about my leaving them because my husband was there too. This I didn't like, of course. I can't recall any other negative experience. Actually, I was offered some jobs, since I am a woman … .

Did they put you onto many committees, which usually happens to successful women?

This didn't happen because I was constantly on the road.

What do you think about the phenomenon that while there are about the same number of male and female students at the undergraduate level, the number of woman professors is a small fraction of the number of male professors?

It may not only be discrimination. It may also be a diversion in women's interests. Opportunities for women have opened up recently in so many new areas. It may well be that women like to spend more time at home, they want to devote themselves to other interests, they may not be so strongly focused on the professional advancement as men usually are. The bottom line is what you observe about the distribution at higher positions, but it is not just the result of discrimination. Also, my experience is that women are more demanding towards women candidates for a position than men are. It may also be that my approach to this whole problem is old-fashioned. I am saying this while I am convinced that intellectually women are at least as capable as men for higher positions in academia and elsewhere. However, society as a whole has not been very supportive toward women in achieving high positions.

What was the greatest challenge in your life?

The lecture to the kindergarten. Scientifically, it was the crystallization of ribosome. Personally, it was the period when my father died.

Are you religious?

No. My parents were religious and the neighborhood where I grew up was religious, but I was not and I am not. My mother left religion as well.

You had a longstanding collaboration with German scientists. Did you feel comfortable in Germany?

I never felt good in Germany as a country, but I liked my working environment. My group was very international. I didn't speak German although I made an effort, I wanted to force myself, but I failed. I had hardly any friends in Germany. I had nothing else in Germany, but my work. I had a car and I drove a lot between Berlin and Hamburg and I slept often in my car. For the first 5 years I had no address (namely no room or apartment) in Germany. When I met Professor H. G. Wittmann for the first time, it was many years ago at a conference in Canada. He gave an interesting lecture (which, in retrospect, changed my scientific life), but it took me days before I overcame my barriers and addressed him. Before, I just couldn't have imagined talking to a German. Of course, this was my problem. At the Institute, the secretaries were all German, and we had very nice working relationship.

You have had a tremendous drive. Where does it come from?

It's luck. I have told you that when I arrived in Berlin, most of the work I had planned had already been done. Nonetheless, when we got the microcrystals, I just immersed in the work because we had this fantastic aim. I could not stop myself, even if I had wanted to which I did not. The vision to understand ribosome was an irresistible intellectual stimulus.

Isabella L. Karle, 2000 (photograph by M. Hargittai).

19

ISABELLA L. KARLE

Isabella L. Karle (b. 1921 in Detroit) is Chief Scientist of the X-Ray Diffraction Section of the Naval Research Laboratory since 1959. She received her M.S. in 1942 and her Ph.D. in 1944 at the University of Michigan in Ann Arbor. She participated in the Manhattan Project at the University of Chicago. She moved to the Naval Research Laboratory in 1946 after teaching chemistry at the University of Michigan. She is a member of the National Academy of Sciences of the U.S.A. (1978) and of the American Philosophical Society (1992). She received numerous awards, among them the Garvan Medal of the American Chemical Society (1976), the Chemical Pioneering Award of the American Institute of Chemists (1985), the Women in Science and Engineering Lifetime Achievement Award (1986), the Franklin Institute's Bower Award (1993), the Aminoff Prize of Sweden (1988), the Bijvoet Medal of the Netherlands (1990) and the United States National Medal of Science (1995). She has been granted honorary doctoral degrees by the University of Athens (Greece), the Jagiellonian University (Poland), Harvard University, the University of Pennsylvania, as well as the University of Michigan, Maryland University, Georgetown University and Wayne State University. We recorded our conversation on May 6, 2000 at the Naval Research Laboratory in Washington, D.C.*

---

*Magdolna Hargittai conducted the interview.

I have read somewhere that you are the world's most esteemed couple in science.

[Dr. Karle is laughing.] I've heard that off and on. It pleases and amuses me, but we don't think about it. It's been a very interesting life. We often travel together to meetings. Also, for much of what Jerome does, he would like to have experimental examples and he uses the results of many of the structures that I have determined. This has not happened by design but it turns out that way.

How did you get into science?

None of my family had any technical training. I didn't know what science meant until I got to high school where I was told that I had to take a science course in order to qualify for admission to a university. Our school offered chemistry, biology and physics. My adviser threw a coin and it turned out to be chemistry. As it turned out, the very first day that I attended a chemistry class, I became fascinated with the subject. I've been fascinated with it ever since.

Did you have other plans before?

My father would have liked me to become a lawyer. It would have been his ambition had he had the opportunity to go to school. My mother thought that there would be nothing better than being a school teacher. She would have done that if she had been able to. I thought that school teaching was very nice. Lawyering I didn't care much for. I was still looking around when chemistry came along.

What was your family background?

Both my parents were born in Poland. They didn't know each other there. They came to the United States when they were still fairly young. They met after World War I in Detroit. When my father left Poland, the country was occupied. He lived near the border between the Russian occupation and German occupation. He never had any formal education except for going to a German school for a little while. On the other hand, his parents, although they were not well off, owned a lot of land. They were part of the landed aristocracy. They managed to have a private tutor for my father when he still lived in Poland. He was the youngest son of his father's second family and with the way inheritance laws worked, it was decided that he might have a better life if he went to the United States. One of his older half brothers had already settled in Duluth, Minnesota, and that is where my father went. This was just before World War I. He enlisted in the United States Army and was one of the very few people who could speak Russian. The Army immediately placed him in a group called the Polar Bear Expedition. The United States sent an army to the very northern part of Russia to try to quell the Communists. He was an interpreter for the officers and he was a censor for the Arkhangelsk newspaper. Fortunately, he personally did not experience much actual combat. After he returned to the United States he spent the rest of his life as a painter, not of the artistic kind.

My mother's family had a lot of children. Her mother died soon after her little brother was born. There was a father with seven children and no wife. He came to the United States and settled in Wilmington, Delaware. His second marriage did not turn out well for the children at all. The three eldest children married, they took in the three youngest one to live with them. The middle boy fended for himself. My mother said that she was happy living with her oldest sister, but they were rather poor. My grandfather apparently owned some farmland in Poland, which he sold and that made it possible for him and the children to come to the United States. If you can believe it, growing up in Delaware, my mother was never sent to school. First, she had to look after her baby half-brother and then, when she was twelve years old, her sister and family moved to Detroit, where she went to work in a factory. She said she was sixteen. However, both my parents could read and write in two languages and were quite knowledgeable in practical mathematics. She led a productive life. When I was five years old my mother taught me to read and write in Polish. My text book was "Polski Elementarz". After I was enrolled in elementary school in Detroit, learning to speak English and reading and writing in English were not a problem.

After you had fallen in love with chemistry, how did you continue?

I graduated from high school in the middle of the school year. Wayne State University in Detroit was a free university and I enrolled in it for the spring semester. In the meantime, there were examinations for the University of Michigan at Ann Arbor, for a four-year scholarship. I took one of those examinations and I placed rather high. As a result, I received a scholarship to the University of Michigan that paid my tuition and had other monetary benefits. I also worked as an assistant to a chemistry professor and my parents helped with some expenses.

I got my Bachelor's and Master's degrees as well as my Ph.D. at Ann Arbor and later was also on the Faculty for a while. One of the reasons for not moving elsewhere was that during World War II it was not easy to go to other universities for graduate work. Besides, I was being well taken care of at the University of Michigan. I had the lowest rank in the faculty, and the reason that I could teach at all was that there weren't enough men around. Many of the male faculty were engaged in war-related research and additional teaching staff was needed. My students were 16 and 17 year-old boys since, as the boys reached the age of 18, they were taken into the armed forces.

When did the two of you meet?

I was a senior when Jerome had come to Michigan as a graduate student. Although he already had a Master's degree in biology, he decided to specialize in chemistry, hence we had some courses together. We married two years later, even before we got our Ph.Ds.

Is it possible to delineate your scientific research?

Jerome and I work together separately. We both did our graduate degree work with the same man, Lawrence Brockway, in gas electron diffraction. Then we spent some time on the Manhattan Project at the University of Chicago in the chemistry laboratory. Our activities differed from those in our Ph.D. programs. I was making plutonium chloride from a crude plutonium oxide and Jerome was trying to make the pure metal. The objective of the Chicago project was to make pure plutonium without any impurities. A number of different paths were being tried simultaneously. I don't know how the plutonium metal was produced eventually, but Jerome succeeded in making some rather pure metal directly from the oxide, and I managed to make pure plutonium chloride by many different paths using very high temperatures in a vapor phase process. The resulting crystals were absolutely beautiful. They were jade green and they grew with large smooth faces.

After we finished our particular projects, we could have stayed in Chicago, but there was no particular urgency since the thrust of the project had moved to Los Alamos.

Isabella Karle at the gas electron diffraction apparatus at the University of Michigan, early 1950s (all photographs courtesy of I. Karle unless indicated otherwise).

Did you have any interaction with Fermi?

Only in that there were weekly lectures and he gave many of them. The lectures were about what the problems were and how the project was coming along. They told us some, though not much, about what was going on in the other laboratories.

I would like to ask you about the situation of women in science. Although there are many women in undergraduate education, as we go higher in the academic ladder, their numbers dwindle.

Let us look at some other countries and consider crystallographers in particular since I know them best. My women colleagues in France always envied me because of the freedom of research that I have had here. There were some very good women scientists in crystallography, but they could not achieve the status that men had at the universities. Women just weren't appointed to higher positions. There are quite a few women crystallographers in Italy, and they always come up to me and ask me about how I managed because they are usually the glorified assistant but rarely the leader of any group. In England, the atmosphere has been better. There are quite a few women in crystallography who are professors and heads of groups. In the United States, it's mixed. It goes all the way from the Italian way to the British situation. I don't quite know about the present situation because I don't have many young colleagues. Of my generation, the late Shoemakers were a good example. Both David and Clara were excellent crystallographers. He used to be a professor at MIT and she was always living off his grants. Not until they went out to Oregon where he became the department head, was she appointed to her own position as a full professor. This was during the last few years of both their lives. Ken Hedberg and his wife, Lise, were both electron diffraction people and I don't think she ever had a real job. She also worked on his grants. There were several other instances of the same kind. The universities wouldn't hire both husband and wife for any number of reasons. Although these women published extensively and did very good work, either they never succeeded or succeeded only in their later years to become independent. From that point of view, research positions in the U.S. Government laboratories were generally a lot better.

Is there any change?

Although more women get their Ph.Ds., they all do not stay in chemical research. Many become public relation representatives of industrial companies or they may become consultants. Research grants are competitive and a woman must be hired by someone first in order to compete for most of the grants.

Do you think that women do science differently from men?

Probably. A lot of them like crystallography because we could bring up children at the same time. Crystallography wasn't something that you had to watch all the time. You could take it home with you, you could think about it while minding the babies. Most of the projects in crystallography, for example, start with an idea or a substance, and there is a possible

Isabella Karle, around 1960.

stopping point when the crystal structure is solved. This can be an isolated procedure. In order to have a research project you may want a number of related experiments that would complement each other, but it is possible to do it in a stepwise manner. In other types of research projects, there maybe a necessity for much more immediate interaction.

How did you manage to have children and stay at the top of your science at the same time?

We were fortunate that after World War II, there were many women of grandmother age whose children didn't want to live on the farm anymore. The mountains of Virginia are about sixty miles from here. Many of the younger people came to Washington during the war and didn't want to return to the country after the war ended. The elder ladies came also and often served as live-in housekeeper/babysitter during the week. They would visit with their own children during the weekends. This procedure worked out very well for us until our children were old enough that we didn't need constant help anymore.

Having children required organization, but I never felt them to be an obstacle for my career. It never interfered with our professional lives. When they got a little older, and by a little older I mean at least seven years old, we always took our children with us for our summer travels to various meetings in Europe, Japan, and the United States.

Did they ever complain about having had a working mother?

Not really. I managed to attend or transport them to the various school functions. They were members of the Camp Fire Girls, had their piano lessons, and practiced their sports like gymnastics. Weekends were reserved for family events or for events with their friends. They especially liked traveling with us in the summer time.

Did your and Jerome's careers influence their choices of a profession?

We never tried to influence them. They were all good students and they all had their own interests, but they all turned to science. Louise, the oldest

Louise Hanson (daughter), (grandson Jeff's girlfriend), Jeffrey Hanson (grandson), Madeleine Tawney (daughter), Nicole Tawney (granddaughter), Jean Karle (daughter), Christine Hanson (granddaughter), Isabella Karle, Michael Tawney (grandson), Jerome Karle, Lionel Skidmore (Jean's fiance), Brian Tawney (son-in-law), celebrating Jerome's 80th birthday, 1998.

one, works at the Brookhaven National Laboratory. The middle one, Jean, lives nearby in Virginia, and she works at the Walter Reed Institute for Research. She designs drugs and uses X-ray crystallography to enhance her research. They are both chemists. The youngest one, Madeleine, studied geology. She worked for the Smithsonian Institution, in their natural history museum, for some fifteen years. She has some young children now and is staying home for a while.

Nowadays, many young women who aim at a scientific career postpone having children.

I have a grandchild who just turned three. He is the son of my youngest daughter who is going to be forty-five this year.

Is there any discrimination against women in science in the work place?

It's different in different places. Some universities have quite a few women on their faculty, others have very few and only at the lowest levels.

Have you met or heard of any reverse discrimination?

It may happen sometimes. A university has to have a woman in their department. I don't know how to adjudicate this sort of thing. In some cases the women are overlooked. In other cases, schools are aiming at a certain percentage of women in order to qualify for their state funds or whomever they get their funds from. There could be some reverse discrimination.

Is this whole women issue an important one or is it being exaggerated?

I don't have enough information to form an opinion about this. I hear many more complaints from women about lack of funding without it being specifically a gender issue.

Often the few high positioned women are overwhelmed by committee memberships and other engagements. Have you had this experience?

One has to be careful about accepting all the invitations that come along simply because you would be the only woman on the committee or on the panel, and that can overrun one's life. There is then something else too. There are not that many women who want to become heads and run things. They want to do their research.

Who are your heroes?

Male, female?

Either.

Haven't thought about that one. Some of my heroes are people who either volunteer or contribute greatly to hospitals for children where the children have terrible diseases, where they need not only medical care, but also loving care, and where the children may not always be in a condition where somebody is there to love them. The people who look after such children are heroic people. They do this sometimes at the risk of contracting the diseases themselves.

Looking back, would you care to single out one or two pieces of your research that you are most fond of or most proud of?

It was certainly pleasing to be able to put crystal structure analysis on a practical basis. Of course, Jerome and Herb Hauptman developed the theoretical work. However, there was a very definite step between having an infinite set of inequalities and having a set of real data with experimental error and limited scope and what you do with transcendental functions. It was very satisfying when I was able to present schemes of operation first for the direct determination of centrosymmetric structures and then for the non-centrosymmetric structures. It's also pleasing that all these procedures have now become so commonplace that there are ten or even fifteen thousand crystal structure publications a year. After that I became more interested in the materials I was working on rather than the procedure. So I spent quite a number of years working on peptides. Of course, peptides are made of the same kind of amino acids as proteins, although not always. The natural peptides that are derived from the low forms of life have a greater variety of amino acids than those that come from plants and animals. This work meant working out their conformations, their folding patterns, recognizing the residues that induce a helix or a beta bend, or a combination of a helix and a beta bend, and trying to see their practical biological activities. This has been a very fascinating area of my research.

Would you care to reflect on the 1985 Nobel Prize?

It would have been nice to have it. On the other hand, I have received many awards of my own, which are significant and which satisfy me.

Isabella Karle with President Bill Clinton and Vice President Al Gore at the White House after receiving the National Medal of Science, 1995.

How did Jerome feel about it?

He had always felt bad about the fact that I was not included in the Nobel Prize. He had been very pleased when I got the Bower Award and the National Medal of Science.

Has he been a very supportive husband?

Yes.

You were both Brockway's graduate students.

Yes, this was so in spite of the fact that Jerome was a few years older than I, but he had had a different path of education from mine.

Did you ever work jointly on a project?

When we first came to the Naval Research Laboratory, we set up an electron diffraction laboratory with our newly-designed and constructed instrument. I did the practical work and he did the theoretical work for the most part. Every once in a while I needed his help. We wrote papers together because both the theory and experiment came together. After he and Herb worked on the phase problem in crystallography, Herb left for other places. At that point, Jerome asked if I would do crystal structure analysis because nobody elsewhere was trying to apply his equations. It was almost five years after they wrote their original papers. So I set up the first X-ray laboratory for our group. We bought or borrowed the equipment and I had to use Martin Buerger's books on how to handle crystals and how to take the photographs in order to identify the X-ray reflections. Most of the people involved in

Isabella and Jerome Karle at the Naval Research Laboratory with molecular models, around 1970.

Isabella Karle with her X-ray diffraction apparatus, around 1955.

Isabella Karle and the formula of dihydrohistrionicotoxin, the poison of arrow frog *Phyllobates aurotaenia*, 1995.

crystallography had learned that sort of thing in graduate school. We published a lot of structure papers together. Again, I did the practical work in solving the structures and Jerome did the theoretical work in trying to figure out which of the formulas might be useful when I ran into difficulties. We work together that way.

Then your interactions extended to many other people.

After a while we were invited to participate in special summer schools where we instructed young scientists on how to use the procedure. These were held mostly in England, Czechoslovakia, Italy, Germany, Poland, Japan and Brazil. However, by that time, Jerome was thinking about other things. I was continuing with crystal structure analysis, especially with a number of people, among them very well known organic chemists who had various problems and decided that maybe I could help them, and I was able to. For example, very potent toxins were isolated from the skins of South American poison dart frogs by Bernhard Witkop, John Daly and their collaborators (National Institutes of Health). Their chemical formulas were quite different from the alkaloids known at that time (late 1960s, early 1970s). I was able to establish the unique formulas of a variety of these toxins, such as batrachotoxin, histrionicotoxins, gephyrotoxin, etc, by solving their crystal structures. In another example, there were a number of strange products that were isolated from photo-rearrangement experiments, for which the chemists had difficulties in determining their structural formulas. These were readily established with crystal structure analysis. In other cases, the total amount of an isolated new natural product was so small that the identification of its formula by chemical means was very difficult. We would grow a crystal and I identified the substance. I went through a whole series of natural compounds that included steroids, alkaloids, terpenoids, carbohydrates, amino acids and peptides, among others. I've been concentrating on peptides and peptide hybrids for quite a while now.

At the beginning of your career you did some pioneering work in gas electron diffraction.

That was interesting in that Jerome felt when we were doing our graduate work that there must be procedures that were superior to estimating the intensities of the diffraction pattern by eye. We had heard about the rotating sector although we never saw one. Finbak used it in Norway. After World

War II we designed a new apparatus with a rotating sector. That, of course, considerably flattened out the steeply falling background pattern, thus enhancing the undulating pattern that contains the molecular structure. Also, Jerome put in a damping factor, so that we wouldn't have a sharp cutoff effect that accompanies data that was necessarily limited in scattering angle. We were able then to perform a Fourier inversion of the data in order to obtain measures of the vibrational motions as well as the interatomic distances from the radial distribution curves. There were not very many people in the field, but about a dozen laboratories sprung up around the world, including Japan and the Soviet Union, and also in Hungary. It was an interesting time and we were establishing the structures of almost all the small molecules. In a few years, the structures of the simpler molecules were established and we got to the point where the molecules were becoming more and more complicated. We investigated rotational motion and it maybe that we would have gone into high-temperature electron diffraction or electron diffraction combined with other techniques. However, crystallography came along, and our interest turned to that.

Do you talk about your actual work at home?

Sometimes.

Do you get advice from each other?

Sometimes.

Does it ever happen that Jerome gets the credit for what you did?

I suppose so.

Does it ever happen the other way around?

Not often.

What was the greatest challenge in your life?

My life has been full of challenges, but I don't know if there was a greatest one. I take them as they come along whether they are scientific or personal.

You told me about your work with peptides and it has continued. What do you do these days in this area?

I've been working with naturally occurring peptides, often with those that form ion channels, and with designed peptides with various folding properties. I've also been working with Darshan Ranganathan, a lady in India who has received the chemistry award from the Third World Academy. She builds hybrids containing both organic moieties and peptide segments. The reason for doing that is to attempt to make pores of various kinds for the transport of electrons or for the transport of metallic ions, or even large molecules. These pores may or may not form passageways through cell membranes. We've had quite an active publication program on these structures. The pores may have applications in somewhat diverse fields, including medical fields, for example, for transporting medication to a proper site in the body. They may have applications in material science as well, for solubilizing non-soluble ingredients and for the study of nanotube formation.

Do you do only what you enjoy doing?

Definitely. I could have retired fifteen or twenty years ago, but I have remained in the laboratory because I do what I enjoy doing.

You have had a spectacular career. Nonetheless, if you could chart it from the beginning, what would you like to change in it?

Not very much. We've traveled a good deal around the whole world. Neither one of us is the kind of person who likes to sit on the beach or to sit in a boat catching fish. We were happy to have children and grandchildren and we keep busy, perhaps too busy.

Would you encourage young people today to embark on a science career?

I think so. There are ups and downs in employment and funding. On the other hand, science keeps changing all the time. Every few years there is something very new and very exciting. If you don't like it this year, wait a little while and something else may come along.

What is more important for a beginner, to have a strong mentor who gives direction or to have independence?

It depends on the person, the young person, that is. Some people thrive very well with mentors; they make excellent teams with mentors. Others are very anxious to do their own work. It is necessary to have both kinds of situations available. Some young people may go astray if they have too much freedom to work independently. Others do quite well when they are on their own. It helps if the young people have a good background in a number of different areas. To be a good mentor does not mean only to be a good scientist, but to be interested in the young people's problems, and in their development. A good scientist, if she or he is not interested, may just not have the time for the young co-worker.

There seems to be a gap of top science in the United States, leading the world, and the general ignorance of much of the population.

The National Academy of Sciences has had a very active committee for the last ten years or so in trying to improve science education from kindergarten through twelfth grade. They have done quite a number of things to make it easier for schools that do not have teachers trained in science to be able to teach their students by making sure that there are textbooks that even teachers who have no idea about science can use with their students. I find with my grandchildren that there has been quite a bit of science introduced from the very beginning. I don't know how many schools they reach. This is a very varied country. There are rural schools and small towns that are poor and don't have the facilities and money for hiring good teachers. But efforts are being made to improve the teaching of science at all levels.

Concerning support for science in the United States, there was a period after the Sputnik when it was enormously increased. Sputnik and Gagarin's flight were very beneficial for American science education as well.

True.

Competition with the Soviet Union had also its beneficial effects. Has that incentive been replaced by something else?

[Long silence] There are all the computer-related activities that fascinate children of all ages. For the most part though, the leading scientists have an inner inspiration for science rather than external incentives or competition.

In addition to crystallography, astronomy seems to be yet another field that attracts women scientists.

There are a number of women astronomers who are members of the National Academy of Sciences, but I do not know why astronomy is particularly interesting to women. I have never considered such matters. Curiously, there are more and more women engineers now.

But there are only two women at the National Academy of Engineering of the U.S.A.

That I didn't know.

Would you care to tell me about your relationship to religion?

I suppose. I was brought up a Roman Catholic. My parents were not particularly religious. I went through the usual schools for Catechism, went through the First Communion and Confirmation, but never considered religion to be of primary importance. Jerome is Jewish and he went through the bar mitzvah, but that was about it. Our children have never been particularly interested in religion.

Do you have any hobby?

I used to like sports, but getting close to eighty, I don't ice skate anymore and I don't roller skate anymore. I still go swimming in the summer time. I like to garden and it is trial and error gardening. Those plants that survive my care stay; those that don't are replaced by something else. I used to do tailoring a lot. I used to design and sew suits and coats, but I don't do that anymore though I still repair things. I like to travel, to go to museums in various parts of the world, and I participate in the activities of the associations that I am involved with. I've been active in the National Academy of Sciences, the American Philosophical Society and the American Academy of Achievement. There are many activities in which to participate.

Has Jerome's Nobel Prize changed your life?

It got busier. We get much more mail, many more requests for all sorts of events, nice things and things that are a bother. We did, and still do, give quite a few lectures to students around the United States and around the world.

Do you lead an active social life?

Yes, because we have a large family. Around Christmas time everybody gathers in our house for two weeks and that is very social. Many of our friends no longer exist. Others have moved away to retirement homes and we don't see them as often as we used to. Our social activities are more concentrated with professional colleagues and the societies that we belong to.

Is it possible to make new friends?

I am surprised, but it is. I have some telephone relationships with young people. There is a young man at the University of Wisconsin, for example, who often calls me up and we talk for an hour. He is an organic chemist and designs molecules, so sometimes it's business and sometimes it's family. There are several others of that nature, not any though, from nearby.

Jerome Karle, 2000 (photograph by I. Hargittai).

20

JEROME KARLE

Jerome Karle (b. 1918 in Coney Island, a part of Brooklyn, New York) is Chief Scientist of the Laboratory for the Structure of Matter at the Naval Research Laboratory. He received his M.A. at Harvard University (in biology) in 1937 and his M.S. (in physical chemistry) at the University of Michigan in Ann Arbor in 1941, and finished his work there for the Ph.D. in 1943. In between his studies at Harvard and graduate work at the University of Michigan, he worked at the New York State Health Department where he developed, as one of his chores, the standard method used for testing the amount of fluoridation that was applied to water supplies. He worked on the Manhattan Project at the University of Chicago in 1943 and partly in 1944. He returned to the University of Michigan to work on a project for the Naval Research Laboratory (NRL). It was in the course of this project that he became familiar with NRL and in 1946, he and Isabella Karle became members of the Naval Research Laboratory as physicists. He is a member of the American Mathematical Society, a member of the American Chemical Society, and he became a Fellow of the American Physical Society in 1960. He is a member of the National Academy of Sciences of the U.S.A. (1976), of the American Philosophical Society (1990) and many other professional societies. He has served as President of the International Union of Crystallography and as Chairman of the Chemistry Section of the National Academy of Sciences. He received the Nobel Prize in Chemistry in 1985, shared with Herbert Hauptman, "for their outstanding achievements in the development of direct methods for the determination of crystal structures".[1] This conversation was recorded in

Jackson, Mississippi, during a Conference on Current Trends in Computational Chemistry on November 3, 2003.*

I would like to start with your family background.

My father was born in 1882 in Poland, in a mining town although there was not much mining going on there. My father's father left Poland and came to the United States when my father was a small child. My paternal grandfather was a painter in the artistic sense. He specialized in decorating ceilings both with paintings and with sculptures. Almost everybody on my father's side of the family had artistic inclinations, except my father and one of his sisters. When I received my Ph.D., the extended family came together to celebrate and they decided that I was the black sheep of the family. It was a fun-loving family. My uncle almost became a teacher at City College, but he could not face the regular hours of such a job. On my mother's side, there was not much artistic inclination. My mother was the fourth child in her family. Her father had to face a difficult situation because his wife died, leaving him alone with his four children. He felt that he could take care of three, but not of all four. My mother was adopted by members of her extended family, who were people who came from Budapest. My mother was very talented. She could sit down at the piano, be handed some notes that she had never seen before, and she played them as they should be played. I also learned to play the piano and would practice at the other end of the house from where my mother usually stayed. When I hit a wrong key, she would know exactly which one it was. She was an absolute expert. My mother was a member of a women's organization called the Eastern Star and she would play the piano or the organ at their gatherings.

I lived in Coney Island until I left permanently at the age of nineteen. It was a good place for a youngster to enjoy. We knew the people who ran the theme parks and I could use them for swimming, other athletics and having fun on the rides.

Having this environment, how could you become interested in science?

When I was seven or eight years old, my mother found out that there was a science museum in a building that belonged to and was used by

---

*Magdolna Hargittai conducted the interview.

Young Jerry Karle, 1937 (all photographs courtesy of J. Karle unless indicated otherwise).

the New York *Daily News*. My mother took me to the museum. It was one of those hands-on science museums and I was absolutely enthralled with all the things that I could do. The museum did not stay for long, maybe a year or two, but it was enough for me to make up my mind to become a scientist. Once, when I was nine years old, I had to write a book report with a detailed description of a garden. I forget what the book was, but what the teacher received from me was something different. For example, I read popularized science books. I read the books by Sir James Jeans in particular. I wrote reports at the age of nine because there was a track system in New York City. They pushed ahead capable students. I graduated from the eighth grade before I was eleven years old.

Then you went to high school?

I went to Abraham Lincoln High School and to City College of New York from which I graduated when I was 19 years old. I could have graduated when I was 18 years old but I felt that I was too young and spent one-half year more than was required in high school and college.

What happened then?

I didn't know much about graduate schools until my last year in college. It was virtually impossible to get into medical school for reasons that are not very complimentary for society at that time.

Were there actual rules against Jews or it just happened?

There were no rules. I don't know whether I should speak about this.

Please do.

I went to Harvard and spent a year there obtaining a Master's degree in biology. I had the illusion that being a good student was all that was necessary to get admitted to medical school. I applied to Harvard and some other places and, of course, I was turned down. I wanted to try again early the next year and I was allowed to have a conversation with the Dean of the medical school. The only thing I got from him was a harangue. He said, "We have enough Jews in Massachusetts, we don't need any from New York City." He was not at all interested in my record as a student. For example, when I graduated from City College I received the first award given at graduation for "excellence in the natural sciences".

What did you do then?

I had applied to various graduate schools just to do graduate work and I was turned down by all of them. So I wasn't doing anything. Then there was just a stroke of luck. In the summer of 1938 I was working in Coney Island and a good friend of mine, with whom I still communicate, told me that exams were forthcoming for civil service jobs in the New York State Health Department. I took the exam and I had the highest grade among those they accepted. There was a rule that after a certain period, perhaps three months, the Health Department could dismiss anyone without an explanation. I stayed for about two years. I learned only later that they had wanted me to leave, along with the rest of the people who arrived when I did, but my boss said that if they tried to dismiss me, he would not accept that and that he needed me for his work. This fine gentleman's name was F. Wellington Gilcrease. He remained my good friend and we kept in touch until the end of his life. I did not know that he saved me from dismissal until I left to go to the University of Michigan. During those two years, I was saving up money as I knew I couldn't get any money from graduate school. At that time someone told me that if I went to the University of Michigan, I would be treated properly. That is why I went there. After the first year, I was funded to continue my education at the University of Michigan. After that I have never experienced any anti-Semitism.

Lawrence Brockway, Isabella Karle, and Jerome Karle.

How did you choose your graduate project?

Both my wife and I became associated with Professor Lawrence Brockway who was well known for his work in gas electron diffraction.

That was also the place where you met Isabella.

I met her the first day that I went to school at the University of Michigan. I was one of those people who, when we had to set up an apparatus, for example, would go and do it, rather than wait until the lab period started. The places in the teaching laboratory were assigned according to alphabetical order. Her last name started with an L and mine started with a K. When Isabella arrived, she saw me and the entire apparatus set up to carry out an experiment. I do not remember what she said, but she was surprised. That is how I met Isabella. The first year I did not see her very often, except in class. We went for an occasional walk in the evening. Once she was not feeling well and I brought material that she had missed in class to her lodging. The next year we started to have lunch together and by the end of that school year we were married.

You have done a lot of work together. Would you say that your interactions produced more than the sum would have been if your research had been done separately?

Perhaps. We collaborated more in the early days than later. Initially, I did about as much experimental work as Isabella did. Eventually, I drifted into more theoretical work. She always made an excellent contribution by carrying out the experimental analysis. That applied both to the gas electron diffraction and the applications that were made in crystallography.

Gas-phase electron diffraction is not a widespread and very well-known technique. What do you consider to be your most important contribution to this technique?

There was a set of papers, a theoretical contribution, that could possibly lead to not only the positions of the atoms and the structures of the molecules, but also obtain information concerning vibrational motion. My wife and I showed that it would come in the package so to speak. So far as our careers were concerned, we received the stimulation from the non-negativity principle that went into drawing the right background lines through the oscillations in a gas diffraction pattern. Later, the negativity principle opened the way to solving the phase problem in crystallography. The question that needed to be answered concerned the necessary and sufficient condition for a Fourier series to be non-negative.

The phase problem in crystallography was pursued at the Naval Research Laboratory. Why did you move there?

We did not move there immediately. After we completed our work for the Ph.D. degrees at the University of Michigan, Isabella and I went to work at the Manhattan Project in Chicago. I joined the project in the Summer of 1943, and Isabella did so a few months later. My job at the University of Chicago was to extract pure plutonium from plutonium oxide. Someone had already set up the equipment to do it, and I was assigned to proceed with it. I tried for a while, but could not make it work. A different procedure finally occurred to me. I spent some time putting together the required instrumentation. In those days, commercial equipment was usually not available. After the equipment was completed and tested, I carried out an experiment with plutonium oxide. The experiment turned out to be successful. Soon after, we both returned to the University of Michigan, Isabella to teach chemistry, and I to work on a project for the Naval Research Laboratory. This is how we both became familiar with the latter institution. After about two years, we both received an invitation to join the Naval Research Laboratory, which we were quite happy to accept.

At NRL, you did some fundamental work with gas-electron diffraction because that was there where you published your papers about determining not only geometrical but also vibrational parameters for molecules. What made you then switch to X-ray crystallography?

I was familiar with the X-ray diffraction field. There were meetings where both electron diffraction and X-ray diffraction were discussed. People from both fields came together at these meetings. What later became the American Crystallographic Association used to be an association for both the gas and crystal diffraction people. I remember reading a book by Martin Buerger. By the way, he had seven children, all girls. After we had three daughters, Martin said to me, "Jerry, quit." At any rate, I read the book, and it made me consider how much data one would get from a copper tube diffraction

Jerome Karle, around 1948.

experiment. I found that there were about a hundred and fifty data per unknown for a centrosymmetric crystal, just from the copper tube experiment, and half of that for non-centrosymmetric crystals. The hundred and fifty data per unknown would correspond to one third of that per atom in a perfect experiment, and half of that for a non-centrosymmetric crystal. As the experiments are not perfect, one gets about twenty data per atom for a non-centrosymmetric crystal. I found this observation of an overdetermination of data to be very exciting and it set my mind to work. This is where my experience in gas electron diffraction and, in particular the non-negativity principle in determining the background line, came in handy. At that time Herb Hauptman was involved in building a low-energy gas electron diffraction experiment. I suggested to Herb to set aside that project and concentrate on the crystallographic problem.

What was the division of work between the two of you?

We worked together and we participated about fifty-fifty.

How long did it take to work out the direct methods?

It was done stepwise. Herb and I had derived the mathematical relations that were necessary for phase determination by about 1955–1956. However, for practical application an additional step was needed. We had to assign a place in the crystal that could be referred to as the origin. This step needed different considerations for each of the 230 three-dimensional space groups and meant a lot of work for Herb and me. Not all the specifications of the origin are different for all space groups but each one had to be considered one after the other. The great virtue in specifying the origin was that it allowed the determination of the phases directly from the intensities. Herb left our laboratory about 1960.

Isabella and I worked out a modus operandi for phase determination which we called the symbolic addition procedure. The first crystallographic structure that Isabella and I worked on alone was a centrosymmetric one that had been attempted at several laboratories. It concerned the cyclohexaglycyl molecule. There were four conformational isomers in the unit cell. This structure established for the first time the coexistence of conformational isomers in peptides and also reliable parameters for the beta-hairpin turn and its hydrogen bond. This paper was one of the most quoted papers in the Citation Index. We solved the first non-centrosymmetric structure by direct phase determination in 1963 and published it in 1964. The substance was arginine dehydrate.

But it was a slow process. Crystallographers did not exactly jump at the opportunity provided by your new techniques.

As we were solving non-centrosymmetric crystals, we found that often the symbolic addition procedure may produce a good piece of the structure, but not the complete structure. There was a formula that came out of the papers that I and Herb wrote in 1956. This formula by itself didn't do the trick, but I thought that I could use that formula with a tangent formula and the partial structure to produce the whole structure. Sometimes just knowing ten per cent of the structure would suffice. Alternatively, if a sufficient number of starts are used, maybe hundreds, maybe thousands, in a fast computer that is available now, but not in the early 1960s, one may arrive at the right solution by using the tangent formula, first formulated in 1956 by Herb and me, which has been and still is the work horse of phase determination and phase extension. Sheldrick has done an excellent job in writing the necessary programs. However, most of the people were never able to apply the symbolic addition procedure as it should be or write a program for it. Isabella and I thought of writing the necessary program to do this, but we have other things to do. The computers have become so fast that the trial and error procedure with the use of the tangent formula has become possible. And this is what people usually do.

I was listening to a lecture not too long ago in which there was a statement that Isabella and I showed that the analyses of non-centrosymmetric structures didn't get anywhere. The truth is that in the 1960s we were the only laboratory publishing non-centrosymmetric structures that did not contain heavy atoms. It was not until 1968, that an Australian paper appeared. Furthermore, our formulas got into everybody's program, but what is omitted is the part of the symbolic addition. That's substituted by lots of trial and error. At present, the results can be obtained quite fast when people use this trial and error procedure. I have no problem with that. But I do have a problem with them calling it the direct method. The statement direct method means that you essentially start with the data, and with very little extra effort, you go directly to the answer. But massive trial and error is not a direct method.

Are you saying that, in fact, crystallographers even today do not apply your direct method?

That's absolutely correct. Massive trial and error is not a direct method.

They don't apply it or they just don't understand it?

Few researchers, besides Isabella and some members of our laboratory, have successfully applied the first part of the symbolic addition procedure. What is done usually, instead of using data plus probabilities, is just trial and error with some reliability measures. This is no credit to the effort except to the people who had invented fast computers.

Could you please tell us a little more about Isabella's involvement?

In the course of the events, when we started solving the first centrosymmetric structure, Isabella collected the data, formulated a modus operandi and solved the structure. A vital additional step was required for many of the non-centrosymmetric structures for which only a partial structure appeared. I provided the ideas for the phase extension with the repetitive use of the tangent formula.

Was there any problem because you worked together?

There was some administrative discomfort because I was supervising my own wife. The solution was that, although she has continued to be a member of the group, she was placed directly under the Associate Director of NRL, just as I am.

Coming back to research, from the start, she has applied the symbolic addition procedure. Isabella and I derived a variance formula to apply probability measures for the analysis of non-centrosymmetric structures. In the 1960s, she solved quite a number of centrosymmetric and non-centrosymmetric structures without the use of heavy atoms. People all over the world followed her scheme and they did solve structures, like she did. Over the years, she has received many compliments from people who did their Ph.D. work following her procedures.

Was it possible at all before to solve non-centrosymmetric structures?

No, it wasn't. Centrosymmetric structures had been solved. If there were heavy atoms in the structure, people used Patterson functions. If there were no heavy atoms present, people just resorted to the methods of trial and error. They would produce a nine or ten atom structure for a Ph.D. dissertation. It could easily require a year or two to solve such a structure.

In the early 1950s, your suggestions for the direct methods were met with ridicule, even in the scientific literature.

Someone wrote a book about 1960 in which he stated that he was convinced that we would never be able to determine non-centrosymmetric structures. He also told me about it personally. [Dr. Karle is heartily laughing.]

While working on your new techniques, did you sense that this was a piece of research of a Nobel caliber?

I can tell you in complete honesty that at that time, I did not think about the Nobel Prize or receiving it.

Wow.

And I had good reason for thinking so. What I did was very theoretical. It seems to me that what happened — and I'm pretty sure that it is true — was that our work became noticed and then was more and more appreciated because of the applications that Isabella worked out and communicated. If you are beginning to wonder whether I was disappointed that Isabella was not included in the Nobel Prize, I most certainly was disappointed. On the plus side, Isabella has received numerous, important other prizes and she is quite satisfied with those acknowledgments of her contributions.

How did you learn about your Nobel Prize?

I was flying home from Europe, from a meeting, and the captain announced it just about two hours before landing in Washington, D.C. It was good to have those two hours because upon arrival, I was taken to a hall where there were fifty journalists with fifty microphones, all asking questions. The two hours in the air helped me to prepare myself.

Did you know whether she was included or not?

I didn't. Only when we met and she told me. I was disappointed.

How did she appear to take it?

There were two times in our life together when I really admired her. She had obviously made up her mind to be happy for me and didn't say a word about how things might have turned out to be different. That was

Isabella and Jerome Karle in 1980.

very remarkable. The other time was when she was potentially very ill and she never gave a sign about the seriousness of her situation. Fortunately, it ended well. The good thing is that she has received a lot of important awards, including the National Medal of Science.

Changing topics, do you ever think about the future of X-ray crystallography?

I do to the extent that we are developing a new field of crystallography. It's called quantum crystallography. It's a combination of structural information with theoretical calculations. We have shown that there is a way to do theoretical calculations on large molecules, such as proteins.

Does the emergence of quantum crystallography point to the diminishing role of X-ray crystallography?

No, it enhances the potentials of the existing techniques as well.

Has the Nobel Prize changed your life?

One thing that I especially appreciate is that it has brought me more into contact with young people who want to make their career in science.

Do you enjoy being in the limelight?

Sometimes it feels like a duty.

Most of your career, you have been working for government institutions. It was the New York State Health Department, the Manhattan Project and the Navy. This must be different from a university setting.

It depends on the individual. All of them may work and work well. I don't have any strong opinion about this.

Do you have a rank?

I am in the Senior Executive Service and I am in category six, the only one in this category at our institution, which is the highest.

Looking back, can you single out any of your research achievements as the most important?

You may have a lot of children and you don't love any of them more than the others. For public health, the most important maybe what we did for the synchrotron data, e.g. my publication in 1980. Then, of course, solving the phase problem for the crystal structures was important. There have been many other projects that are much less known.

What do you do nowadays?

I am working on improving methods for crystal structure determination. My colleagues and I are working in an area, which we call quantum crystallography, a combination of quantum mechanics and crystal structure. One application of interest is the potential improved understanding of the biological mechanisms of living systems.

What was the greatest challenge in your life?

Getting into a graduate school that would accept me. There was this young person who had this wish, who worked hard, and who was denied the privilege.

The Karles and the Hargittais in 1978 (Pécs, Hungary) and in 2000 (Washington, D.C.). Photographs by unknown photographers.

For Isabella, being an internationally renowned scientist with three children, you must have been a very supportive husband. Were you?

We were able to afford having a woman to stay at home during the week just to take care of the children. Also, we always took the children with us wherever we went.

You could observe the situation of women in science at close range. Has there been a change?

It's much better now than it used to be. But there are still big gaps. The main problem is getting tenure and having children. It's impossible to do both at the same time. This matter is being taken into consideration more and more in universities, I believe, thus helping women to have successful careers.

What do you consider most important in a scientist's behavior? Do you have a message?

Ethics is very important in science. For example, other people must be referred to if they had done relevant work. If you will, this is my message.

Reference

1. An interview with Herbert Hauptman has appeared in Hargittai, I. *Candid Science III: More Conversations with Famous Chemists*, edited by Hargittai, M. Imperial College Press, London, 2003, pp. 292–317.

Yuan T. Lee, 2001 (photograph by I. Hargittai).

21

YUAN TSEH LEE

Yuan Tseh Lee (b. 1936 in Hsinchu, Taiwan) is President of Academia Sinica in Taipei, Taiwan. He received his B.S. and M.S. degrees from the National Taiwan University in 1959 and 1961, respectively. He entered graduate school at the University of California at Berkeley in 1962, and received his Ph.D. degree in Chemistry under the supervision of Professor Bruce Mahan in 1965. He did his postdoctoral studies first at Berkeley, then, in 1967–1968, at Harvard University under Professor Dudley Herschbach.

In 1968, he joined the James Franck Institute at the University of Chicago as an assistant professor, and was promoted to full professorship in 1973. In 1974, he returned to Berkeley as Professor of Chemistry at the University of California and principal investigator at the Lawrence Berkeley National Laboratory. In 1986, he received the National Science Medal from President Ronald Reagan. Also in 1986, he shared the chemistry Nobel Prize with Dudley R. Herschbach and John C. Polanyi "for their contributions concerning the dynamics of chemical elementary processes". (Interviews with Herschbach and Polanyi appeared in *Candid Science III*, pp. 392–399 and 378–391, respectively.)

He became President of Academia Sinica after having resigned his positions at Berkeley and moved to Taipei, Taiwan in 1994. He is a Foreign Member of the National Academy of Sciences of the U.S.A. (1979); Fellow of the American Academy of Arts and Sciences (1975); and a member of numerous other learned societies. He has received the Ernest O. Lawrence Award of the U.S. Department of Energy (1981); the Peter Debye Award of the American Chemical Society (1986); the

Faraday Medal of the Royal Society of Chemistry (London, 1992); and numerous other distinctions.

We recorded our conversation in Lindau, Germany, on June 28, 2005, during the 55th Meeting of Nobel Laureates.*

I just returned from a visit to Japan and there much attention was given to the demonstrations in China against the newly approved revisionist history textbooks for the Japanese schools. Taiwan was under Japanese domination for a long time, fifty years. What are the sentiments in Taiwan about the Japanese attempts to falsify history?

In Taiwan, nobody feels so strongly against Japan. When Japan came to Taiwan, they occupied Taiwan and wanted to use Taiwan as a base for their expansion in Southeast Asia. They built up Taiwan in a very nice way. They built many schools, developed an irrigation system, created a railway network, and made a lot of investment in Taiwan. Their presence and operations, however, could not be characterized as fair. The Japanese people drew higher salaries than their Taiwanese counterparts for the same work.

But there were no atrocities as Nanjing, for example?

There were sporadic incidents, but nothing like that.

Would you characterize it as colonization?

Yes, it was colonization. The resentment against the Japanese in Taiwan is not the same as in Korea or as in China.

I knew Michael Polanyi, I talked with John Polanyi and with Dudley Herschbach, the two co-recipients of the Nobel Prize with you. John and Dudley came from very different family backgrounds. John's was a very cultured one whereas Dudley was the first college graduate in his family. How do you compare your background with theirs?

Mine was somewhere in between. My father and my mother were school teachers. My father then became an artist; he did watercolor painting, Western style. He used to teach art in school, but eventually became a famous

---

*István Hargittai conducted the interview.

Yuan Lee at the Nobel Prize award ceremony in 2001; César Milstein (partly hidden) is in front of him; Marshall Nirenberg is on Lee's left; Jerome Karle (partly hidden) is behind Lee, and Richard Ernst is behind Karle (photograph by M. Hargittai).

artist. His paintings were sold for a reasonably high price. We started very poor, but later our family lived very well. I grew up in an environment of a scholarly family. My mother collected literature of authors from the whole world. We also had many books of physics and chemistry.

What turned you to science?

It came very early. Our family name is Lee. Yuan represents my generation; all my brothers have also this name. My middle name is Tseh, which means philosophy. My mother thought early on that I was a smart boy and started asking me all sorts of questions. Whenever she wanted to know something she would ask me questions. So I was being challenged from a very young age. My mother had a Singer sewing machine to make our clothes. It was

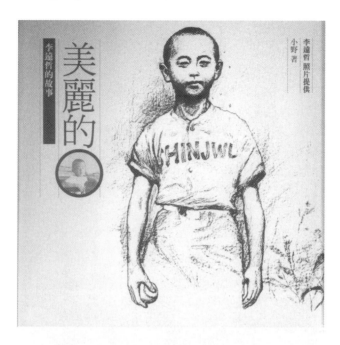

Young Yuan on the cover of a book about him (courtesy of Y. T. Lee).

a very noisy machine, so one day I asked my mother to keep quiet because I was studying. At that point my mother challenged me to fix the machine, to make it quieter. At that time I was in the sixth grade of the elementary school, just before entering junior high school. I took all the tools from my father's toolbox and took the sewing machine apart. I checked where the noise came from, I lubricated various moving parts, and I succeeded in reducing the noise. My mother was very impressed.

My studies were disrupted by the war. We were under Japanese occupation and at the end of the first grade of elementary school B-29 bombers appeared in the sky and were bombing our region on a daily basis. We learned about conservation of momentum from the bombing although I was not familiar with this term at the time. We knew that if the planes released their bombs above us, those bombs would not hurt us. As the bombs were dropped they were moving away because of the momentum. The war had a lot to do with science and I experienced it as a child.

At the end of the war Taiwan was returned to China. At that time China was a very chaotic place, backward, and was considered to be the sick man of Asia. The Chinese people smoked cocaine due to the British tactic of selling opium to the Chinese. Immediately after the Second World

The student Yuan Lee (courtesy of Y. T. Lee).

War, particularly among the young people, there was a movement that only science and democracy would save China. After 1945, democracy and science were the two key words for us. The regime that came from China to Taiwan, however, was very corrupt and the people of Taiwan had an uprising against the authorities after one year following the end of the war.

Then came the civil war and Taiwan became separated from China again. From your autobiography I know that you completed high school and then graduated from National Taiwan University in chemistry. Then you entered graduate school at National Tsinghua University and obtained your Master's degree. You stayed on and did research in radiochemistry and X-ray crystallography. In 1962, you left Taiwan, went to Berkeley and earned your Ph.D. by doing research on chemi-ionization processes of electronically excited alkali atoms. You stayed on in the United States and became an American citizen in 1974. This would suggest to me that you didn't have plans to return to Taiwan.

When I left Taiwan in 1962, I thought that I would return to Taiwan after I received my Ph.D. However, when I received my Ph.D. degree,

I was asked to stay with my research adviser, Professor Bruce Mahan and I did postdoctoral research for one and a half years. In February 1967, I joined Dudley Herschbach's group at Harvard University. In October 1968, I accepted an appointment of Assistant Professor at the University of Chicago. So I was staying on and on in the United States. Before I knew it, 32 years had gone by and I was at the age of 57 when I returned to Taiwan. Actually, in 1972 I took sabbatical leave, which I spent in Taiwan. This was after the first ten years that I spent in America. By then I had been promoted to Associate Professor at the University of Chicago. I wanted to take a look in Taiwan and determine whether I could help. However, my sabbatical stay convinced me that the situation was not quite ready yet for my return. I was not yet well known enough to hope to be able to make a difference. Taiwan was not developing fast enough and the government was not putting enough money into science. I decided that I should be staying a little longer in the United States. In another ten years, by 1982, I had been elected to the National Academy of Sciences of the U.S.A. and also to the Academy of Sciences in Taiwan, called Academia Sinica, as well as to the American Academy of Art and Sciences. So I had become better known both in America and in Taiwan, and I started helping science in Taiwan. Twenty years after I had left Taiwan, I could help Taiwan more effectively. I set up a research institute, for example, at Academia Sinica. I became a U.S. citizen only to facilitate various things, like applying for grants, traveling from and to the United States. I became U.S. citizen when I moved from Chicago to Berkeley in 1974.

Did you ever meet Michael Polanyi?

I met him in 1968 when I took the position at the University of Chicago. There was a conference in Toronto and we met there.

In a way your work was related to his early discoveries.

Dudley and I tried to visualize collisions in chemical reactions. We were concerned about the orientations of the molecules. In 1978, there was a celebration in Berlin called fifty years of dynamical chemical reactions. It was to honor the fiftieth anniversary of the paper by Landau, Eyring, and Polanyi, which was aimed at understanding the energetics of chemical reactions.

Do you think that what is called femtochemistry has overtaken what you had been doing?

Not really. The people doing femtochemistry always say that for studying the molecular beams they have to go to femtochemistry. However, when we do chemical reactions, we already have the rotational period as a clock. In the reaction of potassium and methyliodide what Dudley Herschbach was doing it was possible to see the product bouncing backward in the time period of one rotation. That clock is a picosecond clock. It made it possible to tell how fast the chemical reaction took place. One of the reactions was particularly interesting. It was a charge transfer reaction between potassium and oxygen. At a long distance there is an electron transfer and the oxygen starts vibrating. Then at some point the electron jumps back to potassium. By looking at the angular distribution, it was possible to see the oscillation of electron jump probability based on the molecular vibration. It is a femtosecond phenomenon. In the beam experiments, there is a lot of information provided on a femtosecond timescale. Of course, when you use spectroscopy, you can see electronic excited states and how they decay on a femtosecond scale. However, it won't tell you anything about approach and molecular alignment and other spatial characteristics. Neither will it give information about angular momentum and the conservation of angular momentum.

Yuan Lee and his wife, Bernice Wu, in Lindau, Germany, 2005 (photograph by I. Hargittai).

You married your elementary school sweetheart. Did she also study?

She went to the same university as I did and studied at the Department of Foreign Language and Literature. Then she taught in high school for three years. After that we went together to America and she also continued her education. However, we decided that we should have a family a little sooner. I didn't want to have children when both of us would be very old. Thus she and I decided that maybe only one of us should pursue an academic career and she would take care of the family. She's a smart girl. I always felt that she is smarter than I. Her name is Bernice Wu and we have two sons, Ted, born in 1963, and Sidney, born in 1966, and we have one daughter, Charlotte, born in 1969. They live in the United States.

Did they choose a career in academia?

No. When I was working very hard in the laboratory, our boys said that they wanted to work very hard, very, very hard, when they grow up, but not as hard as I did. They complained that I always came home around midnight and that I worked on Saturdays and Sundays. I said that I was working so much because I enjoyed it. For me it was not work. They said that they also wanted to work very hard but they wanted to enjoy life as well. Our first-born became a journalist and the second a medical doctor.

A large proportion of famous scientists are Jewish and people often ascribe this to their status as immigrants initially and science is a field where immigrants can work hard and achieve. It is also noted often that the people from Southeast Asia, especially the Chinese, have a similar drive and follow a similar path. Do you see this?

This is certainly true to some extent. When you arrive in America, you feel that you have to establish yourself. Science is color-blind, there is opportunity, not too much bias, and you determine your performance and your success. There are also other aspects. Many years ago I visited Jerusalem. The Mayor of Jerusalem had me for dinner in his house. I asked him about his difficulties in his divided city. The Mayor said that scientists are horrible; they always try to solve problems whereas the people of Jerusalem have learned for thousands of years to live together with their problems. That night I asked him the question, "Jewish people seem to be doing well in the scientific area; are there cultural differences, which might work to your advantage in science?" He said something quite interesting. He said, "In a Jewish family when the child comes home from school, the parents ask the child, 'Did you ask some good questions today?' " In Asia, when the child comes back home, they always ask him about his test scores. If he performed to 97, the parents ask him, "What happened to the missing three points?" Then the child says that maybe he misplaced a comma or something like that. Then the parents ask him, "Did anybody do better than you?" If the answer is in the negative then the parents are satisfied. This was an important point. The second key point the Mayor mentioned was that in the Synagogue, when the Rabbi tells a story, he always tells two sides of the story. He tells about the positive side and the negative side and stimulates the students to look at the same problem from different perspectives.

Isn't this also the Oriental approach?

Ancient people for a long time, especially in Japan, Korea, Taiwan, and China, were under the influence of Confucian teaching. It meant that the family paid enormous attention to education. You will find in Asian families, especially among the first immigrants that they use every bit of money to send their kids to the most expensive private schools. Dedication to education is a cultural heritage in Confucian teaching. Two or three generations down the road this may not be present any longer.

Do you observe this change?

Especially our daughter asked us often why people from Taiwan worked so hard and why didn't they enjoy life?

Don't the Americans work hard as well?

They work hard when they are interested. The students interested in science work hard when they are attending college. In high school and in elementary school they don't work very hard. I don't think this is bad and I don't think it's very good when the family makes a student work too hard.

What do you think about the recent explosion in scientific research in China? I edit a journal, Structural Chemistry. *A few years ago we started receiving manuscripts from China. Initially, they were not too god, but they improved very fast and they are often excellent now. How could they make such tremendous progress in such short period of time?*

China decided to invest a lot of money in elite universities. They also persuaded many former students in foreign countries to return to China. People who got their Ph.D. degrees abroad came back and educated some graduate students and they started publishing very soon from China. This was not so hard. Such initial success is not too hard to accomplish. To enter a sustained creative stage will take much longer.

Do you have scientific interactions with the People's Republic?

Yes. I visited China for the first time in 1978 as a member of a chemistry delegation of the National Academy of Sciences of the U.S.A. I visited China almost once a year after that initial visit and I helped their chemical dynamics programs. However, after my return to Taiwan in 1994, that interaction was cut off because of the political relations being not very good between China and Taiwan.

Are there Chinese students from the People's Republic in Taiwan?

There are many postdocs. The interactions continue at the institutes level. One of my former postdocs came to Taiwan with me from America, then he went to China and established himself as a leading scientist.

Living in Taiwan, do you feel that you are under some tension because of the possibility of hostilities?

Yes, certainly. In Taiwan the birthrate has gone down. Part of it is that society has become more prosperous and people want to live a better life. There is then a second aspect and it is this instability of the political situation in Taiwan. Nobody is sure whether there will be a Taiwan in ten years or in fifteen years. China always says that there is only one China.

Many years ago Taiwan used to say the same. Chiang Kai-shek used to say that.

That's true. For the People's Republic, the People's Republic of China was the one China and for the Republic of China, Taiwan represented the one

China. At the present time we have one China, but two regimes, and we should make both sides happy about the situation in a logical way.

Do you think that the Hong Kong model might work?

No. Hong Kong was a colony. Taiwan has maintained independence for quite a long time. We have gradually become more and more democratic and we have not been under the domination of a foreign power during the last fifty years.

You are Taiwanese. Is there any tension between the original Taiwanese people and those who descended from the people who arrived from the Mainland as a consequence of the civil war in 1949?

There has been such a tension, but it has taken a different turn. The Taiwanese were oppressed by the Japanese for fifty years during colonization. After the Second World War we waved the flag and we were so happy to return to the Motherland. But then, the regime that came to Taiwan under Chiang Kai-shek was so corrupt that people felt that it was worse than the Japanese. The difference was that the Japanese ruled by law. It was not good for the Taiwanese people but it was calculable and we knew how we could get by. When the Mainland Chinese came, they were corrupt. The former Japanese properties became national properties but the Mainland Chinese official expropriated those properties. The young socialists were especially upset. They demanded a fair society; they were not card-carrying party members, but people hoped that they would help to remove the Chiang Kai-shek regime. Instead, the Mainland Chinese said that Taiwan could keep its government and military and police so long as you come back to China. Thus people realized that the Communist party served the oppressors rather than the people. The people in Taiwan felt betrayed. This is why so many would like to have independence declared. I don't think though that independence is so important. The people in Taiwan need and seek fair treatment.

How did you feel when the United States changed policy and by recognizing the People's Republic no longer considered Taiwan to be an independent state?

Taiwan is so small compared with Mainland China that Taiwan would not survive without U.S. protection and that has not changed. In this sense, we are helpless. Politics is a very pragmatic activity. China is a very large country and the trade between the People's Republic and the United States has become very important. There is no real choice for any country in the world when the question comes up about choosing which country to recognize, Mainland China or Taiwan. Everybody says that we are a democratic society and they would like to be our friends, but Mainland China forces them to choose between them and Taiwan and they just can't afford not to choose Mainland China. Only a few small countries in Africa and in the Pacific region still recognize Taiwan. None of the G8 and other developed nations recognizes Taiwan anymore. What we hope for in Taiwan is that maybe China one day will become democratic and then unification won't be a problem.

Are things evolving in that direction?

They have to. China faces great difficulties at the present time. There are two problems. One is corruption in China. Socialism and communism, which held people together for the last fifty years have gone bankrupt. People again face a dilemma. KMT or Kuomintang — Chiang Kai-shek's party — was so corrupt that the Chinese threw them out from China to Taiwan. They are still corrupt in Taiwan. They say that the Communist party in China is even more corrupt than Kuomintang was a long time ago. The second problem is that before the socialist revolution there was a large gap between the rich and the poor and the young people resonated with the socialist ideas. They wanted the society to become egalitarian and fair. Now looking at the distribution of wealth, they see an even wider gap than it was before the revolution, so the purpose of the socialist revolution was lost. All the sacrifices appear to be meaningless, in vain. The situation in Mainland China may develop in a dangerous direction. People no longer believe in the socialist revolution and the country is moving in the direction of so-called nationalism and patriotism. The recent anti-Japanese movement is part of this development. Of course, Japan should think about this too; they have never done repentance in the proper way for the crimes they committed during the Second World War. Germany has done such repentance and promised that she would never do it again. Japan has given a lot of money to China but never expressed their condemnation of the war crimes in a proper way. Sadly, however, China seems to need to use the anti-Japanese sentiments to unify people and this is dangerous.

You have been back in Taiwan for eleven years. You had had some expectations of the conditions and yourself. Do you feel that your expectations have been fulfilled?

In some aspects yes and in some other aspects not quite. For instance, when we went back to Taiwan, the economy was moving up and I persuaded the government to spend money on science and education. This part of the budget has gone up ten per cent a year.

What is the percentage of the GNP for research and development in Taiwan?

That's 2.8 per cent. We are shooting for three per cent by next year.

So does the European Union, but there are large differences among the member states.

When I went back it was 1.8 per cent. So there has been a tremendous increase.

Taiwan being a small country, do you try to regulate what people research?

We have a so-called national project; it is info-nano-bio. There is nothing unusual or even new in this because since ancient time, we have always been interested in the creation of the Universe, in the origin of life, in the structure of matter and the forces operating in it. So the info-nano-bio can be considered to be the continuation of previous directions and interests.

China used to be in the forefront in technology in ancient times. What happened?

One thousand years ago, during the Sun Dynasty, technologies were very well developed. Ships regularly went to foreign countries, including Japan. Paper production, printing, navigation and other sophisticated areas were well developed in China. However, the various dynasties that followed the Sun Dynasty thought that it would be advantageous to keep people stupid for maintaining the Chinese empire. The vibrant discussions of the Universe stopped, and they introduced a nationwide examination for civil servants and the entire system of education was subordinated to these examinations. The instructions became simplistic and very formal and superficial.

Edward Teller wrote an article about what he called the "China Syndrome" and he painted a gloomy picture for the United States going the way you just described unless the country was willing to improve its system of education and enhance the appreciation of research and development. The original usage of the expression of the "China Syndrome" was for a hypothetical nuclear meltdown in which the meltdown would be so powerful that the melting would reach from the United States to China through the center of the Earth. But Teller considered the events you just described a more devastating "China Syndrome" scenario for the United States. Can you imagine such a danger for the United States?

I've met Edward Teller but I'm not familiar with this article. It's true, however, that recently an increasing fraction of the research money is being spent on applied science rather than on fundamental research. That maybe a danger.

When you went to America, in 1962, the interest in science was on the increase in the United States, maybe to a considerable extent due to Soviet advances.

At that time America lived in a period when nothing was considered to be impossible. There was also the ideal of freedom, equality and justice. People like us, who went there from repressive regimes and feudal societies, felt truly that we arrived in the land of opportunity. We felt ourselves being liberated. I doubt that people who go there today can quite experience the same feeling. After 9/11, entering the United States has become more difficult; it can't be called a free country anymore. This is very sad.

When you met Edward Teller, what did you talk about?

Chinese history. He was very much interested. I was tremendously impressed because he knew so much about Chinese history. He knew more than I do. This was about 1987 or 1988.

You were still in the United States when the debates about SDI, that is, the Star Wars, were going on. Did you have a position in this debate?

I signed a letter saying that I would not participate in SDI. I was against SDI.

Why?

It could not have solved the problem.

Technologically or politically?

Both. The United States might have spent an enormous amount of money so we might have ended up like the former Soviet Union, spending so much on the military and lagging behind in many other areas. Politically it was very complicated because my position is that people should talk with each other to resolve their problems.

Would you agree that SDI might have contributed to the demise of the Soviet Union because it forced the Soviets to carry out developments that they were in no position to afford? If the expenditure was very large for the United States, for the Soviet economy it meant a much larger percentage.

It might have accelerated the dissolution of the Soviet Union because it was such an inefficient society. If you study Marx and Engels, they certainly underestimated the potentials of capitalistic society to adopt change. The Soviet regime had become inefficient and also corrupt. When I was in high school, I read a book called something like melting down of the snow. I read it in Japanese. It was written by someone in Russia but it could not be published there and it was smuggled out of Russia and was brought to the United States. It described the corruption of Soviet society. It gave me the impression that if a society is moving along those lines, it probably would not last very long. Gorbachev certainly made a tremendous contribution to changing the Soviet Union and by doing so he made a tremendous contribution to the peace on Earth. I met with his science advisor two years ago and he told me that Gorbachev didn't really know how to rule the country and how to direct the economy, and how society functions. The Soviet Union was falling apart and Gorbachev couldn't save Russia.

May I ask you about religion?

Two images of the campus of Academia Sinica in Taipei, Taiwan (courtesy of Y. T. Lee).

I'm not religious. My mother was Buddhist-inclined. She would go to the temple and pray for health and pray for my success in the entrance exams and things like that.

What would be a typical day of your life in Taiwan?

Now I am President of the Academy. We have research institutes that we have to take care of. We have 230 academicians of whom 70 live in Taiwan and the rest live abroad, mostly in the United States. The population of Taiwan is twenty three million on about eighteen thousand square kilometers. Taiwan is the second most densely populated area in the world.

Are you an elected president? How does one become president of your Academy?

When I came back to Taiwan, it was a lifetime appointment. The first five presidents, all died while on the job. The sixth president lived very long and he retired when he was 90 years old. That's when I took over. I thought it was not very good to have lifetime appointments for the president, so I went to the Congress to demand a change of law. I was successful and now the presidency has a five-year term and for a maximum of two terms. This happened four years ago so I could have been president for ten years from the day the law was adopted. However, I disagreed and suggested to count my first seven years as my first term and I suggested serving one five-year term from the day the new law was adopted. That means, I have one more year to go.

What will happen to you then?

I will stay in Taiwan doing research and helping education and science. I have lots of job offers from America from becoming president of a university to returning to research. When I left America eleven years ago, I said I would go back to Taiwan to help people in Taiwan for ten years. So there are people who counted on me coming back to the United States after ten years.

You have made two major contributions, one to physical chemistry, the other to Taiwan. Which is more important for you?

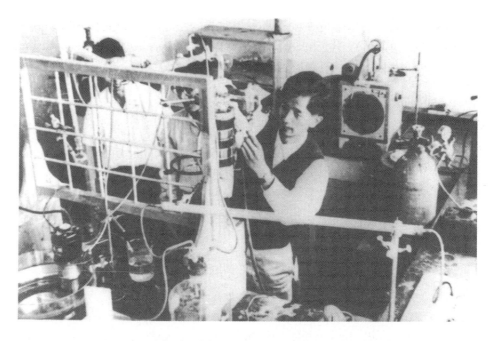

Yuan Lee in the laboratory (courtesy of Y. T. Lee).

If the question is about what is most satisfying for me, I would say that it is that the people who worked with me in research have become very successful. Someone called me up last year from the United States and said that he had done a survey about who produced the most professors in American chemistry departments in a 25-year period between 1979 and 2004 and according to his survey I produced more professors than anybody else. I left eleven years ago, so I produced more professors in that fifteen-year period than anybody else in the longer period for which the survey was made. I produced 13 professors. They are at Caltech, MIT, Berkeley, Cornell, and elsewhere. The second person, Alex Pine, produced 9 professors in 25 years.

How did you do that?

First of all, I had good students to work with. I do believe in one thing. I believe in education. There is always possibility for improvement. When I was in California, I would take everybody who wanted to join me. All my students enjoyed science with me. That's what I feel very proud about. I enjoyed the science I did in the United States. The work I did in Taiwan had more social impact and I have spoken out on political questions. I have spoken out about the need to fight corruption. My stand helped the current president to win the election. He told me himself that I was instrumental in bringing down the fifty-year Kuomintang rule. Of course, in politics if you do something, some interest groups will hate you. So I have earned the hatred of the people who used to be the rulers in Taiwan, but the people who wanted the change like me and appreciate what I did.

Darleane C. Hoffman, 2004 (photograph by M. Hargittai).

22

DARLEANE C. HOFFMAN

Darleane C. Hoffman (b. 1926 in Terril, Iowa) is Professor of the Graduate School in the Department of Chemistry, University of California, Berkeley, CA (UC Berkeley) and Faculty Sr. Scientist in the Nuclear Science Division of Lawrence Berkeley National Laboratory (LBNL). She also serves as senior advisor and charter director for the G. T. Seaborg Institute at Lawrence Livermore National Laboratory. She received her Ph.D. in physical (nuclear) chemistry from Iowa State University (1951). She held positions with Oak Ridge National Laboratory (1952), Los Alamos Scientific Laboratory (LASL, 1953–1979), and as Division Leader of the Chemistry-Nuclear Chemistry and Isotope and Nuclear Chemistry Divisions (LANL, 1979–1984). During this period she spent sabbaticals in Oslo, Norway (1964–1965) and as Guggenheim Fellow at the Lawrence Berkeley Laboratory (1978–1979). She was Professor of Chemistry, Professor Emeritus, at UC Berkeley (1984–1994), and Group Leader in the Nuclear Science Division, LBNL (1984–1996). She served as first director of the Livermore Seaborg Institute for Transactinium Science (1991–1996).

She is a Fellow of the American Physical Society (1986), the Norwegian Academy of Science and Letters (1990), the American Association for the Advancement of Science (1994), the American Academy of Arts and Sciences (1998), and was inducted into the Women in Technology International Hall of Fame (2000) and the Alpha Chi Sigma Hall of Fame (2002). She has received numerous awards and distinctions, including the Iowa State University Alumni Citation of Merit (1978), the American Chemical Society (ACS) Award for Nuclear Chemistry (1983), the ACS Garvan Medal (1990), the UC Berkeley Citation (1996), the U.S. National

Medal of Science (1997), the ACS Priestley Medal (2000), the Sigma Xi Procter Award for Scientific Achievement (2003). She has held visiting lectureships in the U.S. and abroad and was awarded honorary doctorates from Clark University, U.S.A. (2000) and Bern University, Switzerland (2001). We recorded our conversation in her office at Berkeley on February 20, 2004.*

The origin of lighter elements in the Universe is more or less understood but this seems more difficult for the heavier elements.

The current theories postulate r-process nucleosynthesis, that is, rapid neutron capture until you finally get up to the highest reaches of the Periodic Table as we know it. The highest atomic number element that we know of in nature is plutonium-244, which I and some of my co-workers found remnants of many years ago, in 1971. Of course, there is a lot of uranium in nature and it is formed by successive capture of neutrons by lighter elements. It was discovered in 1789 and is the heaviest element found in macroscopic quantities. This is how things were until about 1940. Then when Ed McMillan and Phil Abelson were trying to investigate neutron-induced fission of uranium (just reported by Hahn and Strassmann in Germany) here in Berkeley, they discovered neptunium, which was the first synthetically produced transuranium element.

Your finding plutonium in 1971 is included in your National Medal of Science citation, "discovery of primordial plutonium in Nature".

Yes, that is among the things referred to in the Citation. Pu-244 is the longest lived isotope of plutonium, with a half-life of about 80 million years. Of course, it is a question, whether or not it was formed in the last nucleosynthesis of heavy elements in our solar system. There is also the possibility that it might have accreted from extraterrestrial sources as our earth traveled through the galaxy. But we think that it is more likely primordial.

How did you find it and how did you determine its lifetime?

When we began thinking about a search for plutonium in nature, we looked at the well known properties of plutonium based on studies of other

---

*Magdolna Hargittai conducted the interview.

At the General Electric Co., Knowles Atomic Power Laboratory in 1971. From left to right: Francine O. Lawrence (Los Alamos National Laboratory), Jack L. Mewherter (GE, KAPL), Darleane Hoffman (LANL), and Frank M. Rourke (GE, KAPL). All photographs courtesy of D. Hoffman unless indicated otherwise.

Presentation of the National Medal of Science to Darleane Hoffman by President Bill Clinton, 1997.

plutonium isotopes and we decided to try to find a mining operation somewhere that might be concentrating plutonium-like elements. We found the Molybdenum Corporation of America mine in Mountain Pass California around 1969–1970. They were doing extraction processes for rare earth metals and they used one of the most advanced extraction processes for a mining company. They were extracting cerium, praseodymium, and other rare earths from a pre-Cambrian bastnasite for commercial uses. We chose this operation because the cerium in this ore was already known to be enriched a factor of about half a million relative to its average terrestrial abundance. Cerium in many instances behaves like plutonium because they have the same +4/+3 redox potential. We postulated that plutonium would most likely be in its +4 oxidation state in nature. They were using hydrogen di-2-ethylhexylorthophosphoric acid (HDEHP) as their extracting agent — which was very farsighted for a mining company — so we were able to get, after they had back extracted the +3 rare earths, some of the extractant that should have all the +4s remaining in it from many repeated extractions. So we managed to get a tremendous concentration factor based on this and the 9 liter HDEHP sample we got from them represented the processing of about 260 kg of "as-mined" bastnasite containing ore! Then we divided it into three fractions which were processed separately. Plutonium-236 and Pu-242 tracers were added to each for yield determination and the plutonium was separated from each along with the blanks that were run in parallel with the samples to make sure we had no Pu-244 contamination in our reagents.

At that time we had a collaboration with chemist Jack Mewherter and mass spectroscopist, Frank Rourke from the General Electric Co., Knowles Atomic Power Laboratory (KAPL) in Schenectady, New York. They had the most sensitive mass spectrometer in the country at that time. We sent them our purified Pu samples and blanks. Jack Mewherter purified them some more, and divided them into fractions and then Frank Rourke made the mass spectrometric measurements. They did, indeed, detect Pu-244 in two separate aliquots and after much checking and calculations relative to the added tracers, blanks and yield corrections, the concentration of plutonium-244 was determined to be 0.001 microgram of Pu-244 per g of bastnasite. Then we did a lot of calculations to get its abundance relative to the cerium in the bastnasite and found that it only had to be enriched by modest factors relative to cerium and after much consideration determined that it was probably primordial. We wanted to continue this research collaboration and repeat the experiments on other samples and different ores, but the collaboration was summarily terminated when Dr. Rourke was diagnosed with pancreatic cancer and died a few months later. There was no long term funding for the project, and I and others moved to different positions and the work was finally just stopped. In fact, someone asked me the other day whether they should look for plutonium-244 in various places such as ice-cores or old ores with accelerator-based mass spectrometry which would have much better sensitivity than our previous methods. Of course, I think that would be an excellent research project and should certainly be pursued.

Let's go back to your childhood and your family background.

I was born in a little town with a population of about 340 in northwest Iowa. My father was a school superintendent at the consolidated school there, and that was his first position as superintendent. I am the first of two children, my brother is 5 years younger than I. (He also became a chemist and earned his Ph.D. where I did and then went on to a highly successful career as a professor of physical chemistry at the University of Oklahoma.) When I was in second grade, we moved to another small town in Iowa and I went from third through ninth grades there. Later we moved to yet another town in northeast Iowa, near where my father grew up; he was of Norwegian ancestry. My mother also grew up in Iowa; she was an orphan and a German couple adopted her. Both my father and mother received their degrees from a small college in Iowa and my father went on to the University of Chicago for graduate work.

I always liked art, music, and mathematics. My mother's college subjects were oratory and music and my dad's interest was primarily in mathematics which he taught in high school, but he also substituted as a girls' basketball coach when no one else would do it. When I first went to college, I could not decide whether I wanted to take mathematics or applied art. But I went to Iowa State College (now University) in Ames and if you took applied art you had to enter in home economics. So I entered in home economics and, fortunately, I had to take chemistry. To make a long story short, soon enough I realized that I was not talented in applied art and did not like it anyway; but what I really loved was chemistry. It was taught by a home economics professor, Dr. Nellie Naylor. She was a spinster (an unmarried older woman) and she taught a different course of chemistry for home economics majors from what was taught for science majors. We learned about the various uses of chemistry in everyday life, the environment, and many other things. She was way ahead of her times and I thought this was wonderful. One problem

I had with majoring in mathematics was that I didn't know what I could do after graduation except teach, which I didn't want to do. (Remember this was in 1945!) I saw how poorly women teachers were treated in those days. It wasn't until World War II when they really needed women that they started to treat them a little better because they needed them! But in those days when a woman teacher got married she had to quit her job. I thought that was ridiculous because why should you have to quit your job just because you married? Even in women's colleges in the East women professors had to resign their positions when they married. I didn't want that restriction so I decided to major in chemistry and obtain a non-teaching position to leave the option open for myself and if I wanted to get married I could and still keep my job if I wished and I vowed to still have a normal social life.

What did your parents think about your choice?

I think my father was very happy that I decided to go into science instead of art although he did want me to get a teaching certificate which I never did. My mother was also very supportive; she had never worked outside the home after she married, except for a few substitute teaching positions. But she did insist that besides science I should take some other things such as speech and music in high school as well as in college. These have been especially valuable to me in learning to address large audiences without getting "stage fright". Also, music has always been one of my on-going pleasures.

When you decided to switch to chemistry, how did you go about it?

My applied art teacher, who was also my counselor, was also an "old maid" as we unkindly referred to older single women at that time. She asked me if I thought chemistry was a suitable profession for a woman. I told her that, of course, after all my chemistry professor was also a woman! I suspect that another reason for my choice was my remembrances of Marie Curie, about whom we read in 8th grade — how she and her husband worked together in the lab, and that she had two successful children. As you know, one of them, Eve, wrote a book about Marie Curie. While I was at Iowa State Eve came there to give a talk, so I got to meet her. She also wrote another book, *Journey Among Warriors*, based on her experiences as a correspondent for *The New York Times* in the 1940s. She traveled to North Africa (and later Poland, and even China) along with the military forces, taking with her only a small bag and her heavy old typewriter; it's one of the best non-fiction books I have read.

Going back to my story: in the spring of 1945 I switched to science. I started to take math again and worked at various odd jobs around the college to help with my expenses. A real turning point was in my junior year when they advertized for two undergraduate research assistants at the Atomic Research Institute of the Ames Lab of the Manhattan Project. I didn't know whether I should apply or not but one of my male professors encouraged me so I applied and I got one of the two positions. There I started to work with an inorganic nuclear chemist, Professor Don S. Martin, Jr. One of my first jobs was splitting mica for Geiger counters. Of course, I went on to other things later. It was a great experience and that experience was really responsible for me becoming a nuclear chemist.

When I finished my B.S. degree in 1948 I just assumed that I would go on to graduate school. My professor suggested that he would give me recommendations to California Institute of Technology or wherever I wanted to go but I said, "I'd rather stay here." At that time they were building a new 68-MeV Synchrotron at the Ames Laboratory which would be one of the highest-energy synchrotrons at that time. Dr. Martin agreed to take me as a graduate student and I started research on photonuclear reactions. In the first year of graduate school I met my husband-to-be as he was in the statistics class that my major professor was teaching. Marvin was working at the synchrotron, helping to build it and doing research there.

Is he a chemist or a physicist?

Marvin is a nuclear physicist. We did experiments together; he would run the Synchrotron and I could get my radiations done. That was an important step in my life. I didn't go out exclusively with him but we saw a lot of each other. I finished my Ph.D. in December 1951 and we got married the day after Christmas. I left for a position at Oak Ridge National Laboratory almost immediately. I didn't know it at that time but his major professor said to him that he had made a horrible mistake: he should have married some nice woman who would stay home and took care of him! (I didn't find this out until some 35 years later when I had the pleasure of entertaining him in our home here in Berkeley!) So Marvin stayed in Iowa to finish his degree and I went to Oak Ridge. When he finished in the summer he interviewed in Oak Ridge, but didn't really find a suitable position so he wanted to go to Los Alamos, where he had worked one summer before. He took a position there and after a year, in December 1952, I resigned from my position in Oak Ridge to follow him to Los Alamos.

My husband thought that I had been promised a job there because he'd talked to people about me. I was so young and naive in those days that I didn't even ask what exactly it was. I arrived in December and Marvin was sort of vague about this job; he said, well, I think it was the radiochemistry group in the test division. So I called personnel about it and they said, I am sorry but we don't hire women in that division! This was the first time I had run into such a problem. Of course, I'd observed before how women teachers were treated, but I myself had never run into any overt discrimination because of my sex! I was totally shocked. Then in early January 1953 we went to a party that the director of the lab, Dr. Norris Bradbury, gave for the new hires and I talked to various people. I finally met a man whom I found out after we introduced ourselves was Dr. Rod Spence, Leader of the Radiochemistry Group. He said: "Where have you been? I have been looking for you!" I said, "Oh, I have been here trying to collect my promised job, but I couldn't find it." So there was obviously some kind of disconnect which we solved later — but I have never trusted personnel departments ever since. I always tell young people, if they really want a job with a certain person, make a direct contact first and then do the formalities. He hired me and I thought that everything was settled and I could immediately go to work.

But between Oak Ridge and Los Alamos they had apparently lost my security clearance and spent two or three months trying to find it. Meantime, the radiochemistry group wanted me to come to work and I was just going out of my mind in a little apartment, not being able to go to work as I wanted. Finally Rod Spence suggested starting the security clearance all over and they called in the FBI, who did it in those days. They found my clearance in two days, so I finally could start working on March 13, 1953, more than 2 months after I first arrived in Los Alamos!

The worst thing about this delay was that while I was waiting for my clearance, the Radiochemistry Group researchers were processing the debris from the Mike thermonuclear test (November 1, 1952), the first U.S. thermonuclear test which was conducted in the South Pacific, and in that debris they found two new elements, einsteinium (99) and fermium (100), that no one dreamed would be there. It was a joint discovery by Los Alamos, Berkeley and Argonne Laboratories and I have never gotten over missing that amazing discovery due to administrative errors! By the time I got to work, the discoveries were pretty much pinned down. But they wanted me to develop a new procedure for analyzing multiple samples for plutonium from the nuclear tests and that was my first research project at Los Alamos.

How long did you stay there?

For about 31 years. Our two children were born there; our daughter was born in 1957 and our son in 1959; I worked all the time. We were away for sabbaticals; one was in 1964–1965 in Norway, because I got a senior NSF postdoc and I could take it anywhere. Originally I was thinking of going to Mainz in Germany to work with Professor Günter Herrmann's group there in the nuclear chemistry institute but then Marvin got a Fulbright Fellowship to Norway so I said, it would be nice to go there; I might look up my ancestors. More importantly, there was a well-known nuclear chemistry group at the University of Oslo and also at the Institute for Atomic Energy in Kjeller, Norway where they had a research reactor. Since I wanted to do some research on short-lived fission product separations, it seemed a good choice so we went to Norway rather than Germany. It turned out to be an excellent choice for me and I found Norwegian women were treated much more equally than women in the U.S. at that time. It was all right for women to go out to dinner alone but men also didn't open doors for them, etc. However, if you asked for help it was willingly given. I learned a great deal and developed a much more independent attitude socially and scientifically after our year there!

You said when you had your children in Los Alamos, you never stopped working. How did you manage?

I was very lucky because I always had help in our home. I started out by trying to take my daughter somewhere else, but I decided I couldn't do it. So I got a very nice older woman who came every morning to our home. Her husband worked at the Los Alamos Lab and dropped her off every morning and picked her up every evening; I was very well organized in those days. I planned the whole menu for the week; she would have dinner ready in the evening, and so on. Later when my daughter was about 7 and my son about 5 years old, my mother came to live close to us as my father had died of a heart attack just before my last year in graduate school and had been taking care of her mother who had recently passed away.

How do you look back to this time?

It needed a lot of organization and timing. I enjoyed what I was doing on radiochemical separations and research on new isotopes. We worked odd hours often. The hardest part was that my husband often was on tests in other places, like Nevada or the Pacific. That meant that I was alone in the evening and I could not go anywhere unless I arranged with someone to be there. It was confining, but I managed it and continued to participate in some musical activities.

What would you single out as your most important achievement during your Los Alamos years?

Certainly the plutonium-244 got the most publicity. It was something that I wish I could have carried on and done a more finished job but I couldn't. Also, all the work on short-lived fission products and later on the mechanisms of the fission process itself. We were able to separate very small quantities of fermium-257, which at that time was the heaviest fermium isotope known. Fermium has atomic number 100 and it has a half-life of about 100 days. Isotopes of fermium, as far up as mass 255 were discovered in the Mike debris and also einsteinium, but they didn't discover mass 257 in that debris. We found it later in other tests at the Nevada Test Site and we were able to isolate it and study its spontaneous fission properties. We postulated that maybe atomic number (Z) of 100 would want to break symmetrically into two spherical products with $Z = 50$ and 128 neutrons which is near the doubly-magic atomic numbers of $Z = 50$ and neutron number $N = 132$. So we thought that fermium isotopes might split spontaneously into two fragments near these configurations as the heavier fermium isotopes were reached. And this prediction was confirmed when we discovered symmetric spontaneous fission in Fm-257 recovered from debris from the nuclear tests. Up until that time all the spontaneous fission (SF) that had been observed resulted in two very unequal mass fragments, and was called "asymmetric" fission. At first, the physicists didn't want to believe that we had observed "symmetric" spontaneous fission. I gave a talk on this at an American Physical Society meeting at the University of Washington. I was very excited about this result. Then one of the questions was: do you really think that this is true; why don't you chemists go back and look at your data? I said, no, we have looked at our data very carefully and I am quite sure that we see a significant symmetric fission component. This result was published in *Physical*

Review Letters in 1971. Anyway, it turned out to be a considerable breakthrough and initiated a renaissance of interest in studies of the spontaneous fission process. I consider the discovery of symmetric SF to be one of my most important achievements.

Another thing related to fermium was that we managed to obtain a "larger amount" of fermium-257 produced in the high flux isotope reactor at Oak Ridge National Laboratory. In a joint collaboration with Livermore colleagues, we purified it some more and made a target of only a picogram for reactions at the van de Graaff facility in Los Alamos. It furnished high-energy tritium beams and we did a (t, n) reaction on fermium-257 to make the new isotope fermium-259, which was only a second and a half long. This information, together with the fact that the half-life of fermium-258 was very short, (only a few tenths of a millisecond) indicated why in underground nuclear tests we could never get heavier masses than fermium-257. It was because there was this sharp "fission disaster" as we called it, which just cut off the production of heavier things. It was quite exciting to finally in 1975 produce the 1.5-second fermium-259 — and to find that its SF was primarily "symmetric", that is, it divided nearly always into two equal fragments. The whole systematics of SF in the fermium isotopes was extremely interesting to me and I wanted to extend these studies to

Darleane Hoffman, Jerry B. Wilhelmy, and Joseph Weber with the apparatus at the Los Alamos van de Graaff facility used in the 1975 discovery of 1.5-second Fm-259.

lighter fermium isotopes. I applied for and was awarded a Guggenheim Fellowship for 1978–1979 for these studies and since the lighter Fm isotopes could be readily produced at the 88-Inch Cyclotron at Berkeley I chose to spend my sabbatical there with Professor Seaborg's group. This proved to be another turning point in my life.

In mid-1979 while I was on sabbatical, the leader of the LASL Chemistry-Nuclear Chemistry (CNC) Division was promoted to a new position and I applied for the CNC Division Leader position. There were certain things I wanted to implement in the Division. I had ideas about the kind of collaborations we should have and projects we might pursue, so I thought it would be a good opportunity for me to contribute in a meaningful way to the Division to which I owed so much. I was about 52 years old at that time, somewhat older than most men when they first assume such responsibilities.

Weren't you afraid that you would have to give up some of your research due to the administrative work you had to do in that position?

Yes, but I didn't plan to do this job forever. But there were some things I genuinely wanted to do. For example, we had a nuclear medicine program, we had a stable isotope separation program, we were in charge of the reactor, we had activities going on at various sites of the lab, we had a lot of radiochemical research, pure inorganic and organometallic research, and I just wanted a little broader view of life and I thought that I could help a little with all that. But I remember, the first day I was on the job, I had to fire somebody! One of the group leaders called up and said that he had some very bad news. There was a fellow, one of the technicians, who had been caught stealing, so there was no choice. I thought what a start! But I always had excellent administrative support and I still continued to do some experiments on heavy elements at Berkeley. I held a Division Leader position for five years, first of Chemistry-Nuclear Chemistry Division and then its follow-on, the Isotope and Nuclear Chemistry Division.

Then Seaborg and some other colleagues called and asked if I would be interested in coming to UC Berkeley as a full professor in the Department of Chemistry. I said, probably. One of the things I wanted to make sure about before considering such a move was whether I could keep my University of California retirement plan — Los Alamos operated as part of the University of California system and this was worked out. In a way I felt badly about leaving, but on the other hand, much had been achieved while I was there. About 18% or so of our staff were women and we had done well with our

Darleane Hoffman with Glenn T. Seaborg in the Heavy Element Nuclear & Radiochemistry Group radiation detection laboratory in the early 1980s.

nuclear medicine program, and many other things. I felt that if I stayed, I would be doing the same thing all over again and I would prefer to go on and try to help in training some new nuclear and radiochemists. I felt some obligation in this respect — never mind that I said, I would never teach! By that time my husband decided that he was ready to retire and he told me I would be a fool not to take advantage of this opportunity so we moved here in 1984. I was also interested in the fact that here we could readily produce heavy elements, such as fermium, einsteinium, californium, and eventually maybe even superheavy elements. This time, however, I was cleverer than I was in the previous position at Los Alamos. I asked from the personnel people for something in writing saying that I could keep my retirement when I moved here. It was a good idea to do so because later when I wanted to switch from the UC Berkeley campus to LBNL each summer they wanted to terminate my retirement and health benefits — so at least I had learned something! Thus I came here to Berkeley to become professor of chemistry and to take over the heavy element nuclear and radiochemistry group. Glenn Seaborg by that time was no longer taking students and he was very helpful to me in all possible ways.

Please, tell me something about Seaborg.

He was a wonderful resource. He would have a brown-bag lunch once a week and he would remember everything because he wrote everything in his journal. More than that, every day he talked into a tape recorder even when he was gone and when he came back his secretary would transcribe it. He started his diary when he was very young and kept doing it all his life. I wish I had done that! Many of his books were based on his journals. He had volumes of his journals. When I was here on sabbatical, in 1978–1979, I remember the brown-bag lunches in his office, when people would tell what they were working on and we would discuss this in detail.

When I moved here as a professor, my husband told me that I should not be too surprised if I didn't get too many graduate students because maybe some of the young men would not want to work for a woman. I said, don't be absurd! Then when it was time to go down for the first party of incoming students, Glenn told me, let's go together and see who is there. I am very short but he was very tall, about 6 ft. 6 in., and he would look around and introduce himself. Of course, at that time, all the students knew who he was — so I had no trouble in attracting students — even if not for my own appeal! He was very helpful. And he continued to have his famous brown-bag lunches with my group. He traveled a lot. Helen, his wife, whom I also know very well, is a wonderful woman, too. She almost always went with him on his extensive travels and that must have made it easier for him. She accompanied him when he had to go to Washington on many occasions because he was the former chairman of the U.S. Atomic Energy Commission (1961–1971) and subsequently was often asked to Washington, D.C. to give his advice. Whenever I needed to find out about something, I just ran over to his office in the next building, and he often said, I can look this up in my journal to make sure. He could tell me everything I had done, what I ate on a certain day, as well as everything else I wanted to know. He had a fantastic memory which, of course, was reinforced by his faithful recording of his daily activities in his journal each day, a practice he began when he was 12 years old! He was also always extremely enthusiastic about students and he had a remarkable way of communicating with them.

Eventually you became director of the Seaborg Institute...

I retired from active teaching in 1991 and Seaborg was very upset with me. What happened was that in 1991 they had an incentive package for

In front of the Seaborg Institute at the Lawrence Livermore National Laboratory, March 1994, on the occasion of the announcement of naming element 106 as Seaborgium. From left to right: Helen and Glenn Seaborg, Darleane Hoffman, Duane Sewell, Carol Alonso (one of the co-discoverers of element 106), and Christopher Gatrousis, principal co-founder of the Livermore Seaborg Institute.

people who were in the California Public Employees Retirement System which meant that I could retire with more than the salary I was currently earning! But I took a different option so that if something happened to me my husband would get the same amount I was receiving, so I took a little less. But it was one of those one-time financial incentives that you could not turn down. But I never saw Seaborg so upset as when I told him about this. I told him that I would continue working the same as I did before but he said, it wouldn't be the same without the power and authority. However, when I retired I took the position as first director in 1991 of the Seaborg Institute at Lawrence Livermore National Laboratory on a half-time basis. Together with Drs. Chris Gatrousis, Tom Sugihara, and Patricia Baisden, I had helped to found this Institute and write its Charter. I retired as Director in 1996. Now I continue to serve as an advisor.

What was your main area of research here (at Berkeley)?

When I first came to Berkeley we started to study reaction mechanisms for production and then the chemical properties of the heaviest elements. At that time no aqueous chemistry had yet been done on the element 105, then known as hahnium and since 1991 as dubnium. My group performed the first aqueous chemistry on element 105. At that time Ken Gregorich was finishing his Ph.D. with Dr. Seaborg and he became my first postdoctoral student at Berkeley. I also had arranged a Miller Professorship for Professor Herrmann (Mainz, Germany) so he could spend some time at Berkeley. He said, we know about the chemistry of tantalum and niobium in Group 5 and suggested that perhaps some of their chemistry could be applicable to element 105, if it is the heaviest member of that group. Based on the chemistry of these lighter Group 5 homologs, Ken developed what we called the "glass chemistry". We produced the element here in bombardments at the 88-Inch Cyclotron and then collected the atoms via a helium gas-jet system on glass cover slips. Of course, it is much more difficult to do this with the heaviest elements than with the homologs because the heavy elements primarily decay either by alpha emission or by spontaneous fission. So you can't count a thick sample of these because it absorbs these radiations, and we had to do the chemistry extremely fast. We found that we could do it in ~50 seconds. That was fast enough because the half-life of the longest known isotope of element 105 is about half a minute, so we could do it fast enough to have enough left for definitive measurements. Thus we investigated the chemistry of element 105 and we did show that, indeed, it behaved like tantalum or niobium so it belongs to Group 5 of the Periodic Table. But it behaved more like the lightest homolog niobium than like the heavier homolog tantalum in the extractions we performed and this we didn't understand. When we do chemistry with the heaviest elements, we always measure their nuclear decay properties, which is how we can positively identify them. Alpha decay is particularly nice because you can measure the whole chain and the energies and time correlations. If only SF is measured, we cannot be so sure that we have what we want because it results in two fission fragments that cannot be convincingly related to the fissioning nucleus on an "atom-at-a-time" basis. Later, we also did gas-phase chemistry on element 105. These experiments started a long-term international collaboration with the Germans, Swiss, and many other scientists in other countries. This rejuvenation of "atom-at-a-time" chemistry was my second big contribution after the ground-breaking spontaneous fission research. Another frontier research project was on electron-capture delayed fission of heavy elements. My first graduate student to receive his Ph.D., Howard Hall, developed new instrumentation and initiated these experiments in 1988. This research provided the first direct proof of this process in americium and subsequent students extended the studies to other actinides.

This field is obviously a very competitive field; who finds and publishes a new element or isotope first. This then brings about the danger that you might rush with a publication before you can be absolutely sure of your findings.

Exactly. We also fell into that trap with our report of production and identification of the element 118 decay chain, except that in this case we were victims of fraud.

Fraud?

Yes. It was perpetrated by one of our accomplished and trusted co-workers whom we brought here from the Gesellschaft für Schwerionenforschung (GSI) at Darmstadt, Germany to help build the instrument known as the Berkeley Gas-filled Spectrometer (BGS). We thought that we had observed three decay chains from element 118 produced in the reaction of lead-208 with krypton-85 and the results were published in 1999. Then we continued with additional experiments to try to do an excitation function for the reaction and weren't able to repeat it. We looked very carefully at the data tapes that had been saved and couldn't find evidence for the originally reported events on them. None.

Why did he/she do that?

If I could answer that I would be a psychiatrist instead of a nuclear scientist. He had also apparently similarly fabricated events reported in the initial discoveries of elements 110 (the second event) and 112 (the first event) at GSI before he came to us to help build the BGS. By the way, his work on building the separator itself was outstanding and with subsequent improvements it is among the best in the world.

But you asked him what happened ...

Of course, and he looked me straight in the eye and told me not to worry about it — that we would find the events in the data. Naturally, he was fired after a committee was appointed and reviewed all the data and also could find no evidence for these events on the data tapes, although one tape remains missing to this day. I think that probably because of his childhood background and the various things that had happened to him he no longer could think clearly. As I said, he also seems to have duped the German team at GSI and fabricated data on elements 110 and 112 as they later found out after we warned them of our experience.

Then what happened? You had to retract your paper?

Yes. You see, all results have to be verified by another group before they can be accepted. We published our retraction quickly and then after a series of additional attempts to reproduce the results, we published a comprehensive paper in 2003 setting an upper limit on the cross section for production of element 118 via this reaction of less than a picobarn. But the strange thing about this was that when he did this to the Germans he made up the first event about element 112 but later they saw two more events that apparently were genuine. He didn't quite guess the right alpha-energies to put in for the first event so they even had to postulate an isomer to accommodate this. Why does somebody want to do something like that? When we questioned him about this, he denied it, he said that somebody changed the data. Anyway, the lesson of this story is that first of all you have to publish your evidence to the best of your ability and it is very difficult to uncover fraud by a trusted team member. But a result also has to be verified by you or others — that is the scientific method. It worked here and ultimately the truth will be found. I think that it is an essential requirement. The field is very competitive and the work gets harder and harder. The experiments are more and more difficult.

When we started out with element 105 we were getting a few events per hour. Now we are lucky if we get one or two in a week or even longer. There are not many places where you can do this kind of research which makes confirmation of results difficult. We can still do it here, you can do it in Dubna in Russia, and the Japanese are currently working in this field. They are trying to make element 113 directly and have seen a couple events at the femtobarn level. The different element 113 isotope that the Dubna/Livermore collaboration has reported was not made directly but rather by the alpha-decay of element 115. Based on our unfortunate experience, I cannot emphasize strongly enough the necessity for confirming all results before they are accepted as discoveries.

What are you involved with currently?

I am doing quite a lot of writing, various book chapters and reviews and give many invited lectures. My last graduate student finished at the end of 2004 and he stayed as a postdoc. He is working with elements 110 and 111. In addition, we wanted to use a different reaction to try to produce element 107, bohrium, with larger cross sections in order to perform more extensive studies of its nuclear and chemical properties. I am also a co-principal investigator on a project to help train the next generation of scientists in nuclear and radiochemistry required in the U.S. for both applied and fundamental energy security in the broadest sense. As professor of the graduate school in the Department of Chemistry, I also counsel undergraduate transfer students in chemistry and mentor postdoctoral students. I continue to serve on many review committees for university, national laboratory and government organizations.

Please, tell me a little about your present family.

Our daughter, Maureane R. Hoffman, who is our oldest, is an M.D./Ph.D. She wanted to be a veterinarian as she has always loved horses. Because we lived in New Mexico, we could have horses. She finished high school at 16 and went to New Mexico State University and she wanted to go to veterinary medicine school. She got a degree in animal science and the professor with whom she was doing research suggested (based on her research ability) that she should go to the University of Iowa to get a Ph.D. rather than try to go to Veterinary Medicine School since there was none in New Mexico and she did that. In her first year there they advertized for people who wanted to do an M.D. and a Ph.D. concurrently, so she applied and got this Fellowship which paid all her tuition and fees plus a small allowance. Then she did a residency at Duke University, and after that went to the University of North Carolina to head the Blood Bank and then returned to Duke University where she is now on the Medical School Faculty and Head of Pathology at the Veterans Administration. She is very engaged in research on blood coagulation and related topics and gives seminars and travels a great deal both in the U.S. and abroad.

Our son Daryl thought that he wanted to be a lawyer, so he went to UCLA to study political science, but he decided during his second year there to switch to psychobiology and then to biological science. He was admitted to medical school at the University of New Mexico and then did his residency in surgery, specializing in plastic surgery at Stanford University and is now a plastic surgeon in the Palo Alto/San Jose area of California. He married a Los Alamos girl, Dr. Susan Fraser (M.D., internal medicine) in 1985 and they have our three grandchildren, Sarah (born 1989), Daniel (1991) and Michael (1995). Marvin and I greatly enjoy being close enough to enjoy their activities, and especially in taking them to their tennis tournaments, a long-time sports interest in my family.

Are you religious?

Yes and no, I guess it depends on how you define it. I was born and raised as a Methodist and was extremely active in the Church, especially as a Choir Director and vocalist for a long time, but I am no longer participating in this way. I believe that humans need moral guidance and a religious faith to undergird their concepts of right and wrong and of living for a higher calling than just gratifying their own selfish desires.

Are you involved with trying to get women into science and trying to keep them there?

I have been very involved over the years. I was very proud of the fact that while I was division leader at Los Alamos I could do something positive about this. One of the most pleasurable management tasks that I had occurred during the last year I was there. The laboratory had some lawsuits concerning women not receiving equal pay to men in equivalent positions and so each division was given an amount of money to bring the average of women's salaries up to the average of men's salaries in the same position. We distributed this money according to merit among the women and I felt that this was one of those win-win situations. Over the years I have been involved with many women's conferences and women in science. I think that overall I had about 30% women among my graduate students at Berkeley. When I first went to Berkeley in 1984, only about 18% of the graduate students in Chemistry were women, and I was only the second tenured woman professor in chemistry out of a faculty of about 40. At that time I felt that if we increased the number of women Ph.Ds. in Chemistry, the faculties of the major research universities would also increase in a commensurate manner. However, according to statistics from the American Chemical Society, this has certainly not been the case. Now some 50% of the B.S. degrees in chemistry go to women and more than a third of the Ph.Ds. in Chemistry are earned by women, but (2005) only about 8% of the tenured full professorships at our major universities are held by women. We need to investigate the university climate and the concepts of tenure and why so many women choose not to even apply for positions at the major research universities in the U.S.

Anything else you would like to add?

Whatever things I have accomplished in science were possible only because I was very fortunate in having wonderful help at home (my mother and others), a supportive and understanding husband who was also a colleague in nuclear science, and many outstanding colleagues, students and collaborators. My love and heartfelt thanks to all of them.

Richard L. Garwin, 2004 (photograph by M. Hargittai).

23

RICHARD L. GARWIN

Richard L. Garwin (b. 1928 in Cleveland, Ohio) is IBM Fellow Emeritus at the Thomas J. Watson Research Center, Yorktown Heights, New York. He did his undergraduate studies at Case Institute of Technology and received his Ph.D. degree in Physics from the University of Chicago in 1949. He stayed with the University of Chicago for three years and then joined IBM Corporation in 1952. He was Director of the IBM Watson Laboratory and Director of Applied Research at the IBM Thomas J. Watson Research Center. He did consulting and had part-time appointments at the Los Alamos National Laboratory. He is Adjunct Professor of Physics at Columbia University and was Professor of Public Policy at Harvard University. He was a member of the President's Science Advisory Committee 1962–1965 and 1969–1972, and of the Defense Science Board 1966–1969. From 1997 to 2004 he was Senior Fellow for Science and Technology at the Council on Foreign Relations, New York.

Dr. Garwin is a member of the National Academy of Sciences of the U.S.A. (1966), of the American Academy of Arts and Sciences (1969), of the American Philosophical Society (1979); a Fellow of the American Physical Society, a member of the National Academy of Engineering (1978), of the Institute of Medicine, and other learned societies. His numerous awards and distinctions include the Enrico Fermi Award (1996) and the National Science Medal (2003).

We recorded our conversation with Dr. Garwin in his home in Scarsdale, New York on August 15, 2004.*

---

*István Hargittai and Magdolna Hargittai conducted the interview.

First we would like to ask you about your family background.

I was born in Cleveland, Ohio, in 1928. My father, Robert, had come to this country when he was two years old; he was of Ukrainian-Polish extraction. He lived in Chicago with his three brothers, mother and father. His father operated a shoe store in Chicago with a partner. I was told that when my father was seven years old, the partner shot my father's father, thus orphaning my father, together with his three brothers. His mother eventually moved with her children to Cleveland where there was a better orphanage for Jewish children. She kept the oldest boy and the youngest with her and she put the two middle boys, my father and his brother into the Jewish Orphan Asylum. They graduated high school there, and then my father went to Case School of Applied Science for a degree in electrical engineering. He taught high school electricity and in addition he had a job nights and weekends as a motion picture projectionist. He never worked as an electrical engineer because of prejudice and restrictive hiring practices against Jewish people. His family name was Gawronsky and he and his brothers all changed their name to Garwin in 1921 just because it was a simpler name. My father did well; we lived in a two-family house where the other part of the house belonged to one of my mother's sisters and her family. We moved to University Heights, a suburb of Cleveland, in 1940 into a new house. My mother's sister Irene was divorced by then and she and her daughter came to live with us. I went initially to public schools in Cleveland and I graduated from Cleveland Heights High School in 1944, somewhat accelerated because I'd skipped a year or two earlier on. Besides, during the war, they had school in the summer time. I entered what was still the Case School of Applied Science but soon became Case Institute of Technology by the time I graduated in 1947, and more recently it became Case Western Reserve University.

My mother was born Leona Schwartz; she moved with her parents and her many brothers and sisters from Hungary to Cleveland when she was twelve years old. She was born in 1900 and my father was born in 1898. My mother's mother never learned much English and spoke Hungarian all the time. My mother left school in the tenth grade and had a job in a department store, where her boss told her once that, "You are a very good worker, but we can hardly understand you when you come in on Monday. By the time you leave on Friday, your English is quite good." After this she never spoke Hungarian again. She died in 1996 with full command of her faculties except for the last few months, but she couldn't

Richard Garwin with sons Jeffrey and Thomas in the backyard of their home in Scarsdale, New York, probably 1957 (all photographs courtesy of R. Garwin unless indicated otherwise).

recognize even a word of Hungarian because she had repressed it so. My mother had initially told me that she was born in the United States and so had my father. In all the forms I had filled out for the U.S. government I had them native born. At a very late age, when she was about 85, she told me that was not true, and some years later she told me that my father also had come to the United States as an immigrant.

What made you interested in physics?

My father, in addition to being a graduate electrical engineer, was a motion picture projectionist, and was interested in training other motion picture projectionists who had to take an exam. It was a union job, quite restrictive. He set up optical instruction equipment, he had big demonstration lenses, and an optical bench. He made beautiful demonstrations. When sound movies came in 1928, he took upon himself to instruct the motion picture projectionists in Cleveland how to handle the sound.

He had the sound equipment around the house, too. There he had a shop where he and his younger brother, Joe, created a business, the Garwin Theater Equipment Corporation. It was mostly installing and repairing motion picture and sound equipment for schools and industry in Cleveland. It was a small organization; just the two of them and never more than three or four other people working for them. I would work with them when I had free time, even after I had graduated from the University of Chicago. My brother did the same; he was born in 1933 and he has also a Ph.D. in Physics from the University of Chicago. He went there just before Fermi died; he was a graduate student of Val Telegdi. He has been at SLAC in Stanford for 35 years.

When I was growing up, there were all kinds of technical books around the house and fascinating equipment to find out about. I read all the books and I would repair equipment and build better amplifiers. I found out about the fundamental limits of noise and devised means for reducing the noise. You can't get rid of the noise, but you can push it to bands where you don't care about it, something which stood me in good stead at multiple points in my career later on. My father and I built a glass-working bench from surplus equipment and marble slabs from toilet stalls. We had some machine tools so I could build equipment myself.

I didn't want to work for anybody so I wanted to become some kind of research person. My father wanted me to become an engineer because he had an outdated idea that it was the engineers that built things with their hands. By the time I went to college, engineering was a theoretical profession; the people who actually knew how things worked were scientists.

How did you get to do physics at the University of Chicago?

My physics mentor, Robert Shankland, encouraged me to do physics in Chicago. I got a fellowship from Chicago to go there. Some of my work in Chicago is covered in the talk I gave in September 2001 at the Fermi Remembrance Symposium at the University of Chicago. I titled my talk "Working with Fermi at Chicago and Post-War Los Alamos". After I'd been at the University of Chicago in 1947, after I'd attended my graduate courses for six months or so, I took the initiative to go to Professor Fermi and volunteered to help him in the laboratory. He welcomed me. He had a large laboratory, 20 meters by 10 meters.

He was working there with Leona Marshall on some experiments with positronium, using Geiger tubes. Their innovation was to mount a little cotton string inside the Geiger tube envelope that they had soaked in a very dilute solution of sodium-22, which is a positron emitter. The radioactive material would decay, neutrino would come out undetected, and the positron would come out. It would slow down, come to rest, and after some time it would decay having formed a positronium, the hydrogen-like atom of a positron and an electron rather than a proton and an electron — and with half the binding energy of the hydrogen atom because of the smaller reduced mass. They were interested in measuring the decay time of positronium, and they found two decay times, one for the singlet and one for the triplet state. They had competition in the form of Martin Deutsch at MIT, and Deutsch scooped them because he was using end-window photomultiplier tubes, which had just been pioneered by RCA. Fermi learned his lesson and got half a dozen of these phototubes, which then they used in their further experiments.

Jack Steinberger was in another corner of the laboratory. He had a pile of layered carbon and brass Geiger tubes and he was using cosmic ray muons, probably measuring lifetime or their spectra. I didn't pay much attention to that.

I decided that the Fermi–Marshall experiment was very much behind the times. Everybody was using coincidence circuits left over from the 1930s that had microsecond resolving time so they were limited to small sources strengths for their experiments. On the other hand, with nanosecond circuits, a thousand times more intense sources could be used. Much more detailed experiments could be done much faster than our competitors, and so on. I devised such circuits and our measurements became orders of magnitude more sensitive. My circuits were then widely reproduced in Chicago and used in the elementary particle physics field for about twenty years.

With these fast electronics, I began my work for my Ph.D. thesis, which was the first experiment to look at the angular correlation between the beta-ray from radioactive decay and the ensuing gamma-ray from the excited nucleus. This followed on gamma-gamma correlation done by Martin Deutsch and others in which a radioactive nucleus would decay with a beta-ray and there would be two successive gamma-rays. The correlation between the two gamma-rays would tell you something about the angular momentum of the nucleus, what state it had been left in. However, it didn't tell you anything about the initial nucleus whereas beta-gamma correlation would tell you about the product nucleus as it was created by the beta-decay.

It was a great experience to work with Fermi in his laboratory. He had a lathe and a milling machine and he delighted in making things himself. He could design things, of course, and then have the university shop build them, but he complained that they built everything too accurately and therefore it took them too long. No matter how he would try to mark the drawings so that they should not fine machine the surface, it did get fine machined so that the machinists could take pride in their job. But Fermi wanted to get the experimental results so he often resorted to making things himself, which was, of course, my strength as well.

At lunch one day Fermi wondered whether Robert Mulliken's work at Chicago in molecular structure energy levels could be enhanced by an analog computer; they did not have digital computers in those days. Then there was the entire problem of nuclear shell structure of the regularities in the nuclear properties, especially the binding energy as a function of the atomic number and atomic mass. Fermi maintained that he could solve only six problems. He was an expert of the hypergeometric function but that was not much help when you were faced with an arbitrary potential function in the spherically symmetric compound nucleus model. He wanted to know what the energy levels in the nucleus were assuming various shapes for the potential energy function. There were people who were postulating such things. He worked to develop some intuition himself. Rather than numerically integrating differential equations that he was very good at (using an accountant's pad, a slide rule for multiplication and division, and a Marchand electrically driven mechanical adding machine for addition and subtraction), he thought he would make an analog computer. He planned an analog of the Schrödinger equation, he would have a bar magnet suspended on a torsion wire, and have a solenoid around it with horizontal axis, so the restoring force on the bar magnet to its neutral position would be proportional to the current in the coil. If the current in the coil could be made proportional in time to the potential in the Schrödinger equation in space, then the time behavior of the angle of the magnet would mimic the evolution of the wave function in space There were all kinds of variation of such an experiment and I told Fermi that he had better things to do with his time and I would build him an electronic analog computer, which I did, and published it in the *Review of Scientific Instruments*. It was a rather monstrous thing, but it worked well. Fermi was the only one who ever used it, aside from some trials by Clyde Hutchison after Fermi's death.

So I went into physics not because I wanted to find out about the Universe, but because I wanted to occupy myself. When I got my Ph.D. degree, I was asked by the University of Chicago to stay on as an Instructor and then as an Assistant Professor and that was from December 1949 until hydrogen temperature. But I did that with long stainless steel rods, which came in at an angle. I designed also at the same time some nuclear explosives of the same principle that could be carried by airplanes, and they built them — named Jughead. The Atomic Energy Commission built five or six of these; they were ready also in 1952. They could be flown over the Soviet Union and destroy targets there. That was no mean feat because they had to have liquid hydrogen, liquid deuterium, refrigerators, and things like that.

Did you patent anything?

No, it does not pay to patent; the work was done for the government; they owned all the rights; so I couldn't have made any money out of it. Also, it would not keep other countries from doing it.

Why is the Ulam–Teller design still secret?

I don't know. I think the paper should be declassified. It's really a very good paper.

Do you think that other countries might use it?

That's the only reason for maintaining it classified, but I don't think it's a good judgment in this case, because the basic idea has been declassified, and the rest is details. There is no design to copy in the Teller–Ulam paper.

Do you know if the Soviet hydrogen bomb design was basically different?

Let me first finish my answer to your previous question. I do have one patent from the nuclear weapon program and that's for a pulsed neutron source. My co-inventors were the late Ted Taylor and the late Carson Mark. This was also tested in October 1952 and has been used on tens of thousands of nuclear weapons. That's still secret also, so I can't talk about it anymore at all.

That was 1951 and 1952. The design for Mike was all done and it was being fabricated. I designed some other things in nuclear testing as well. For Greenhouse George, I introduced the idea that stable isotopes could be added in various places in nuclear explosion tests and these would be intensely activated either by thermonuclear reaction or by the neutrons. So we put a little bit of stable isotopes here and a little bit of another stable isotope there. After the entire explosion and the mushroom cloud, the airplanes go out to sample. Then in the tiny samples that you get, you could look for a specific radioactivity and have detailed information about what is happening on a centimeter scale inside the nuclear explosive. I devised some of those things and they were used in George and elsewhere. I also designed other tools for diagnosing radiation implosion. The radiation implosion concept was later declassified by the Atomic Energy Commission. They revealed officially that our hydrogen bombs are two-stage weapons in which there is a preliminary nuclear explosion called the primary, which exploded inside a metal case of uranium or lead. Most of the energy released in the explosion of the first device goes into soft X-rays that fill the radiation case. That radiation then assembles and heats the secondary charge which may or may not have enriched uranium, and this is how our hydrogen bombs work these days. They can release energies over a wide range, from two kilotons, that is, one-sixth of the Hiroshima bomb, to multi-megatons, yet they are all very similar.

In 1952, I was working with Marshall Rosenbluth, one of the people who did the real calculations for Mike, and he and I were spending a couple of weeks at the National Bureau of Standards in Washington, D.C., using their primitive digital computer. He would run his computer calculations and I would write on the back of an envelope my analysis of what the answer would be and we would come pretty close by the simple concepts of what's going on in these radiation implosions.

I did more work at Los Alamos; I was there every summer through the 1950s, until about 1958. Then I spent a year at CERN, and I was again in Los Alamos half the summers in the 1960s.

By then you had been with IBM for some time where the jobs are for twelve months rather than the nine-month university appointments.

Yes, at IBM I had a twelve-month job, but when they built this new laboratory to study condensed matter physics, Emilio Segrè had been asked by IBM to head the laboratory. Segrè was on a sabbatical at the University of Illinois at Champaign-Urbana so I went down to see him before going to New York to work in the laboratory. He was considering directing it and we had a good discussion. I eventually agreed to go and Segrè did not. Instead, he went to Berkeley.

The laboratory continued to be headed by Wallace J. Eckert, who was the astronomer who introduced the punch card into scientific computing in the 1930s. During the Second World War he did war-related computing work (producing the Nautical Almanac) and after the war he returned to Columbia University as Professor of Astronomy. He was also the head of the IBM Watson Scientific Computing Laboratory. He was an excellent director. When I agreed to work for them it was with the proviso that they would give me one third of my IBM time to work with the U.S. government on matters of national security. I would turn over to IBM any money I would earn from the government, like consulting fees or travel expenses, but IBM would pay my regular salary and would not ask what I was doing. That arrangement worked extremely well for the forty years that I was with IBM. That covered not only my work at Los Alamos but later when I was on the President's Science Advisory Committee for two terms in the 1960s and 1970s, and with the Jason group of consultants to the U.S. government.

Whereas there was general agreement among the physicists about the necessity to build the first atomic bombs, the presidential decision for the development of the hydrogen bomb was preceded by rather sharp disagreement among them.

Many people who opposed the hydrogen bomb ultimately worked on it. It was a matter of formulating policy. First of all a lot of people felt that it was not practical and that Teller had worked all these years and if he hadn't made progress until 1949, then it wouldn't work. The General Advisory Committee of the Atomic Energy Commission was asked by AEC for advice on this and they advised unanimously against pursuing the hydrogen bomb. There was a minority report by Fermi and Rabi and it said that it is inherently evil in itself because it yields unlimited power and there was no limit to the destruction it could cause. However, after Truman's decision to proceed with the hydrogen bomb, most of these people, including Fermi and Bethe, went to Los Alamos to participate in the efforts. Fermi had not opposed it publicly because he was on the General Advisory Committee, but Bethe had led the opposition to it, that it was not necessary. They also argued that if we would work on it, the Russians would certainly work on it, and we would have a net loss of security. Bethe continued to believe that, but he as an émigré from Germany, a refugee from the Nazis, and all the others felt that so long as the government took these decisions in the appropriate legal fashion, it was their responsibility if they had special talents, to support this work. So Bethe went from opposing

Meeting of the Advisory Board to General Dynamics in 1953: On the right, Theodore von Kármán (second from the back), John A. Wheeler (fifth from the back), Edward Teller (partly hidden, third from the front); and on the left, Eugene P. Wigner (partly hidden, second from the back), George Gamow (fifth from the back), Richard Garwin (second from the front).

the bomb to heading the theoretical work on the hydrogen bomb as a part-time consultant at Los Alamos. Fermi, once the decision was made, contributed whatever way he could to building it. I did not have strong views one way or the other; I looked into it technically, whether it could be done and how it could be done later on.

Herbert York wrote later that it was not necessary.

I know Herb York very well. We spent seven weeks in La Jolla this summer and we saw him several times. Herb York has done a real service by writing his books; the most relevant is *The Advisors: Oppenheimer and Teller and the Superbomb*. He sent me the manuscript in draft for my comments twenty years ago, before it was published. That's when I first learned that the AEC had actually made viable and deliverable versions of Mike according to my design. He wrote that we would not have been disadvantaged in national security had we not built it and the Russians, the Soviets, had. I think that's true, but I think it would have been very bad for the scientific community, in those days of Senator Joseph McCarthy and others, to have opposed the hydrogen bomb. The Russians would have built something, which is ten or a hundred times as powerful as we had, despite the fact that the nuclear weapons that we had were entirely capable of destroying Soviet cities. I think that we could A), not have gotten the Russians to avoid working on it given the relations we had with them at the time and with which I am quite familiar. And B), yes we could readily have caught up and whether we would have caught up or not would not have made any difference for our national security.

But at that time York was a supporter of developing the hydrogen bomb.

Yes, York worked very hard on it. Incidentally, I met York for the first time in 1951 when he was taking a principal role in the diagnostic experiments for the 1951 tests, for the Geenhouse test and others. Then, of course, he was the first director of Livermore [the second weapons laboratory], which was formed in the fall of 1952. When Edward Teller astonishingly was disaffected with Los Alamos — I was at the meeting at Los Alamos in the late summer of 1952 when Edward said that this was not proceeding fast enough and we need another laboratory. How anybody in the world, any laboratory, any system, could have taken Mike from concept to test in less than sixteen months, I don't know. It was something that Teller was fixated on; he needed another laboratory because they were not working hard enough in Los Alamos on hydrogen bombs, but that was no longer true, once they found out how to do it.

But let me return to your question about the Soviet hydrogen bomb. Their first nuclear explosion in 1949 was a copy of our plutonium Nagasaki bomb. It was also the one that was tested a few weeks before Nagasaki on July 16, 1945, in New Mexico. The reason that they had that copy was that Klaus Fuchs had transmitted that information to the Soviets. Kurchatov, who was the head of the Soviet nuclear program, was the only one who was provided with the intelligence by the KGB. He kept this material in the safe and when people on his team would come and tell him how they were building their nuclear weapon, he would say, "Have you considered this or that?" and he would guide them to something, which was exactly what we had built. The Russian scientists were unhappy about this, for they realized that what we had tested was more conservative than what they were doing. They tested their version next year, in 1950, and it gave considerably higher yield, it was more efficient, and more advanced. Instead of using a solid ball of plutonium as our bomb design did, it used a different arrangement in their high explosive implosion. Their first thermonuclear yield came in 1953 and that was not a radiation implosion. Sakharov said in his memoirs that he had three ideas; one of them was radiation implosion, another was the so-called layer cake in which the nuclear fuel of lithium deuteride is mixed more closely with the fissionable material. I was on the small team that analyzed this Joe-4, the fourth Soviet nuclear explosion — the first one with thermonuclear yield — it was the so-called Bethe Panel, which analyzed this explosion. Hans Bethe, Fermi, myself, and Lothar Nordheim — a theoretical physicist from Duke University — were on this panel. The Soviets soon gave up this dead end approach, which merely provided an augmentation to the yield of the fission weapon and they also went instead to radiation implosion. The Russians often maintain that they had the first deliverable hydrogen bomb with this layer cake approach in 1953 and we didn't have our deliverable hydrogen bombs until 1954. However, that's not true because we had our deliverable liquid deuterium hydrogen bombs in 1952. In any case, the Russian nuclear designers were very competent, as are the Chinese. People in general do a very good job if they are allowed to do that.

You write that there had been about a thousand U.S. tests...

Of course, after 1963, the United States had no more nuclear tests in the atmosphere. Up to that time, the United States and the Soviet Union mostly exploded three hundred some megatons, officially, in the atmosphere. There was a lot of carbon-14 produced also from the excess neutrons from fission and fusion.

Bear with me for what sounds like a digression. A lot of things were not well known when the test program started and in 1968 there was some question in regard to the supersonic transport aircraft, which I studied for the President's Science Advisors and published on April 1, 1968. It was not classified, but it was a document that was not released until later by the government under the threat of a court suit. That was a one-month-long analysis of the U.S. supersonic transport program. Just at that time, people made charges that the nitrogen oxide from the exhaust of the SST in the stratosphere could catalytically decompose ozone in the stratosphere — thus is what would be the electron distribution about this velocity of the μ meson if the muon spin did not change during the couple of microseconds it was waiting to decay? Many people began this experiment and they all got results, but they got bad results and when they double scanned they found that they had different probability of detecting this forward emitted electron compared with the backward emitted electron, which was much more visible than the forward emitted one.

Professor Chien-Shiung Wu from Columbia University had come to me in August or September 1956 because I was making a demagnetization refrigerator and she asked whether we could work together to demagnetize and therefore cool cobalt-60, so that it would be highly polarized in these magnetic fields. I talked to her about it and told her that I had recently created a superconducting computer program at IBM and I was in charge of a hundred people in several laboratories, and I couldn't do that right away. I thought that she ought to be working with the people at the National

Richard Garwin with the g-2 experiment for the muon around 1957; the picture showing Garwin with part of the shielding at the Nevis cyclotron of Columbia University that was used as a source of muons.

Bureau of Standards. I was involved in the discussions of Lee and Yang's paper. Then in January 1957 Miss Wu and others at the Bureau of Standards were getting results, and the results were in the form of electron counting, and they were reproducible. However, they were not very striking results.

At this point, Leon Lederman called me one evening, January 4, 1957, a Friday. He said he had had an idea, and that is that in the cyclotron at Columbia, but any cyclotron would have done, we could do another experiment to prove parity non-conservation. This experiment involving pions and muons has become another famous experiment. All we had to do, in principle, was to move our detectors and measure the amount of muon decay electrons as a function of angle. It was a very simple experiment which we could do during a weekend although there were misadventures on the way. This experiment showed a very strong effect of parity non-conservation. In this experiment I could utilize a lot of experience of my previous work in elementary particle physics and even in magnetic resonance.

This was not only a very decisive experiment for proving parity non-conservation, but it also started a new area for me because I had been doing condensed matter physics and now I had to have somebody take over my superconducting computer program at IBM. For a couple of years

Richard Garwin peering at the camera through a quadrupole beam-focusing magnet of the g-2 apparatus, 1957.

I worked in this new field made possible by the parity non-conservation in the pi-mu-e decay. I led a group at CERN, which built an 80-ton magnet to store muons for 6 microseconds in order to determine the deviation of the muon g value from the value it would have of 2 if it were a Dirac particle. This was a very nice experiment and all the people who worked on it loved it. They included Nino Zichichi and Georges Charpak, and others.

You also know Val Telegdi.

Of course, I know him; I knew him back at the University of Chicago; he is a very fine physicist. We were good friends with Val and Lia. We were happy to have our sabbaticals at the same time at CERN in 1959–1960.

Originally I went to CERN for my sabbatical wanting to do nothing else than to read and to write, having worked very hard for years for IBM and for the U.S. government. I had a small office at CERN, shared with Giuseppe Fidecaro. Leon Lederman had been at CERN in the previous year and we had discussed some experiments on which he had been working while he was at CERN, and I felt induced to take up some of the challenges of these experiments.

Richard and Lois Garwin with Polykarp and Edith Kusch around 1957.

I may have not mentioned before that one of the reasons I left high-energy physics and went to IBM in 1952 was because I didn't like the sociology of high-energy physics. At the Chicago cyclotron, in order to get time on the cyclotron you needed to work with a group of five or six people and tell them six weeks ahead of time what it was what you wanted to do. I wanted to do what I wanted to do, stay up all night and do something the next morning. That was not possible. Now, of course, in particle physics, you have to work with six hundred people and tell people six years ahead of time what it is you want to work on.

I was fortunate in January 1957 because I had good friends on the Columbia faculty; they were using all the equipment that I had devised, so I knew exactly what was going on. It was easy to jump back in; it was no telling anybody, Leon Lederman and I just did the thing and had the result before anybody knew it. But over here, at CERN, you had to demonstrate

The g-2 CERN group, which Richard Garwin headed. Left to right: Antonino Zichichi, Hans Sens, Valentine L. Telegdi (not part of the g-2 group, but worked with it at the beginning), Georges Charpak, Francis J. M. Farley, F. Sauli (Charpak's assistant), Richard Garwin, and Leon M. Lederman (also not part of the CERN g-2 group, but helped start the experiment), in the CERN administration building.

a lot of things in advance to convince people that what you wanted to do was feasible. It was also a lot of fun devising all those techniques.

So instead of just sitting in the library, I was pressed into service to lead this group with such wonderful co-workers as Nino Zichichi and Georges Charpak and others. I listened to everybody, but I had to make all the decisions. We had meetings to be sure, but at the end, it was I who had to decide, just as when we were working on the hydrogen bomb. People were very enthusiastic and it worked out very well.

Georges Charpak was so enthusiastic that he said it spoiled him for any other research in physics. He was just about this time to go into medical instrumentation, where he stayed all these years whereas I kept urging him to continue his work on detectors for physics as well as for medical purposes. That experiment at CERN was probably my biggest physics experiment.

My last physics work at IBM was to look for gravitational radiation. Joe Weber in 1969 had published an article on the detection of gravity waves. He made very nice instrumentation to do this. But he was a terrible experimenter and could not interpret his data. He found large numbers of gravity waves, so many that the Universe would have run out of rest energy in a mere hundred million years creating these gravity waves. So it was dubious to begin with.

He had found these by coincidence detection in his so-called Weber bars, multi-ton aluminum bars. When Weber published his paper, Luis Alvarez said that you could get any answer you want in the way Weber analyzes these things with strip charts and marking exceeding amplitudes and then estimating what his resolving time is, two seconds, and subtracting random coincidence rate that would be estimated from the observed singles rate. Alvarez said that physicists had solved that problem long ago, they just measure the coincidence rate and they do a delayed coincidence. You take exactly the same coincidence chart and you take 97 seconds off and you see what signal is there. So you don't have to estimate the resolving time.

Weber said that he had worn out his eyes in six weeks or six months with his strip charts. When I visited him at the University of Maryland, I tried to get him to do better experiments, but he would not acknowledge the deficiencies.

Jim Levine and I planned an improved experiment in our laboratory at IBM and said that if we don't find anything then we've shown that the gravity waves don't exist. If we do find something then of course we have to do further experiments. We ran our experiment for a whole month and we found much better analytical tools. Unfortunately, the program manager at the National Science Foundation, which supported Weber, wanted to be a hero for having funded Weber's discovery, and NSF were not happy at all with our results. There were then other people who reproduced our negative results. Nobody has ever detected a gravity wave. There is no doubt that they exist; Joe Taylor and Russ Hulse have a Nobel Prize for looking at the spin-up of double stars by the radiation of gravity waves. But their coupling to matter was never strong enough to be detected in a Weber bar. Now we have the laser interferometer gravity wave observatories (LIGO) and so on and they are still not sensitive enough to detect such things. That was my last physics interest.

You have done more for national security than the hydrogen bomb.

That's true. I did a lot of national security advising. When I went to IBM in 1952, one of the first things they asked me to do was to spend a year or two away working on a program for extending the radar air defense environment to the sea-lines of approach to the United States and Canada, but I told IBM that it was not what I came to IBM to do. But I did work half time on it and it was very interesting. It was in Cambridge, Massachusetts.

Thus I met the leaders of the LAMP LIGHT study — Jerome Wiesner and Jerrold Zacharias — who were then on the President's Science Advisory Committee (PSAC). I became a consultant to the Committee. Now I worked on the effect of nuclear weapons and what nuclear weapons would do to societies and how to counter them. I told Wiesner and Zacharias that while we are working on defending our country against the airplanes approaching us from the sea, the Soviet Union will have missiles carrying their nuclear weapons. Zacharias said that we first solve this problem and then we'll solve the other problem. He had some other aphorisms, like "Don't get it right, get it written". That meant that when you are working with other people, you could take a long time to perfect some calculation or understanding, but what's really important is to get it down on paper so that other people can judge whether it is right and they can use it in their work, and they can build on it. That's so true.

I was a consultant to the President's Advisory Committee, particularly in matters of intelligence and military technology. PSAC worked on other things too, for example, John Tukey, I believe, was a well known statistician, who headed a panel on pesticides and insecticides; Edward Land, the inventor of Polaroid, polarizing material and also the Polaroid instant photography system, worked also in intelligence. There were 18 PSAC members; they

Richard Garwin (on the left) and Luis W. Alvarez (second from the right), with two other members of the PSAC (President's Scientific Advisory Committee) Military Aircraft Panel (RLG chair) and others on the flight deck of an aircraft carrier, probably 1964.

met for two days every month in the old Executive Office Building next to the White House, and had at any time ten or twelve panels each of which had ten or twelve people on it. I chaired the Military Aircraft Panel for ten years or more and a Naval warfare panel, and I was a member of the Strategic Military Panel that looked at the problems of the missile attack.

Did SDI make sense?

No, but we have had long experience with SDI. That is, in the 1950s when satellites were first discussed, RAND Corporation and RCA had proposed some what they called Bambi, Boost-phase Anti-Missile Ballistic Intercept. That would be a series of satellites that would observe the launching of ballistic missiles and attack them when they were still in boost phase. In 1983, when President Reagan gave his speech, it was mostly space-based ballistic missile defense.

Were you there?

No, I was in California, getting an award for interdisciplinary science. I remember, I was getting dressed for my acceptance speech, it was a formal occasion. It was 5:30 p.m. in California, but 8:30 p.m. on the East Coast, when President Reagan gave his speech. I turned on the TV and here was President Reagan giving his standard support for the defense budget, but in the last couple of paragraphs he was asking "scientists who had given us nuclear weapons now to give us the means for rendering them impotent and obsolete". Reagan may have thought that the only way to deliver nuclear weapons was by missiles, but it was by no means true. The problems with SDI were that nobody in the government, no technical person believed that it could be done. Reagan had contacted just a couple of people. Even Admiral Watkins, Chief of Naval Operations, and Robert McFarlane of the National Security Council, had their doubts.

What was Edward Teller's role?

He was pushing people, all over the Pentagon and in the Congress for doing these things. Of the government people, even George Keyworth, President Reagan's science advisor was initially against it, but then he became a convert. Just before the speech, Keyworth had asked two people, the late Solomon Buchsbaum from Bell Laboratories and Ed Frieman, both of whom were on the much diminished President's Council of Advisors on Science and Technology. Back in 1973, Nixon demolished the Committee and later it was brought back but in a diminished role and it met only a few times a year and it didn't have nearly the power and the staff that PSAC had in its heyday.

But Frieman and Buchsbaum were asked to go over the speech and to make some sense out of it, so it was much worse before. The same day that President Reagan gave his speech the head of the Defense Advanced Research Projects Agency, DARPA, was in Congress explaining that his powerful lasers would take many years if ever to develop into any useful kind of weapon. There were people in Congress who were lobbied by Teller and by Lowell Wood of Livermore. There were two aspects of SDI; these people were in competition with one another. There was the Heritage Foundation High Frontier folks and there was Lieutenant General Daniel O. Graham, who wanted to have Bambi, that is, these modest rockets in space, which would attack missiles or warheads. Then there were the other folks, the directed energy weapon people, the neutral particle beams or lasers and some people at DARPA wanted rail guns that made absolutely no sense at all.

But the big problem with SDI is the vulnerability of these things in space and the susceptibility of countermeasures. This is still a question, because the Republicans really want to have a defense against ballistic missiles whether it works or not. They're putting interceptors in the ground in Alaska, in California, but they do not have a chance of working because North Korea could get an ICBM equipped with countermeasures and they would use up more interceptors than we have. As for deploying multi-billion-dollar laser systems in space, these are all vulnerable to tiny little space mines that

The 40th wedding anniversary of Robert M. and Evelyn Frank, White Rock (Los Alamos), New Mexico, probably 1982. Richard Feynman is at the center of the photo. On his left is Evelyn Frank and directly in front of her is Nick Metropolis. Further right in the picture, the person seated with white hair is Isidor I. Rabi with Helen Rabi on his left. The tall person seated next to Feynman is Robert Frank and the third person on Feynman's right is Norris Bradbury (sitting with shoulders hunched). Lois and Richard Garwin are in the back row. Going from Garwin to the right, the fifth person with his head lowered is J. Carson Mark, the next man is Harold Agnew, and the next is Rudolf Peierls. Hans Bethe is in the lower right corner with Luis Alvarez behind him.

would come up and sit next to them and be ready to explode at a moment's notice. All of this I have been saying since 1982.

Don't you think that the SDI program might have contributed to the collapse of the Soviet Union?

No, I don't. In addition to my work with the government on national security and intelligence, I had a lot of influence amplified by these committees in which we could see what was right and what was wrong. In the 1960s, our military aircraft panel was a very strong advocate of what became GPS, the Global Positioning System and the use of homing systems, GPS homing on missiles and bombs, and also laser-guided bombs of which we used 25,000 in Vietnam between 1969 and 1974. We used about 9000 laser-guided bombs in the recent war in Iraq.

I've been involved with the Pugwash group for a long time. In association with Pugwash, Paul Doty of Harvard University had an informal group of six or eight people that would meet with the Soviet scientists in conjunction with the Pugwash meetings. It was a kind of cover for Soviet scientists for having bilateral meetings with American scientists. It was during the Pugwash meetings, so they would not need additional travel authorization. I've got to know the Soviet scientists and engineers very well, some of whom were of very high quality.

Richard Garwin with Joseph Rotblat, President of the Pugwash Movement, in New York, 1996.

Eventually these activities in the 1960s and 1970s were supplemented or taken over by activities at the National Academy of Sciences. It was CISAC, the Committee of International Security and Arms Control, which was headed first by Marvin Goldberger, then by Wolfgang Panofsky, and for the last eleven years by John Holdren from Harvard. We were created to meet twice a year with the Soviets and we discussed all kinds of things, ranging from command and control of nuclear weapons to the number of nuclear weapons, to defenses against them. In 1982–1983, we were discussing a lot on directed energy weapons and how you can distinguish a laser on a Soviet space probe whose job it was to knock off a little bit of the Moon in order to do time-of-flight mass spectrometry on it, from a laser whose job it was at a thousands-of-kilometers distance to destroy some missile. We had a whole session for several days in Washington just before the Reagan speech, when we were totally unaware of the upcoming Star Wars speech of March 23, 1983. The Soviets were quite concerned that we had known about it and had not told them and misled them. We assured them that this was not the case.

If SDI was not going to be effective, why were the Soviets concerned about it?

Richard Garwin with John Paul II and Vittorio M. Canuto (partly hidden) at the Vatican in 1985 during a meeting of the Pontifical Academy on weaponization of space.

This is what Henry Kissinger always asked. Because when somebody is doing something, like the Americans were doing Star Wars, the Soviets thought that they might be missing something. That was the general response. The Soviet generals and the Soviet scientists rushed in and wanted support, they wanted money for the same sort of thing. Of course, they had particle accelerators; they had lasers, and very good ones at that, so they asked for support for doing the same kind of thing. They too don't look at this thing overall, what countermeasures can be introduced, what vulnerability these things have for their destruction.

Gorbachev was concerned briefly and then we talked more with his advisors who Gorbachev had tapped when he first took office, before the bureaucracy could form around him. He, like President Eisenhower, was very suspicious of the bureaucracy. Eisenhower relied greatly on scientists as advisors and so did Gorbachev. His main advisors were our counterparts to CISAC — Evgenii Velikhov, Roald Sagdeev, Georgii Arbatov, and Evgenii Primakov. We talked to them and then they advised Gorbachev and then Gorbachev announced that if SDI went forward, he could counter it with means that were asymmetric and that would cost maybe one per cent as much as SDI. I've published a considerable number of papers showing in detail how even if you have the so-called Brilliant Pebbles, the very small interceptors based in space for countering missiles, these things can

Richard Garwin with President Jimmy Carter in Plains, Georgia, in 1985.

be destroyed one by one much more economically because you can take your time doing it. You don't need anything that goes to orbit, just to the orbital altitude, which is probably by a factor of twenty times cheaper. No, SDI makes no sense at all.

But the United States defeated the Soviet Union in the final account.

Yes, Gorbachev defeated the Soviet Union.

What are the hopes for defeating the terrorists?

That's different, unfortunately. The terrorists do not have a leader, like Gorbachev. Gorbachev was a person who wanted to reform the Soviet Union. He did it in the wrong order. Instead of introducing first *glasnost*, which destroyed people's faith in the Soviet system, and then hoping for *perestroika*, after he had eliminated his power, I think (and I told my Soviet colleagues at the time), he ought to do it in the other order. First he should get people refrigerators and roads and trucks instead of military equipment, and then introduce the openness as to the legitimacy of the origin and behavior of Soviet society. But it has been peaceful. CISAC also deals with China and we meet with them one and a half times a year. The Chinese, to my mind, have adapted a much better job of transformation from the communist society because they put the economics first. After the dissolution of the Soviet Union and the Warsaw Pact they were panicked.

What do you do these days?

I retired from IBM in 1993, but I still have an office there as an IBM Fellow Emeritus. IBM paid for my secretary for five years; now I pay her. I write things. I am on various panels. When anything comes up, I may get involved. In 1989, when cold fusion was claimed in Utah, I got IBM to do some of the experiments within the next week or so. I have analyzed what would be required for deuterium fusion to take place in metals; it just wasn't going to happen. After a while after my retirement from IBM I took a half time job on the Council on Foreign Relations staff in New York as a Science and Technology Senior Fellow. I had two goals. One was to influence the members of the Council and the other was to do some work. However, I did not find it helpful because I have so many connections of my own that I can be more effective if I am working on my own, so I am no longer with them since March 2004. Now I am

Richard and Lois Garwin in their home, Scarsdale, New York, 2004 (photograph by M. Hargittai).

working on the new version of my book with Georges Charpak, *Megawatts and Megatons: The Future of Nuclear Power and Nuclear Weapons*. Our colleague Venance Journe is translating the book back into French and expanding it, as well as putting it on a sounder basis. I am also trying to do something about space weapons again, publishing two papers, one in *International Security* and one in *IEEE*, the electrical and electronic spectrum magazine. I am also working on a *Scientific American* paper on missile defense, SDI and such things.

Do you regret that you were so instrumental in creating the hydrogen bomb?

No. I wish hydrogen bombs were not possible; I wish the fission bombs were not possible. I do believe that we are lucky not to already have had a terrorist nuclear explosion in one of our cities. I confidently believe that we will have one within the next few years.

Would you favor negotiations with terrorist leaders?

No. We can't negotiate with a bin Laden so long as he wants modern society to disappear.

Do you wish you had done something different in your career?

I could have learned more mathematics. Sometimes I felt like conducting guerilla warfare in using mathematics whereas a more structured approach might have been helpful. But I am happy the way my career turned out. I have had more recognition than I could have expected. Of course, I wish that the world had turned out better.

Donald A. Glaser, 2004 (photograph by M. Hargittai).

24

DONALD A. GLASER

Donald A. Glaser (b. 1926 in Cleveland, Ohio) is Professor of Physics and of Neurobiology in the Graduate School at the University of California at Berkeley. He was awarded the Nobel Prize in Physics in 1960 "for the invention of the bubble chamber". He received his B.Sc. degree in physics and mathematics from Case Institute of Technology in 1946. His thesis research was on thin films as studied by electron diffraction. He did his graduate work at the California Institute of Technology on the momentum spectrum of high-energy cosmic rays and mesons at sea level. He received his Ph.D. degree in 1950. From 1949 till 1959 he taught at the University of Michigan, rising to the rank of Professor in 1957. He has been at Berkeley since 1959. In addition to his title of Professor of Physics, in 1964, he was given the title of Professor of Molecular Biology. His recent main interest has been in the construction of computational models of human vision, his goal being the understanding of its physiology and anatomy.

Donald Glaser is a Member of the National Academy of Sciences of the U.S.A. (1962); of the American Philosophical Society (1997); and of the American Academy of Arts and Sciences (2003). We recorded our conversation with Donald Glaser in the Glasers' home near Berkeley, California, on February 19, 2004.*

---

*István Hargittai and Magdolna Hargittai conducted the interview.

You have been involved with three major and quite different areas in science, namely, physics, molecular biology, and lately neuroscience. How is it possible for you to move so easily from one field to another?

It goes further back than that. I began as a musician and then, for about two months I was an engineering student because neither my parents, nor my teachers knew the difference between engineering and physics. It took me two months to realize that physics as a profession was different from engineering. Then I worked seriously in physics. I've been thinking about what I should tell you to answer a question like that. The answer is that there is a thread through my life and it is escaping from big science. I like to work with a group of students and postdocs, but I don't like to be in very large research teams. I began with a thesis in cosmic ray physics, which is a very solitary thing; I did my own little project. Then I was offered a job — when I was first looking for a position — to work with cyclotrons at Columbia or MIT or at some place else where there would've been a large group. Instead I took a position at Michigan, which is a good university where I was promised that I could do whatever I wanted to do. I decided that what I really wanted to do was to find out more about the so-called elementary particles of physics.

At that time the only really productive method was the large cloud chambers. Sometimes high in mountains and sometimes at sea level you just waited hour after hour, day after day, to get by chance a picture which showed some novelty that could not be explained by the current understanding of the particles. Sure enough, there were things, which began to be called V particles because they made tracks of the shape V. Then the theorists decided that these particles were illegal according to the current understanding and so they began to call them strange particles because they did not obey the theoretical structures. The quest was to look for more of these particles like pieces in a jigsaw puzzle that might make enough to give a global theory of what these particles were and what their role was in describing the dynamics and the structure of the Universe. But it was painstakingly slow and I decided that I would try to device a method for increasing the rate of getting information about this family of particles that we were beginning to discover and by we I mean the world collectively.

My fantasy was that if I was clever enough, and had a good enough instrument, I could sit by myself in a little cabin on the top of a mountain and gradually collect more information. In those days one could get one interesting picture a day perhaps. Then with a slide rule, we could calculate a little bit of relativistic dynamics that you had to do to measure the mass and charge of the particle. We didn't use computers and it wasn't necessary. Later, when computers were available, of course, that made it a lot easier. I went through a series of inventions, which were particle detectors of various kinds and finally, the bubble chamber, which increased the rate of collecting information by about a thousand fold. But a very disappointing thing was that it was no good to use on top of a mountain but was ideally suited to use at big accelerators. So I was trapped.

In order to advance the subject that I was interested in, I was forced to work at the big accelerators. I did this for a number of years, but they got worse and worse, and each experiment in those days cost twenty or thirty million dollars, so you could not have an impulse and go to the lab in the morning and tell your students about an idea and do something about it. Instead, it was committee after committee and weighty decisions and so on. Finally, we did generate large numbers of pictures, which contained interesting events. They were so many that we couldn't analyze them and it became a question of automating the pattern recognition and diagnosis software and hardware. We built scanners and software and studied the question of how humans can recognize patterns so readily and computers can't. In those days, pattern recognition was very unsuccessful. I sent pictures all over the world and a lot of universities were looking at our pictures and finally we had to meet in Geneva in order to agree on a final draft of a paper by 23 authors. At that point I decided that that was it.

Luckily, at that time I got the Nobel Prize, so I could quit working in that field without being fired and it wouldn't matter if I wouldn't publish anything for two or three years. That is the reason why I left high-energy physics. It has become so ponderous; by now there are five hundred or a thousand names on each paper. Enormous sociological, economic, and political issues are involved. Though the field remains as intellectually fascinating and critical as ever, the style of working is such that I just could not do.

I had always been interested in biology largely due to Max Delbrück. I was a graduate student at Caltech when he and his group were learning about the genetics of bacterial viruses. I would go to their seminars now and then as an outsider, but I never met Delbrück; I just went there to listen. When I got my degree and was about to leave, I went to see him and I told him that I was interested in what he was doing because it had an enormous appeal. The issues were stated simply enough that a physicist could understand what was happening and could design an original experiment and get the answer within a few days, which was way beyond what was happening in physics, which was much more elaborate.

In fact, the early days of molecular biology attracted many physicists. A lot of fundamental work was done by physicists or, more accurately, by people who had been trained and had earned their Ph.D. in physics. There were many reasons, but one of the commonly quoted reasons is that physicists have been so successful in their science generally that they have an arrogance to believe that they could solve any problem. I would call it confidence rather than arrogance because they deserved their confidence and it continued to work. The question is whether you could be that rigorous and tough in insisting on truth and rigor and quantitative predictability in biology that we were used to in physics. Many of the biologists thought that Nature was so beautiful, all we can do is describe it and many people believed that we couldn't even hope to understand it, like we understood the orbits of planets.

Eugene Wigner said something similar about physics in his Nobel lecture, that physics does not endeavor to explain nature, and the great success of physics is due to a restriction of its objectives in that it only endeavors to explain the regularities in the behavior of objects.

I didn't know that quote. Wigner made a number of interesting statements. One of them was that he proved that life could not exist. He published a paper on it and the essence of the paper was very clever. First you wrote down a wave equation and a wave function, which putatively described the state of a living system. Then you imagined some operator that in physics is used to describe the dynamics of such a system. He started with an initial state and then subjected it to the operation. As the system would go to the state of the next moment in time, the number of transitions, which produced something living would be a tiny fraction of all the transitions. In other words, almost anything you do to a living system kills it in terms of quantum mechanics. Therefore, nothing can live. Of course, it was a huge joke, but it wasn't a joke. I interpreted it to mean that physics is a wrong tool to describe living systems. As we come further along, I will explain why I agree with that.

Anyway, I went to Delbrück and asked him if I could be a postdoc in his lab and work in his field. He was a tough character. He turned to me and said, "What's the matter, can't you get a job in physics?" I was terrified and never went back. I went on and took a job in Michigan and started to do physics. That explains the first part. I left physics because the way of working was something that I didn't like. When I came here, to Berkeley, I was doing physics for a while. I worked at the Bevatron, but soon the Bevatron was no longer at the cutting edge and I had to go to the Argonne National Lab in Chicago or to CERN and I didn't like to be traveling all the time in order to be able to do science. I liked to be at home and think hard and read and work that way. So I had many reasons for leaving that kind of physics.

The other field that interested me enormously was cosmology, which is now a very, very exciting part of physics. But instead, I went into molecular biology. Like everyone else, I was very much interested in what causes cancer and what kinds of mutations cause what kinds of changes and I decided to join the rest of the world in trying to run *E. coli* into the ground. As you know, *E. coli* is the most studied organism on earth. All of us had the idea that it was the simplest autonomous living thing and we really wanted to understand what life was that would be ideal. If all of us would work very hard about it, sooner or later we would be able to see what life was. I was very impatient with the standard business with Petri dishes and so on that everybody used. So I used my knowledge of computers and automation and all the stuff that I had learned in high-energy physics to build a big machine to automate many of the procedures of molecular biology. I called it a dumb waiter because it had large trays of agar, which moved up and down and was equivalent to I don't remember how many Petri dishes. It could carry something like 10^8 or 10^9 colonies. I developed a technology, which was very efficient. So again, off we went and we isolated a very large number of mutants.

The strategy was that as it went through the machine, each colony was recorded by television cameras and you could administer penicillin, you could add amino acids, and you could change the temperature. You could get a dossier of each one of the 10^8 or more colonies and begin to do a genetic archive describing all of the abilities and fragilities of each organism. Again, we had way more data than we could handle, so I sent mutants everywhere. People are still trying to understand — many years later — all of these mutants. What we did was the easy part; the hard part is the detailed kinetics and biochemistry of each organism. It's a little bit the current DNA sequencing business: now you have the Human Genome, but the real problem had just begun with trying to understand what each gene did, what was its protein, how it functioned, and that's going to be a very long detailed biochemical, genetic, physiological study.

When you were sending mutants all over the world, was it altruism? People usually like to interpret their data first.

I knew that I couldn't do all of it. That was the same as in physics that I had developed new methods that generated so much information and I wanted to see what the answer was. I knew that there was no chance doing it myself during my lifetime. If we worked together then we could get much closer to the goal. It was that simple.

Then bad things happened and that was partly my fault again. Some of these organisms were valuable because they made an antibiotic or because they were useful in some other way, and I realized that I could use these methods to do large scale screening for useful organisms, useful in the commercial sense. But you can't do that in a university, so I started a company with some friends, Cetus. That was the first biotech company in the world. It didn't use the methods I used on the campus, but it did use the idea that you could do things on a large scale. I was very careful never to do anything at the company that was also what we were doing at the university. We had quite different goals.

At the company we did things, like taking an organism, which was being used commercially by a very large drug company to make a very important antibiotic called gentamycin. It's an antibiotic, which you don't hear of much because it's very dangerous and it can only be administered in a hospital. For a patient who is very sick and nothing else is working then they can use gentamycin because they have to monitor the blood concentration very accurately to get the desired effect without killing the patient. The company that was making this stuff gave us their organism after a lot of legal arrangements. They were planning to build a factory in Puerto Rico and another one in Ireland, but we told them that we could double the yield of this organism which was grown in huge baths of hundred thousand gallons. We did this, we doubled the yield by methods involving automation and so on and they didn't have to build those two factories because the existing factory could satisfy their market. That's the kind of thing we did, and finally we got into the development of real biotechnology and it has a long history.

At the same time, molecular biology had become more and more sophisticated and more and more depended on real skill and knowledge in biochemistry, which I don't have, or, rather, which I'm not good at.

When you got into biology, did you take courses or read books, or how did you make this transition?

Of course, I read books; then I went to seminars. I did not take any courses although in retrospect, I should have. That's a more efficient way. Anyway, I just started and I picked things for which the genetics and biochemistry were simple. The problem was to get enough information to see the overall pattern. But as soon as it got to problems like the mutants I described, I was not competent to analyze any of these mutants in enormous detail, to track down which gene did what and what was the biochemistry. I did not have any training in that. I didn't really try. The one exception was when we were studying a disease called zerodermopigmentoso, which is a skin cancer that people have who lack repair of DNA damage caused by ultraviolet light. If these people go out into the sunlight, they get skin cancer. If they never go out during the day, and only go out at night, they live a normal live like everybody else. The reason is that when an ultraviolet photon is coming in, it hits the DNA and can cause the thymine to dimerize. This may happen where there are two thymines opposite of each other in the DNA double strand. It can cause a linkage between them, which is not normal and which inhibits cell division.

It's amazing that we have several enzymes that do nothing but correct that kind of UV damage. One of them is sort of like calipers; it goes right into a rail, like a railroad track, finds a bulge on the track, it stops, and it calls in another enzyme. This other enzyme cuts the offending part and then yet another enzyme comes and puts in a new good part copying the strand, which remains. Finally another enzyme comes, an endonuclease and seals the ends of the new piece. What we did was, we used Chinese hamster ovary cells, which is a mammalian equivalent of *E. coli*; it's a cell, which can be grown in culture in large amounts, which is not quite normal, but not fully cancerous, sort of in between. It's used a lot as a model. We used those cells to identify those enzymes, but we did not do anything about their enzymology because that was beyond my competence. We were not alone; there were other people working on it. But that was one of the projects that we did all by ourselves. We didn't send the problem out because we could handle that.

That sounds to be a very successful project.

It was and produced a number of papers.

But you just said that bad things happened.

The bad thing was that biotechnology had also become big business and now you find many authors on those papers because a number of different

Magdolna Hargittai conducting the Glaser interview, in the Glasers' home, 2004 (photograph by I. Hargittai).

technologies are required in the laboratory and sometimes one lab can't have all those competencies. So it again became a sociologically very complicated business. I resigned about two years ago from industry altogether. But we had 64 major contracts with other pharmaceutical firms because they had patent number 1, we had patent number 3, and somebody else had patent number 7 and so on. We had two thousand relationships with individual molecular biologists at universities. To manage that socio-technical network, it got to be not much fun. It's an enormously complicated administrative thing. In the laboratory — unless you really were an expert in biochemistry and in various other technologies — you really couldn't contribute very much. The time in which physicists could help a lot in biology was over unless they really became biologists and many of them did, Wally Gilbert, for example and Seymour Benzer. That's what I meant when I said that bad things happened.

But didn't you have trained biologists and biochemists in the company?

Sure we did.

How many people worked in your company?

A few thousand and the company spread over many companies in several countries.

Wasn't it your company where Kary Mullis discovered polymerase chain reaction?

Exactly. He claims that I was on his Ph.D. committee when he was a student. I don't remember that but he must be right. He was a student at Berkeley and then he worked for Cetus.

Did you get involved with his dispute with Cetus?

No.

Several Nobel laureates have singled him out as a person who shouldn't have received the Nobel Prize because of his unbecoming behavior. Don't you think though that his discovery was the perfect achievement for the Nobel Prize?

I have a somewhat different view of it, which I would not like to see published. I just tell you that he could never have done it except for two other people at Cetus who were really competent. He had the idea, but it would've meant nothing without them being able to do something in the lab that showed that it worked. I felt that it wasn't fair that these other two people were not included. He gets credit for the idea and there is no doubt about that in my mind. There have been a lot of discussion that Kornberg had thought of it and Khorana had thought of it. But when the literature was really examined, both of them, and particularly Khorana, backed out. Khorana said that he didn't do it, he thought of things like it and he could kick himself, but he didn't. Kornberg actually got involved in some of the lawsuits, which is unfortunate. But the thing I think is wrong there is that the other two men [at Cetus] should've been part of that Nobel Prize and they weren't.

Did you know that when Kary Mullis was a graduate student in biochemistry, he published a paper in Nature *(**1968**, 218, 663–664) about the cosmological significance of time reversal?*[1]

I didn't know that, but that's interesting because that was one of the topics I was working on in physics. I didn't publish on it because we didn't get any sharp enough result.

Coming back to your path …

That combination of circumstances led me to leave molecular biology simply because I didn't have the skills that were required at that stage of the development of the field. I know you guys are chemists, but I don't enjoy chemistry very much.

The next thing that interested me was neurobiology. It takes a certain arrogance to switch fields like that. On the other hand, it has to be fun, it has to be interesting, and it has to be important for me to want to do it. I had a close friend, Werner Reifhart, a German physicist, who worked in neurobiology and he tried many times to interest me in neurobiology. The field attracted me because there didn't seem to be big money in it, it didn't require large groups, and the problems were extremely interesting. I focus my work now on human vision. I do two things. One is psychophysics, a term invented by Helmholtz, and it means the use of physically defined stimuli to measure the response of people to those stimuli. The psycho part is that people are being observed and the physics part is that the stimulus is well defined. It's a huge field with an enormous amount of publications.

My wife thinks that psychophysics was invented in Hollywood, it sounds like it, but it's really an experimental science, it's experimental psychology. I have about a dozen undergraduates in my lab. They serve as observers and I show them complicated patterns on the computer and I ask them, whether they are going to the right or going to the left, and whether A is closer or B is closer; these are standard experiments, which are widely done everywhere in the world. We are collecting quantitative data. Mostly I formulate the tasks but I don't do the experiments. I have a very talented colleague who was a graduate student and is now a postdoc, his name is Kumar and he runs the psychophysics side of things and we talk all the time. We have a huge lab, painted totally black. At one end there is an undergraduate and at the other end there is a computer. They are looking across and pressing buttons and telling what they see.

The other side of our work is making computational models. What I mean by that relates to something I said earlier. I don't think that traditional mathematics is a suitable tool for trying to understand the brain. Instead, I make models which is sort of graphical picture models and when I do write down the equations, which are often non-linear differential equations, so non-linear that they even have Brillouin terms in them, that is, decision branches, which is standard stuff in computer algorithms.

For example, I represent the cerebral cortex — the gray matter which is two or three millimeter thick as an interacting array of very small computers, like tiny computers of a checkerboard. Each little computer is a neuron in this model and I postulate how these neurons talk to each other. I don't make it up, I take it from the physiological literature. I put in as much as I can of what's securely known about the properties of neurons and the interactions of neurons and the nature of the signals between them. I can put that into the model as a computer command without really using mathematics. I can write down the equations but they're very complicated and non-linear, so I know they can't be solved.

The thing to do is to run the computer model, something like playing a chess game for which you can't write down an equation, which describes the outcome of a chess game. The only way to find out what's going to happen is to do it. That's true of this kind of model, which is closely related to a mathematical construct called the cellular automata. These are, however, not cellular automata. In a cellular automata each node either is zero or one, a very simple, binary thing. In our model, they can have any value. The result of all this is to predict brain waves of certain patterns, which correlate with perceiving certain stimuli. That has been successful so far in understanding the perception of motion and the perception of depth. Now we are working hard on texture.

I have uncovered some really surprising optical illusions. The first one was invented not by a scientist but by a French painter named Leviant. I have a copy of that picture here and it's really shocking. When you look at it you see things happening, which aren't real.

Donald Glaser during the interview, 2004 (photograph by M. Hargittai).

This is a color pattern and it moves or gives the perception that it does. Would that have the same effect in black and white?

We're doing many experiments like that and I can't answer that as I don't have the data yet. You can see the motion in this pattern, but not everyone sees this. We were at a Valentine's day party on Saturday; there were mostly young people and some older people. It was a wild party with a lot of noise. Some people were drunk and some people got stoned on marijuana. I had a chance to show this pattern to people in various conditions and I even took data. I have a more elaborate version of this, which is on the computer. There was great individual variation. This French artist published a similar pattern and even wrote an article about it in the *Proceedings of the Royal Society* in 1996. He described how he made the pattern. He called his pattern Enigma and the one I am showing to you I call the Spiral Enigma.

Is there any relationship to the Moiré patterns? They are created by superimposing infinite planar patterns and they are also very dynamic.

I don't think so although I haven't studied them in detail. Anyhow, to give you some idea of how the work goes, I'm now struggling with trying to decide between two very different explanations. One is that the eyeball is dancing around and I would have to tell you why that might work. The other is that there are tessellations of flickering at the cortical level. What we know is remarkable that the back of your head, in the occipital lobe, the gray matter there carries a literal map of the world of what you're looking at. There is a famous picture in which a monkey is staring at a bull's eye for archery and you can measure by a lengthy autoradiography method the pattern of excitation of the neurons. And it's a bull's eye, literally. When we try to make some model of how the brain or the visual system maybe working, it's very encouraging that we have something to start with that we really understand. If we only had all kinds of crazy patterns, with no intuitive insight, it would be very difficult to build any kind of a logical analysis.

We now know that there are fifty different regions of the brain that are specialized; one region recognizes faces, another one is for motion, another one is for color; we don't know what most of them are, but those we know. The notion is that we start with a literal image and then through a mechanism like my theories or some others, you modify the map through a series of transformations. Finally, you end up with a form of it, which is simple and compact enough that you can put it in memory. Then at some future time, if you see a similar thing, you can then do the same process and you can compare the simplified versions. It's sort of like encoding a picture.

But obviously you don't remember everything. There are a bunch of books behind you at which you had looked a few minutes ago. If I asked you now without looking over there, you could make some guess of how many books there might be just from the dimensions. Or you look at a forest and may not remember the leaves. We do remember some fraction of what's there and it's limited by how much space we can give in our memory. That's determined by this process of boiling down the image to something simpler. I'm now trying to understand the sequence of processes, which could produce what the psychologists call n-gram, which is a form of information, which is recorded in some permanent or semi-permanent way, something you'll remember your whole life. Other things you remember for ten minutes. There is a whole series of different memories.

What's going on here, in this pattern, could be motion here, but it could be tessellations or jitter or jiggling or some kind of transformations in the brain. At the moment, my prejudice is that it's going on in the brain and not in the eyeball. Luckily, there have been a lot of measurements in the oscillations of the eye. I'm just learning that literature and I have to find out whether the jumps are big enough and frequent enough to correspond to the pattern of motion and I suspect that they are not going to be. Then the other game is a statistical game and that is to predict how many of these components are there in the pattern. And the next question is why some people at this party didn't see any of them at all, even if they weren't drank or stoned? Even a 25-year old young woman couldn't see it. I have an Indian student in my lab who is very bright, but he doesn't see it.

I should say that all the theories that I'm working on require very special properties. In other words, if you change one of the parameters a little bit, it won't work anymore. I suspect that different people might be slightly differently tuned. Now we're going to play games; we're going to modify this until there is a form of it that the Indian guy can see and the rest of us probably won't. It's a new kind of tuning of the brain, and it maybe all fantasy, but that's the result of our research meeting yesterday. The guy told me that the model was fine but you had to have just the right parameters. This is what prompted me to think that maybe they are tuned to the individual.

Leon Cooper is also interested in vision. He also comes from physics. Do you have any interactions?

We see each other at meetings. Usually he speaks and I speak or I speak and then he speaks. We are friends and we talk things over but he is going at it in a very different way from what I am.

He also has empirical components of his work.

Yes. The deprived cat was one of them.

Do you discuss with Francis Crick your brain research?

Yes, but not very often. When I see him, we have interesting conversations.

Is this such a complex field that all these different approaches have not reached a point of intersection yet?

I don't know if there is any profound reason. We are all friends, but I have quite a close relationship with a man at MIT; he and I in some sense resonate. He is Tony Poggio. He is a mathematician and works in an artificial intelligence lab at MIT and we talk quite often, but we don't co-author papers. We stimulate each other, but it hasn't come to cooperation. It may change though. It may also be a logistical problem; I don't like to travel very much and we don't visit each other very often.

There doesn't seem to be too much overlap.

That's true. The history of my work is that I've been regarded as crazy, many times. When I asked for 25 hundred dollars from a government agency for my work on the bubble chamber, I got back a letter saying that it would be an irresponsible use of public funds to support this research. That research earned me my Nobel Prize. Then, when we started this first biotech company, Cetus, I and my partners went around to all the big pharmaceutical firms. We told them that we had something exciting, they should know about it, and they should invest in our company. They told us that they were doing fine, they told us that they trusted us and knew that we were respectable scientists, they knew about DNA, but they thought that it had nothing to do with the pharmaceutical business. It happened over and over again, people thought that I was crazy, or at least irrelevant. They were not angry with me, they didn't think that I was critically crazy, but they thought that what I was doing was not relevant to the field.

People are not prepared to absorb truly groundbreaking ideas. This is why it often happens that research reports that eventually earn Nobel Prizes maybe difficult to get accepted for publication.

The first time I wanted to publish a paper about the bubble chamber, I used the word bublet to mean a little tiny bubble, like droplet. The paper was rejected because the word bublet is not in the dictionary. I had to change back and then it was OK.

Which journal?

It was *Physical Review*.

Could we go back to the bubble chamber? Looking back, it seems to be a straightforward logical idea to develop the bubble chamber for high-energy particles after the Wilson cloud chamber that had been around and worked well for low-energy particles, did not work well enough for high-energy particles. How did the discovery actually happen?

The principal idea was to increase the rate at which interesting things happen. In the cloud chamber you always had a metal plate. A particle coming in from cosmic rays did something in the metal plate and things came out. But you couldn't tell exactly because it really happened in the metal, which is opaque. What I wanted was something, which was transparent. Ideally, a transparent lead brick in which you could see tracks, that was the fantasy. I tried three things.

The first one was Dacron, a polymer of acrylonitrile. Acrylonitrile is soluble in water if I remember it correctly. I made a solution of acrylonitrile and the idea was that an ionizing particle would break bonds and could lead to polymerization to form Dacron, which is insoluble in water. The fantasy was that something would come in, it would make a big spray of particles and then that would form a little plastic Xmas tree, I would pull it out, and I would measure all the angles. That was the first attempt. It worked beautifully. What happened was that the solution turned brown in radiation, which was the polymerization that I was looking for. But it did not make big enough structures to achieve rigidity. So essentially what I invented was a dosimeter. It was a very sensitive way of measuring radiation dose, but I didn't even publish it because I wasn't interested in it.

The next thing I did was that I took a pair of glass plates that had been covered with silicon fluoride. It was a conducting coating. I wanted to have a strong electric field so that if a particle goes through it, it makes a spark. I could then photograph the sparks. My fantasy was to have a whole stack of such glass plates and I could follow the track of particles. That worked well except that the conducting coating degraded after a while, and the whole thing turned into a neon sign. That was the progenitor of the spark chambers, which are now used everywhere, and that was Charpak's Nobel Prize. For me, however, that didn't work because I was not as clever as he was in the electrical engineering of it.

At the same time, at Brookhaven National Lab, a group headed by Ralph Shatt was building a high-pressure cloud chamber. They had the same goal I did; they wanted to have a high-density material. They had cloud chambers at some 20 atmospheres, 300 PSI. That means heavy engineering. They took a picture with an enormous bang and they had to wait for 5 minutes for the system to settle down thermodynamically because otherwise there were strong convection currents within the volume of the chamber. It was a ponderous heavy big thing. I saw that it didn't do because it produced a picture every five minutes whereas the accelerator produced a burst of particles every five seconds. They were off by a factor of 60 in the use of the machine.

Then I had the idea of the bubble chamber, which is based on a very simple notion, which is that every particle detector is an energetically unstable system. Even if you're looking at a photographic film, if you immerse the film in the developer in the dark room, and the film had never been exposed to light, nothing happens. That's a chemically unstable system. When a photon comes in, it makes an initial nucleation of a silver bromide unit, and that sets off a chemical reaction. It happens without adding energy because the energy is there in the instability. In the cloud chamber, the instability is in that you make a supersaturated vapor, which wants to condense into droplets, but if you're careful, there is no place for it to start. The particle going through in the right moment will nucleate the transition. The spark chamber is electrically unstable, it's charged, it wants to make sparks, and the particle going through initiates that.

The question is, can I make instability in a liquid, which will be triggered by a charged particle? Then I thought about super-cooled liquids that would make crystalline particles when the particle came in. That was ridiculous because I wouldn't be able to reverse it except by huge pressure. So the bubble chamber remained and the question was whether it would work?

I made a calculation of the growth of a bubble in a superheated liquid above its normal boiling point. It was complicated. It was a little, tiny bubble, 10^{-6} to 10^{-8} centimeters, getting down to the angstrom level, to the molecular sizes, or maybe a hundred molecules across. The question was also whether the bulk properties of liquids that you find in tables were accurate enough to describe such a tiny thing?

I had little Carnot engines running and little pistons running into the bubbles; I did a lot of thermodynamics, and finally I decided that it would work if I could take, for example, diethyl ether whose boiling point is around 35 degrees centigrade. The theory said that I had to get it up to 132 degrees centigrade. That seemed ridiculous; getting a liquid so much higher than its boiling point, it would explode. The idea was to make a super-pressure cooker and if you can take off the cover from the pressure cooker very fast, you would get an explosion. But if you are very careful and the glass is very smooth, and there is no dirt in there, then there is no nucleation center, and you generate a situation of enormous energetic instability, which is the paradigm of any detector. So I wanted to do that but I also looked at the literature.

I was really lucky to find a paper of 1924 in the *Journal of Physical Chemistry*.[1] In that paper, they were trying to see how much they could superheat diethyl ether, and I used diethyl ether! I had chosen diethyl ether because I thought that for this substance I could trust the handbooks and I need not have to purify it to make my own measurements. I just asked the chemistry stockroom guy for some pure liquid and that was what they had at a reasonable range of temperatures and pressures. So this paper was remarkable. They raised the temperature higher and higher in a simple experiment, and once they got up to 130 to 135 degrees centigrade, it became very erratic. Sometimes it would boil right away when they pulled the lid off; sometimes it wouldn't, so they gave up. They said that they could overheat it but that it would only last a second or two. Then they did a thing, which you wouldn't be allowed to do nowadays in a publication. They said that to show you how bad the experiment is here is a list of 30 waiting times in seconds. You pull the lid off and you wait two seconds and it explodes, then you repeat it and the waiting time is one second and the next time it is eight seconds. It's not a chemical explosion, just a physical explosion, a boiling burst.

I made a plot of their data and that was a perfect Poisson distribution. That meant that it was some kind of a random event. I asked myself, what random event do I know about? Cosmic rays! Then I showed that the rate, at which they saw these things going in an apparently random distributed way, was twice as high as you would expect from a cosmic ray. We now know that at sea level the background radiation is about half cosmic rays and half radioactivity from bricks and cement and all kinds of building materials. So there was perfect agreement. All I had to do was reproduce the experiment with a known source as cobalt-60, a gamma-emitter, and I could show immediately that it could make it explode right away with this source. So I saw right away what those guys were seeing, but in 1924, hardly anybody knew anything about cosmic rays. It's no reflection on them that they didn't see this. Nowadays you wouldn't be allowed to publish such a set of data showing how bad your experiment is. They would tell you, "Look, this is a serious journal, you have to reproduce your data better." Luckily, in those days, you were allowed to do that. Then it became an engineering task and I built bigger and bigger chambers.

The next question was whether I was getting a volume affect in which the phase changes over a big volume or it is really localized tracks. I was able to borrow a fancy movie camera from the engineers, which made three thousand pictures a second. I made a little bubble chamber of the size of my thumb and connected it to a hand piston, I cranked down the piston that lowered the pressure and I started the thing and I got movies showing that I got tracks.

Then Fermi invited me to come to Chicago because they were very excited about this possibility. I went and gave my talk. I had never met him before and he was a very courteous, friendly man and he asked me, "Why did you believe this would work?" I said that I had made a theoretical analysis of it. He said, "Really, what did you do?" So I went into all the theoretical stuff and he kept cross-examining me and I couldn't understand why the great man would care about this theoretical thing when it clearly worked. I was very young then, 26 or 27, and I asked another young physicist who was sort of my size, why did Fermi give a dam about the theory. It turned out that Fermi had this idea too and he proved that it couldn't work, and he put it away in his desk. When I told this story to a friend of mine, he told me that Fermi had a book on thermodynamics and I didn't know that. I went and looked at Fermi's book and there was a place where he dealt with capillary rise and capillary depression, water goes up and mercury goes down and it depends on the contact angle. There was a minor error in the analysis of that diagram; he forgot to put in the weight of area outside of the column of liquid in his equations. He got the wrong answer for the conditions inside of a potential bubble;

at least that is my interpretation. I never went to ask him. I was so lucky that I didn't know about Fermi's book in thermodynamics because then I probably wouldn't have noticed the mistake and I would've decided that it wouldn't work.

A wonderful story. Alvarez writes in his book Alvarez *about your meeting in Washington in 1953 during the meeting of the American Physical Society. You were scheduled to be the last speaker and you were worried that nobody would stay to the end to attend your talk. Alvarez was also to leave before, so he asked you to tell him about your talk and you described to him the bubble chamber and even showed one to him. That night Alvarez and his colleagues discussed your idea and decided to build a bubble chamber with liquid hydrogen.*

I haven't read his book, but what you're describing is quite accurate. The American Physical Society is extremely democratic. Anybody who says they're interested in physics can join. Once you're a member, anybody who is a member can give a paper. I don't know if that's still true, but it was in those days. So there are a lot of crazy people who give papers. The biggest meeting in the year was the one in Washington and they preserved the last half-day, a Saturday afternoon, to the crackpots. I was in the crackpot session because everybody thought I was nuts. The reviewers, whoever they were, thought, it was craziness. On one of the previous nights I was sitting with my young physicist friends. Some of them were from Berkeley, so Luis came over and I told him about it. None of the famous physicists came to my talk, but Luis sent his postdocs, and they took notes in detail.

In my original paper, I said that the ideal target is pure hydrogen, but at that time there was no liquid hydrogen in Ann Arbor where I was teaching and nor were there any accelerators. Short of liquid hydrogen, I used hydrogen-rich things, like diethyl ether, simply for this funny reason. Later I used propane and various other things. Then, a lot of other people at Columbia, at Berkeley, at Geneva, started building bubble chambers. They had big engineering groups and I didn't have anything like that and I couldn't really compete and continue in the direction of hydrogen, which was the important direction. I went in the crazy direction; I went in the direction of xenon. The point is that in the hydrogen and in the propane and so on in these bubble chambers you can only see charged particles, but there are many events, like the production of a neutral pi meson, which decays into two gamma rays, which you can't see in any detector. It was only postulated. I wanted to see whether you could really see these things.

What I needed was that famous transparent lead brick again and the closest I could get was tetraethyl lead, which is used as an anti-knock agent in gasoline. I looked up tetraethyl lead whether I could get it hot enough. The answer is that if you bring it much above the boiling point, it explodes spontaneously in the absence of oxygen; it's a very dangerous material. The only material that worked and seemed reasonable was xenon. I built a xenon bubble chamber and xenon is very expensive, and my bubble chamber needed close to a million dollar worth of xenon. I had to do the engineering very carefully not to lose any xenon. I got all of it from Germany and the United States. The Russians would not let me have theirs. The Atomic Energy Commission was willing to pay for it because they were quite excited. We had enough so we could discover the pi-nod and we had two pi-nods and we had four gamma rays. Each gamma makes a positron-electron pair and that makes it easy to identify it. I could do a little original stuff, but I couldn't go in the main direction for practical reasons. I continued my own experiments with my propane chambers. There were several meetings where Luis's group was there and they were all looking for parity violation and time-reversal violation. They were looking for certain asymmetries in the V particles. We had pretty good evidence for it, but it wasn't quite statistically strong enough to publish. Luis also did that. This was asymmetry of the pattern in the production of the so-called strange particles.

When was this?

Must have been around 1956.

Was it parity violation or CP violation?

Simply parity violation because we were looking at geometrical asymmetry. We were also looking for time-reversal variance in another experiment. But this would not have been the first detection of parity violation. This would've been showing parity violation in a different physical system. This was in the case of the strange particles to show that they also had this effect. We didn't have quite enough data, Luis didn't have quite enough data, so we put it all together, and we published it together with a lot of students. In that sense we collaborated but only at the pulling data stage.

The bubble chamber figured strongly in Alvarez's Nobel Prize.

Oh, yeah. That's what it was for. He really gets the credit for developing large hydrogen bubble chambers, which made possible experiments, which led to a lot of data. It was a major administrative-engineering success to do that. He had come from experience at Los Alamos where they were building hydrogen bombs. They had an enormous amount of technology of how to handle liquid hydrogen safely. He was able to borrow engineers and even equipment that had previously been classified. That gave him the ability together with the remarkable machine shops at the Berkeley lab to do a project, which not many people in the world could've done. CERN could do it later on and so did other places.

Talking about particle physics, if we suppose that there will be no larger accelerators, except the one at CERN expected in around 2005, how do you see the future of particle physics?

I don't really have a right to an opinion because I haven't been active in the field for a number of years now. I'm very impressed with what little I know about the attempts to observe very large cosmic-ray events. But those probably are going to teach more about the origin of the high energies that you get in cosmic rays and therefore about cosmological phenomena than about the particles themselves.

Do you think that there are even more fundamental particles than the ones in the Standard Model?

The claim is that the String Theory not only explains but also requires gravitons, which were causing trouble with the Standard Model. But that's beyond the level of my understanding. I'm so overwhelmed keeping track of the literature in neurobiology that I don't read professional things but only popular things in physics nowadays.

You were very young when you received the Nobel Prize. How much did it change your life?

It allowed me to quit working in high-energy physics. I would've done it anyway. But it did. For a while I had to give graduation speeches at universities, there are increased social obligations. It gave also opportunities. People were interested in me; they were inviting me to meetings or to meet people that I might've not met otherwise. From the point of view of social interactions, it enriched my life. It didn't effect my scientific work very much because I had always worked by myself or with students only and deliberately picking small science so as not to be involved in endless committee meetings. The Nobel Prize didn't have any effect on that. It's

Donald Glaser and István Hargittai in Stockholm, 2001 (photograph by M. Hargittai).

interesting that people took me more seriously when I proposed crazy things. They listened to me more, but they wouldn't really accept my crazy proposals. Luckily, I could go ahead and do these things without anybody's approval. In an environment in which one couldn't have that freedom, it would've been distractive.

We would like to ask you about your family background.

My parents both were born in southern Russia and I'm not sure whether it was the Ukraine or Georgia and they came to this country when they were very young, 10 or 12 years old. They came independently; they didn't know each other. They were forced to work almost immediately and they could graduate high school only by going to school at night while they were working during the day. My father was one of twelve and my mother was one of seven. They were not formally educated people, but they read a lot and were wonderful parents.

Jewish?

Yes.

Did you ever experience anti-Semitism in American academia?

I wasn't aware of that. I've read about it. I went for my undergraduate studies at Case Institute of Technology, which is now Case-Western.

Cleveland was a multi-ethnic city.

That's right, but I was never aware of ethnicity. I didn't know that word. I had heard about anti-Semitism, but I didn't notice any in my own life. It may also be that I'm insensitive. After Cleveland, I went to graduate school at Caltech. It was a wonderful experience; the level of teaching and the commitment to science were uncomplicated by any political or other things. Life was really in an ivory tower.

How far did your parents witness your success?

My father died of Alzheimer's at age 55 or so. Although people in those days did not recognize Alzheimer's but the level of senility at such an early age was so high that the diagnosis was clear. My mother committed suicide just a few years later. She was tremendously broken up. So they didn't witness my success.

What language did you speak at home?

English.

Not Russian, not Yiddish.

My grandmother spoke Yiddish and my father spoke Yiddish with her, but I had almost no contact with my grandmother.

When did you first think that you would turn to science?

As a child, I always was interested in how things worked; I built a lot of model airplanes and all that sort of things. In high school, I loved mathematics; unfortunately, the physics instruction was awful. I went to Case as an undergraduate, quite young, when I was 16. I didn't do anything heroic in high school; it was just a series of strange events. When I was in the second grade, they wanted to send me to a school of retarded children. My mother got a letter from the Principal of our school saying that I was no worse than the other little boys, but I couldn't keep up with the other children. They were not polite in those days, there was no learning disadvantage, it was retarded children. I remember a lot about that early time. I remember how we were supposed to learn how to do division, but it was silly because

I would just look at it and see the answer. I simply refused to learn their system. Then we were supposed to sit there and practice penmanship, but I felt that nobody would care how I would write if I could write, so I refused to do that. I'll never forget the turning point of my intellectual life, which was in the fifth grade. Somebody said that I had to divide an eight-digit number into a twenty-digit number. I could get the first three digits right, but I couldn't get all of it without learning that damn system. But I had little confidence in the educational system. We were given a map of the world and we were supposed to color the British Empire red. Why did we have to do that? We were supposed to learn that the Sun never sets on the British Empire. That was a famous slogan that little kids were supposed to know. Next day we were supposed to color green all the countries that exported copra.

What is copra?

What is copra? Nobody knew. The teacher didn't know. I thought this was dishonest. We looked it up and it was dried coconut. Why does anybody care what they use dried coconut for? Shampoo. The educational system did not inspire confidence.

Have you had any challenge at any point in your career that was an obstacle, a hurdle to overcome?

Sure. Or, maybe I shouldn't say so quickly. A major thing was that I had invested an enormous amount of effort and time in building this large automated machinery for isolating interesting mutants for studying *E. coli* and Chinese hamster cells with the support of the National Institutes of Health, but then at a certain point they said that they had a policy that they don't do big science, and they cancelled my grant. They told me that there was nothing wrong with what I was doing, but it was against their policy, which is a huge joke in the light of the DNA project.

How much money was involved?

A million a year, not much more.

When did this happen?

In the early or mid-1960s. They told me that instead of the machinery, I should spend the money on postdocs. But I had six postdocs and they were extremely productive and they were more productive in using the big machinery than as if they had worked individually or in a small lab. We were also doing automated medical diagnosis. We analyzed urine for sterol. If you have a urinary infection, it's dominated by one organism and the question is what organism and what is its drug sensitivity. And this machine could do such diagnosis well and fast. Now, with bio-terrorism this machine would just be what you need.

It's no longer around?

No, we tore ours apart. The space was needed for other projects. By now serological methods are probably better. But at the time, there were a lot of medical applications, and it was silliness to withdraw their support from that work.

Can you single out one person or a few who made a strong impact on your career?

I've noticed in some of your write-ups that you sometimes ask a question about heroes.

That will be my next question.

It's hard for me to say because I had such good teachers at Case and at Caltech. There is a tiny story when I was taking a course in the physics of X-rays as an undergraduate. At a certain moment I was supposed to cut a piece of film to fit inside the X-ray camera. There was a pair of scissors and one of the blades was broken and was shorter than the other one. I went to the professor and asked him if he had any good scissors and he told me, Glaser, don't be so helpless, go to the machine shop and grind down the other blade so they are equal. That made a big impression on me. Otherwise, I just have a general admiration for the teachers. I have been an independent person, not defiantly independent, I don't give big speeches, but I like to think things through myself.

Also, I am a fundamentalist, not in the religious sense, but I don't like to work on problems until I really understand the fundamental issues. Very often I find out that they are a little unclear and there is a large element of common culture that everybody believes a certain thing and then they do lovely and sophisticated stuff, but the question is, Is the fundamental assumption right?

I've made a very serious mistake in my career in which a very distinguished person, Sydney Brenner, who just won the Nobel Prize, who was a good friend, figures. I was trying to do an interesting experiment having to do with the reproduction of DNA and he said, one of the big questions is that the DNA as it reproduces, makes a reproducing fork and it may go this way and may also go that way. He told me that it would be good to look into this question. So I did some experiments with some precursors, which had high molecular weight and I found that it was more or less equal, the two directions. But Brenner asked the question, "Which direction was the preferred one?" John Cairns had published a picture showing a partially reproduced circular chromosome of *E. coli* and the question was, "Which direction is it?" And I published a paper saying that statistically there is a slight advantage for this direction, and that was wrong. I should've published what I really saw, which was about equal. But I accepted what a truly high-quality scientist had told me what the question was. I have failed to go to the fundamental question, which should have been to ask, "Is this really the question?" It was the most serious error I have ever made.

Coming back to your question of who influenced me, I can't point to a single person, but I could make you a whole list of really good teachers.

So how about heroes? Living or dead.

I noticed that one of the physicists said Newton and Einstein and of course they are everybody's heroes. In general, the hero tends to be a powerful theorist among physicists. If asking this question about experimentalists, it reminds me of a time when Dick Feynman and I were both on the same platform talking to a packet of audience. I don't know what led him to do this, but I was impressed when he said, you know, I could've invented Special Relativity, but I could've never invented General Relativity. That really got to me.

Any hero among biomedical scientists?

There isn't much in the way of real theory in biology. In physics, there are laws, which are very powerful and general and if somebody does an experiment that violates one of those laws, it's a very serious matter, and is called a paradox. With some luck, it leads to a better law. Relativity is an example on top of Newtonian mechanics. But in biology, there aren't any laws; there are mysteries; the solution to a mystery in biology is a miracle.

This means that somebody finds a new molecule or a process, something unexpected, not predicted by any theory. So physics has paradoxes and biology has miracles. There are great thinkers in biology, who don't work at the level of abstraction in physics. You can go back to Mendel. Francis Crick is one of the outstanding conceptualizers. Sydney Brenner is another one. Hermann Helmholtz was a physicists who was one of the first ones who did serious work in biology and covered an enormous range both in physics and biology. I have great respect for his contribution. I haven't tried to make a list.

Personally, you don't have heroes.

I have people for whom I have great respect. To me a hero is somebody who is outstanding in military combat. That's the usual use of the word.

I never use it in this sense.

I don't either.

I suppose, you are not religious.

I'm not.

At the start we enumerated your three major areas of activity. The bubble chamber contributed greatly to fundamental knowledge, your work in molecular biology was more practical and brought you wealth, and your neurobiology so far is a risky project. Is this a correct if oversimplified characterization?

All three were risky businesses. I started my academic career as an Instructor, and I was an instructor for three or four years because I was trying all these different ways of detecting particles and if something hadn't turned out, that would've been the end of my career. I was too naive to know what a risk I was taking. I wasn't doing and publishing respectable standard stuff like a scientist is supposed to do. I'm not claiming that I was brave, I was just naive. It's also true that I was trying to do something that was important, but if it had failed, it would have been nothing. Instead, I could've made a lot of measurements of cosmic rays, joined a cyclotron group, and published papers. So it was risky in terms of picking an ambitious but what I thought extremely important problem.

Looking back, if you could chart your career now, how would you do it?

The word chart is overly optimistic. A career is a result of a combination of lucky events, unlucky events, external circumstances, so if I had done exactly the same things and NIH was enthusiastic about large-scale diagnostic instrument, then I probably would've stayed with that for some years longer. I don't think that the sociology of high-energy physics could've gone in any other way because high energy requires big machines and big money and large groups. Another possibility might have been that instead of going into biology, I might have gone into astrophysics, which is much closer to physics and which would've been using my physics background, and I really enjoy and love physics and mathematics very much.

A companion question: if you were 20 years old today, where would you go?

I would have a serious look at cosmology. There's a really exciting development now. However, it's big science. I forgot to tell you that I was also involved with the space program very early. At a certain moment, NASA called me and asked if I would join the artificial intelligence group at MIT to look at the question of what should we do if we're going to get to the Moon. It was at a very early stage, we didn't have any satellites, the Russian satellites hadn't gone up yet, maybe they had, but we were just three kids in that group. We had to report back to NASA. What should we do? First we thought we needed a television camera to look around. The second thing we had to do is that we had to have a lot of on-board computing ability because we would have to do a lot of pre-processing of data before communicating them back to the Earth. Then I said, we didn't want to overlook the obvious, we should have the ability for the television camera to look at all directions and then make an enormous noise and look again and see if anything changed. Then make an enormous bright light and look again. Then we talked about looking for samples of organic material and so on. The NASA people were furious when they got our suggestions about the big bang and about the big light, they thought it was a big joke and we thought that we shouldn't overlook the obvious. That was fun and it led to some of the initial planning, but I didn't do anything after that. It was just one meeting of a couple of days and we wrote the report.

Please tell us about your children and your wife.

We have two children; they did not follow my path professionally. My daughter is a pediatrician and has two daughters. My son got a triple degree in physics, math, and computer science; he is a computer guy, he is at Cornell, and has a start-up company. I have a wonderful second wife, Lynn. When we first met, she was a professional musician; she performed and taught harpsichord at Caltech; we met through music; I was a violist and we played together.

Do you think that your children may have been intimidated by your fame and success?

I tried so hard not to let them.

How did you do that?

They didn't know that I won the Nobel Prize for many, many years.

Glaser poster in the street in Berkeley, California, 2004 (photograph by I. Hargittai).

You walk around in Berkeley and your huge portraits are hanging on street lamps.

I refused to sit for that because I didn't want to see my pictures all around. So they just pulled a photograph out of the archives. Finally, one of my colleagues, a wise old guy, Dan Koshland …

A former editor of Science …

Yeah, a biochemist, said, "Look, if it hadn't been you, it would be football stars." It didn't occur to me, so I didn't mind. Lynn didn't know that I was a Nobel laureate during the first six months that we were going together. In any case, I tried for my children and I don't know if it intimidated them or not. They're both bright and talented people so they had nothing to worry about in the sense that they could do well.

Did you know any of the famous Hungarian scientists?

Once we had both Teller and Szilard in our home. At that time we lived in a small house on the hillside and you had to manage about eighty stairs to get there and I didn't know that Teller had a wooden leg, but he didn't complain. What Teller and Szilard were fighting about on that occasion was that Teller was claiming that a little bit of radiation was good for you. The basis for it is that the life of certain flies was longer if they were exposed to a low level of radiation. But it was known that the reason for that was that there was a parasite or bacteria or fungus that was very sensitive to radiation and Szilard was claiming that Teller knew that. You can imagine what kind of fight that led to. They weren't yelling and screaming, but it was a very intense discussion. I knew Teller when we confronted each other in highly classified meetings and both of us were advising the government. We were almost always on the opposite side. Then by pure chance, I was sitting next to him on an airplane going to Washington. We were having a very nice and interesting conversation. One of the things he wanted to do was to teach physics in Berkeley.

At that time we were trying very hard to be nice to the undergraduates. It was pressure from the legislature. Every department had a course, which was called Blabla 10; so I taught Molecular Biology 10, which got to be known as Molecular Biology for Poets. I had a hundred kids in there, who knew nothing about science and they were mad at me because I was expecting them to learn something about biology. They could take

I know of them only at a popular level, from seminars. I can imagine that tools like that will lead to remarkable advances, but I can't name any particular one. In neurobiology, it's even more complicated; it's called neuroscience on our campus because it goes all the way from computer engineers to neurochemists, and everything in between. As long as the money keeps coming, these researches are not limited and progress can be fantastic.

References

1. For the Kary Mullis interview, see, Hargittai I., *Candid Science II: Conversations with Famous Biomedical Scientists*, edited by Hargittai, M. Imperial College Press, London, 2002, pp. 182–195.
2. Kenrick, F. B.; et al., *J. Phys. Chem.* **1924**, *28*, 1297.

Nicholas Kurti, 1994 (photograph by I. Hargittai).

25

NICHOLAS KURTI

Nicholas Kurti (1908 in Budapest as Miklós Kürti–1998 in Oxford) was Professor of Physics, Emeritus, of the Clarendon Laboratory, Oxford University, when we recorded our conversation at the headquarters of the Royal Society in London in October 1994. He had his secondary education in the Minta-Gimnázium in Budapest. Anti-Jewish laws in Hungary prompted him to seek higher education abroad. He studied at the University of Paris and the University of Berlin (Dr. Phil.), under F. E. Simon. Subsequently, Kurti worked as Simon's assistant at the Technical University of Breslau (then in Germany, now Wroclaw in Poland) in 1931–1933. When Hitler came to power, they both went to England and continued their joint work at the Clarendon Laboratory in Oxford. Kurti worked there in 1933–1940. He participated in the U.K. Atomic Bomb Project in 1940–1945. After World War II, he worked at the University of Oxford, rising to be Professor of Physics in 1967–1975 and was Professor Emeritus until he died. He received many honors and was a member of many learned societies. In 1956, he was elected Fellow of the Royal Society (London). He was Vice-President of the Royal Society in 1965. In 1973, he was appointed Commander of the British Empire. In addition to his numerous publications in physics, especially in low-temperature physics, he was also known for his culinary art, called "gastrophysics" by him, culminating in the book created jointly with his wife, Giana Kurti, *But the Crackling is Superb* (Adam Hilger, 1988). Below are some edited excerpts from our conversation.*

---

*István Hargittai conducted the interview.

Please, tell us about your family background, education and the beginning of your scientific career?

I was born in 1908 into a middle-class Budapest Jewish family. My father died when I was three years old. Originally, my ancestors on my mother's side settled in Abony, about 100 km from Budapest in the late 18th century, coming probably from Eastern Europe, from Galicia. I have no information on my father's side. My father was a banker and died very young, at the age of 42. I went to the so-called Minta [Model] Gimnázium [Gimnázium = high school] in Budapest. I went to the Minta together with many other young men — women didn't count in those days — of the Jewish middle class in Budapest. *Numerus clausus* was in effect in Hungary at the time of my graduation from high school. Jews were about 5% of the population and they were admitted only in the same proportion to the universities. Many of these young men were bright or worked very hard, so most of them should have gotten into the university, but could not. There did not seem to be much future for Hungarian Jewish boys in Hungary, so this was the reason for emigration in the 1920s. I was lucky because, although my father had died, the bank gave my mother a pension and, moreover, they agreed to pay half of my expenses for education. I also had an uncle, József Pintér, who was Vice President of Egyesült Izzó, the Tungsram Works in Budapest. He was an electrical engineer and he also helped with my education. Some years ago I learned with obvious delight that a street was named after Pintér, near the Tungsram Works in Budapest.

I had one more great luck. If you look at Neumann, Wigner, and Teller, these great émigrés had studied chemical engineering after high school. I also wanted an academic career, but I also knew that I was not all that bright. So I wanted something practical. Of the various engineerings, for me too, chemical engineering appealed the most. It was when I was about 16 years old that my mother happened to meet, at a tea party at the house of József Pintér, the then senior physicist of Tungsram Works, Jakab Szentpéter. Mr. Szentpéter convinced my mother that there were too many chemical engineers, and if her son wanted to make money, he should become an applied physicist. It was a great foresight in 1924. I was also advised by him to go to Paris and study at the Sorbonne. I had a letter of introduction from Felix Ernhardt, a physics professor of Vienna University. He was a distant relative and an unhappy man. He was the first to use what eventually became known as the Millikan drop method. However, his experiment was not a very clean one and he then suffered for the rest of his life that it was Millikan who got the Nobel Prize and not he. Anyway, he knew the famous French physicist Professor Paul Langevin quite well and he gave me a letter of introduction.

At the age of 18 I went to Langevin in Paris. Fortunately, I had learned French by then. Langevin told me that first of all I must get a *Licence ès sciences physiques* degree. This meant passing a rigorous examination, first a written test and then an oral one. I succeeded doing this within two years. I studied chemistry, physics, and mathematics. I had the most famous professors in both chemistry and physics. For the continuation of my studies Szentpéter's idea was to go to Berlin. For this I had a letter of introduction to Michael Polanyi who was at that time in Berlin. Polanyi suggested to me to do one year of postgraduate work and then to do a doctorate. The field I chose was low-temperature physics and Professor Franz Simon was my supervisor. He was one of the founders of low-temperature physics in Germany. Those three years, between 1928 and 1931, in Berlin were the most fantastic. As a city to live in, Berlin did not appeal to me. What I missed most was the Quartier Latin of Paris where I used to live. Walking up and down the Boulevard Saint Michel was the best recreation I could ever have. Berlin was different. Compared with Paris, it was a soulless city. It was all right though because I just wanted to work hard. Still I managed to do a few good things. For example, a few weeks after the premiere of the *Dreigroschen Opera* by Bertold Brecht, I went to see it four times.

The most important thing though was the *Physik Kolloquia*, organized by Max von Laue in the Physics Department. These were not colloquia in the present sense of the word. They were more like the American journal clubs, just one two-hour session every Wednesday. A few people simply reported on recent publications from the literature. It was characteristic that in 1929 or 1930, Max von Laue could have an overview of the whole physics literature by looking at the *Proceedings of the Royal Society, Physical Review*, and *Physikalische Zeitschrift*. If you went regularly to this colloquium, you could know what was going on in physics. Then you could keep up with everything. Laue would ask the audience about papers as he was looking for volunteers to review them for next time. It was regarded as the thing for graduate students to volunteer. Just think of it, you were reporting about a recent paper by a famous physicist and there was the audience, in the front row, Planck, Schrödinger, von Laue, Gustav Hertz, Haber, Nernst, about 6 or 7 Nobel laureates or future Nobel laureates. Behind them were Wigner, Szilard, and others. It was a very interesting experience. It was also wonderful to see that every now and then the great men could also make some silly mistakes. I remember when once Schrödinger suddenly stood up in the middle of a discussion of the spectra of triatomic molecules and suggested that the calculations could be simplified if you assumed that the three atoms are in the same plane. There was a silence, followed by laughter.

I was doing low-temperature physics and almost blew up the laboratory. I was the first person to produce very strong magnetic fields by magnet coils cooled in liquid hydrogen. I was doing my first experiment and everything looked all right and I pressed a button which sent a current of 40 Amps through a copper wire of half a millimeter diameter. So I pressed the button and there was a huge bang and my Dewar blew up. Fortunately I was standing about two meters from it and nothing happened to me. What happened was that I had this wire in liquid hydrogen, and in order to save hydrogen, I precooled it with liquid air. At that time we didn't have liquid nitrogen, just liquid air. My mistake was that I didn't wait until all the liquid oxygen evaporated. There must have been a spark, a bad contact and that did it. Eventually I got my experiment right and got my doctorate too.

Then I had a private assistantship with Simon. In the meantime, he got a full professorship in Breslau at the Technical University, so I followed him there. Then Hitler came to power and Simon decided that it was not the place to remain especially since he had two small children. Although Simon was Jewish, at the beginning he was not under the anti-Jewish law as he had served in World War I and was awarded the Iron Cross, First Class. He used to mention many times that he was among the first gas casualties on the Western Front. Simon knew F. A. Lindeman [later Lord Cherwell] and it was arranged that Lindeman invited him to England and he brought me with him as his assistant. This was in the Spring of 1933. Simon decided not to make a big fuss out of his departure. He just resigned his Chair. He called me into his office and told me that he just signed his letter of resignation when he received a letter from the *Notgemeinschaft der Deutschen Wissenschaft* which was the German funding agency for scientific research that time. Simon had applied for a big grant six months before. Now he received this letter telling him that all his requests had been accepted, including the complete refurbishing of his laboratory. Simon left nevertheless. This is how I came to Oxford.

Where did you go to work then?

Simon was invited by Lindeman to work in the Clarendon Laboratory in Oxford, and I was to be his assistant. I arrived in Oxford on a Saturday evening, and it was a muggy autumn evening. I missed my boat-train and arrived late, everything looked dismal, and all the buildings were gloomy black. Everything changed though overnight. The next morning was a beautiful sunny morning and I got a fantastic impression of this city. The buildings were exquisite and the streets were clean and the lawns green. I thought to myself, why should I ever leave this place, and I never did. Those six years then until World War II were very exciting in the Clarendon Laboratory. Much has been said about Lindeman; he had been a doctoral student of Nernst's but he was not a great scientist. He always maintained that the Clarendon Laboratory has become what it did primarily due to the great luck of bringing in the refugee scientists from Central Europe. When Lindeman became Professor in 1921, the Clarendon Laboratory was completely unknown. There was practically no research there at the time. Other Oxford laboratories had already gained fame in chemistry, in physiology, and other fields. Lindeman decided to convert the Clarendon Laboratory into a great research place. He brought in a few brilliant scientists whom he met during World War I when he worked in the aircraft industry. G. N. P. Dobson was one of them who eventually became a great name in atmospheric science. Another was Derek Jackson, an atomic spectroscopist. So there were about four people, and all of them had private funds, and they were the nucleus to start something. Then in 1933 the refugees joined the Laboratory.

How long did you work for Simon?

Until he died in 1956.

By then you were an established scientist yourself. It was also the year of your election to the Royal Society.

This happened six months before Simon died. However, the building up of the low-temperature laboratory happened in the 1930s. Then the atomic energy project started at the beginning of World War II. It was almost entirely started by refugee scientists.

Why by refugee scientists?

When the war started, the Government didn't want to run a risk and employ ex-aliens in sensitive and secret areas. Therefore none of them went into the radar project, for example. However, the refugee scientists also wanted to do something. First I did some work on applying magnets for pulling out metal splinters of the brain. Then Peierls started thinking about the atomic bomb. He knew Simon quite well so they got together in the early 1940s. Although they had been refugees, by then they were British subjects and so was I. This is why we were not interned.

When people talk nowadays about the possibility that Germany might have built the atomic bomb, it is nonsense. There is no trace of it that there was anything in any German document approaching anything like the two fundamental studies that made Britain and eventually the United States realize that we could build an atomic bomb. One was the famous Frisch–Peierls memorandum. They sat down and worked out what an atomic bomb should be like. All right, they made a mistake in the estimation of the neutron capture cross section of uranium-235. This mistake did not interfere though with the decision to go ahead. The other important thing was the separation of the isotopes. Simon was not an engineer but became very interested in this particular question. The first experiment he did was with some simple kitchen sieves and soda water. Eventually, Imperial Chemical Industries (ICI) was also brought into this project and finally it was worked out how big the plant would be and how much energy it would need in order to produce 1 kilogram of pure uranium-235 per day. When then the plant was built in about ten years, he proved to be correct within a factor of three or four which is remarkable.

These were the two studies that made all the difference, and the government decided to spend some research and some manpower on the bomb. All those who started the work were refugee scientists, Frisch and Peierls, Simon, Cohn, myself, and others worked on it, all foreigners. The head of the committee was G. P. Thompson. It's also important what made the Americans work on the bomb. Of course, first it was the Einstein letter to President Roosevelt, but what led to a concerted effort came following a visit by two emissaries from the U.S., who came over in 1941 and looked around and reported back that there is a country which is being at war and yet is devoting some effort to developing the atomic bomb. This report was very important to make the Americans make up their minds. In fact, in 1942 it was decided that it was impossible to build the bomb in Britain. On the other hand, a number of people have gotten involved by then in the U.S., such as Eugene Wigner, Edward Teller, to some extent Leo Szilard, and, of course, Enrico Fermi. So it was decided that the main efforts should be made over there. It was also decided though to continue the work on the diffusion membranes in Britain. The theoretical people, including Frisch and Peierls, went to Los Alamos. I went only once to the United States for setting up a membrane testing apparatus and spent a total of four months in Columbia University, at the end of 1943 and beginning of 1944. Otherwise, I continued this work in Oxford until 1946, when I went back to peace-time research.

What was it then?

It was still low-temperature physics. One of the things I was interested in was the orientation of atomic nuclei. The question was, what happens to radioactive emissions, gamma-ray emissions if we try to orient them? They are usually isotropic if you have a piece of material. So the question we asked was what happens if, instead, all the nuclei which emit radiation get aligned in one direction. It was easy to calculate how strong a field was needed to produce a sizable orientation of the nuclei at the liquid helium temperature. It turned out that it was tens of millions Gauss. Various methods have then been developed, but by then we could also go to much lower temperatures. At about 1948–1949, we could produce temperatures in the order of a few thousandths of a kelvin, that is, a few millikelvin. This was achieved by so-called magnetic cooling. We put the paramagnetic material in a magnetic field, orientated the electronic spins with 10 to 20 kilogauss, at the temperature of one kelvin, and then isolated the material from the surroundings, removed the magnetic field, and the substance cooled in the same way as if having a compressed gas, isolate it thermally, expand it, and the gas condenses. So this is how we did it, first producing hundredths of a kelvin, that is, say, 10 millikelvin, applying a 10 or 20 kilogauss field, and then produced sizable polarization. But even that was a little bit too complicated and further methods were developed making use of hyperfine interactions. Hyperfine interaction is the interaction between the electronic spin and the nuclear spin. The electronic spin exerts a magnetic field of a few hundred kilogauss in the vicinity of the nucleus. All you have to do is to make sure that all the electronic spins are pointing in the same direction. They will then produce a local field making the nuclei point in the same direction if the temperature is low enough.

This is to replace an external field.

That's correct. The first experiments already showed that if the atomic nuclei are oriented parallel, they emit a gamma-radiation that is anisotropic. This anisotropy depends on two things, on the temperature and the interaction field. Thus, for example, if you know the temperature and if you know the radioactive decay, then you can calculate the interaction field. This first experiment was carried out in 1951 in the Clarendon Laboratory in Oxford. The other importance in orientating the nuclei is that if you orientate them, and then reduce the field, you can do a cooling by reversing the nuclear spins rather than the electronic spins.

Let's use the analogy with gas expansion again. The lowest temperature the gas reaches is the temperature of the boiling point. Once it boils you can expand the gas and by decreasing the pressure you can decrease the temperature of the boiling point. Now, what defines the boiling point? It is the van der Waals forces. The weaker these forces, the lower the temperature you can reach by expanding the gas.

It is the same with the magnetic phenomena. What limits the cooling? It is the interaction between the magnetic moments, associated with the spins. You can calculate this for a solid and find that this interaction corresponds to a temperature of a few hundredths of a kelvin, possibly a few millikelvin. So if you want to do the cooling with electronic spins, you can't get very much further than a few millikelvin. However, the nuclear spins have a much lower magnetic moment than the electronic spins, a thousand times lower. Accordingly, the limiting temperature will also be much lower. In that case, however, you must use a strong external field because you can't reduce the interaction field between the electronic spins and the nuclear spins but you can reduce the external field. Again in Oxford, in 1956, we did the first experiment to show that a temperature of a few microkelvin could be reached by the adiabatic demagnetization of nuclear spins. All this has further developed since and nowadays people reach these temperatures more routinely.

When we talk about these very low temperatures, in the order of microkelvin, and then of nanokelvin, then we refer to the temperature of the nuclear spin system. The rest of the material is at a higher temperature. But it is not cheating because the nuclear spin system is completely autonomous. The nuclear spins are in energy interaction with each other, they are on speaking terms with each other, if so to speak. In fact, my name was once in the *Guinness Book of Records* as the creator of the lowest temperature in the world. It was about a hundred-thousandth of a degree above absolute zero.

Are these local low temperatures?

No, the whole substance is characterized by the low temperature. Imagine the following analogy, which I worked out in the good old days of the Cold War. The Soviet soccer team Dynamo comes to England and plays Arsenal. Dynamo wins and afterwards there is a cocktail party. The players mix but the English players don't speak Russian and the Russians don't speak English. All the Russians are exuberant, they won and they are enjoying themselves, and all the English are very sad and quiet because they lost. They are then like the nuclear spins and the electronic spins. The electronic

26

HERBERT KROEMER

Herbert Kroemer (b. 1928 in Weimar, Germany) is Professor of Electrical and Computer Engineering and of Materials at the University of California, Santa Barbara. He graduated from the University of Göttingen, Germany in 1952. He worked at different non-academic research laboratories, such as the Telecommunications Laboratory of the German Postal Service in Darmstadt, RCA Laboratories in Princeton, New Jersey, the Philips Research Laboratory in Hamburg, Germany, Varian Associates' Central Research, and the Fairchild Semiconductor R&D Laboratory in Palo Alto, CA. He joined the faculty of the University of Colorado in Boulder in 1968. Since 1976 he has been at the University of California in Santa Barbara.

Dr. Kroemer was co-recipient of the Nobel Prize in Physics for 2000, sharing half of the prize with Zhores Alferov[1] of the Ioffe Physico-Technical Institute of the Russian Academy of Sciences, St. Petersburg "for developing semiconductor heterostructures used in high-speed and opto-electronics". The other half of that Nobel Prize went to Jack S. Kilby of Texas Instruments "for his part in the invention of the integrated circuit". He is a member of the National Academy of Engineering of the U.S.A. (1997), Fellow of the Institute of Physics, London (2000), and Member of the National Academy of Sciences of the U.S.A. (2003). He has received numerous awards and recognitions, among them the Order of Merit of the Federal Republic of Germany (2001), and the IEEE Medal of Honor (2002). We recorded our conversation in his office at UCSB on February 8, 2004.*

---

*Magdolna Hargittai conducted the interview.

A recent article called you the "information age's father". How do you feel about that?

Certainly, it is not the way I would consider myself; the information age probably had many fathers. To a certain extent I contributed to the technology — but calling me *the* father would discriminate against others who deserved to be that equally well.

What made you interested in physics?

That started back in high school. It's hard to tell. I was generally interested in both math and science. In the German high school science started with chemistry and then came physics. Physics appealed to me more than chemistry; it required less memorization — you laugh at me — it relied on the acceptance of certain fundamental principles, and you proceeded from there. I enjoyed that. I also enjoyed math, so it was a natural combination. By the time I finished high school there was no doubt in my mind that physics was what I wanted to study.

Which university did you attend?

I went to the University of Jena, which was in East Germany. I stayed there only for one year. During the Berlin blockade, which was in the summer of 1948, I worked in Berlin at the Siemens Company as a summer student, and at the end I decided not to go back to Jena. At that time the political pressure in East Germany was rather terrible. Then I went to Göttingen; I was very lucky that I got accepted there. That is an interesting story in its own right. I had written to various universities from Berlin for acceptance. They did not accept me in Göttingen but, fortunately enough, the rejection letter never reached me. I had told to one of my professors in Jena that I would like to get into Göttingen. He suggested that I should talk to Professor König and give him his regards. I went there and looked him up; he was one of the junior professors there at that time. He told me that all admissions were closed by then. Still, he took me to Professor Becker. Professor Becker talked with me about various things and then he took me to another professor — I think it was Professor Paul — and eventually it became clear to me that this was more than just a social discussion. I never forget the question that Paul asked me, "Well, Mr. Kroemer, you know that a mirror interchanges left and right, why does it not interchange top and bottom?" I was stunned for a few seconds and then I gave him the answer that he probably was waiting for. Finally, I was led back to Becker and he said that although all the admissions were closed, two of the people whom they admitted went elsewhere, so they had two openings. They had a meeting next day and I got accepted. This is how I got into Göttingen. It was an absolutely wonderful place. First of all, it was one of the few cities that had seen practically no destruction during the war. It was an intellectually very exciting environment. Many refugee scientists from the Eastern provinces had basically been collected in Göttingen, to some extent by the British military authorities, and accommodated in various positions, basically in a holding pattern.

I was in my third semester by that time. I attached myself to a junior faculty member, Privatdozent Hellwege. He held a seminar in which every student was given a certain scientific topic and they had to report on that. This was much more interesting than the formal courses. I remember once he gave me a paper by Zernike, who was Dutch and the paper was in Dutch. I told Hellwege, "I can't read Dutch!" He said, "Mr. Kroemer, you know German, you know English, in high school you probably learned some medieval German; you will be able to figure it out!" Of course, it was easy for him because he came from Friesland, which is close to the Dutch border, but for me it took two weeks just to get through the first two pages.

Eventually I signed up with Hellwege to do my diploma work. But then I had given some talks in theoretical seminars, which were run by Sauter. He was that time the acting director of theoretical physics while Becker was on a one-year sabbatical leave. Sauter suggested that one of my talks would be a suitable topic for a theoretical thesis. I had a talk with Hellwege about this and he said I should take Sauter's offer because I would be finished with Sauter before I would reach my turn on Hellwege's waiting list. So this is how I became a theoretical physicist. Later I also did my Ph.D. thesis with Sauter, again in theoretical physics.

What do you consider yourself today?

I am more theoretical than experimental, but I am not one who just computes something. I rather like to look at concepts; my idea of theory was always stressing concepts and analyzing their consequences and not on creating fancy formulas. Through much of my university career my theoretical background was used to guide experimental dissertations by the students and that worked well.

Did you stay at academia after having gotten you Ph.D.?

No. My first position was at the German Telecommunication Service; it was part of the postal service. They had a telecommunication research laboratory at Darmstadt and that's where I went to work. They had a small semiconductor group and they were looking for a theorist. It was an interesting job. I was supposed to be available to answer theoretical questions if they came up, but I was not allowed to do any experimental work. That had an interesting side effect: my experimental colleagues did not see me as a threat. But I was encouraged to talk to them as much as I wanted to. So I did that, I asked what they worked on, what the problem was. Also I had to give lectures regularly on topics I thought important. Nobody learned more from these lectures than I did. I had to look into topics that I had never before heard about. So, for example, I had to learn about phase diagrams that I never learned about before, and later in my life this became very handy. It was an enjoyable job; I had a lot of freedom but I very quickly realized that if I wanted to become a really active participant in physics, I had to come to the United States. I had to come where the action was.

How did you do that? Applied for a position somewhere?

No. It was in 1953, the German Physical Society had its annual meeting in Innsbruck, and at that meeting, for the first time, several invited speakers came from the United States. One of them was William Shockley. Another was Ed Herold from RCA Laboratories, and he was talking about their transistor work. I found it very interesting because his talk confirmed something that I had been speculating about and about which my German colleagues did not believe that I was right. It turned out that I was completely right. So I started to talk with him and I asked him, "Why don't you come and visit us?" He happened to have a day available so he came and we talked. He also talked with my boss and it was an enjoyable visit. There is one thing I remember rather vividly. In those days they were making what we call *pnp* transistors, basically consisting of a slice of germanium with indium alloyed on both sides. I had been interested in the question whether one could make *npn* transistors. So I asked him whether they had ever tried that. He said yes. I asked what they used as the alloy metal? "Lead," he said. He did not hesitate to tell me that. "But lead is not a donor, you must have added some donors." He hesitated a little and then he admitted that it was antimony. That was something I would have used. Then I asked him, "How much antimony and at what temperature did you alloy?" That he refused to answer. Then I told him what I thought the answer would be. I told him I would use 9% antimony and the temperature would be around 600 degrees. His jaw completely dropped.

How did you know that?

Because I knew the phase diagram! Then I asked him if I could get a job at RCA and he said that it could be arranged. So I spent three very happy years in Princeton.

Your vita shows that you spent the first about 15 or 16 years of your career at different companies.

Yes, that's right. From RCA I went back to Germany and worked at the Philips Research Laboratory in Hamburg for a few years. Then again I came back to the United States and worked at Varian Associates for a number of years and also at the Fairchild Semiconductor R&D Laboratory, both in the Bay area. After these years finally I decided that I really want to go to a university.

Why?

Well. More intellectual freedom, first of all. I have been asked this question many times before. I remember two answers; one of them is sarcastic: I needed a new set of frustrations. The other one, which is a more serious one: the university is the one place where you do not have to be a boss in order not to have a boss. I was really not interested in being a boss but certainly I did not want to have a boss whose judgment I did not necessarily respect.

But the university setting also involves teaching duties. How did you feel about that?

I love teaching. Teaching introduced one major difference compared with research laboratories. In research your hours are rather flexible, while with teaching they are not at all. But it did not turn out to be a problem.

What subjects do you teach?

Quantum mechanics, statistical thermodynamics, electromagnetic theory, solid-state physics. Basically theoretical physics subjects.

You said earlier that you consider yourself a theoretical physicist. You teach such subjects. Still, the heterostructures, for which you received the Nobel Prize for, sound like a very practical topic.

Indeed, it is. You can call me an applied theorist if you want to. Heterostructures are not something that has a deep background in math, it is concepts. The idea was that a variation in the energy gap of a semiconductor introduces gradients into the energy band edges; these gradients represent true forces. True forces that are not simply electrostatic forces in the form of a gradient of a potential; they are basically quantum-mechanical. This is a concept that I recognized how powerful it was, and applied it. First it was applied to a transistor, the so-called heterostructure bipolar transistor (I did not use the word heterostructure at that time). This was the first application, back in 1954, and it took many years until the first of these transistors were built. It was done by others. I tried it but that was not very successful. I did some experimental work while I was at RCA. I tried to build a transistor made of a germanium base and a germanium/silicon alloy emitter. I saw that the technology was not ready yet and so I decided to get out of transistors and not get back to it until the technology was ready. And, of course, then came the laser and it gave the idea for the double heterostructure laser.

I would like to ask you about the heterojunction concept. When did it start?

It did not all happen at once. The important conceptual contribution ranged over a number of years with big gaps in it. 1954 was one year, 1957 another, then 1963. I was working at the German Postal Service laboratory and we were making the very first junction transistors. Those things were hideously slow. I was trying to see how one could speed them up. The key idea was that if you could somehow incorporate an electric field into the base; an electric field that would reach from the emitter to the collector and would speed up the carriers, that would speed up the device. The initial idea was to incorporate an electric field by using a gradient in the doping. We had a higher doping concentration on the emitter side, and at the collector side there had to be a force because the bands had to be sloped. That was the original idea and it dates back to 1953, but it was not yet a heterostructure.

Was it you who first thought about this idea?

In that form probably yes. I get back to this in a moment. Remember, the idea was to have not a discontinuity but a gradient. I always viewed a discontinuity as a special case when the gradient is very steep. This doping was my idea; I gave a talk about it at a Physical Society meeting in 1953. That was an interesting experience. I predicted it would take about 10 years to do this. Well, I was wrong; it took only 9 and a half years. Anyway, the great Walter Schottky sat in the front row during my talk, nodding and expressing approval. So I was very happy. Then somebody else, who was also from that company and who was in charge of transistor development, but did not have any background in semiconductor physics, decided to take me apart during the discussion. It wasn't just that he challenged the physics, it was very clear to me that he did not understand the physics. It was a sort of an *ad hominem* attack. I tried to explain what was going on and finally I said, "Well, if what you say is true we all do not understand semiconductor physics." Which was obviously an insult because it meant that he did not understand it. I intended it as an insult. At that point some people laughed, others applauded and then the chairman of the session decided to end the discussion. Then came an intermission during which the chairman came to me and said, "Dr. Kroemer, I don't know whether you are right or not right but I liked the way you handled Dr. ***. When can you start working for me?" I told him that I was about to leave Germany to work for RCA. That was the history of the drift transistor.

But back to heterostructures. While I was working on the drift transistor, around '53 or '54, I realized that another way to incorporate a force acting on the electrons was a gradient in the energy gap. (This I described in a German paper, which appeared in an obscure German journal that nobody reads. It was a leading journal in theoretical electrical transmission technology at that time.) Then the person who was our technologist and who would have to do it, said, "Dr. Kroemer, I see no way how I could do it. The most I could do, he said, is to put an emitter with a different energy gap but that clearly does not introduce a field ..." On the way home, I tried to think about what would be the consequences if we did try to do that and I realized that this would have an altogether different benefit of its own. Thus came the more or less abrupt transition from the emitter to the base. I was totally unaware of the fact that this had already been mentioned in Shockley's patent — I do not read patents. Certainly I did not at that time. This work was done around 1954 and the Shockley patent was filed in '48. Shockley there had the idea that one could improve the injection by using a wider energy gap in the emitter. Clearly he had a very abrupt transition. I had basically rediscovered that. But Shockley never elaborated this into a general principle. I think, there is no indication in Shockley's patent that this is something that you might want to do even if you didn't have to do. Then, in 1957, I wrote up the idea of the wide gap emitter in a paper in the Proceedings of the IRE, and it was actually a referee of my paper who called my attention to Shockley's patent. I also wrote a paper in RCA's house journal, the *RCA Review* (here in parentheses, I would like to mention that I can only advice young and ambitious scientists NOT to publish in obscure journals) and there the principle of what I call quasi-electric fields is completely clearly spelled out. I pointed out there that this could be used as a principle to accomplish things that could not possibly be accomplished with ordinary electric fields. I gave as an example the two modifications of the transistor. But they were not terribly convincing examples. They became important later on and they played a role in my getting the Nobel award. But to me it was not a totally new idea; it was more as an improvement of what you could do already.

Then in 1962 the first semiconductor lasers were reported at conferences, particularly at the Device Research Conference, which is an annual event to which I regularly went. There was one session completely on semiconductor lasers. These were nothing but ordinary Ga/As *pn* junctions driven very hard so that they emitted light. I was totally uninterested in semiconductor lasers. Of course, I knew the theory but I was working on other things. In fact, Alferov went through pretty much the same experience. He also was not interested in lasers.

Around March 1963 a colleague of mine, Sol Miller, was asked to give a talk on semiconductor lasers at Varian, where I then worked, and he gave a beautiful talk. In that talk he described what had been done up till then but he also carefully described the limitations. These systems worked only at low temperatures, and only for short duty cycles; one microsecond on, a millisecond off, something like that. Our Research Director, Ed Herold (he had also come from RCA to Varian), then said that although this was very nice, in order to become practically important it had to work at room temperature. The question was, what are the chances? Miller said that this has been looked into and it was fundamentally impossible. Herold was not about to accept the comment that something was fundamentally impossible without being given an explanation, so Sol Miller presented the argument. It was basically that there was no electron confinement and as soon as you injected electrons on one side they would flow out on the other side. Having worked on heterostructure transistors I said that this was a pile of crap! All you have to do is to put wider gap regions on the outside. So the idea of the double heterostructure laser grew out of this discussion and arose the moment I was aware of the existence of the problem. There was the problem and here was the solution. And, of course, this was a spectacular solution. Of course, it took many years until it was technologically realized. And I was not allowed to work on the technology.

What do you mean by not allowed?

You see, that was industry and in industry you do not have the degree of freedom that you have at a university. First of all, there were some people who questioned the physics. That I did not take particularly seriously because I had been through that sort of argument before and I was quite convinced that I was right. Another argument was that there was no technology and that was true. But then came the killer: that there is no point in developing the technology, because this device could never possibly be useful. The reasoning was that in order for a laser to be useful there has to be a very clean spectrum, like the helium-neon laser, or it has to have large power, like the carbon dioxide laser, so that you could drill holes through bricks. This one did not have a clean spectrum nor was it ever going to have a high power, so it did not have a hope for applications. I wasn't told that our company was not in this field; I was told they were no possible candidates for applications. And that was nonsense, of course, because the principal applications for any sufficiently new and innovative concepts were always applications that were created by the new technology. Not just improvements of something that has already been around and you made them a little better. And so these new applications have come around. It took a few years but they have come. It started with the laser and then the light emitting diode, etc., our whole opto-electronics.

You already mentioned Alferov, your co-recepient. I was wondering, when did you first learn about his or generally of the Russians' involvement in this field?

I don't remember when I first heard of them. I do remember that the first time I met Alferov was in 1972. That time he spent the better part of the year at the University of Illinois as a visitor. I got a call from a government agent whether I was interested in receiving a Russian visitor. Of course, as someone coming from behind the Iron Curtain, the FBI had a close watch on him and on what he was doing. I was not particularly interested in receiving a visitor I have never heard of before. But then he mentioned his name, Alferov, and I said, sure, of course — by that time I *did* know of him and of some of his publications in heterojunctions. But I certainly was not aware of his publications in which he got ahead of the Bell Labs people.

Weren't you surprised that they were allowed to publish their results in this field and that with all that information they were allowed to travel?

Well. They were allowed to publish. But the key patent that Alferov submitted, almost exactly one week before I submitted mine, that patent was never published. It was an author's certificate or something. We once compared dates and he was a week ahead of me. But mine got published. In fact, I think that my patent is a better paper than the regular journal paper I published on that. So coming back to your question, they had been publishing general studies on *pn* junctions in gallium arsenide and also on some early heterojunctions. One of the reasons I think they were allowed to publish was that they did not appear to have any apparent applications. The matter of visits is, of course, a different one. Alferov was one of the very few people who were allowed to travel. Not with his family, of course. And politically he was probably trusted; he was a devoted communist.

What do you think of Russian physics today?

In my field the theoreticians are very good and also the technology is excellent, so I have a high respect for it. The theoreticians are among the best in the world; I don't know why, maybe this is the effect of Landau and his school. On the experimental side, the Ioffe Institute, for example, where Alferov is, is superb on an international level.

Do you have a lot of patents?

No. There is one on the laser, it was filed around 1963.

Do you get a lot of money for it?

Yes. I made more money on this than anybody else. I got one hundred dollars.

How much did Varian make?

A negative amount; they carried all the expenses. You know, the problem was that the patent expired by the time the technology was ready. A patent runs for 18 years and you cannot renew it. So it had expired. I have an interesting story about this patent. Shortly after this patent got issued, I went to the annual IEEE Electronic Devices Meeting in Washington, DC. During the cocktail party a person walked up to me and said that he was the examiner on that patent. He congratulated me on how wonderful a patent it was. I had had some correspondence with him because there was a figure in the patent about which he said that that particular configuration would not work. I explained to him why he was wrong and he immediately understood it.

Would you care to tell me something about your other research topics?

Heterojunctions have been the dominant topic, of course. Another topic that goes back to my Ph.D. dissertation is the behavior of the electrons under high electric fields, so called hot-electron effects. That topic has always interested me and if you count the number of years that I had actually spent with heterojunctions and with this topic, this would probably be more. Then during those years when I was not allowed to work on heterojunctions, I worked with the so-called Gunn effect. That is a phenomenon where, if you take a piece of Ga/As with ohmic contacts and apply a high electric field, that would break spontaneously into oscillations of microwave frequencies. I worked with that topic from about 1963 until I came to UCSB in 1976. Then I decided to get out of it; I don't like to spend too much time on the same thing. At a certain stage in a physics research when you had already solved the most important problems, I don't think it is worth to spend 90% of your time on the remaining 10% of the problem. I rather spend 10% of my time on 90% of a new problem.

You wrote somewhere that you like to work on problems that are one or two generations ahead of established mainstream technology. What do you mean by that?

Herbert Kroemer and Alan Heeger, two of the University of Santa Barbara's Nobel laureates, 2004 (photograph by M. Hargittai).

are active in the whole spectrum of modern electronics. We are not active in electrical power.

If this discussion that you had back in Colorado with Dr. Stear in 1974 had happened today, what area would you suggest to pursue?

I would not suggest anybody to get into compound semiconductors. We have taken care of that.

Obviously. But what would be the topic, two generations ahead, so to speak, to pursue?

I would probably suggest what many others would suggest: bioengineering. Trying to find what electrical engineers can contribute to the biological sciences and how then can work with them. I do believe that this is a coming field. This is far more obvious today than compound semiconductors were at that time. I believe that engineering and the biological sciences are two complementary fields.

I interviewed Freeman Dyson who predicted about 20–30 years ago that by the year 2000 physicists would be involved seriously with the biological sciences. My interview took place in 2000 and he was somewhat disappointed that this happened slower than he had thought it would.

If I had an idea that could be applied in biology, I would jump right away even at my age. But I don't want to get in there just for wanting to be in it. I often call myself an opportunist. I know that this is supposed to be a dirty word but I don't look at it that way. For me an opportunist is someone who is looking for opportunities. As simple as that.

I am working on some ideas that sort of grew out of the heterostructure project. I have spent a considerable amount of time looking at what others have been doing in adjacent fields. So I would not hesitate to get involved with something if I could say that I have a meaningful contribution.

What are you doing these days?

I do collaborate with colleagues and their graduate students, although I do not have graduate students of my own any more. I am interested in the hybrids of semiconductors and superconductors, I have spent quite a few hours on this. There are a number of unsolved problems there.

I have recently become interested in the topic of negative refraction that many others are also getting into. The Bloch Oscillator is also a topic that interests me. The idea is that if you have a semiconductor superlattice it should be capable of being a source of electromagnetic radiation at high frequencies. This is a field where you can merge the ideas of heterostructures and the ideas of high electric fields. The idea is to push things to higher and higher frequencies and much of my work has been connected with pushing devices to higher frequencies. And, I have always been interested in didactic problems.

Please, tell me something about your family background.

My parents were working class parents. My father's father worked at a tile manufacturer and my mother's father was a plumber. Neither of my parents had gone to high school, but both of them felt that their children should get the best education. I have two brothers.

What was it like to grow up in the Germany of the 1930? How much do you remember?

I remember a lot. Of course, we were under the influence of the regime, where critical opinions could not be expressed publicly. My father, whatever his private thoughts were, did not want me and my brothers to get into trouble. He knew that I would not keep my mouth shut if he were openly critical. So he was never critical in my presence, which had the side-effect that I did not become critical of the Nazi regime until the Nazi regime was over. I was a believer. That changed very quickly afterwards.

Was it a shock to you to learn what had happened?

It was a terrible shock. I think the shock hit me in full force already about January 1945. I remember that I was at a railway station in Weimar, in my hometown, and I saw on one of the tracks a train with open carriages with people. I suspected that they were being taken to the concentration camp in Buchenwald. That was when it hit me. Later, of course, the growing up in East Germany under the rapidly very Stalinistic regime was also something that I felt was not all that much better.

When you moved to West Germany did your family stay back in East Germany?

Yes. Both my parents lived there until the end of their life.

James W. Cronin, 2004 (photograph by M. Hargittai).

27

JAMES W. CRONIN

James W. Cronin (b. 1931 in Chicago) is University Professor of Physics at the University of Chicago. He received his B.S. at Southern Methodist University in 1951, his M.S. at the University of Chicago (1953) and his Ph.D. (1955) at the University of Chicago. He was at the Brookhaven National Laboratory, from 1955 till 1958, from where he moved to Princeton University, where he spent 16 years till 1971. He has been at the University of Chicago since 1971. Professor Cronin shared the Nobel Prize in Physics in 1980 with Val Fitch,[1] "for the discovery of violations of fundamental symmetry principles in the decay of neutral K-mesons". He is a member of the National Academy of Sciences of the U.S.A. (1970), of the American Academy of Arts and Sciences (1967), and of the American Philosophical Society (1999). He received the John Price Wetherill Medal of the Franklin Institute (1975), the Ernest O. Lawrence Award (1977), and the National Medal of Science (1999). We recorded our conversation in his office at the University of Chicago on September 2, 2003.*

Please, tell us something about your family background.

My father was professor of classical languages; he received his Ph.D. in 1936 from this institution, the University of Chicago. There was no hope to find a job at a university, so he went to Alabama, where he taught at a small girl's school. Then he got a job in 1939 at the Southern Methodist

---

*Magdolna Hargittai conducted the interview.

University and that's where he stayed for his entire life. He and my mother met at Northwestern University where they both were undergraduates and both studied classical languages.

What made you interested in science?

I cannot tell you that exactly. I know that my father encouraged me to study science. When I went to do my undergraduate studies at Southern Methodist University, my father suggested to me to study general science, so that if I wanted to be an engineer, I could do my graduate studies in engineering. I graduated in physics and mathematics. I can't say that there was any particular influence on my turning to science, only that I had always been fascinated by numbers, data, and measurements. That is something I enjoyed and still enjoy. If I do an experiment and get the data and finally can sit down and start analyzing them; that is my greatest pleasure.

How did you become a physicist?

I was good at math and the sciences as an undergraduate and I was accepted to several graduate schools, of which I chose the University of Chicago. I came here in 1951 with a teaching assistantship. I spent four years here and got a degree in theoretical physics. This was in 1955, when the golden age of particle physics was just beginning. Then I had an opportunity to go to the Brookhaven National Laboratory where I worked with the newly-built Cosmotron.

Coming back to your graduate years at the University of Chicago; you must have had very famous teachers here at the beginning of the 1950s.

Oh, yes! Fermi, Teller, Gell-Mann, Goldberger, all were here. On the chemistry side there were Urey, Libby, and many others. This was quite remarkable, but I don't think it made much of a difference for me because it was a time when the passion for physics was felt by everybody.

Still, can you mention anything about these famous people, their teaching style, their personality?

I just finished editing a book on Fermi. His style was extraordinarily good; one felt that one understood everything that he said in class. At the end of each class he assigned a problem to us orally, that we had to take down

James Cronin lecturing at a conference (all photographs courtesy of J. Cronin).

and turn in the next class; classes met three times a week. While we felt we understood everything in class, it was not always easy to work out these problems and we had to use books to help us.

From here you went to Brookhaven ...

Yes. I was there for three years and I would say that if there was any influence on my attitude or success in physics or science, it was during those years. Not at Brookhaven but it happened like this: while I was at Brookhaven, the Cosmotron, their new 3 GeV accelerator broke down due to some magnet failure. Therefore, the group that I worked with, led by Rodney Cool, moved the experiment to Berkeley, to their Bevatron, that could go up to over 6 GeV. This is where I met two other physicists, William Wenzel and Bruce Cork. I learned from them that one has to be extremely bold in making the experiment. Certain measurements require a certain size of apparatus, a certain sophistication of the experiment. I didn't get that sense from my sponsor or from the people at Brookhaven, but I did get it from those two physicists. I also mentioned this incident in my autobiography in the Nobel book. I feel that they were my real mentors and after that I was not intimidated by any size apparatus that I needed.

Next you moved to Princeton where, as I read in your autobiography, you worked with the spark chamber. What is that?

The spark chamber is a particle detector that was invented by two Japanese scientists and I saw it as a beautiful instrument to do particle physics with. We used it for a number of experiments between 1960 and 1970 to photograph the tracks of the particles and measure them. I didn't develop the spark chamber but I sensed how important it was for a certain period of time in physics instrumentation. The spark chamber was very important in developing the apparatus that had the sensitivity to measure the K_{long} to π^0 effects, that lead to the discovery of CP violation. Instrumentation is always extraordinarily important in all sciences.

Let us talk about the K-meson decay experiments that led to the Nobel Prize.

Well. Its implications were extremely important, the measurement, as things go, was relatively simple. We just had to set up an apparatus with some spark chambers and magnets and show that a long-life K-meson decays to π^+ and π^- meson and this is a clear signal that there is a violation of symmetry of parity and charge conjugation. That, in turn, is a symmetry that should exist between matter and anti-matter. Now the argument is a little bit complex because you can say, well, a K_{long} decays to π^+ and π^-, what is asymmetric about that? Well, it is forbidden, since it is its "colleague", the short-life K that goes to π^+ and π^- but K_{long} should not. We did the experiment in the summer of 1963, and that means that we took the pictures, the photographs of the particle tracks and we then had to analyze the results, so we didn't get the answer right away, there is no eureka moment in these experiments. At the end, it was very clear that this phenomenon did exist. And the consequence was inescapable that we had an asymmetry between matter and anti-matter. That is a very significant result and it is fascinating. Just imagine, here we are, doing this particle physics experiment; we were working on the floor of an accelerator with no air-conditioning, with not too-well built apparatus that did not work all the time, but nevertheless, we did the measurement and we knew very well its extraordinary importance. We were just lucky in some sense to make the measurement, with no sense, whatsoever, that we would see this phenomenon. The consequence was pretty amazing. This was in 1964 and the Swedes didn't recognize its importance, although I think we did. Not in that context but it was really important because it had implications about the development of the early Universe and the development of dominance of matter over anti-matter, and the evolution of the Universe. That was recognized very quickly. The most prominent person who recognized it early was Andrei Sakharov in about 1967. During the 1970s people were dragging on this and eventually the Swedes figured out that there was probably something in this because, as you know, the Prize is not given for smarts, it is given for discoveries.

CP violation was a great surprise to everybody, wasn't it? Already parity violation, when it was suggested and soon after proved in 1954, was a surprise but that people accepted. Still everybody thought that CP symmetry still should hold.

There are two reasons to this. I think the most fundamental one is that if you can design an experiment and push it to the limit, in this case by a factor of 30 or so, you might get something. Originally there was no effect, it was one event in about 10 thousand and we managed to change that to about one event in 300 and this was a big improvement! We were further stimulated because there was an experiment in a bubble chamber that suggested a so-called anomalous regeneration and we wanted to check that, too.

Were there others as well looking for CP violation?

No. The reason is that the development of the spark chamber — I didn't invent it but I developed it as a useful tool for particle physics — permitted us to build an apparatus that was much more sensitive than others.

Why did it take so long for the Nobel Committee to recognize the importance of this? This is especially strange since the Nobel Prize to Lee and Yang was one of the fastest; one year after the discovery.

I don't know. I don't think that people recognized that this had something to do with one of the most fundamental aspects of nature, with the origin of the Universe. I think that it took for a while to realize this. For me this was actually a good thing, I was much too young at that time to deal with such a thing as the Nobel Prize.

What has happened in this field since 1980?

We made our measurement that said that K_{long} decays to πs with a certain branching ratio, one in 300. That initiated a whole long series of experiments until about 3–4 years ago to finally understand the nature of the K-meson decay. Of course, in the meanwhile it was discovered that there are different quarks, and there are other K-meson-like objects, which are made of beauty quarks or bottom quarks. The K-meson problem was more or less understood. Then the B-meson topic came up and two specialized machines were built, one in Japan, the other in Stanford just to study the CP violation in B-meson decay. They also found the phenomenon in the B-meson decay. They also found that it can be traced back to a certain complex number in a 3×3 matrix called the CKM matrix. Now we know that the bulk of the CP violation happens because of this complex number in the CKM matrix. I don't consider that an understanding at all! It's a phenomenon. I am unsatisfied that the reason is just one number somewhere — but we don't know more about it yet. But it is clear that the CP violation is an essential ingredient in understanding how the Universe evolved into matter dominating. Nothing would exist if we had not have CP violation. If we accept this then its magnitude, maybe one part in 10^9 is all you need, because if you have a mixture of matter and anti-matter, a very slight unbalance, such as one part in a billion is enough, all the others annihilate into photons and that's what we call the microwave background, which is the debris of all the matter and anti-matter annihilating and what's left and what the Universe is, is matter. You could even go so far as to say that our very existence is a strong argument for CP violation in the early Universe. That is just turning it around. When you have such huge energies compared to the masses you are going to get equilibrium between particles and anti-particles, unless there is something to break that equilibrium, which is the CP violation.

Now of course, astrophysics is booming and there are so many magnificent experiments being done, and so many mysteries, maybe we eventually get more insight for CP violation. Not from machines but from measuring the microwave background radiation, or dark energy or expansion of the Universe, so there are so many logical possibilities.

Are people looking for CPT violation?

Oh, yes, whenever they can. There were a series of beautiful experiments when you have annihilation between a proton and anti-proton, you can measure all kinds of details of the decay. But nothing yet. It is very hard to quantify when something isn't violated, what the level is of it not being violated! But certainly, there is no evidence for CPT violation at all. Some of the more manifest ways would be a difference between the mass of proton and anti-proton, or difference in the spectral lines of hydrogen and anti-hydrogen. CERN is beginning now to make anti-hydrogen by positron capture of anti-protons. CPT violation if it exists, ought to be proved by experiments but the bar to prove that would be very high.

Symmetry violation is a very fundamental phenomenon in Nature. Just think about the importance of chirality in biology, for example.

Oh, yes. Why are all the proteins left-handed and the sugars right-handed?

Is there a connection between the two; I mean the symmetry violation at the molecular and at the subatomic particle level?

A lot of work was done in this shortly after parity violation. For example, experiments were being done in racemic solutions being bombarded by beta rays and checking if any evolutionary imbalance happened — but to my knowledge there was never an experimental proof that there was any direct connection between parity violation and the chirality of molecules. But it strikes me almost like CP violation when we could almost say that our existence demands it. We might almost say that, perhaps, in the biological material the existence of the total right-handedness is due to some fundamental symmetries. But once you have the predominance of one type of chirality, then it is understandable that it has to prevail. But here, it is more likely just like tossing the dice.

Why did you leave Princeton?

I left Princeton with a great sadness. But being a particle physicist and also wanting to have a reasonable family life, I had to. That time the new accelerator was being built at Fermi National Laboratory. Also, I have been a graduate of Chicago, so when I got an offer to come to Chicago, being close to the accelerator was a great attraction and thus I decided to move. People were somewhat shocked at Princeton, because they think that when you move there you buy your burial plot and never leave. But there was also the influence of my wife; we had two daughters who were about 10 and 13 that time, and my wife wanted to finish her studies in sociology, fulfill her intellectual desires, so we moved. If you have a happy wife, you will be able to do a lot better work and that is a key point. I think it is harder now. When I was young, my wife had to take care of the children, and raise them, and she did that with a good grace, but it is not so easy now, I think. I was lucky that during the most important years of my scientific career, we split up the duties in a way that would not be considered proper now; but it really made a difference to me.

What have you been involved with recently?

Around 1986–1987 I got tired of particle physics. In the golden age of particle physics, you could put together a group of people of reasonable size and do the work. But with time it got more and more complicated. As I told you that I learned from my colleagues at Berkeley that a certain measurement takes a certain scale and thus when the scale gets too big

Some of the participants of the Meeting of the Superconducting Super Collider (SSC) on September 17, 1985, at the Lawrence Berkeley National Laboratory. The participants included, from the left, first row: Robert Wilson; Robert Hofstadter; Maurice Goldhaber; Burton Richter; Paul Reardon; Robert Hughes; Robert Diebold. Second row: Boyce McDaniel; Gerson Goldhaber; Stewart Smith; Robert Palmer; and the rightmost two in the second row: Stanley Wojcicki; Larry Sulak. Third row: Karl Berkleman; Frank Merritt; Leon Lederman. Fourth row: Michael Smith (on the left); James Cronin (in the middle); Jim Peebles; Maury Tigner; Murray Gell-Mann; Sidney Drell. Courtesy of J. Cronin and the Lawrence Berkeley National Laboratory. We thank Val Fitch for his assistance in identifying the persons in the picture.

you need hundreds of physicists. But I was proud enough that anything I did I wanted to leave a mark on, and that is quite impossible if there are so many people involved. So in the late 1980s there were reports about cosmic rays coming from point sources in our galaxy and, since I was always fascinated by cosmic rays, I decided to switch. Together with my colleagues, we got some money and built a rather large detector of about half a km radius in Utah. We showed that all these claims about these point sources were wrong and they did not exist. We also did some positive things but this made me interested in what the real problems were in cosmic rays and I considered the highest energy cosmic rays. This means macroscopic energy in a macroscopic particle. Cosmic rays have been measured by now well into the 10^{20} eV and that's like 50 of joules energy in one elementary particle and that is quite significant and one doesn't have any ideas about how Nature produces these things.

Cosmic rays are protons and nuclei coming from the Galaxy and perhaps from outside the Galaxy and they are striking the earth making radiation. They interact in the upper atmosphere and they make showers of particles that we can detect very easily on the ground. In one square meter we have about a hundred particles per second coming, they are going through us right now. It is a natural phenomenon and an enormous amount of work has already been done with cosmic rays. But the most fascinating are the highest energy cosmic rays because there is no understanding, no theory even how objects of nature, even astrophysical objects can produce cosmic rays of such high energies. They are so energetic that the magnetic fields that we know about in the Galaxy don't bend them at all. They most probably are extragalactic and the magnetic fields are too weak to bend them.

In order to study the cosmic rays we have to build a very big apparatus. I said earlier that I didn't want to be part of those particle physics experiments that involve hundreds of physicists. This I wanted to be part of because of the passion to understand why this is happening. We are now building the first such instrument in the Southern hemisphere, in Argentina.

Why there?

First, of course, we had to have a site-survey. There were three places that were possible in the Southern hemisphere: Australia, South Africa and Argentina. We need large flat lands, also lots of sunshine and no clouds because these are optical instruments. It was also important that the place welcomes and encourages you and has a scientific infrastructure. Argentina was surprisingly better than Australia in this regard. Australia is extremely powerful in radio astronomy and they thought that this was some kind of bastard astronomy. We ended up with Argentina. It is already working.

We are also starting to build the detectors in the Northern Hemisphere, possibly in Utah or Colorado. I started to get involved with this in 1991 and my fear is that I am not going to live long enough to see the whole thing done. But somebody will.

You said that the primary cosmic rays reach the atmosphere and there they collide with particles of the atmosphere and produce secondary rays, showers, which we observe. I was wondering, doesn't the nature of these showers depend on what atoms or molecules the primary rays hit?

This is an extraordinarily good question! We know quite a bit how these particles interact and the most crucial part is when protons hit nitrogen atoms and one knows from accelerator studies, which are at somewhat lower energies, that the atmosphere behaves like a converter, it converts the single energy of the high energy particle into the energies of lots of particles. It turns out that the exact detail mechanism of how this works is not all that important. The key point is that when this shower develops, it grows; it reaches a maximum and then decays. By the time it gets to the maximum the rest is pretty well known. The most uncertain part is how deep in the atmosphere is the maximum. But even independent of that, the number of particles you get on the ground, particularly far from where the original particle was going is quite insensitive to what one assumes about the interaction. It is like thermodynamics. One particle converts into 10^{11} particles and then you have thermodynamics and statistics that save you, because you are averaging over all.

Is there any idea about where the primary cosmic rays come from?

Many. I did a survey and found that over the last few years one article per week has been published on the topic. All sorts of interesting and even crazy ideas. But nobody understands anything about this. But eventually Nature will tell us what it is; if we build a good experiment we will have a good chance of finding out. There must be many different origins, depending on the energy. This is a mystery and the only way for us to find out is not theory but through measurements. And my critics are forgetting how we got where we are now. We built bigger and bigger machines, producing amazing results that we've never anticipated.

Is studying cosmic rays still your major occupation?

Yes, completely. I retired formally from teaching in 1997, when I was 65. The university is very kind to me and they let me keep an office and I still have some grants.

Which of your work are you most proud of?

I don't know. I am not sure if that question has an answer. I've followed a line, which is just research, research and teaching. I was often asked to be chairman, dean, etc., but I have never done that. I have been accused of being irresponsible for that but my passion is whatever I am doing at the time. There are a few eureka moments in life. One of them was when I was working with the spark chamber and I figured out that if I put a small voltage on it to clear out the ions, the spark chamber could live and see one particle out of 10^6 per second going through. Such things, or making a big mistake and then finding out the error. Like I said enjoying working on the data, the day-by-day work.

Do you have any heroes?

It probably sounds bad to say no but no, I don't. There are people I admire but if you look back in physics, there were extraordinary scientists whom I admire. But if you look at this Nobel business when you take the whole array, I don't know how to make a scale but there is probably 6 to 10 orders of magnitude difference between the lowest man or woman on that and the highest. Look at what the pioneers of physics did in the 20th century; I can't even imagine! I have maybe one talent: to choose the right experiment to do and find the good ideas to do them, what is extremely important. Ideas are the most precious things in the world; I don't care where they come from. The range of quality of the people who receive this Swedish prize is just enormous and I think that some people think that when they are put in that category, they are all of the same level. Never!

Please, tell me something about your present family.

We have three children; two daughters and a son. We have four grandchildren. Our son lives in California and wants to be a screenwriter but now he lives on reading scripts for movie companies. Our older daughter is the publisher of the Harvard Business Review, she has done very well, and our younger daughter has her own business on environmental assessment for pipeline construction; there are about 70 people in the company, it is in Minneapolis.

What do you consider success in science?

Enjoying it. It's got to be more than that. Clearly there are some minimum things, experiments to propose, get approved, make the experiment and make a difference. To be successful in science is quite a bit of luck, clearly some skill.

Has the Nobel Prize changed your life?

No, I don't think so. It has given me the opportunity of doing things that I otherwise could not have done, such as putting together this big international cooperation to build the cosmic ray detectors. I am shameless in using the Nobel Prize to get access to presidents, prime ministers, groups in the United States and Argentina to arrange things. That's what it is very good for.

Don't you think that being a Nobel Prize winner and thus having a larger visibility, gives you a certain responsibility in promoting science or other similar issues?

Probably I should do that more than I do. I didn't ask them to give me the Nobel Prize. I chose to accept it. I don't see that there is any logical connection that should make me more responsible for communicating with young people. As a matter of fact, in this Auger project, we are in a small Argentine town and we have a large visitor center, lectures, programs, so we do this. But I don't like this "Nobel oblige" — they didn't have to give it to me. It's not that I am against popularizing science, it is just that I don't feel that I have this obligation.

There is no such obligation. What I meant was that the Nobel Prize has far the largest prestige among all prizes and probably the only one that even the general public appreciates.

I've spent a lot of time talking to young people but again, that prestige of the Nobel Prize is just unrealistic. I give you an example. I was invited to Yale to talk about the Auger project. These days I can't travel much because my wife is not terribly well, so when they asked me, I told them that I could not go but they should invite my colleague, Paul Mantsch, who is the project manager of Fermilab and is really the person responsible for getting it built, and he has the same passion for the project as I have. But they would not invite him. This is really terrible and they will really be missing something because they won't hear about this beautiful project. Also when I travel to other countries, they make so much fuss about this Nobel Prize, I am set aside from my colleagues, and all that; it really is embarrassing.

Earlier you mentioned that the 1950s were the golden age of particle physics. What do you think of its future?

Intellectually it is very good. The sociology has changed a little bit; the problems that are being attacked have changed, and the new machine at CERN, with its 7 TeV will certainly produce interesting results. The real beauty and challenge in particle physics is designing detectors. The theory is somewhat stuck between the very effective phenomenology and string theory. When one has lived through science and asks himself the question, it is amazing to me that there is anything at all. And I don't know if we'll ever get the answer, the answer in some ways is that Nature, or God, or whatever has tried trillion things and one worked out and that's the only one we are sensitive to. These are amazing things.

What was the greatest challenge in your life?

Oh, gee, why do you ask these questions? That's kind of a Nobel-oriented question. What's the greatest challenge? I really don't know.

Is there anything else you would like to convey?

No, I don't think so. I think I am one of your duller people but that's OK. I haven't had any great challenges and there were awards and prizes that came unanticipated but I appreciated them very much.

Reference

1. An interview with Val Fitch appeared in Hargittai, M.; Hargittai, I. *Candid Science IV: Conversations with Famous Physicists.* Imperial College Press, London, 2004, pp. 192–213.

Wolfgang K. H. Panofsky, 2004 (photograph by M. Hargittai).

28

WOLFGANG K. H. PANOFSKY

Wolfgang K. H. Panofsky (b. 1919 in Berlin, Germany) is Professor and Director Emeritus of the Stanford Linear Accelerator Center (SLAC) at Stanford University in Palo Alto, California. He received his Ph.D. from the California Institute of Technology (1942). He has had positions at Caltech, the University of California at Berkeley, and he has been at Stanford University since 1951. He organized and oversaw the creation of SLAC, and was its director from 1961 till 1984. He is a member of the National Academy of the U.S.A. (1954), the American Academy of Arts and Sciences (1962), and foreign member of the French Academy of Sciences (1989). He has received many awards and distinctions, among them the National Medal of Science of the U.S.A. (1969), the Franklin Institute Award (1970), the Enrico Fermi Award (1979), and the Leo Szilard Award (1982). He has served on the President's Science Advisory Committee, the White House (1960–1964), as consultant to the Office of Science and Technology, Executive Office of the President (1965–1973), and on the General Advisory Committee to the President, the White House (1978–1980) among many other similar activities. He has also served on numerous national and international committees on nuclear energy and disarmament, such as The Ad Hoc Group on Detection of Nuclear Explosions (1959), as Vice Chairman of the U.S. Delegation to Geneva (1959), and on the Department of Energy Panel on Nuclear Warhead Dismantlement (1991–1992). We recorded our conversation in Dr. Panofsky's office at SLAC on February 11, 2004.*

Dr. Panofsky, after having prepared for this interview I don't find it easy where to start. You have accomplished so much and in so diverse areas.

I have had a complicated life.

You come from a highly intellectual but not science-oriented family. Still, both you and your brother became physicists.

That's correct. My father, Erwin Panofsky, was a well-known art historian and he always called us, the two scientists, die Krempeln, the plumbers. He was not interested in physics, although he had some interest in mathematics, he wrote a well-known paper on perspective in geometry. But anyway, we did not follow in his footsteps. In fact, I have had rather little interest in pictorial art.

My family background was rather harmonious. My parents were more or less supportive of what we were doing but they were not intensely involved with our upbringing. Not only my father but also my mother was an art historian. They used to have seminars in our home and both were very busy people. My father wanted me to go to the Gymnasium in Germany but I never finished that. I was still 3 years away from finishing it when we left Germany in 1934 because of Hitler and emigrated to the United States. Our emigration was quite routine. We left early and my father had already been a lecturer at New York University and he had, so to speak, one foot across the ocean so it was not that difficult to pull up the other foot. Our move was not that painful and difficult. We came to Princeton and instead of continuing high school, I went to Princeton University at a relatively young age. One reason I went to science was that my English was very bad and at most humanities courses you had to write long essays in English that I could not do, so I started taking courses in engineering and the sciences.

After graduating from Princeton, you went to Caltech to graduate school. Why there? At that time Caltech wasn't yet such a famous school as it is today.

---

*Magdolna Hargittai conducted the interview.

No, it was not, you're right. Since my undergraduate grades were very good, I applied to several places. What persuaded me to go there was that I received a long, 5-page letter from Robert Andrews Millikan, the president of Caltech, explaining all the virtues of the school. So it seemed as a reasonable thing to do — although, to be frank, I did not really know what I was doing. I did not know anybody there but it seemed like an interesting adventure. I was only 19 years old and I went to California for the first time. I traveled there on a freighter through the Panama Canal, so it was a very complicated way to get there.

Caltech was a very harmonious place. All the graduate students lived there together at the Atheneum, and all graduate students were involved with teaching, so the graduate students were very much involved with the intellectual community of the institute. There was also an interesting aspect to Caltech in that the number of undergraduate and graduate students was about the same; they had a relatively small number of undergraduates.

There I worked with Jesse W. M. DuMond, who also became my father-in-law. He was a well-known physicist, and I liked him very much because he was a person, who believed very much in the integrity of all of physics. He was a good machinist, a good experimentalist, he was very much interested in detailed mechanical design, and was a reasonably good theoretician, so he was not compartmentalized. The war was beginning to start and my life became complicated because he went to Washington to work on some military problems during my graduate years and I was more or less in charge of a fair fraction of the activity there. I was quite independent and given a lot of latitude in what to do. I also worked with Carl Anderson and we wrote a textbook together on electricity and magnetism, a lower-division textbook, because the ones we used were not very up-to-date. There was much activity. It was also that time that I started to get involved with war work in what was called the National Research Defense Council.

Were you an American citizen by that time?

No. Actually it was quite amusing. I started war work while I was still an enemy alien so in order to get security clearance I had to fill out the longest government form I have every seen, it covered the whole living room floor. That time there was a complicated situation because quite a few foreign scientists participated in American war-related work. Enrico Fermi, for example, was a non-citizen when he started work on the atomic project. My war work was together with DuMond. First on a fairly straightforward problem, on improving the accuracy of anti-aircraft fire. Then later I became involved with work done at Los Alamos through Luis Alvarez. He was charged by Robert Oppenheimer to determine how to measure the explosive power of nuclear weapons. He read some of the papers I had written together with DuMond and others on how to measure shockwaves from supersonic bullets. Alvarez said that we had already done what he was supposed to do and he didn't want to do it again. So he got in touch with DuMond and me and eventually I became the linkage between Caltech and Los Alamos in developing a device to measure the yield from nuclear weapons. I converted instruments that we developed at Caltech and they were used at Los Alamos. We flew over the test over Trinity in Alamagordo but the weather was too bad. I was in a B29 airplane over the first nuclear explosion but we could not release our devices because the weather was too bad. So we just made observations. But the device that I designed was used over Hiroshima and Nagasaki as the means to measure the power of the explosion. I still have one of these gadgets here, in my office. The actual one that was used is now in the Nagasaki Peace Museum.

Did you know that time that Alvarez was to fly with the Hiroshima mission and that he took with him a letter addressed to a Japanese physicist?

Yes, he told me about it.

Beforehand?

Yes. He told me that he had been planning to do that. I also have a copy of that famous letter. You know the story?

Yes.

It was a new way to communicate with the enemy. He addressed it to Professor Sagane, who was a Japanese physicist, who earlier worked at Berkeley. The letter was taped to the device that was dropped. It was found and delivered to Sagane, who then delivered it to the Japanese high command. But there is no evidence that it did or did not influence the Japanese decision to surrender.

It was a very bold move to write this letter. It stated that there were many more such bombs available and to be used if they don't stop the war, and, of course, there weren't.

That's right. There weren't. There were materials for maybe one or two more but not actually in the pipeline. This whole story is rather dramatic. It was published in the United States in *The Saturday Evening Post,* but it did not really receive much attention even though it was quite a dramatic event. It was an extremely bold move and, to my knowledge, it has never become known whether the letter did or did not influence the Japanese to surrender. I mean the time sequence was there but it probably did not.

What is your feeling about the bombs dropped on Japan?

I was a kid that time. I was very young but I could see that people were dying daily in large numbers so initially when I was doing this work, I had not really given it any particular thought. For instance, I went in the B29 over the Trinity explosion and people asked me how I felt in this dramatic moment and my honest answer was that I fell asleep immediately afterwards because we had a deadline to meet to get ready with it and we did not sleep enough and everybody was exhausted. It was not a time to think profound thoughts. But afterwards, when the war ended, I was very much concerned. When I went to Berkeley, I joined what was called the Northern California Association of Scientists and went around giving speeches to labor unions and professional associations about what the atomic bomb was and how it differed from all previous weapons and all that. I felt that the public coverage did not sink in, the fact that nuclear energy increased the energy that you can put into a certain size by a factor of more than a million somehow did not penetrate. So I went around giving public speeches, which did not help much either. I was always interested in that and from then on in, I sort of pursued a double career, one in physics and the other worrying about the nuclear bomb.

I would like to come back for a minute to the fact that you already participated in war-related work while still a foreign citizen. You did that in California where they had the famous Enemy Exclusion Act. Did that concern you too or was that only against the Japanese?

That was rather paradoxical because, indeed, it was aimed against the Japanese and as such, was quite racist. I had to register each year as an enemy alien and it showed the stupidity of some of these things. I was required not to go more than five miles from my place of residence, which was Caltech that was doing secret work, so I was restricted to stay there. I remember that I liked to go hiking with my wife and we had a map with the five miles radius indicated on it. When we were hiking in the hills behind Pasadena, we were worrying not to cross that sacred circle.

Then I was giving lecture courses to military officers without being a citizen. They were refresher courses on physics. I remember, I was just a kid and I was being rather mean with them, making them work out problems at the blackboard and so forth. Then came Pearl Harbor and they were ordered to wear uniforms and it turned out that they were all generals, with all those stars on their uniform. It was funny. Then I also gave evening courses on physics to engineers and physicists in the local military industry and once, while I was giving the lecture, a military policeman came in and asked why I was not yet in bed because there was a 9 o'clock curfew for enemy aliens. It turned out that I just got my citizenship one day before. He then congratulated me and left. But the people at the class were wondering what was happening. So the enemy exclusion act other than the five-mile radius and late-night curfew and all these regimentations did not keep me from doing my work.

How did you get to Berkeley?

I became well acquainted with Luis Alvarez during our war work. He, during his later years at Los Alamos, conceived the idea to reuse the surplus parts that were left over from the radar work done during the war. Alvarez used to work at the MIT Radiation Laboratory before he went to Los Alamos, and he knew about these radar sets and he judged that they could be used as cheap sources of radio frequency power, to power the accelerator at Berkeley. He recruited me to become his sort of right-hand man, to help build the 32 MeV proton linear accelerator, which at that time was a very ambitious project. I went to Berkeley as his deputy, but he had many other things to do so I mostly ran the project and became the chief operating guy for building the proton linear accelerator. It was interesting because the whole process was to design it around the surplus radars that was being made available. But the main process of designing and building the machine was to throw these away and replace them with equipment more suitable for the job. By the time we actually built the machine there was essentially nothing left of the main reason for building it. It became a very good machine and a lot of research was done on that. At that time at Berkeley, there was a very flexible atmosphere and people who helped build machines were also given the privilege to do experiments with them. There was no distinction between accelerator builders and experimental particle physicists. So in addition to working on accelerator design and construction I did experiments on the proton linear accelerator and also on the other machines at Berkeley.

What were these experiments?

The experiment that turned out to have the largest impact was an experiment I did on the 184-inch cyclotron on what's known as the absorption of negative pi-mesons in hydrogen and deuterium. This was a completely new approach of doing physics at the time. It consisted of putting a high-pressure hydrogen vessel very near the target, inside the 184-inch cyclotron and then measuring the gamma ray spectrum emanating from that secondary vessel. From that spectrum one could learn a great deal of what happened when a pi minus absorbed on a proton and became either a neutron and a gamma ray or a neutron and a pi zero. We determined a lot of things. We determined the mass difference between the negative pion and the neutral pion. When we did the experiment in deuterium we determined what is known as the intrinsic parity of the pion. We nailed down a lot of unknown parameters of the pi-meson, so it was an extremely productive experiment.

Then I worked with Jack Steinberger on Ed McMillan's electron synchrotron and we did the experiment that discovered the neutral pi-meson. There were also other experiments on meson-induced fission, on the half-life measurement of the pi-meson, and so forth. So I kept going back and forth between building the machines and using them.

I did the calculation that set the energy of the bevatron to make it possible to measure the anti-proton because that was a relativistic calculation and it was not obvious what energy was needed to do that. So I did a bunch of things. I also got recruited into teaching. I taught at Berkeley and organized an upper division lab for experimental techniques in physics. I taught a graduate course in classical electrodynamics and prepared mimeograph notes for that. It was quite successful so people suggested that I should publish it but since it is a lot of work I wanted to have a co-worker. I was looking around for one and then E. U. Condon proposed that I should see whether Melba Phillips could join me. Do you know who she is?

No.

She was a theoretical physicist who worked with Oppenheimer and she got into trouble with the House Un-American Activities Committee and got fired from her job at New York University. But she was a very good theoretical physicist. So I wrote to her and she said, she would be happy to collaborate on the book. We did the book together while I never met her; the post office obviously worked very well. I only met her in the last year of writing. The textbook, on classical electrodynamics, became very successful. Later on she became very prominent, she became the president of the American Association of Physics Teachers. She is now 93 or 94 and we had some correspondence lately. The book is out of print and is now being reprinted by Dover Publications.

You mentioned the House Un-American Activities Committee. I read that you also "bumped" into problems with the so-called "loyalty oath" controversy.

That was the reason I left Berkeley. Basically the loyalty oath was a way in which the University of California wanted to preempt actions by the legislature to actually forbid any professors to engage in "un-American" activities. I had signed the oath because I had all sorts of security clearances. In fact at the time I was working on a project called MTA (Materials Testing Accelerator), in Livermore. At that time — this is a sideline — there was a belief that with the Korean War starting, the United States would be cut off its uranium supply, which at the time came from Africa. Then Lawrence proposed that plutonium could be made by irradiating depleted uranium with an accelerator beam. It was a complicated story but as a result a secret pilot project started with the name MTA, which was just a cover name. I became the chief engineer of that. There was no room for that project in Berkeley so Lawrence found an abandoned site that used to be a naval airfield at Livermore. We built the pilot project there.

The machine got very high current but in the meanwhile people looked for uranium and found a lot of it in the United States, so the economic motivation disappeared.

Then came the loyalty oath. As I said, I had signed it but some of my mainly European colleagues did not and they lost their jobs. I got very unhappy about that and told Lawrence that I was going to leave because of that. He took me to see John Francis Neylan, the Chairman of the Board of Trustees, who lived here out in Woodside. He asked, "Young man, what is bothering you?" I said that people who have different values should be respected. He said, "Listen to me, young man," — and he talked to me for two hours. I never said anything else further on. I went back and decided to leave. I had several job offers but then Felix Bloch and Leonard Schiff came to Berkeley and twisted my arm to go to Stanford. So by the principle of least resistance I decided to go to Stanford, although I had offers from Princeton and Columbia University as well. By that time I already had a large family and moving just across the Bay was the easiest. But, again, I was very fortunate because I certainly did not know what I was doing. So I went to Stanford. Luis Alvarez tried to talk me out of it. He said, "Nobody ever does anything at Stanford, you could kill yourself."

Does that mean that nothing really was going on at Stanford that time?

Oh, no, on the contrary. There was a fair amount going on there. Of course, Felix Bloch was here, the famous theorist, who had measured the spin of the neutron and other things. But it was a very tightly knit department. At that time William W. Hansen, a Stanford physicist, a young man — who went to the MIT Radiation Laboratory and did many major contributions to radar — invented, what he called the rhumbatron as an idea to make high energies for accelerators. Then he died tragically of beryllium poisoning at the age of 40. Then Edward L. Ginzton, an applied microwave physicist, had gotten support from the Navy to build a one-billion eV electron linear accelerator here, at Stanford. Robert Hofstadter was recruited to do experiments with it. But somehow the chemistry here was such that Hofstadter worked on preparing for experiments and Ginzton worked on building the accelerator and the two did not cooperate much. They really did not adapt very well. The accelerator had lots of problems, so Schiff and Bloch came to me and said that they wanted somebody to come to Stanford who had experience with both the accelerator and with doing experiments.

So I came to Stanford and found a really terrible mess in the accelerator in the sense that the machine did not work very well. Ginzton was involved with lots of other applied work, mainly related to the Korean War. Hofstadter did not interact with the accelerator people. They had to promise the government that the machine would reach a billion volts but it needed more length. So they added some but then there remained only six inches between the end of the machine and the end of the wall, certainly not enough to put detection devices there. So overall the whole thing was in a bad shape. I became the head of that laboratory and I designed some magnetic systems to clean up the spectrum of the machine so that Hofstadter could use it. Much of the machine was re-engineered, we enlarged the lab space, and added the detection devices. I started a whole bunch of graduate students to do other experiments. But basically I founded a bridge between the experimental people and the machine people.

How old were you that time?

It was in 1951, so I was 32. I also kept being interested in nuclear weapons problems and at that time the Atomic Energy Commission (AEC) got worried about what is now a popular subject, namely nuclear terrorism. Robert Oppenheimer was asked in congressional testimony how would one detect a nuclear bomb in a crate smuggled across the boundaries of the United States. His answer was, "With a screwdriver." He meant that you take the crate apart. Then Hofstadter and I were commissioned by the AEC to write what became known as the "Screwdriver Report". That was about how you use radiation detectors and accelerators to detect what's inside a box that you can't open. That became a sort of classic reference on how you detect smuggled nuclear material. The basic physics has not changed any since then. So I was involved with that, parallel to trying to make the accelerator run here. Then I became the head of what's now known as the High Energy Physics Laboratory (HEPL). Ginzton ran the microware laboratory because he was interested in microwaves but not in what you do with the machines. Hofstadter was interested in using the machines, so HEPL was extremely successful and it led to the experiments with Hofstadter on electron scattering. We also did a lot of other things. Several graduate students and I worked on pair production of muons and on measurement of the radiation length in hydrogen, there were experiments on inelastic electron scattering, and so on. The machine, we called it the Mark III accelerator, became a very productive machine and it demonstrated that high-energy electrons could really contribute to fundamental particle physics, while before the field had been dominated by proton accelerators. The success of that machine was basically the background for this laboratory.

How did the idea of organizing SLAC come about?

That was one of those interesting things. I think that the first proposal to build SLAC came from Hofstadter. He was the one who said, "This machine is great, why don't we build a bigger one?" Of course, he was not interested in the "how" only in the "what", he simply wanted to have a machine with higher energy so that he could extend his scattering experiments. He had the courage, if you call it that, to bring this up. We then had a number of informal meetings with the senior staff from the High Energy Physics Laboratory, the Microwave Laboratory, and the Physics Department. Some of these meetings were held in my house, others in a beer joint nearby and they eventually led to the proposal to build this laboratory. I have a copy of that proposal here. This was in the old days, there was no bureaucracy. You can see that this was a typewritten proposal for what is now over a 100-million-dollar unit. We collaborated on writing it. At that time support of science was growing rapidly and there was relatively little competition for new initiatives. We made the proposal to three agencies of the government: the Defense Department, the National Science Foundation, and the Atomic Energy Commission. This was in 1957. But the proposal was generated by a very informal collaboration of local people. I already mentioned the laboratory people but there were also others. For instance, some of the members of the Board of Trustees ran large architectural and engineering firms and they volunteered some analysis of the geology of the site here and checked the effect of earthquakes. After we submitted the proposal, four years of politics followed.

There was a diversion in 1958 when I was drafted to negotiate with the Russians about a nuclear test ban. That was, again, a complicated thing. The science advisor of the president appointed a committee to look into the technical means of detecting nuclear explosion in outer space. I became chairman of that committee and I had the great distinction of chairing a committee, which was joined by such scientists as Hans Bethe and Edward Teller. The distinction was that a unanimous report was produced, which was signed by both these gentlemen. After that the powers to be thought that I was ready to negotiate with the Russians.

That conference in 1958, called the Conference of Experts, and chaired by Bethe, was about whether or how we should talk with the Russians about stopping nuclear tests. Its report became, however, somewhat undone by Teller bringing up some ideas about the ways to cheat. Then two new conferences, called technical working groups, were organized, where the Russians and the Americans would meet to look at technical means to avoid cheating. I chaired one of these.

What do you mean by "ways to cheat"?

Well, Teller had proposed that you could secretly test nuclear weapons by testing in outer space by sending out one rocket to carry the nuclear bomb and another rocket to carry the diagnostic instruments to measure the radiation from the nuclear bomb. That would take place somewhere between ten thousand and a hundred thousand miles in outer space and nobody would know. In addition one of Teller's colleagues invented what was called an underground cavity. That means that you test underground and make it in a hole, which is big enough so that the acoustic shockwave from the bomb would be sufficiently weak. And it would not give that much of a disturbance. This way you could decouple the explosion and so you could not detect it so easily with a seismograph. Those two technical working groups dealt with underground detection of nuclear weapons and detection in space.

Has either of these suggestions become effective?

No. They were purely theoretical evasion possibilities. The fact that you can technically cheat is not sufficient reason to cheat. In effect, if the Russians really wanted to go to all that trouble to send two rockets to space to do nuclear tests maybe America would be better off if they spent their efforts doing that than doing something else bad. The resources needed for evasion had to be factored into the overall judgment whether the evasion was a real possibility or not. Some other people had taken the absolute position that if there is any technique in which you can cheat that's bad. Never mind if you have to use half the treasury in order to do the cheating.

Anyway, we talked to the Russians and we issued a report. Essentially the outer space report was the basis of the 1963 treaty that Kennedy finally negotiated and which forbade testing anywhere except underground. So outer space got ruled out and it never happened. That was a sideline.

Coming back to starting SLAC…

There was complicated politics. Eisenhower made a speech in about 1960 supporting SLAC because his advisors supported it. But at that time there was a Democratic congress and a Republican president and he forgot to consult Congress before making that speech. So they decided not to go along with that and not to authorize SLAC. Then the next year, in 1961, Congress did authorize it in a rather amusing way. There was a project in Congress to get nuclear power from the plutonium production reactors at Hanford, Washington, and the Democrats liked that because there was a lack in power there. Since Eisenhower had supported SLAC, it was a Republican accelerator, so they paired Hanford and Stanford — I used to say because it rhymes — and authorized both projects at once.

In the meantime we have carried the design quite far and we were ready to break ground in 1961. Then Ginzton, who was originally the co-director, had to become the chief executive of Varian Associates (it is a complicated story why) and he could not be both, so he resigned. Thus, more or less by default, I became the director of SLAC. It is interesting that there is no record anywhere of me ever having been appointed, although I think I probably have been the longest serving director of any national laboratory. It just happened. Ginzton by virtue of this problem at Varian needing a leader — and he was not interested in particle physics anyway — sort of left it to me but there was never any official act that officially gave me that dignity.

SLAC was built in a remarkably short time, in four years, on schedule, on budget, and its performance exceeded what had been planned. All this is documented; before that accelerators were generally not documented properly. I was probably the right person to do it because I was deeply involved with particle physics, had done a lot of experiments, but also had experience in experimental design and electrodynamics. One thing that I felt very strongly about was that even with a project of that magnitude the head should be a technician first and administrator second. That means someone whose objective is science and not making money. The leadership should be interested in science and not in management or administration. We gathered people who shared this philosophy. We had a very able head of the business department but we called it the "business service department", just to make sure that doing a clean job and managing money and keeping the budget balanced was not an end in itself but the end was the science. Similarly we had "administrative service department". We felt that in order to have scientific productivity all the people who are in various key positions should fundamentally be technically competent and resourceful and put the technical achievement ahead of everything else and be able to contribute to the technical solutions.

You were also very successful in recruiting top scientists.

Yes, we were very successful in doing that. It is a long story. That had to do with the fact that in negotiating with the government we were very successful in having Stanford University policies preempting governmental policies, whenever they were both applicable. This Laboratory had its own Faculty, which is part of the University Faculty. The scientific leaders here are faculty members and they have the dubious privilege to serve on university committees. They can supervise graduate students and they are part of the academic structure of the University.

This was probably very important in order to attract good scientists.

Yes, it was, although there were lots of tensions. On the one hand, there was pressure from the University Physics Department that this Laboratory should simply be adjunct to the Department and serve the Physics Department. I said that it would be impossible. This Laboratory is too big and we have to serve the whole international community. But then there was pressure from the other direction as well in that the Laboratory should be what they called that time a "truly national laboratory", meaning that it should not be connected with any university but should be run by a university association of some kind. That I felt didn't make any sense either because then it would lose its academic identity. But I went through the motions. I asked Caltech and Berkeley whether they would join in management of this place. They said, "No, we are busy with our programs," so that didn't work. Eventually SLAC got organized as a national facility which is halfway between a national laboratory and a strictly university organization. We are totally open to all outside users, nationally or internationally. Anybody can make proposals. Today it is utterly wide, we have over 3000 outside scientific users and the language on the third floor is broken English. Sometimes we are even confused where people are coming from but we have a very bureaucratic system, in which people make proposals, we have committee structures to evaluate whether the proposed experiments are feasible, how much they cost, what is their scientific merit, and so forth. Even though the scientific leadership is members of the Stanford faculty, we are completely accessible on an equal opportunity basis to people from all over the world.

Do you consider SLAC your most important achievement?

I don't know. But I suppose I have to say, yes. I think that some of my early experiments at Berkeley were very important. Of course, when I started being the director here, for the first year I participated in one of the experiments but then I decided that the young people are too smart, so I stopped. Your question is hard to answer.

What is the most exciting thing going on at SLAC at the moment?

Well, of course, I am twice retired. I retired in 1989 and my successor retired since then. But the present director is very good and he maintained the tradition that the leadership is technical person first and bureaucrat second.

The most exciting thing now is actually two things. One is the so-called "B-factory", which is an electron-positron storage ring where there are 3 GeV positrons against 9 MeV electrons, making collisions and forming b-mesons.

The decay of the b-mesons can be recorded and from that — and it is a long and complicated thing — one can test one of the fundamental symmetries of Nature, the so-called CP symmetry. One of the hard things to understand is why the world is not balanced equally between particles and anti-particles. This experiment gives at least part of the answer to that question.

The other interesting thing going on is that since accelerators on Earth are limited in size we decided to diversify in two more directions. One is into cosmology because today there is a convergence between the investigations into the very small and the very large. That convergence is caused by the fact that in the very first moments of the Universe, the energies were comparable to the energies we have here. So we are now creating a Cosmology Institute, which is on the SLAC campus and we have some very good people who joined us.

The other direction is — and this became clear in the middle of my tenure here — that when electrons go in circles, they radiate X-rays and those X-rays are many orders of magnitude more intense than what you get from an X-ray tube. That is called synchrotron radiation. Its utilization became sort of a cottage industry. We created the Synchrotron Radiation Laboratory, which has made contributions in many fields, among them biology and materials science; in fact all fields where extremely intense X-rays are useful as an investigatory tool. So, in fact, there are three major fields carried out these days here, the "B-physics", cosmology, and synchrotron radiation — and all three of them are in good shape. We also maintained the tradition that we have four or five major ongoing construction projects and we have never exceeded the budget. We have never exceeded the scheduled time of a project either, so we have a sort of a hallow in this respect, which is a perishable commodity.

The most important work right now is, in fact, the work we are doing in the B-factory. It's again ahead of schedule and already has gotten important results both in understanding the b-meson system and the systematics of that, and also in the fundamental studies of CP symmetry.

I saw on the website of SLAC that NIH is one of your supporters. Is that because of the synchrotron radiation facility?

It's only a minor supporter. Since everybody is in need of money, the people who do molecular structure studies on the synchrotron radiation source persuaded NIH to contribute to constructing some of the beam lines of the synchrotron radiation source and they also became joint supporters of the latest improved project of that facility. But that is only a fraction of the synchrotron radiation work, which in turn, is only a spin-off of our other work. Still it is making major contributions to biology. This was definitely a development that was not predicted by anybody.

Several Nobel Prizes were awarded for work done at SLAC. You yourself never received the prize. How do you feel about that?

I am relaxed. And I have lots of other prizes. I mean the Nobel Prize is both a great thing and maybe a not so great a thing. It is a particularly difficult thing in a field like particle physics, where if you look at a paper, there are a hundred authors on an experiment. Sometimes it is not that

Wolfgang Panofsky receiving the National Medal of Science from President Richard M. Nixon in 1969. Herbert C. Brown is in the background (courtesy of W. Panofsky).

easy to determine who the prime movers and shakers are. I am happy that Burt Richter, Marty Perl, and Dick Taylor got the Nobel Prize. They are great guys but there is a certain amount of serendipity involved in how it works. We find, for instance, that it is very difficult, when we make new faculty appointments, for the leadership of the laboratory to really find out from publication records and even from lectures and speeches, who would be the best candidates for the future leaders of the laboratory. The good thing is that there is a huge amount of collaboration both national and international, but the bad thing is that the individual leadership is sometimes difficult to determine.

You mentioned that particle physics is very much teamwork and a large number of people work on any one particular project. Is there then room for individual creativity?

Oh, yes. There is a tremendous amount of room for individual creativity. There is creativity in instrumentation and there is creativity in diagnosing why something works or doesn't work. As you push these devices to the extremes of their performance, you run into all sorts of barriers and then you have to be able to diagnose how to fix them. Then there is creativity in terms of simply thinking of new forms of acceleration. On the experimental side, there is creativity in instrumentation but also in data analysis. One thing that is absolutely astonishing to most people who do not work in this field is that when you do experiments in, say, the B-factory, you record, literally, hundreds of millions of events, when particles are colliding. Then you have to be inventive to design tests, what we call cuts, to find the needle in the haystack. For instance, with the big discovery in the CP symmetry in the B-system, the first paper was based on about 70 events out of a hundred million. How do you know that in the process of massaging these hundred million events you didn't, in fact, fake the events simply by throwing away everything that did not look like what you were looking for. You have to make inventions criteria analyzing tools which can't throw out things that have objective reality against the background of many things that used to be interesting. After all, yesterday's discovery is today's background. So there is a tremendous amount of invention in data analysis. And, there are, of course, practical applications. Just because particle physicists are impatient with the limitations of their tools, they invent. For example, the world wide web was a by-product of particle physicists.

Was it really?

Oh, yes. SLAC had the first broadband network established with China. That, of course, is not invention, that is just impatience with limitations.

Then, of course, on the theoretical side there are great puzzles there. There is a whole bunch of theoretical work in string theory, in which we have several practitioners here. It paints a self-consistent picture, which unifies gravitation and quantum theory but thus far has zero implications in anything you can see. So there is still reasonable doubt whether it is Nature or just a theory. There is then the present struggle in trying to understand cosmology in terms of the rest of physics. There is a tremendous amount of room for innovation everywhere.

You are also famous for your work in arms control and disarmament. How did it all start?

I am extremely worried and to put it mildly, extremely unhappy and critical about what's going on at the present moment. Of course, it is a problem where science and politics intersect. Basically, scientists have managed that fewer and fewer people can kill more and more other people. First of all, nuclear energy concentrates more and more destructive or constructive power into a given volume or weight. Biology makes it possible to understand more and more of our life processes and therefore there is more and more knowledge how to manipulate life processes. So unless we get our social house in order, the combination of scientific advances and our lack of getting our social order in good shape are clearly destructive. I am trying my best to communicate this and to work with many other people to find constructive moves. Nuclear weapons, of which I know a lot about, is a particular example. In a way it is extremely remarkable that between 1945 and 2004 they have never been used, notwithstanding the fact that there have been well over a hundred major armed wars. Still, nobody has used these most destructive devices. That's the good news. The bad news is that there is no assurance that this will continue and if all countries in the world and certain sub-national groups have nuclear weapons, the world is going to be even harder to manage than it is now. So one of the great achievements in nuclear arms control has been the so-called Non-Proliferation Treaty, which is a very complicated bargain between the nuclear weapon states and non-nuclear weapon states. That bargain has been now under threat to come apart in two directions. One is that some of the non-nuclear weapon states are cheating and trying to get nuclear weapons, and also because the nuclear weapon states have not lived up to their side of the bargain. To some extent our government, to be frank, is being hypocritical in that we are all for non-proliferation as long as it does not effect us.

I have been trying to work on several things. Here is the draft of a long study that we are doing for the National Academy of Sciences on how to extend arms control to all nuclear weapons. Mr. Bush and Mr. Putin signed with great fanfare a treaty, the so-called Strategic Offensive Arms Reduction Treaty, which is, really, if anything, a step backwards, rather than forward. It limited only the strategically deployed offensive weapons that are becoming a minority of the total number of weapons in the inventories of both Russia and the Unites States. It has no means of verification of any kind and has no steps to get there and expires at the date when it is supposed to achieve its goal. Extending arms control to reach all nuclear weapons is one of the big international problems. You can police strategic nuclear weapons because it is pretty easy to track submarines, strategic bombers, and missiles by looking from satellites. But nuclear weapons that are carried for short range are much harder to track. It is even harder to track the materials that make nuclear weapons. We have a waste excess of nuclear explosive materials, both the Russians and the Americans. The Russians have enough of them to make over a hundred thousand bombs and we have comparable amounts. We made some progress in trying to get rid of them but that has come to a grinding halt. There has been a disdain for the international binding treaties just because there have been violations. But to me that is very troublesome, it is as if we were saying that we have to legalize murder because not all murderers are caught. The fact that somebody is cheating is not a reason to abandon the constraints but to try to understand the reasons for the cheating.

My main work is on this. There is a committee on international security and arms control at the National Academy, which is doing two things. It is working on studies on how to do a better job. We also have regular meetings with people from India, China, and Russia, with like-minded people, generally also affiliated with the academies. They are either scientists or military people. In a funny way military people currently are more sensible on some of the things than the political leadership. This is partially because in the Unites States military people, right now, understand science better then the political leadership. I cannot claim any raving success as a chair of a panel of the National Academy that wrote a fairly influential report on how to get rid of excess plutonium and several things of that kind. There is a long list of committees that I have been on. But this is something I passionately feel that we could do more progress than we actually do. Unfortunately, since most of it is not science but politics, I feel fairly frustrated.

You mentioned earlier nuclear smuggling. How small can a hydrogen bomb be nowadays?

They are not that small. There was one hydrogen bomb that could be handled by one soldier but it still weighed about 150 pounds. They are not that small but they are small enough so that you can stick them in a crate or put them on a truck. Carrying it in a suitcase you ought to be awfully strong and it would set off all sorts of alarms. But it is a very real problem. Senator Nun said that the limit of damage a terrorist can do is limited by the tool at his disposal. The reason why we have not used nuclear weapons is deterrence. Namely, had either the Russians or the Americans used nuclear weapons it would have been suicidal and that applies to all the nuclear weapons states. But when you talk about subnational groups, when people believe that life in Heaven is better than life on Earth, then deterrence does not work. So there is a real risk there. The idea of detecting smuggling is a multilayered thing. You have to detect smuggling but the most important thing is that you have to control the inventories at the source. Terrorists cannot make plutonium or uranium. They have to swipe it or buy it or bribe somebody.

Isn't in this respect Russia (or rather the former Soviet Union) an especially dangerous place?

It has gotten better. Right after the end of the Soviet Union, there were many problems of demoralization of the guards and underpaid troops. I once had a talk with the head of the nuclear regulatory commission of Russia and he said, "Dr. Panofsky, our nuclear weapons used to be guarded by elite troops. Now they are being guarded by a babushka holding a cucumber and she does not even know how to use that cucumber very well." But it has gotten better. There has been some cooperation but it is underfunded. Still well over half of the facilities in Russia are not guarded well. In fact, it is not guarding but the counting and book-keeping that is still not done in a reasonable standard. And there is so much of that material! To make hundred thousand nuclear weapons! It is a little bit like arguing whether the cup is half full or half empty. Roughly half of the Russian installations are now in good shape and the other half are not. Is it good news or bad news?

Edward Teller has died recently. What do you think of him?

He was a very complicated person. I knew him quite well, I met him several times, I debated with him, and I was together with him on committees. I don't pretend to know him; I am not a good enough psychologist. He had an obsessive concern with the communists. He had an excessive confidence that weaponry can protect us. He was a very troubled person in many respects because since the Oppenheimer case many of his friends broke off the relationship with him and he was a very social person. He has been wrong in terms of his predictions quite frequently. I remember I once had a ride with him in Washington, in a taxi, and we made conversation. He was complaining about the lack of more extensive science education in America and I said, "Edward, this is one subject on which we agree." And he said, "If you understood the other ones you would agree with me, too!" — and we rode the rest of the time in silence to the airport.

He has been wrong in many of his predictions. He was wrong on ballistic missile defense, he was wrong about the usefulness of peaceful nuclear explosions and, many other things.

How about the hydrogen bomb project, was that also wrong?

The question at that time was not whether we should pursue it or not but how rapidly. I think he was wrong in making it a crash project. What Oppenheimer and other people said that it was very difficult and that fission bombs already had so much explosive power that even if there is a hydrogen bomb, you could not use it any more than a fission bomb because of deterrence, so then what's the rush. The way it was usually publicly presented was whether we should build the hydrogen bomb or not. But what the real question was: should we make it a matter of high urgency because the Russians may think of it first. At that time fission bomb reached the megaton range in explosive power and so if the first hydrogen bomb would have had 4–5 megaton power, well, you can be just so dead. It would not have upset the military situation in any drastic way for quite some time. So the main issue really was whether to make this a crash program. But then, in the other direction, the alternate issue is how seriously you pursue arms control. Clearly, simply racing one another in these destructive things cannot lead to a constructive end. So at some level, arms control has to work whether it is before or after the catastrophe. I think that his pathological urgency of pursuing the hydrogen bomb when the atomic bomb had already demonstrated its destructiveness was wrong, and the efforts to control the atomic bomb really should have been given much higher priority. There I disagreed with him. But there were many personal complicated backgrounds between Oppenheimer and Teller.

I think that Teller has done a lot of damage in some respects. The political leadership is crucially dependent on scientific advice and if scientific advice is not balanced — and scientists do not agree with each other on many things, as I am sure you found out in your interviews — that is a problem. The politicians have to get consensus and responsible scientific advice. You cannot have science channeled through your pet advisor. Teller had done some damage during the time of Reagan when Reagan rather uncritically accepted some of the rather dubious scientific conclusions.

Are you referring to SDI?

Yes.

Could not this be looked at in such a way that pushing the Soviet Union into this arms race, he helped to bring down the Soviet Union?

Again, I am not expert in the subject. I think the Soviet Union sow the seeds of its own destruction. About SDI, as you know the Russians had built a defense system around Moscow, using nuclear explosive warheads, but it did not work very well and now, although it is still in existence, it is in fairly bad disrepair. I once had a meeting with some Russians and they told me, "You people are so awfully good, so if you are pushing this so hard, you must know something that we don't." And we didn't. Certainly, some of them had an exaggerated idea of our technological powers but I don't think that was a major component. They were certainly driven to their destruction by any number of factors. Their economy worked very badly, productivity was low, the military spending was too high and a lot of their military activities were uncritically done. So I am sure, the SDI contributed but in my view it was not a controlling component.

Earlier we talked about Teller and his testimony at the Oppenheimer case, which hurt him very badly. But — as his coauthor of his memoirs, Judy Shoolery told us last night — he felt that he cannot avoid testifying because he was brought up in having to absolutely obey the law and since he knew a lot about the circumstances, he felt it his duty to testify even if he did not want to do so.

I read his memoirs and the record is quite clear that he certainly did not push to testify. What he actually said was something like he would prefer the security of the nation in other hands. But the main point is this: security clearance — and I have had lots of security clearances, even as an enemy alien — is not something that should depend on differences in judgment. People do have major differences in judgment. Teller had certain judgment and Bethe had certain judgment, and all the rest of us had certain judgment. We also have to understand that even if scientific facts become more and more confirmed over time, during their evolution they can very easily be belittled or be bases for disagreements. These should in no way interfere with security clearances because otherwise the country gets trapped into essentially preselecting advice. There is a difference between disagreeing on which way things should go initially because it is a mixture of technical reality and political judgment as to what people are going to do and not giving people access to classified information. Why people were so deeply disturbed about the Oppenheimer hearings was that the subject was not about what advisors the government should select but who should and should not have a security clearance. And there was no evidence other than the earlier transgressions about not being sincere about the Chevalier case that Oppenheimer himself had in any deliberate way compromised information to the Russians.

But there were also others who had the same opinion as Teller, such as Alvarez, Lawrence or Wheeler. Still, it is only Teller who is blamed.

That's true and that is unfair. I know Alvarez very well, we collaborated a lot. He was a very outspoken men and I remember going home after work and telling my wife, "Today is a red-letter day because Alvarez didn't call me a traitor!" Alvarez was a guy who agreed or disagreed, but we still could work together harmoniously on lots of things. We worked together on the MTA in Livermore to regenerate the supply of plutonium. Alvarez, and to a certain extent Teller, also was an idealist. In 1956, the barriers went down in high energy physics to talk to the Russians and there was a first meeting of high energy physicists in Russia in Dubna. Alvarez and I went there together on a plane and talked a lot. He was of the opinion that war is now impossible and nuclear weapons have saved us and we would have a peaceful world from now on. Of course, I am paraphrasing him only. But he had an idealistic view in some respects, that humanity was sensible enough that the very existence of nuclear weapons have made war impossible. But, of course, it is not so and we are in a very dangerous situation today.

I did not know Oppenheimer very well but to me he was a very strange person. I knew him way back when I was in Caltech and he used to go back and forth between Berkeley and Caltech. I used to go and listen to his lectures and understood nothing. He always had a slightly mystical aura around himself, so I never understood him. But I didn't understand Teller either. Oppenheimer had this self-created mystical aura around him and Teller was very doctrinal. When we were on committees together usually he was rather unwilling to listen and participate in an informed debate, he was not very good in that. We once debated on arms control before a committee of Jewish physicians and Teller was impossible. He didn't pay any attention to the length of time and he just didn't stop. The chairman wanted to stop him and give some time to me but it just didn't work. Engage in a mutually respectful argument with Teller was very difficult. But they were both very good physicists. I admired Teller more as a physicist than I do Oppenheimer. Oppenheimer was an interpreter but he hadn't done very much physics. He introduced European physics to America but his own contributions were really not all that profound. I don't pretend in any deep sense to understand either of them. I am too much of a down-to-earth character.

You met many famous Russian scientists as well. Whom would you single out as the most interesting or most important?

I met Sakharov several times. He certainly impressed me very much. But the only Russian scientist I knew well was Kapitsa and also his son, Sergei Kapitsa. Sergei translated my book on electrodynamics into Russian. There was a funny occasion. Once I got a letter saying that the next time I go to Russia they could pay me some rubles. It turned out that they translated my book without any authorization from anybody. So when next time I went there, Sergei Kapitsa met me and took me to an accounting house. There were many ladies with abacuses and they said that they owed me 336 rubles and 35 kopeks. So they gave me a bunch of single rubles and I had to count them and, of course, I counted them wrong and they were very upset that I disagreed with them because they thought it was an indication that I considered them dishonest. So they counted them again in front of my nose. Anyway, I got those rubles and I bought an oboe. You could not take the rubles out.

I knew Kapitsa quite well and we had many good discussions. Once I went for a long walk with him out in the countryside in Russia and there were a bunch of cows there and he asked me, "Can you tell the difference between the communist cows and the capitalist cows?" He asked because some of them were owned by the collective and others by the people in their dachas. Of course, I couldn't tell the difference. He was a very worldly character. His son didn't do very much by way of physics but he ran a popular science program on the radio and got me to talk on that a couple of times.

I used to know Igor Tamm. Once we went for a walk in Geneva and we were climbing the mountains and he was terribly worried that we were going outside Switzerland into France since his permit from Russia was only for going to Switzerland. We talked a lot about arms control.

Your political views must be very similar to those of your parents or at least your father.

Yes, they are somewhat similar.

When I interviewed John Wheeler, he told me a story. Apparently you and they lived next door and when the FBI came to check him out during the H-bomb project, they went to your father and asked about him. Your father said, "Oh, they are not subversives, they are mass murderers. We are the subversives!"

I heard that story secondhand but I didn't know that Wheeler knew about it. My father, of course, didn't understand physics. He didn't like secrecy. Interestingly, Teller didn't like secrecy either. There was one interesting thing: both Teller and I felt that there was over-classification and over-secrecy on many things, even if for slightly different reasons. Teller felt that the Russians knew it anyway and may even be ahead. I felt that most of the things that were kept secret were in fact known to most intelligent technicians in the field.

My father was probably more liberal than I. Of course, I don't even know for sure what that means. I feel very strongly that many of the problems we have today have military and non-military solutions and we are simply not working intelligently enough on the non-military solutions. But again, my differences from Teller and to a certain extent also from Oppenheimer may come from the fact that they were both theorists and, to quote my father, I am a plumber. So you have to temper your desire of some tool being needed with a practicality and you have to balance more intelligently than we are our political, diplomatic, economical, and human tools with the use of force. On that score, I guess, you can call me a liberal.

About your brother. Was he also a physicist?

He was a meteorologist and he died over ten years ago. That's also an interesting story. He graduated as an astronomer and then the war started so he decided to do something for the country and he enrolled in the air force as a meteorological observer. They went over his papers and said that you can't join this course since you are not a citizen. But you can teach it! So instead of taking those courses, he taught them and that converted him from an astronomer into a meteorologist. He became very prominent in his field; he worked on atmospheric turbulence.

I would like to ask you about your present family.

I married the daughter of my thesis supervisor, in 1942. Everything happened in that year, I got my citizenship, I got married, and I got my Ph.D. We got married when she was 18. We have five children, 11 grandchildren and one great-grandchild. None of my children is a physicist. Among my grandchildren one is studying to be one. My children have a wide variety of professions, for example, two of my daughters are classical musicians. We are a very close family. My wife just turned 80 and we had a great chamber music performance in her honor. She is a self-educated paleontologist. It turned out that when we were excavating for this laboratory, we found a rare fossil. My wife spent 21 years on restoring this fossil. It is now

Wolfgang Panofsky and his wife in the 1990s (courtesy of W. Panofsky).

out in the Visitor Center. She often goes on paleontological expeditions. She is also very much a practical, hands-on person.

How did you have time for family life with all your activities?

I managed. We went for hikes often when our children were small.

Are you religious?

No, I am agnostic. I am not religious but I respect the role of religion in our cultural heritage. Sometimes I get into discussion with religious people and I know more about religious matters than they do but I am not religious.

Have you ever experienced anti-Semitism?

No, not really. Not since I was in school. That was, of course, acute. We left Germany when I was 15 years old and the last years there, 1932–1933 were very unpleasant. But not in this country. I was very puzzled about racial attitudes when I went to Princeton, it was an essentially all-white institution and also all male. Once I asked a colleague of mine what would he do if a black man would sit next to him in the cafeteria. He said, he would leave. I asked him why and he said, "Superiority." Just like that. That has always puzzled me.

Do you have any heroes, role models?

Not really. There are many people I admire but not above anything else.

What makes a good scientist?

Damned if I know. It's very hard to tell. Mainly he has to be curious and not accept obvious explanations but go deeper. But I really don't have much wisdom on that. Luck is an important component.

Anything else, you would like to add?

Not really. I am 85 and retired. The problem scientists have created essentially is that more people can do harm, and fewer and fewer people can do harm to more and more people. That's a fact. The social organization really hasn't figured out how to cope with that. The biologists are facing now the same crisis as the one the nuclear people faced when fission was discovered.

I listened to a lecture by Fermi way back, perhaps in 1939, when he just calculated on the blackboard how one could make a nuclear explosive, what the critical mass was, etc. Now one lucky thing is that nobody knows how to make a nuclear explosive unless you have either uranium or plutonium. So at least there is a narrow channel to control. But in biology, they are now facing the problem that when there is new dangerous information how you restrict that information. If you do restrict information, you can't channel the restrictions in as narrow a way because there are so many things that can get you into trouble. There is a tremendous amount of debate going on now about how to deal with that. There were several such committees at the National Academy, too. They are in a real state of crisis about how to balance security and unrestricted evolution of knowledge. Of course, the burden has to be on the social and political structure. You can't deal with that within the narrow scientific community.

But as I see you feel that a scientist has social responsibility.

I feel tremendously that we have a responsibility and that's why I write poison-pen letters. But I myself went through a tremendous evolution. Right after Sputnik, Eisenhower created his presidential science advisory committee and I was a member of it from 1959 to 1964. I knew Eisenhower quite well. He met with the scientists all the time and we had really open debates and open committees. What was important is that the President was science literate; of course, he was not a scientist but he calibrated our strengths and weaknesses pretty well. But that ability of the top administrators has eroded steadily. Kennedy understood it more or less. I met with Carter several times and he, in a funny way, overestimated what technical people could do. For example, I wrote a chapter on nuclear reactor safety and we had a one-hour appointment with the President where he briefed me and at the end, when we had only two minutes to go he told me, "Dr. Panofsky, can you explain to me the difference between the sodium fuel cycle and the uranium fuel cycle?" He was the President of the United States and surely he had strengths and weaknesses but he was a curious man — but of course, no human being could explain in detail these things in two minutes. So I sort of chickened out. He listened to too much detail in science so he could not see the forest for the trees — but at least, he listened. But since then ...

How about Clinton?

Wolfgang Panofsky in the President's Science Advisory Committee. President Dwight D. Eisenhower sits in the middle; George Kistiakowsky is on his right and Isidor I. Rabi is on the right in the picture. Standing in front is Jerome Wiesner, first on the left; Alvin Weinberg is third from the left; and Wolfgang Panofsky is fifth from the left. Glenn T. Seaborg is in the middle of the back row, just behind Weinberg, and John Bardeen is behind Panofsky to the right (courtesy of Lawrence Berkeley National Laboratory).

Clinton was not that good. I was not directly involved with the President's council on science and technology but knew several people who were. They mainly talked to Gore and not to Clinton. Gore, again, understood science but he thought he knew the answers, so was not a very good listener. Clinton understood the value of science and he was technically literate but he did not take the time to really understand some things. He was better though than the current administration. But it was Eisenhower who was very good in appreciating science. Of course, he was a university president so he knew the academic animal fairly well. Johnson had essentially no communication. It's not a matter of being a Democrat or a Republican; it's very much on the individual style and character of the particular gentleman. But anyway, that's the way it is.

Burton Richter, 2004 (photograph by M. Hargittai).

29

BURTON RICHTER

Burton Richter (b. 1931, New York) is Paul Pigott Professor of Physical Sciences Emeritus at Stanford University and Director Emeritus at the Stanford Linear Accelerator Center (SLAC). He received his Ph.D. at the Massachusetts Institute of Technology (1956). He moved to Stanford in 1956, first to Stanford University (1956–1963), then to SLAC, where he was Associate Professor (1963–1967), Professor (1967–present), Technical Director (1982–1984) and Director (1984–1999).

He received the Nobel Prize in 1976 together with Samuel Ting "for their pioneering work in the discovery of a heavy elementary particle of a new kind". He is a member of the National Academy of Sciences of the U.S.A. (1977), the American Philosophical Society (2003), the American Academy of Arts and Sciences (1989), and the European Physical Society, and is a Fellow of the American Physical Society (of which he also served as President) and of the American Association for the Advancement of Science. He received the E. O. Lawrence Medal of the Department of Energy (DOE, 1976). He has served on the boards of directors of several companies. He was President of the International Union of Pure and Applied Physics (1999–2002), and is a member of Le Haut-Commissaire à l'Energie Atomique Visiting Committee (CEA) and chairs the DOE Nuclear Energy Research Advisory Committee, Subcommittee on Accelerator Transmutation of Waste.

We recorded our conversation in Dr. Richter's office at SLAC on February 17, 2004.*

---

*Magdolna Hargittai conducted the interview.

Let's start with your family background. I would also like to know how you became interested in science and how you started your life in it.

I am the first of my family to have been born in the United States. My father came to the United States from Austria when he was eight. My mother came from the area near the Poland-Russia border when she was about five or six years old. Things were not so good for Jewish people in the old country, and both families felt it would be better if they left. My father's family settled upon the lower east side of New York. They were relatively poor, and my father had to work to help support the family. He ended up going to law school at night. He wanted to be a physician, but he couldn't go to medical school at night. My mother's family was better off; they lived in Brooklyn. I don't really know a lot about their background because my grandparents on my mother's side didn't speak English. On my father's side they did, so naturally my sister and I talked much more with our paternal grandparents than with our maternal ones. I got interested in science as a child, and my father, because of his background, wanted to make sure that his children could do what they wanted to do, that they didn't have to do things to support the family, and that they didn't have to go into the family business. He and his brothers set up a business and it was relatively successful, but my father never liked it; he was not happy in that kind of work; and he wanted my sister and me to do things that we enjoyed.

I had a rather extensive science laboratory at home when I was a kid. A friend of mine, Micky Wolf, and I were science freaks. Micky lived in an apartment, so he didn't have any room for science gear. We lived in a house with a basement, and that's where our laboratory was. At that time we could do things that kids now can't do. The chemistry outfit we put together would now be regarded as much too unsafe for any young person to play with. We did a lot of exploration in biology, too, with a cheap microscope. For some reason I recently looked at the prices of microscopes students could use these days and the equivalent to what I had now costs three hundred dollars. The one I had was a used one and it cost five dollars back in the early 1940s. Maybe today's microscopes are fancier, but we could do an awful lot with that cheap microscope. There were also a lot of other opportunities. One other thing that occurs to me often, and I talk about it when I talk to young people about science, is that I had something to intrigue me that they don't have: I could see the stars. Nobody today who lives in an urban area has ever seen the

Burton Richter (center) with Stanford University Physics Nobel laureates Robert Hofstadter (left) and Felix Bloch (right) at the party celebrating Richter's Nobel Prize in 1976 (photograph by R. Muffley, courtesy of SLAC Archives).

sky in all its glory. They see a rather dull thing, but this blaze across the heavens is invisible unless they go to Yosemite, or camping in the mountains, or do something similar.

So I was an eclectic science student. I went to Far Rockaway High School, but I did not know Richard Feynman and I did not know Baruch Blumberg; the three of us are Far Rockaway High School's three Nobel laureates. Feynman was much before me and I don't remember when Blumberg was in high school there. Far Rockaway was a very peculiar place: it was almost idyllic and was as far away from Manhattan as you could get and still be part of New York City and not be on Staten Island. Far Rockaway during the wintertime was like a small town, with a population of ten or twelve thousand. During the summertime, people moved out from the city to get away from the heat and to have a sort of vacation. We had this relatively small-town life, but you could get onto the Long Island

Railroad and head to New York and be in the heart of Manhattan forty minutes later. When I was a teenager, we could go to Greenwich Village to jazz clubs, and still have our small-town life.

I moved from Far Rockaway High School to Mercersburg Academy when I was a junior in high school. My father was worried that Far Rockaway High School wouldn't be good enough to get me in to where I wanted to go. I already knew I wanted to go to MIT, but I didn't know whether I wanted to do physics or chemistry, and I didn't figure it out till I actually got there. The Mercersburg Academy was an absolutely terrific place with very good teachers and broad education, not only physics and mathematics, but English and literature. I was a disaster in foreign languages, and I still am. I went up to MIT in 1948.

At that time MIT was still recovering from the Second World War. I entered as a freshman and we lived in the barracks that had been built for the influx of people from the military to get technical training. The first room I lived in was this great big barracks hall, where 24 of us lived. We rearranged all our lockers so that we would have a study area. It was an interesting life. That's where I confronted the issue of physics or chemistry. When I took my first physics course, it became clear to me that I wanted to do physics and not chemistry. I had a not very sophisticated vision as a child: I wanted to know how the universe worked. When I got to MIT and started doing both chemistry and physics, I very quickly came to the conclusion that doing chemistry would teach me how a lot of things worked and how you could create all sorts of wonderful materials, but not how the universe worked. For that, I had to do physics.

My experience at MIT was slightly atypical. First of all, I was lucky that in my second year, I was among the first to be moved into a brand-new dormitory that'd just been finished. That dormitory was simply gorgeous compared to the barracks. The courses were interesting, but I went through some troubling times and frustration as a student about whether I really wanted to do science. In the summer of my junior year I prevailed on one of my professors, Francis Bitter, to give me a job doing research in his laboratory. Bitter was the developer of very-high-field magnets that were used for research. He had a ten-tesla magnet and a six-tesla magnet in his laboratory. At the time these were among the most powerful magnets in the world. I became a gofer, "go for things", for Francis Bitter. I also met some very interesting people at that time including Professor Martin Deutsch, who was doing his experiments on positronium in Bitter's lab. I became Deutsch's extra set of hands because I knew where everything in the lab was when Deutsch and his people came into Bitter's lab to use one of his big magnets. Deutsch was an interesting person; he was rather kind to me, in spite of my being a junior, because I was not supposed to know all sorts of things that his graduate students were assumed to know. He was tough, and sometimes even mean, to his graduate students. I think Deutsch perhaps considered me to have a moldable young mind, and he helped me learn a great deal of physics. Generally, however, I found my junior year at MIT rather depressing, and another teacher, Francis Friedman, was very important in pulling me out of that. After I got my bachelor's degree, I decided to stay at MIT. I was one of the few very lucky graduate students whose support came directly to the student rather than through a professor. That meant that I could do whatever I wanted to do.

I started off in Bitter's magnet lab continuing some work I had done when I was an undergraduate. That work actually led me to change fields. We were working on determining the magnetic moments of mercury isotopes. There is a whole string of these mercury isotopes, and the question was, what was the nuclear physics that varied the nuclear magnetic moments of these isotopes as the number of neutrons in the nucleus was changed? One of my jobs — a very interesting job — was a sort of reverse alchemy, to make mercury out of gold. I took gold to the cyclotron that we had at MIT, and bombarded the gold with deuterium from the cyclotron. Once this was done, I rushed with the sample to a quartz apparatus and distilled the mercury out of the gold to get the relatively short-lived mercury isotopes. Then I rushed the distillate over to Bitter's lab, where we would use the mercury in an apparatus that I'd helped build. It was essentially a scanner which used the normal Zeeman Effect to make a variable-frequency light source, and we would measure the hyperfine structure of the spectrum.

Eventually, I found what was going on at the cyclotron to be more interesting than the stuff that I made. When I was in my second year of graduate school I decided that what I was doing was not for me. At that time I was an NSF Fellow, so the money came to me and I did not have to find a faculty member who would use his research money to allow me to experiment in a new area. Bitter was very nice and helped me find my way. I talked with Professor David Frisch, whom I had known from course work and from hanging around the cyclotron. Frisch arranged for me to spend three months at Brookhaven National Laboratory to see whether I liked high energy physics. I spent a three-month period at Brookhaven: January, February, and March. January and February are the worst possible times to be there. I knew all about it because I'd grown up on Long Island, but this was my opportunity to find out about high energy physics. After those three months I came back convinced that particle physics was the right thing for me. I looked around to find a professor to work with, and found Professor Louis Osborne. I spent the next two and a quarter years with him in particle physics, completing my Ph.D. Along the way, besides particle physics, I learned a lot about accelerators. The MIT synchrotron where I did my research was a relatively small machine (it had a diameter of about a meter) and four graduate students ran it. I was responsible for the magnets and the correction system. One of my colleagues was responsible for the control system, the third was responsible for the radiofrequency accelerating system, and the fourth was responsible for the main power supplies. The four of us had to maintain that machine and keep it running, and we had to learn about accelerators enough to do this. I never forgot it and accelerators and high energy physics became central to my career.

Did you also learn machining and building apparatus?

At MIT, in the junior laboratory, you had to build things. Then, as a graduate student, and in Bitter's laboratory in particular, you built things. I learned to use machine tools. At that time an experimentalist had to become a fairly good engineer because one designed and built a large fraction of the apparatus that one used. Today's graduate students are not good engineers because the apparatus has tended to become larger and more expensive, and is designed and built by professionals. However, they are wizards at computers. I am not a wizard at computers although I can use them. The skills you learn as a scientist depend on the science you're doing and the time when you're doing it. I learned to be not only an engineer, but also an electronics expert. When I came out from MIT to Stanford, I was more of an electronics expert than anybody in the laboratory. Anyway, I got my Ph.D. under Lou Osborne; I met a lot of people in high energy physics at MIT, and I learned a lot about accelerators.

How did you decide where to go after your Ph.D.?

I had been thinking about physics a lot as a student. At that time quantum electrodynamics was not really understood. Feynman, Schwinger, and Tomonaga hadn't done their work, and there was a real question about the validity of quantum electrodynamics at very short distances. I had job offers from Stanford, Chicago, Columbia, Cornell, and I think there was one from Princeton, too. At MIT they didn't want to take their own graduate students. I had an experiment in mind that I wanted to do and Stanford was the best place for it. So I came to Stanford, and never left. The time was 1956, and this place was gorgeous. People would talk about the beauties of the Santa Clara Valley and the Vale of Kashmir in the same paragraph. I decided if I could get my research funded, I would be crazy not to come out here. I got the funds for the research I wanted to do, so I started out to do electron-electron scattering at a large momentum transfer. But the apparatus didn't belong to Wolfgang Panofsky who had hired me; it belonged to Robert Hofstadter. Hofstadter had his own ideas on what he wanted to use the apparatus for, and he didn't want to spend a lot of time on electron-electron scattering. One of the graduate students, a theorist, J. D. Bjorken, and one of the professors, Sidney Drell, were working on a book on quantum electrodynamics, and Drell said that it was possible to find out about short-distance behavior by looking at large-angle pair production: the study of electron-positron pairs produced by gamma-rays, where either the electron or the positron comes off at a very large angle relative to the gamma ray beam. I decided to build an experiment for this, which I designed, and I also modernized the electronics of what was called the High Energy Physics Lab. I was shocked that I was the most sophisticated electronics person there as far as detector electronics was concerned. Their previous electronics guru had left to go someplace else. I did the experiment which became at the time the most stringent test of quantum electrodynamics that had ever been done.

I was invited to give a paper at the meeting of the American Physical Society in New York and my father came to attend his son's talk. It was an invited session; it was held in the biggest ballroom in a big hotel; and there must have been three thousand people sitting in that ballroom. Every seat was taken, but not because of me; I was the second speaker. The first speaker was Bob Hofstadter and he was just announcing his results on the structure of the proton for which he received the Nobel Prize later on. He gave his talk, there was the usual discussion, then the chairman said, "I now call for the second paper," and two thirds of the people got up and left. My father was furious, "How could they do this to my son?" I was rather impressed that a third of them stayed. I soothed him afterwards and allowed him to take me to an especially good lunch.

Along about that time came colliding beams. G. K. O'Neill of Princeton had been thinking about colliding beams. Instead of smashing a beam into a stationary target, he proposed colliding one beam with another beam moving in the opposite direction. For a beam striking a stationary target, the center-of-mass motion meant that the available interaction energy only went up by the square root of the beam energy. So if you took the then 30-billion-electron-volt energy of the accelerator at the Brookhaven National Laboratory, then the highest-energy accelerator in the world, and you put that onto a stationary target, you would get roughly the equivalent of 7 or 8 billion electron volts of interaction energy. If you used a second 30-GeV accelerator beam as the target, you would get 60 billion volts of energy available. So if you build a second machine of the same size as the first, and collide the beams, you would get eight times the collision energy. Twice the cost gives eight times the bang. People at the University of Illinois had been thinking about this also, and they and O'Neill were the pioneers. However, nobody could figure out quite how to do it. One of the reasons was that we didn't really know enough about beam dynamics back then compared to what we understand today.

O'Neill had the idea of doing the experiment with an electron machine instead of a proton machine like that at Brookhaven. What made life relatively easier for an electron machine is that the electrons radiate synchrotron radiation, losing energy in the process, but that energy can be restored with a radiofrequency acceleration system. This gives rise to a phenomenon called radiation damping. You can think of an electron oscillating around an equilibrium orbit. The energy it loses is lost in the direction of its motion, while the energy put back with the radiofrequency system is only the straight-ahead component. The result is that the transverse component of the motion shrinks and the particles form a small beam at the center of the accelerator's vacuum system. A lot of the problems present in a proton machine which would not have radiation damping would be non-existent in the electron machine. Of course, it's not quite that simple, but this was the general idea. So O'Neill came to Stanford because he thought that the Stanford linear accelerator was the best machine for his ideas. He convinced Panofsky, who was the laboratory director, who then recruited me, a postdoc, Walter Barber, a staff member, and B. Gittelman, also a postdoc then and now at Cornell. We built what became known as CBX, the colliding-beam experiment at Stanford, which was a joint Stanford–Princeton project. Panofsky convinced the Office of Naval Research to fund the project. It was at the time the largest-ever physics experiment, costing eight hundred and fifty thousand dollars. It started a revolution in accelerators, and now almost all high energy physics accelerators are colliding-beam machines. Ours was the first; we had a lot to learn; and we learned a lot.

I became the person responsible for the whole magnet system; Gittelman took on the vacuum system; Barber took on the injection system; and O'Neill himself took on the radiofrequency system and some special features of the injection system. This was a bold adventure. This machine was supposed to be not only an experiment to demonstrate the feasibility of colliding beams, but to do physics. The beam current we achieved circulating in that machine was six tenths of an ampere, an intensity not exceeded for more than twenty years. There were other people at that time playing around with storage rings and colliding beams, but I think it is fair to say that ours was the first to confront all the issues that had to be understood to create colliding-beam facilities that could do advanced high energy physics experiments.

Our machine worked, but I had begun thinking about what I was sure would be a better way to do physics, an electron-positron colliding-beam machine. A group of postdocs and graduate students had asked a theorist, J. D. Bjorken to give us a course on calculating with Feynman's new quantum electrodynamics. I always believed that if you couldn't do theory, or didn't at least have an understanding of it, you would be only a technician. I always was interested in theory. When I started my graduate studies, before I settled in with Bitter, I actually did a theory problem with Francis Friedman to see whether I liked experiment better or theory. So I was thinking about the possible outcome of the electron-positron annihilation experiment. I thought that I could figure out what I called the relative structure of the strongly-interacting particles. I could produce pi-mesons and K-mesons for example. Their production would depend on what their structures were, and this way I could do something for these unstable particles like what Hofstadter had done for the proton. I believe I was the first to realize this but I never did write it up for publication. Others eventually had the same idea and began to build such machines.

The Russians at their lab in Novosibirsk and the French at their lab in Orsay started building machines that were capable of making real physics measurements. At Frascati in Italy, a higher-energy machine was planned that eventually came into operation in 1968. I was interested in building a high-energy electron-positron colliding-beam machine. I started working with Professor David Ritson of Stanford University, who was also interested in accelerators, and we began the design of a high-energy machine capable, in principle, of measuring my structure functions. To me and to Ritson this meant two things: to have enough intensity to do the experiments and to have enough energy to get beyond threshold effects to be able to see what I called relative structures. The Italians were using too low energy (1.5 GeV), and we decided to use 3 GeV.

At the time (1963) I was an Assistant Professor at the Physics Department at Stanford University and Panofsky made me an offer I could not refuse. He offered an Associate Professorship at SLAC, and also offered me the direction of the high-energy colliding-beam program. I was 32 years old at the time, I was one of the world's real experts, and I had this whole

From left to right: Gerson Goldhaber of the Lawrence Berkeley Laboratory, Martin Perl and Burton Richter of SLAC meeting in late 1974 in the control room of SPEAR (photograph by W. Zawojski, courtesy of SLAC Archives).

program in my hand. If I look at what we do with the younger people now at 32 years of age, it would be very unusual to have that kind of responsibility. One reason I suppose is that the costs have gone up. The question today is are you going to give somebody without much experience a project costing hundreds of millions of dollars? Also, experiments have become much larger and more complex. So can you put a 32-year old in charge of a project with 600 collaborators, including a considerable number of full professors? Perhaps my experience is hard to duplicate today, but the laboratories should pay more attention than they do to bringing along the next generation of leaders.

Our high-energy electron-positron collider was called SPEAR. It took a long time to begin because it was very difficult to get funding for it. The first proposal was submitted in 1964. By 1969 I was close to giving up. I wanted to do physics, and although I was doing some experiments with the SLAC machine, it was a frustrating period. I decided to make one more try, and, by divine bureaucratic intervention, the Comptroller of the Atomic Energy Commission became fascinated with our project. The problem was getting a line into the budget of the Atomic Energy Commission for a construction project. That required specific Congressional approval and, before that, you had to have specific administration approval. The Comptroller asked Panofsky a leading question: was the project a facility or an experiment, and Panofsky's response was that it was an experiment. He asked a second question: was there a special permanent building for this experiment, and the response was that there was not. At this point it was determined that there was no need for a specific line in the budget because it was not a construction project, and the regular budget could accommodate it. I never met John P. Abeddesa, but I will never forget him.

We had approval in October 1970, and twenty months later the machine was ready for operation. The team was terrific; John Rees came out here from Cambridge to be my partner. The Italians produced early results from their ADONE storage ring in 1969, and four different experiments gave inconsistent results. John Rees and I visited Frascati, and looked at the data, and I told John that their experiments disagreed with each other by a factor of two and disagreed with theory by a factor of one hundred. I knew that we could straighten out the situation only if we built a detector that would see everything coming out of each interaction. Budgets were tight and the government started putting pressure on us to reduce costs, but I told Panofsky that we could not give up the magnetic detector.

The two "November Revolutionaries", Burton Richter and Samuel Ting, during a symposium in 1984 at SLAC (photograph by J. Faust, courtesy of SLAC Archives).

We built both the machine and the detector in record time and started our experiments. Then came what is still called the "November Revolution" (1974), and the rest is history. First was the discovery of the ψ (psi) (J at Brookhaven) and all the particles that went with them. These particles were not allowed by the old three-quark model, which even then was known to be incomplete. There were many competing explanations of how it might be improved and the results of these experiments eliminated all but one possibility. With that the Standard Model began to take real form. We had just begun to straighten out the Standard Model when Martin Perl, using data from the same experiments, found the *tau* lepton, and that said that if our new Standard Model was right, there ought to be a third family, which was eventually found at Fermilab and Cornell, and here we are.

How did you know at what energies to look for particles?

This is a wonderful story. What we did was first a sparse scan in energy over a broad range. At each point we did many runs, and found that at one particular point the results did not seem to be repeatable. We chased the apparent problem for months. A lot of people of the group were involved with analyzing the data. Finally in November the machine was scheduled to run for Bob Hofstadter who was to do some experiments that were interesting but not earth-shaking. At that point Gerson Goldhaber, one of my senior collaborators on the experiment, came to my office and told me about an excess of K-mesons in the funny data points. I immediately rescheduled the program of the machine; decided to run the machine over the weekend to check this out. We started the run on Friday; by Sunday morning the data were still pouring in, and I did something I'd never done before, I sat down and started writing the first draft of the paper. I knew what was happening and so did everybody else in the group, 28 of us. I wrote the first draft and I don't remember who wrote the second draft, it may have been Goldhaber or G. Trilling. In the meantime, the word had gotten out that a new state with properties not allowed by the standard theory had been found. Information sometimes seems to travel faster than the speed of light, and people started phoning in. By evening the paper was written and the only thing missing was some drawing. I didn't know about Sam Ting's work at that point until Sam showed up the next morning for a committee meeting at SLAC and told me that he had some wonderful news to tell. I told him we had some wonderful news, too. It turned out that we had both seen the unexpected and unallowed particle in different experiments done 3000 miles apart. This was the reason for the instant acceptance of our findings. They came from two independent experiments, one done at Brookhaven and one at Stanford using two totally different techniques. Our two papers were published in the same issue in *Physical Review Letters*.

We continued our experiments. Our machine was capable of scanning the energy spectrum and we turned a big IBM computer into a real-time data analysis machine, the first time one of the big computers had been used in that way. We also did some model calculations to predict the energies where one might look for the next particle. We looked, and the particle was there. There immediately followed an intense interaction between the theorists and the experimenters. SPEAR was a particularly flexible machine so different hypotheses could be tested rapidly. Within a few months the leading explanation was that we had found a bound state of a new quark and its anti-particle. If that was so, there should also be bound states of the new quark with some of the old quarks. In 1975, Goldhaber found the so-called D-meson in the data, which was one of the predicted states. Experiment and theory systematically confronted the collection of possible explanations and the only interpretation that remained valid was the fourth quark.

So the Standard Model was greatly strengthened by your discovery.

I would say that there were several experiments at the foundation of the Standard Model. One of them was a CERN experiment on neutrino interaction in CERN's heavy liquid bubble chamber. Initially the experimenters themselves were skeptical of the explanation of their data in terms of what are called weak neutral current interaction. It took a lot of analysis and a long time before they accepted it. There was the deep inelastic scattering experiments of Taylor, Friedman, and Kendall at SLAC which showed that quarks seemed to be real; the psi/J experiments by Sam Ting and by us, showing the bound state of the quark; and finally the discovery of the *tau* lepton, showing that there was a third generation. All this happened very fast, between 1970 and 1975. Since that time, the Standard Model remains unshaken, driving all of the theorists and experimentalists to frustration. The situation is not that different from that at the time of the psi/J discoveries.

Burton Richter receiving the E. O. Lawrence Award of the Energy Research & Development Administration, 1976 (photograph of ERDA, courtesy of SLAC Archives).

The Standard Model itself is known to be incomplete. The theorists make wilder and more obscure explanations for what is beyond the Standard Model. New data are needed and we all hope that with the LHC at CERN soon to come on, something dramatic will come along that will prune away all the weeds and leave us with the one flower instead, that is, with the New Standard Model.

So you also believe that there must be something more fundamental beyond the Standard Model.

There has to be. The simplest Standard Model, when extrapolated to higher energy than is now accessible, is not consistent with some fundamental limitations. For example, it predicts certain reaction probabilities to be larger than one, an obvious impossibility. My own feeling is that we are never going to come to the end of the quest for a final answer to everything, because philosophically, we can never prove a negative. You cannot prove that there is nothing beyond the present knowledge when one reaches regions that are not accessible to experiments today. We still have much to learn. I have a lot of trouble with people who invoke the anthropic principle. It would say that all of the constants of Nature have to be what they are because if they weren't what they are, we wouldn't be able to exist and to speculate about them. This is an observation, not an explanation. It has no power to tell you why something is so. Our goal before theorists became frustrated with their inability to unify quantum mechanics and general relativity was to explain everything in the universe in terms of a small number of equations and a small number of arbitrary constants. Most of us have kept to that goal, but haven't made any progress in a long time because of a lack of data. There is nothing known about what is beyond the Standard Model other than knowing that there is something beyond the Standard Model. For me the anthropic principle with the many-universes model is a surrender, saying that we can't explain anything anymore.

Will the Large Hadron Collider, LHC, provide new experimental proofs?

We certainly hope so. A lot of the theorists believe in a model called supersymmetry, which predicts new, observable phenomena. Personally, I think it is a silly model because it sharply increases the number of fundamental entities rather than reducing them and simplifying the picture. A lot of

Burton Richter in 1984 (photograph by J. Faust, courtesy of SLAC Archives).

the theorists believe that LHC will have enough energy to produce the particles that are predicted by the super-symmetry model. I hope it will have enough energy to tell us what is beyond the Standard Model, whether it's super-symmetry or something else. If we come up empty, I think the LHC may be our last big accelerator. The LHC costs five billion dollars to build and the next would cost even more.

Couldn't astrophysical observations bring in some information?

They're bringing a lot of information in now, which makes the questions more complicated, but they haven't given us the answers.

What is dark matter and dark energy?

These are exactly the questions that we don't have the answers to. If the LHC turns up super-symmetry particles, they may be the answers. I personally wish people wouldn't be quite so certain that there is a phenomenon called dark energy. I would like them to be a little bit more questioning about it, but it's very exciting. I would like to see the results from the proposed experiment called the Joint Dark Energy Mission, JDEM, to look in greater detail at the supernovas that gave the first hints of dark energy. This will be an experiment of much higher precision than the ones at Berkeley and Harvard that showed the first evidence that the expansion of the universe was speeding up. This will be a big job involving NASA, DOE, and possibly the NSF. Given NASA's budget troubles, I don't know when it will be off the ground, but it will be an important experiment.

Coming back to your discovery, the Italians and others had missed out something big.

They did indeed. As soon as the word got out about our experiments, the Italians and the Germans cranked up their machines to higher energies, and our observations were confirmed around the world. But the editors of *Physical Review Letters* refused to print their papers in the same issue that carried Ting's paper and ours as the discovery papers. They were published in the next issue.

The Nobel Prize came very soon after your discovery.

It came very soon, just about as soon as it could come.

How did it change your life?

It's hard to say. The first year was frantic. Nobody takes the advice of previous laureates, but my advice would be, you must learn to say no, otherwise you get consumed by requests. Then, after things calmed down, I asked myself the question, what mountains are there left to climb? If I hadn't got the Nobel Prize, would I have switched to science administration? I don't know. Two Nobelist friends of mine, Charlie Townes and Bob Hofstadter, never switched to administration. I did.

Why?

It's fair to say that I am one of the pioneers in the development of the colliding-beam technique, not only with the machines, but with the detectors that collect the data. Our apparatus at SPEAR was the first to fully surround the event and thus be able to see all the particles produced in each event. That is now the standard way of using these machines. However, experiments were getting bigger; groups were getting bigger, and I didn't just want to run a huge group of people who were collaborating on an experiment. I wanted to build things that would illuminate the important issues of physics, but it got to a point where the machine builders, the detector builders, and the data analyzers were separate groups. That was not my style. So in 1984 I agreed to become Panofsky's successor as Director of SLAC. It was a job where I was responsible for finding the funds for

Burton Richter about to swing at a pitch during the annual Theory versus Experiment Softball Game, Stanford, 1990 (courtesy of SLAC Archives).

and running an operation that let others do the physics they thought important. But it also gave me the opportunity to advance things I thought were important. For example, in 1978 I had done something with two other people, Alexander Skrinsky of Novosibirsk and Maury Tigner of Cornell University. We'd met at a seminar on the possibilities and limitations of very-high energy colliding-beam systems, and discovered that we were all thinking about the same thing: what would come after the colliding-beam storage ring? When I was on sabbatical at CERN in 1975–1976, I wrote what I think was a very important paper on the scaling laws for electron-positron storage-ring colliders. In those laws I set up the parameters of what would become the LEP, but I also concluded that storage rings for electrons and positrons had costs that scaled as the square of their energy. At the seminar I mentioned, the three of us set out the basic design equations for a new type of colliding-beam device called a linear collider. This kind of collider had a cost scaling law that went as the first power of the energy. Back at Stanford in 1978, I started the design of a Linear Collider based on the existing SLAC linear accelerator. As lab director I could, and did, make it happen. I am glad to say that the machine worked as predicted and the world high energy physics community is now pursuing a linear collider of much higher energy that would be a complement to the LHC at CERN.

Did you achieve your goals as director?

I like to think so. As director, I had to worry about three things: What is the Lab doing now? What's the Lab going to do in five years? And what's the Lab going to do in ten years? The Lab has to be on the frontier, has to be a center of excellence, all the time. What we are doing now has to be truly excellent if we are to get the support to begin preparing for what we will be doing five years from now. That in turn has to be inspiring now and done well in the future to support the R&D necessary for the work we will be doing ten years down the road. That was my philosophy for the 15 years that I was director. I stepped down in 1999. I was fortunate in that SLAC was an integral part of Stanford University. Many of the ideas that are the basis of our science program today came from other university faculty. SLAC has enabled their research and their research has enriched SLAC's program. This is a case where the whole is more than the sum of its parts.

How do you see the future of high energy physics?

It has to answer the questions about the fundamentals of matter and energy. The biggest questions facing physics are about dark matter and dark energy. To answer these questions we need space experiments. These are non-accelerator experiments. There are important experiments involving accelerators. They may answer questions like — just to give you an example — whether neutrinos have CP violation or not. For this we need neutrino beams that are ten times the power that we have now, requiring accelerators that put two- to four-megawatt proton beams onto the target. These would be much cheaper accelerators than the LHC, costing only a few hundred million dollars. The LHC, as I mentioned earlier, is the most important facility right now. A large linear collider would be its companion. With the LHC and the linear collider we expect to see what's beyond the Standard Model. If the LHC does not turn up anything new, high energy accelerators are probably finished because the cost of an accelerator much more capable may be too large for society to fund. There may of course be better acceleration techniques, lasers or plasma wake-field for example. Colliding-beam technology revolutionized high energy physics. There is no way the interaction energy of the LHC could be achieved with the old fixed-target technique. Young scientists are getting on with R&D on new types of accelerators. I hope they succeed.

Is particle physics the same as high energy physics?

It's the same. The people who call themselves high energy physicists are particle physicists who live on a moving energy frontier.

High energy physics involves the work of large teams. How does individual contribution figure in the teamwork?

If you look at how the work actually gets done, it takes the team to run the facility. But in the analysis of the data, it is small subgroups of the larger collaboration that do the work. Today, author lists are typically done alphabetically by institution and alphabetically within the institutions. At SPEAR we followed the practice of having the people who did the analysis in the front and then the rest of the authors in alphabetical order. The exception was if a paper was a graduate student's dissertation. Then the first author was always the graduate student. However, nobody else followed in our footsteps. I would like to see all the two thousand people on the first four papers coming out of the LHC experiments because it took the efforts of all two thousand to build the apparatus and make it work. After that people should take responsibility for their own actions. Those who do the analysis should be on the paper with an acknowledgement to the rest. This would make it possible for invitations for talks to go directly to the individuals rather than to the speakers' bureau typically associated with a large collaboration, as the practice is now. Even more subversive of the present order, I would like to see all our data put in to the public domain, like NASA does. In that case anyone could work on extracting the physics, even things that are there but not on the priority list for analysis.

Please, tell us about your family.

My wife and I have been married for almost forty-six years. We have two children in their forties; one of them became a scientist (our son got his Ph.D. from Stanford in Applied Physics and went into industry).

Do you have heroes?

As a physicist, my biggest hero has always been Fermi. He was the last person who was both an experimenter and a theorist. After Fermi, life became too complicated. Dick Feynman is also a hero.

Are you religious?

No.

Did you ever experience anti-Semitism?

Sure. Anybody who grew up in the 1930s and 1940s did. It was a very different time from now; there was a lot of visible and active anti-Semitism. I went to a very peculiar elementary school in Far Rockaway, where there were three groups, the blacks, the Jews, and the Catholics. It's a funny way to describe the situation as one racial and two religious groups, but that's the way it was. In the schoolyard, the two minorities, the blacks and the Jews, had to be allies because otherwise the Irish Catholics would have given us much too hard a time. After school, we never saw the blacks because they lived in a different region. We had enough of the young Catholic kids whose minds had been poisoned by bigoted parents and priests, so we had a fair amount of trouble. When my father was trying to get me into a prep school for my junior year in high school, I found out about the quota system. There were only a certain number of spaces for Jews, and if the spaces were filled, it was impossible to get in. My father felt this rejection from some of the schools rather keenly. I cannot know whether Mercersburg Academy was just better than the others or whether their quota wasn't filled, but I have to say that I didn't see anti-Semitism at Mercersburg although there was a big Christian chapel there.

It's hard to miss that there are so many Jewish Nobel laureates among your generation. Any comment?

I explain this in terms of what was important to the parents. The Jewish community has always regarded scholarship as very important. The rabbi is a scholar. "My son, the doctor," says the Jewish mother in the joke. It's serious. The question is what was open to Jews during historical times, when Jews were not allowed to do many things in Europe, or even in the

United States. They went to scholarship. I see in young Asian kids here in California the same family attitude toward the importance of scholarship and education. The valedictorians in high schools in California are usually Asian girls with the Asian boys not very far behind. So, returning to your question, I don't think it's a Jewish thing, rather, it has to do with family life and what families regard as important for their kids because that's what the kids will think is important.

Edward Teller died recently. He lived nearby. Did you know him?

Edward and I talked regularly when he was here at Stanford. But, this was after his active days. He wanted to keep up with what was going on in physics and would call me periodically. I would go over to Hoover Institution or to his house spending a couple of hours with him on such visits. He liked me and he didn't like many people. I read his autobiography and I found it fine up until the Oppenheimer hearing after which it becomes self-serving. His tone changes completely and he becomes defensive. He lost many of his friends of his own generation because of his attitude toward Oppenheimer, but many younger scientists worshipped him. What he says in his memoirs about what he said at the Oppenheimer hearing doesn't ring true.

Do you think his testimony was important for Oppenheimer's fate? According to many, Oppenheimer would have lost his clearance in any case.

It was certainly his testimony that gave a bunch of people the cover they needed to declare Oppenheimer a security risk. I thought even back then when I was much younger that Oppenheimer had no business as chair of the General Advisory Committee to put his personal morality in front of his advice. The chairman of the general Advisory Committee has to answer the question posed to the committee. If he wants to say at the end that something should not be done, that's all right, but I thought it was very wrong to put his moral objections to going to the Super, as the hydrogen bomb was called, in front of giving frank and honest advice. The final decision had to be based on considerations like what are we going to do if the Russians are going to have it? I thought even at the time that they should have just not reappointed Oppenheimer, but they shouldn't have taken his clearance away. He was not a security risk.

Looking back, do you think it was right to push for the Super, given the fact that the Soviet Union was already building its own?

Going back in time, there are so many what ifs. I think the Russians would've done it anyway, even if the United States had decided not to develop it.

To conclude, when you were a kid, you opted for physics because you wanted to know how the universe worked. Did you figure it out?

Did I figure out how the universe works? No. Did I learn a lot about it? Yes. Looking back, I have no regrets.

Samuel C. C. Ting, 2004 (photograph by M. Hargittai).

30

SAMUEL CHAO CHUNG TING

Samuel Chao Chung Ting (b. 1936, Ann Arbor, Michigan) is Tomas Dudley Cabot Institute Professor of Physics at the Massachusetts Institute of Technology (MIT) and at the same time spends considerable amount of time at the European Organization for Nuclear Research (CERN). He received his Ph.D. at the University of Michigan (1962). He spent 3 years at CERN as a Ford Foundation Fellow, after which he went to Columbia University (1965–1969). He spent a year at the Deutsches Elektronen Synchrotron (DAISY) in Hamburg (1966). Since 1969 he is at MIT.

He shared the Nobel Prize in Physics with Burton Richter of SLAC in 1976 "for their pioneering work in the discovery of a heavy elementary particle of a new kind". Dr. Ting is a member of the National Academy of Sciences of the U.S.A. (1977), the American Physical Society, the Italian Physical Society, and the European Physical Society. He has received many awards and distinctions, among them the Ernest Orlando Lawrence Award (1976), the Eringen Medal from the Society of Engineering and Science (1977), and the DeGasperi Award of Italy (1988). He holds honorary degrees from many universities. We recorded our conversation in his office at CERN on May 15, 2004.*

---

*Magdolna Hargittai conducted the interview.

I read in your autobiography that you were born in the U.S. more or less by accident. What were your parents doing there?

My father was studying engineering and my mother studied psychology. They wanted to go back to China because at that time the war between China and Japan just started and most of the Chinese students who studied abroad went back home to help. I grew up in China in World War II, and because of the war I did not go to school. After the War, in 1946, we moved to Nanjing, which was then the capital. I started to go to school there but then the communists came. In December 1948 we moved to Taiwan and that's where my education really began; before that I always went to a school and after a short while we left. From 1948 to about 1956 I studied there. Then, since I was born an American citizen, I went to do my Ph.D. at the University of Michigan. First I studied engineering for one year, but by then I saw that I didn't understand that.

Did you start engineering because of your father?

No. One thing about my parents: they always let me alone; they never tried to push me in any direction because of tradition or other reasons. I guess this was because my mother was a psychologist. So at the beginning of the next term I had an interview with my advisor and I told him, I could not understand engineering drawings, so I better change. Then I started to study both physics and mathematics, and in 1959 I got a degree in physics and also one in mathematics and received my Ph.D. in physics in 1962. My thesis was on particle physics. The University of Michigan was extremely supportive of me; I got scholarships all the time and I didn't take any general courses, only the ones that were important for my degree.

Before continuing with your life stories I would like to ask about your mother. She went to college in the 1930s and eventually became a professor of psychology in China. Was that natural for a woman in China at that time?

No, not at all. It was very unusual. My grandfather, her father, was one of the earliest Chinese students studying in Japan. After he returned to China he participated in designing the Trans-Asia railroad and then he decided to join the revolution. Then somebody betrayed him and he got caught and was killed. That time my mother was only about 3 years old; they were living in a village. My grandmother, who was left with the child alone, decided to go to school — she was about 30 years old that time. She finished all the necessary courses and became a schoolteacher and that is how she could support my mother. But still, they lived under very difficult circumstances. Eventually my mother went to college and then to the University of Michigan to graduate school. After they returned home she became a university professor in China, and that was very rare at that time, especially because she came from a very poor family.

Let's go back to your science. What did you do after your Ph.D.?

First I came to CERN. After spending about six years in the United States, I thought it would be important to see what Europe was like. I came here in March, 1963 and was here for about a year. After that I went to Columbia; my thesis advisor, Martin Perl, wrote a letter of recommendation and with that I got there. The University of Michigan is a very good school, but what was most famous about it was its football, not its physics. While I was there, I was more interested in that than in physics, so I missed many classes because of that. The consequence was, of course, that when I came to CERN, I really didn't know much. There was a colleague, Giuseppe Cocconi, from whom I learned a lot. Then I went to Columbia. Its physics department was really very, very good; Lederman, Schwartz, Steinberger, C. S. Wu, T. D. Lee were all there. I watched what people were doing, and at that time there was an experiment trying to measure the size of the electron. They found that the electron actually has a size but according to the modern theory of quantum electrodynamics, the theory of Feynman, Schwinger, and Tomonaga, the electron should not have a size. So I decided, maybe I should repeat this experiment.

I went to Boston, to the Cambridge Electron Accelerator, and talked to the people there. The leader of the group suggested to me to join his group and repeat the experiment. However, I said, no, I want to do it in a different manner. That time I was just a young faculty member, nobody was interested in supporting me — and they should not have because I had absolutely no credentials. Fortunately, earlier at CERN I worked with many people, and two German colleagues suggested to me to go to Hamburg, where they just finished building their accelerator. I talked to Lederman, and he told me that I was crazy because I would never be able to do that experiment; these experiments are very difficult and people spend all their lives to learn such experiments. "You would never be able to do it. In fact, I bet you 20 dollars that in three years you are not going to have any results," he told me.

I went to Hamburg nonetheless. I found there a good group of people, young researchers, including a quite famous professor, Stuart Smith from Princeton — and we had this experiment done in six months! We showed with this experiment that the electron actually does not have a size. After this experiment I gradually started to measure electron pairs. I proposed a new experiment to CERN and also to Fermilab but it was rejected at both places: I was looking for a new particle. However, they rejected my proposal for two reasons: one of them was that at that time most of the theorists thought that the particle I wanted to find could not exist, and the other was that most of the experimentalists thought that this would be a too difficult experiment. The required sensitivity for observing the particle over background for one particle was over 10^{10}. After that I went to Brookhaven, we did the experiment and we found the new particle that is called the J particle.

Recently I visited Burton Richter at SLAC and he told me about their experiment with — what they called — the psi particle and that eventually you visited them and the two of you found out that you were talking about the same particle.

Three Nobel laureates at CERN: Carlo Rubbia, Samuel Ting, and Georges Charpak, 1992 (courtesy of CERN).

I was on their program committee. We found the particle in October, but we kept looking at higher energies and related phenomena. Then I visited him in about November and that's when we found out that we have the same thing. We found that, by doing two quite different experiments; we found the same particle and that's when we announced this together.

Did you expect the Nobel Prize for this?

You should never expect the Nobel Prize. Besides, I know many truly great scientists, who have never received the Nobel Prize but always think they should. And that makes them miserable. Also, as you know, from the day of the discovery to the day you actually get the Nobel Prize 20 or 30 years may pass by.

But in your case it was just two years. How come?

I don't know. I think they got overexcited about it. By the way, if you think you have to get it, it's better to get it early. I was 40 years old and for an experimentalist it is considered to be quite young. And it was very helpful to me.

In what way?

I remember, I received a phone call from Isidor I. Rabi. He said, look, Sam, from now on, people will support you more. That turned out to be true. It is easy to get collaborators, and there is more access to funding agencies. Experimental results in particle physics come from proposing new ideas and if you already have the Nobel Prize people tend to think, maybe this guy's ideas are not totally crazy.

Beside the fact that this particle you found was a new particle, what was the importance of this experiment?

After this particle was found, people at DESY and SLAC found a family of particles. Of course, there are hundreds of particles, so why is this particular group of particles so important? It is like if you suppose that the average life of people is, say, 60 years. Then, suppose you find a village somewhere, in which the average life of its inhabitants is 60 thousand years. You start to wonder, what's happening to these people? The idea of having only two or three quarks was not correct; there should be four. But once you found the fourth, you started wondering, isn't there a fifth? That was also found, and now we have six. Most people say that there should be only six — but of course, if you don't look for more, you will never know, so you have to look. These fit the Standard Model. But still, we should still look, are there more, does the quark have a size, does the electron have a size? We found that they do not have a measurable size; their size, if they have it, is less than 10^{-17} cm. When I did the first experiment looking for the electron size, we could say that it cannot be larger than 10^{-14} cm. During the past 30 years, we could move the lower limit of their size by 3 orders of magnitude.

Is there another, deeper layer of physics beyond the Standard Model?

When a new particle is discovered, many eminent physicists say that now we understand everything, while, in fact, we were just opening a new door. I think that a hundred years from now our understanding of the Universe may be totally different from what it is now. And if we don't do experiments, we will never know.

They are now building here, at CERN, the Large Hadron Collider, a larger than ever accelerator. That will probably make it possible to make many new discoveries. But what will happen if people won't be able to build even larger accelerators, maybe because there won't be enough money for it or for other reasons, what will happen with the experiments? There will, of course, be theoreticians, but a theory always remains a theory until it is proven with experiments...

You are absolutely correct! Until you measure something you can't know for sure.

So what will then happen with the experiments?

There are three possibilities. One is to find a new technology for acceleration. Today in these machines there are thousands of people working with them. The consequence of thousands of people working together is that somebody with creativity won't devote his energy to that. You pointed out a major issue of particle physics today. I would think that for traditional accelerator research we need to develop accelerator technology, to make them smaller, cheaper, and producing larger energies. There are then the underground

Alfa Magnetic Spectrometer (AMS) on the International Space Station (courtesy of S. Ting and NASA).

detectors and, finally, experiments in space. No matter how large the accelerators we build we'll never produce particles with such high energies as the ones in space. In the last ten years I have been doing experiments in space.

Could you, please, elaborate on this a little?

Over the last 40 years there have been many discoveries in space by measuring X-rays, gamma-rays, microwave background and by measuring neutrinos. Light-waves and neutrinos are particles without charge and without mass. But in space beside chargeless and massless particles there are particles with charge and mass; protons, electrons, positrons, different nuclei. These have never been measured. Of course, to carry charge, they have to have a mass, but if they have a mass, they are absorbed by the Earth's atmosphere. As the simplest way, you can use balloons, they go up to about 80 km from the Earth's surface and they do not stay there for a long time.

In the last 10 years we have been working on an experiment that puts a precision particle detector into the Space Station. We measure very precisely all the charged cosmic rays. We decided to put a magnetic spectrometer in space. A magnetic spectrometer in space, as you can easily imagine, is quite difficult because when you have a magnet it has north and south directions, but if you are in space, you are in trouble. So we managed to design a magnet in such a way that the field is entirely on the inside.

How large is such an instrument?

The first one we did was 5 tons, the second one 7 tons and its size is 3 by 3 by 3 meters. The first one was put on the Space Station some time ago, and the second one was to be put on the Space Station in 2006, but because we lost the Shuttle it will probably be 2007. We are planning to measure the whole charged spectrum very accurately.

Talking about theory, we can ask many questions. One of them is that, of course, everybody knows that a very large part of the Universe is made up of dark matter. What is dark matter? Nobody knows. Most of the theorists think — and they are not necessarily right — that they are neutrinos. If neutrinos are really the dark matter, then the collisions with them can produce an excess of charged particles, such as positrons. So if we suddenly see an enhancement of charged particles in the measurement then we know that we understood dark matter.

Another thing, that we can look for with these space experiments, is anti-matter. The Universe comes from the Big Bang and after the Big Bang there must have been equal amount of matter and anti-matter. The question is where the Universe is made out of anti-matter? We look for anti-particles, such as anti-helium, for example. We can also study the properties of cosmic rays and look for beryllium-9 and beryllium-10. This is not an easy experiment; in fact, from all the experiments that I have done, the most difficult was the Brookhaven J-particle experiment and this one is similarly very difficult. If it were done on the ground it would be very simple. But in space, once you realize that you lost one component, you lost the whole experiment; you cannot fix it. You lose a fuse and the whole experiment is gone. Also, during the Shuttle lifting from the ground the acceleration is 3g and therefore vibrations are very large. Once in space, the temperature goes from +40 Celsius to −40 Celsius every 90 minutes. We flew only once but that turned out to be very successful. During the first flight we discovered many new phenomena. They are new phenomena because nobody has ever been there before and the circumstances were so different from what we have on Earth.

What did you find there?

Many things. For example, helium isotopes. It has been known from balloon experiments that the distribution is 90% ^{4}He and 10% ^{3}He. What we found in space was that ^{4}He is distributed everywhere but ^{3}He is only distributed above the equator, about 400 km above the equator.

Why?

I don't know, it is very, very strange. We also found the ratio of electrons and positrons. You would expect that there were equal amounts of electrons and positrons. What we found was that actually there are four times more positrons than electrons near the equator region, up to very high energy. Probably this means that we do not quite understand the Earth's magnetic field.

How many people are working on this project?

498. We need different specialists, but about 90% of them are engineers.

We talked about anti-matter. We might suppose that equal amounts of matter and anti-matter were produced in the Big Bang. What happened to all the anti-matter?

There have been very few experiments concerning this question. In 1967 Andrei Sakharov postulated that if there is a very strong CP violation, then the proton, which is a stable particle, actually decays. If proton decays then you can predict that the anti-matter disappears (this is the so-called barion genesis) — but for that you need proton decay. But nobody has found the strong CP violation and nobody found proton decay. But still most of the theorists believe that this is the way to explain the disappearance of anti-matter. I have a different approach; I want to look for it.

Fitch and Cronin proved CP violation. Wasn't the amount they showed enough for this?

No, it was just a small amount; maybe one part in a thousand and that's not enough for proton decay. You would need much more than that. This is why they are all looking for this at the big accelerators at CERN, SLAC, and in Japan.

Please, tell me something about your family.

My wife is a psychologist. I have a son who goes to university this year. I have two daughters from a previous marriage; they are much older.

Do you live in the United States?

Not only. We have two houses, one in France, close to Geneva, and one in Boston. My family lives in Boston but for the summer they come here. I move frequently between the two places.

Which of your work do you consider the most important?

I would say three. One of them is the systematic measurement of the size of the electron; we never found a size. Another, of course, is the J-particle, and also a work that I did in Beijing, it was the discovery of the so-called gluon jet. That was basically when you have an electron-positron collision and only produce quarks, you have only two jets, but if you produce gluons, we found three jets. The fourth work that I consider important is the work I am doing now in space, looking for particles with charge and mass. It is extremely difficult. Every day I wake up I start worrying; did I do this right, did I put enough redundancies, did I do enough tests and many such questions. Because what we are doing here nobody has ever done before, so there is nothing to read about, nothing to compare with. We are trying now to put a superconducting magnet in space. This is a very international project: we have people from France, Italy, Germany, Switzerland, Spain, Portugal, China, Taiwan, Mexico, Russia, altogether 16 countries, supported by different European countries and NASA provides us the Space Shuttle. The use of the Shuttle is a safety issue and I talk with their people very often. They have exceptionally good engineers.

Do you have connection with Chinese scientists?

Yes. They are very eager, their economy is coming up and they are very interested.

China is an enormous country and there must be many good talents there. Still, even if there are many famous Chinese scientists, especially physicists, none of them became famous for work done in China. Why is that so?

The reason is obvious. There was the cultural revolution, a whole generation disappeared; there was no good support for science. But I believe that now changes have started to happen, if the political system will be stabilized, and the economy will continue growing, soon they will have their share in contributing to science. People in China always ask me the same question. I always tell them that long ago, before the 13th century, China was a very advanced society and made many contributions to science, all through the 16th and 17th century. I hope that with governmental support and with the growing economy, they will soon make their contribution to world science proportional to their population. I have talked with important people there, with the president of the Academy, the Minister of Science, the Prime Minister; they all are really, genuinely, interested in science.

Do you go to China often?

Quite frequently, two maybe three times a year.

How do you feel yourself, as a Chinese or an American?

I feel myself American. Before I first went to study at the University of Michigan, my father talked with me. He said that most of the students in the U.S. go through college in such a way that they support themselves. Therefore, he gave me only a hundred dollars. Of course, he didn't say that most of these students spoke English! I did not. But when I went there, they gave me a scholarship and that supported me. Basically I was educated at the University of Michigan. I feel at home in Ann Arbor. I go back there every year, visit the University and go to see a football game. Even though I grew up in Taiwan, I spent some time in Europe and in Boston, the place I am most attached to is Ann Arbor and the University of Michigan.

Do you have heroes?

Michael Faraday; he was a truly great experimentalist. I read about his life and he is the scientist I most admire.

Would you like to mention mentors who helped your career to shape?

I wouldn't call it mentor but if you go through your career, there are many people who help you along the way; who believe in you and who support you. I give you one example: the former head of NASA, who was the head of NASA for ten years. When first I went to visit him and described to him what I would like to do, in spite of the fact that I had no experience in space whatsoever, he decided to support us, and NASA invested quite heavily in this project.

Is there anything that you feel you missed or regret that you didn't do?

I don't like to regret things because blaming yourself does not help.

If you started your career today, would you do the same?

No. I probably would go into biology. I don't understand it but from my conversations with my daughter, I think that is a very interesting field and one in which there are great opportunities. I would not like to start particle physics now because of the social structure that we talked about before. It is too large, too expensive, and too many people are involved, and physics itself is not the dominant factor any more. When I first started my experiment in Hamburg there were only 4–5 people. Then you can know everyone involved, you can talk with them and exchange ideas. Now, with the experiment in space there are 498 people. This way you have to spend quite a bit of time with managing and that takes away time from science.

What would be your advice to young people who are interested in doing research in physics?

If you are interested in physics, you have to make up your mind first, what is the most important thing in your life. Everything else is secondary, because it is very, very difficult to do physics and to be a musician or to do art at the same time. The people I know in experimental particle physics, they live in total concentration of their subject.

That brings us to another subject and that is the possibilities for women in physics. What you just said implies that a woman should choose physics as her field only if she does not want to have a family.

No, I didn't say that. There was a lady-physicist, Madame C. S. Wu who was very successful and also had a nice family. It all depends on how you weigh things and how you manage things.

What are the elements of success in science?

I would imagine that part of it is intelligence, and that part you cannot control, it comes from your parents. Then, a sense of curiosity and not listen always to what other people tell you. Before I do an experiment, I always discuss it with colleagues, listen to their views and opinion — but I do not necessarily do what they say. But you need a perspective.

Is there anything else you would like to add?

No, I think I had a full confession today.

Martin L. Perl, 2004 (photograph by M. Hargittai).

31

Martin L. Perl

Martin L. Perl (b. 1927, in New York) is Professor at the Stanford Linear Accelerator Center (SLAC) in Stanford, California. He received his B.S. in chemical engineering at the Polytechnic Institute of Brooklyn (1949) and his Ph.D. in physics at Columbia University. He worked at the General Electric Company (1948–1950), at Columbia University (1950–1955), and as assistant and associate professor at the University of Michigan (1955–1963). He moved to Stanford in 1963 and has been at SLAC ever since. He served as Chair of the High Energy Physics Faculty at SLAC from 1991–1997.

He received the Nobel Prize in Physics in 1995 together with Frederick Reines of the University of California at Irvine "for pioneering experimental contributions to lepton physics" and in the case of Perl, "for the discovery of the *tau* lepton".

He is Member of the National Academy of Sciences of the U.S.A. (1981), of the American Academy of Arts and Sciences (1997) and was a member of the CERN Scientific Program Committee from 1991 to 1997. He received the Wolf Prize of Israel in Physics in 1982. He has honorary doctorates from the University of Chicago (1990) and from the Polytechnic University (1996). We recorded our conversation in Dr. Perl's office at SLAC in February 2004.[*]

First I would like to ask you about your family background and about the beginning of your career.

---

[*]Magdolna Hargittai conducted the interview.

I'm a first-generation American; my parents came to America at the beginning of the 1900s, in 1910 or so from the Russian part of Poland. My father came from a town called Pruzany, which is now in Belarus. It was part of the usual Jewish immigration to this country; to avoid anti-Semitism over there; it was a big family, and the sons wanted to avoid being drafted into the Russian army. My father was the most successful among his brothers and sisters; he went into the printing business, and by the early 1920s had his own printing business. I was born in 1927. My parents did pretty well, and by the 1930s, we were in the middle class, although the money was always tight, but for a while in the early 1930s we even had a live-in maid for a few years. She came from the coal regions of Pennsylvania. I was raised in the middle class in Brooklyn, New York, with good public schools. Living in Brooklyn was easy and safe; when I was ten or eleven I had a bicycle and could go anywhere, even to Manhattan.

I was a very good student and my parents were very practical people; they always thought that I should get into the professions. I didn't want to join my father's business; I found the printing business hard and messy. In high school I was interested in science, but the idea of becoming a scientist just never occurred to me; we didn't know what research was; we didn't know any scientists and knew of Madam Curie and Einstein only. So I decided to become an engineer. I graduated from high school in 1942 and although I took only one year of physics, I received a physics medal. However, I went into chemical engineering at the Polytechnic Institute of Brooklyn; now it is called the Polytechnic University. I could live at home, which was an advantage. By that time my father was doing some printing for companies in the chemical industry, so there was some connection.

I studied chemical engineering for a few years. Near the end of the war, when I became 18 years of age, I wanted to join the war efforts, but my parents thought that I was too young for that, but they did allow me to join the merchant marines. Actually, for a while I trained to be a cadet officer and went on a ship, a training freighter. When the war ended, I was drafted into the United States Army, and I served there for a year. The moment I was discharged from the army, I got back to Brooklyn Polytechnic and got my degree as a chemical engineer in 1948. Upon graduation I went to work for General Electric Schenectady, and stayed with them for a couple of years. I enjoyed my work as a chemical engineer, but I needed some more courses in basic science, and for that there was a very good school, the Union College. I met a wonderful physics professor at Union College, Vladimir Rojansky, whom I got to know reasonably well and one day he told me that I was really interested in physics rather than chemical engineering.

So in 1950, I applied to a number of schools; I had very good grades; and I went to Columbia University in New York City. It's amazing that I had had hardly any physics, just one year as an undergraduate and some physics at Union College. I was at Columbia University from 1950 to 1955; I was a good student, but not the best student. I went to work for Nobel laureate Isidor Rabi in molecular beams, and produced a good thesis. When I finished, Rabi convinced me to leave molecular physics and go into elementary particle physics. He suggested this partly because he was interested in it; he was one of the founders of Brookhaven National Laboratory; and partly because he always liked to tease his fellow professors by giving such directions to his students. Rabi was a very intense advisor; he taught me a lot. He was not a good experimenter by himself; he never worked on apparatus; we always went to another professor for advice, Polykarp Kusch. Kusch knew a lot, but was not a kind man, he didn't hesitate to yell in public at the graduate students.

Rabi had a way of thinking very deeply and even though he was not very good in techniques, he always insisted on working on his own ideas; he didn't like to work on other people's ideas. This is what caused his conflict with Leo Szilard. Rabi told me that he was apprehensive of Szilard's ideas because he wouldn't work on something that Szilard might have suggested to him even if Rabi himself had also thought about the same idea. When I first started out with the atomic beam experiment with sodium, we used a very obsolete detector with which it was very hard to collect data. When I talked to him about it he said that he had had a student who was now at MIT and who was moving into electronic detectors. Rabi knew nothing about electronics, but he knew what was going on, so I went up to MIT for a few days and learned how to do it. I learned a lot from Rabi, especially I learned from him how to be independent.

Rabi was also very careful. When I finished my thesis, I was eager to publish my results and get back to a job — I had a wife and child — but Rabi told me that I had to check my results using some other techniques. There were some people in France doing similar work and Rabi sent off my results to them to see if we were in agreement. When word came back that it was all right, Rabi told me that I could publish my findings. Rabi was a deep and original thinker, but also very careful, and wanted to make sure that everything was all right. I have tried to follow Rabi's example in my research.

When I left Columbia University, I decided to go into particle physics, and I went to the University of Michigan.

Please describe your research at the University of Michigan and why you moved to Stanford University.

I was at the University of Michigan from 1955 to 1963. Working with colleagues Lawrence Jones and Michael Longo, I carried out experiments on elementary particles at the Brookhaven National Laboratory in New York and the Lawrence Radiation Laboratory in California. I studied the strong, also called nuclear, interactions of protons, neutron and pions. The experiments themselves were a pleasure to build and carry out, but I wanted to study a more basic part of particle physics: the nature of the very simple particles — the electron and muon. The electron and muon along with their associated neutrinos are called leptons — the term lepton indicating that these particles have small masses compared to other elementary particles.

In 1962 I was invited by the great physicist, Wolfgang Panofsky, to join the Stanford Linear Accelerator Center at Stanford University (SLAC), where the world's highest energy electron linear accelerator was to be built. This was a great research opportunity for me, high energy electrons and muons would be copiously produced at SLAC. And since I was a boy in Brooklyn I had always wanted to live in the California of the movies. In 1963 my family and I moved to Palo Alto, the beautiful city next to Stanford.

How did the tau lepton discovery happen?

For my first few years at SLAC I still carried out experiments on strong interaction physics and I also used the linear accelerator for some experiments that I hoped would elucidate the nature of the electron or muon. But I didn't get any insight from these experiments. I began to think about the possibility of the existence of leptons other than the electron and muon — particles that could be called heavy leptons.

Meanwhile in the late 1960s, a SLAC group led by Burton Richter began thinking about building an electron-positron collider, and I realized that an electron-positron collider would be a good way to search for heavy leptons! So we got involved, myself and my group, with building a detector for what was called the SPEAR electron-positron collider. There was a bit of luck in all this; there was trouble in getting funded, and other people got ahead, and built their colliders. By the time we could begin building our collider, and it was 1970, we had a much-improved design, and it went to higher-energy and higher-intensity luminosity. The group had made a proposal

There is a lot of competition between different detectors; then, in a funny way, there is a lot of competition to get no results: to prove that something does not exist or to prove that some interaction has a smaller energy than what others had reported. That's considered to be a better experiment, but if there is no theory, it isn't very meaningful. So there's a lot of competition for no results.

Has the Nobel Prize changed your life?

The Nobel Prize is wonderful because then you can do what you want to do. You can also give a lot of talks and go to every meeting or you may choose to stay at home.

Do you have heroes?

I don't really have heroes; I have good colleagues. There have been some very exceptional people, like Newton and Einstein, maybe Fermi, but the rest is just very smart people who were working on the right subject when the subject was being developed. Other very smart people may have worked on some problems for which the data had not developed yet, so they did not accomplish what they might have under different circumstances.

Would you care to tell us something about your family?

Burton Richter, Martin Perl, and Richard Taylor, three SLAC Nobel laureates in 1995, on the day when Martin Perl's Nobel Prize was announced (photograph by P. A. Moore, courtesy of SLAC Archives).

I was married for a long time; my wife and I were divorced, but we are still friends. I have four children with ages from the late thirties to the early fifties; none of them went into scientific research. Their careers in order of their ages are: art critic and writer, accountant, medical doctor, and computer scientist. My former wife is in math education; she is well known in her field; her name is Teri Perl. We both were very much involved in our careers, and probably paid less attention to our children when they were growing up than people do these days.

Coming from a Jewish family in Russia, do you know what your original name was?

I don't know.

Have you not been interested in it?

I have been; I should ask my sister.

Are you religious?

I am not religious, but I am a very strong Zionist, very pro-Israel, and I have been very active in supporting Israel.

Did you ever experience anti-Semitism?

Yes, in Brooklyn. After school, we went to study in a Hebrew school, and walking back we used to pass a Catholic church, and we used to have fights with the boys, who hung around and who went to the parochial school. I have not had any such experience since I grew up.

In your career?

None, but I have been in a field that's dominated by Jews although not too recently. The first director of SLAC was half-Jewish, the second fully Jewish, the third director, also Jewish, but there are very few young Jews in particle physics now; they go somewhere else, maybe to biology and business.

Do you have any hobbies?

I collect mechanical toys, do woodworking, swim, and read a lot.

What was it that you liked to play with in your childhood?

Martin Perl in 1994
(courtesy of M. Perl).

I loved playing with Erector sets that were similar to Meccano sets in Great Britain and Märklin sets in Germany. I have collected hundreds of different versions of construction sets, and sometimes I build things from my sets. I constructed my own mechanical construction set for children. It was made out of wood and plastics and there were big nuts and bolts that the children could tighten with their hands. The idea was to return to the basics and build houses and wagons, chairs, all sorts of things. We tried to have these things made in China to make it cheaper, but it did not become a marketing success. I tried to patent it, but it was not granted because it was claimed that there had been some previous attempts at similar things. I also have another hobby, gardening. We have a nice place in Palo Alto; we have a lot of fruit trees and other trees.

What was your greatest challenge ever?

My greatest challenge is right now, how to keep doing new physics.

What makes one a good scientist?

You have to be smart, but you also have to have dreams; you have to dream that you can find something different. I still have such dreams, like finding dark matter particles, and it may be crazy, and I am even a little embarrassed to talk about it, but dreams are important for being a scientist. A good scientist has to think outside of what other people are doing.

Carlo Rubbia, 2004 (photograph by M. Hargittai).

32

CARLO RUBBIA

Carlo Rubbia (b. 1934, in Gorizia, Italy) is currently Professor of Physics at the University of Pavia, Italy. He graduated from the Scuola Normale in Pisa; he worked at Columbia University in the early 1960s, then moved to the European Organization for Nuclear Research (CERN) in Geneva. He was also Higgins Professor of Physics at Harvard University from 1970 to 1988. He served as director of CERN from 1989 to 1993. From 1999 till 2005 he was the President of ENEA (Italian National Agency for New Technologies, Energy and the Environment).

He received the Nobel Prize in Physics in 1984, together with Simon van der Meer, "for their decisive contributions to the large project, which led to the discovery of the field particles W and Z, communicators of weak interaction". He is a member of The Pontifical Academy of Sciences, The Royal Society, The Austrian Academy of Sciences, The Polish Academy of Sciences, The National Academy of Sciences of the U.S.A., The Russian Academy of Sciences and many others. He has received numerous honors, among them, the "Cavaliere di Gran Croce" (Knight Grand Cross) of the Italian Republic in 1985, the "Officier de la Légion d'Honneur" of the French Republic in 1989, and the Polish Order of Merit in 1993.

We recorded our conversation in his office at CERN on May 15, 2004.*

---

*Magdolna Hargittai conducted the interview.

When did your interest in science start?

I suppose it was there from the very beginning. Science has always been important to me, no question about that. It's the way I wanted to be and the way I succeeded to be.

I read in your autobiography that you did your thesis work at the Scuola Normale in Pisa already on cosmic ray experiments.

You are absolutely right. That was the physics of the time. That was the time when cosmic rays were still conflicting with accelerators. Later on, accelerators took over. But that was a period of time, especially in Europe, before the construction of big accelerators, when cosmic rays were very useful. Normally one would just go to the mountains and detect cosmic rays. Many things were learned from these experiments — in fact, many of the early discoveries in elementary particle physics have been cosmic ray discoveries. Accelerators took over only after a while, when the power of the equipments became sufficiently high to get real physics from them.

After your Ph.D. you went to Columbia University…

Yes. That, in fact, was the first time when I met accelerator physics. Columbia University at that time was a very exciting place — it still is but then it was truly exciting. Most of the high energy physics events happened there. There were a huge number of very relevant people. Not only that, but there were also many people there who later on in their careers became very relevant. As an example, at that time Steve Weinberg was a young postdoc. Mel Schwartz was completing his Ph.D. and participating in the search for and subsequent discovery of the mu-neutrino, together with Leon Lederman and Jack Steinberger. On the theoretical side, T. D. Lee was there. From low energy physics, Chieng-Shiung Wu, who proved experimentally parity violation for the first time was also there. So all in all, that was probably the best place to be. For me it was a very exciting time because I came from a faraway Italian environment to the most challenging and appealing place in the world.

You also did some work on parity violation, probably a little later than those famous experiments that proved it.

We tried to do something more complicated and different. Parity violation was already discovered before I went there. But it was not yet shown in muon capture. As you know, there are three channels: muon decay, beta decay and muon capture that are the three basic constituents of weak interactions. Somehow the area of muon capture was left untouched by others. So we tried to produce polarized muons and find the asymmetry. We did find it; so we showed that parity was also violated in muon capture. Later on there was much work on muon capture but at that time muon capture was a less vital subject than some others; especially because muon capture comes from the nuclei and therefore you have to go through nuclear physics in order to understand what's going on. Later, when I came back to CERN, we did muon capture in hydrogen and that was more direct physics because hydrogen was without the problem associated with nuclear structure. We did it in a bubble chamber. But that was a few years later.

Have you done experiments on CP violation?

I am surprised by your question. You see, CP is something that was considered to be unquestionably valid for a good number of years after the parity revolution. In fact, it was only in the mid-sixties that Fitch and Cronin found that K_{long} and K_{short} were both decaying in the same 2π channel, a forbidden channel for both in the case of CP conservation. In fact CP had been considered to be a sort of underlying necessity. I must say that when Fitch and Cronin came along with the discovery of CP violation, it was a big discovery. It was, however, a rather strange effect because CP violation was very tiny while parity violation was as large as you could think of. We, at that time, at CERN, found CP violation extremely exciting and did a lot of experiments on it with the CERN accelerator. We measured both the channel into π^+, π^-, the channel into π^0, π^0. A lot of very careful work allowed us to understand on one hand the CP violation effect and on the other side another phenomenon, the so-called Pais–Piccioni effect, namely unique quantum mechanical coherence effects which may occur only in the regeneration of neutral K particles, whenever K_{long} can be transformed into K_{short} when interacting with matter. This was rather exciting because it represented the possibility to study quantum mechanics with extremely massive objects. We are talking about several hundreds of kilograms of copper blocks separated from each other by several meters. Even in such a macroscopic situation you could observe strong quantum mechanical interference phenomena. So this was a very good school for us, to learn to systematize all this. This work has been done together with Jack Steinberger. There was also John Bell, an incredibly sharp theorist, who strongly contributed to the theory of these phenomena.

Is the origin of CP violation known by now?

CP violation is a long story. In fact that explains why Fitch and Cronin did not get the Nobel Prize immediately, but only after many years. Now we know that it is produced so weakly in the case of the K_{long}–K_{short} system is not due to the fact that CP is intrinsically weak, but comes instead from the fact that the quantum mechanical amplitudes necessary to build up a CP violation are several, of which some are very weak. In fact, we later have been looking at another system — the Beauty, anti-Beauty system — very similar to the K_{long}–K_{short} system. We did a lot of early work that time with the Collider and found that the B-Bbar is also violating CP and, in fact, it has a much stronger CP violation. Andrei Sakharov had postulated that at the time of the Big Bang, at extremely high energies, under the presence of CP violation and in non-equilibrium conditions, matter may emerge over anti-matter. Most likely, the absolutely fundamental result of Cronin and Fitch is just the early occurrence of an effect capable of producing very gigantic differences and consequences at the cosmological level.

Press conference announcing the discovery of the intermediate *W* boson, 1983. The first and second from the left are Carlo Rubbia and Simon van der Meer (courtesy of CERN).

But CPT symmetry is still true.

We also looked into that. Of course CPT is assumed to be theoretically correct because that's how you write down local equations. But the question of CP violation and correspondingly T violation are questions of fundamental importance. It is a very exciting area in which all kinds of symmetry phenomena can be studied.

Let us get to the work that brought the Nobel Prize to you and to Dr. van der Meer. Professor Ekspong said in his presentation speech that "van der Meer made it possible, Rubbia made it happen." Please, tell us your side of the story.

This event is a small part of a major development in the history of elementary particle physics, which was the development of the Standard Model. That was the main thrust of the elementary particle community during the 1970s and the 1980s. But in order to perform accurate experimental tests of theoretical ideas, new types of instruments had to be developed. One kind of these instruments was the accelerator. We had to go from the situation of the fixed target experiments of the early 1970s to colliding beams. A first step was the proton-proton experiment with the Intersecting Storage Rings (ISR), followed by proton-anti-proton experiments with the proton-anti-proton collider. We had to introduce very important developments in accelerator domains. We also had to eliminate a few fundamental problems like the Liouville theorem obliging us to conserve the space charge volume of a beam. That was done with electron cooling in Russia by Budker and with the stochastic cooling by Schnell and van der Meer at the time of the ISR. It was a development that did not seem to have great value at the time, but in the end it turned out to be fundamental. On the other hand, there was a big development in detector technology, thanks to Georges Charpak. CERN was not only building the physics of the day that time but it was developing the physics of the future as well. There was truly a collective impulse to develop new methods and new ideas that made all these discoveries possible.

Other works of yours you consider important?

Well, I normally don't think about the things I am doing as important. But there is no question that from the early days of cosmic rays all the way down to the last experiments done with the big accelerators, it's been a very exciting period. We felt that we were the right people at the right

Participants of the seminar at CERN in 1999, honoring Carlo Rubbia's 65th birthday: left to right, Gerard 't Hooft, Georges Charpak, Alan Astbury, Luciano Maiani, Klaus Winter, Carlo Rubbia, Arthur Kerman, and Val Fitch (courtesy of CERN).

place. There were a number of results and discoveries, which I think represent a life dedicated to science. I am perfectly happy with what I have done.

Why did you come back to Europe after spending exciting years at Columbia?

First of all, I did not go back to Italy, I came to CERN and there was an idea behind it. The idea was that Europe would work together and build this very significant center of science. So I did not come back because I did not like Columbia; I came back because I felt that my role would be to be associated with the growing science here at CERN.

I spent quite a few years here and then, when I decided to couple research with teaching, I felt that the best place to do it was Harvard, so I went there. I spent 18 years of my life as a professor at Harvard, and these were very profitable years. It is a perfect place to build up science-based education. The reason why I left Harvard was that I was asked to be Director General of CERN and that was not compatible with being a professor in the United States at the same time. I was director general for five years and that gave me the opportunity to contribute to CERN in a different form; not only as a simple scientist but in the role of a person responsible for science policy.

Anything special you would like to mention in this respect? What was accomplished during your mandate?

It was a very exciting time. We had a number of very important things. LEP (Large Electron Positron Collider) was just about finished by my predecessor

Erwin Schopper but was not yet started. So I could witness the early startup of LEP that has been a great success. It has produced a tremendous amount of physics and covered an enormous number of precision experiments for almost a decade. It has provided for a systematic verification of the Standard Model, after the earlier discoveries of the proton-anti-proton Collider. At that time the foundations for the LHC (Large Hadron Collider) were also laid down. LHC occupies the same tunnel as LEP, in fact it uses many of the facilities of LEP. It is also of truly international nature. In addition, I was much involved with opening CERN to new countries; Hungary, for example, joined CERN when I was the director general. So I might say that CERN when I left was not the same as when I came. Its transformation into a truly worldwide laboratory, with countries like Israel and Russia being members also happened then.

There are large hopes for the LHC that is to be ready in a few years...

I think it is more than hopes; we have to make sure that it will be built in time. Technologically it is an incredibly difficult challenge. We should not forget that, for several reasons, including the complexity and cost, America's similar machine, the SSC (superconducting supercollider), was cancelled by the Congress. That left Europe alone with the burden of the problem that if neither the SSC, nor the LHC would be built, things in accelerator physics would come to a rest. We thus have a responsibility, and also the commitment to finish the work. LHC is an absolute *must*; if it does not work, we have a lot of problems. But I don't see why it should not work, even if it is extremely difficult because we are talking about technologies on a very large scale — 27 kilometers of circumference is a hell of a lot pathway. It is the ultimate limit of what technology can manage today in this field.

Does this mean that the future of particle physics depends on LHC?

Well, you never know what will be the next step. The next discovery may come from anywhere; for instance there are lots of results coming from underground physics, which is just the continuation of the old cosmic rays we talked about earlier. There are now laboratories in the world, one of them at Gran Sasso in Italy, which have the capability of producing high-quality physics. Nonetheless, it is true that the main road of accelerator physics today is LHC. That's why we carry a huge responsibility toward the whole high energy physics community of the world.

You travel back and forth continuously between Geneva and Rome. What do you do in Rome?

At present I am doing there something different. I've decided that I would no longer do accelerator physics. I've spent my lifetime doing it and I need to do also something different, something new and exciting. There are two things on which I am working. I am working on a rather unusual experiment in the Gran Sasso tunnel, which has all kinds of fundamental interests both in astrophysics and in elementary particle physics. We are looking for dark matter, we are looking for symmetries, we are looking for proton decay; all kinds of very important questions that are likely to be discovered and which is certainly worth looking for.

The highly predictive physics that was developed at the time of the Standard Model, permitted people to think that if they could predict something, most likely it was true. Parity violation has been as it was predicted to be; CP violation was difficult but after a while we understood it; W and Z were produced where they were supposed to be; the top quark came out with a collider experiment at Fermilab and it was more or less expected. Probably even the Higgs might come out eventually and complete the system. But I think that the level of unexpected surprises has decreased with respect to what it used to be in the early 1960s, when nothing was known and everything was possible.

So the interest may move from accelerator physics to non-accelerator physics, and there are important fundamental questions which are vastly unknown and which, if true, might change the world. And these are things that could be done. Certainly, we have to work hard, but we don't know if the answer is yes or no. Let me give you two examples of these important questions. One of them is proton decay: the fact that on a cosmological scale protons can decay. Glashow used to say that diamonds are not forever. It should be a process inverse to the process of the Big Bang, when energy was turned into matter. Everybody is convinced that this process should be possible, and that, in fact, protons are not stable particles. But the question is: what is its lifetime? At present we have reached the lifetime of 10^{33}–10^{34} years for proton decay. That, in fact, represents an immense number, the product of two huge numbers: the lifetime of the Universe and the Avogadro number. Which means, that within the Universe's lifetime, 15 billion years, maybe just one out of the Avogadro number of atoms has decayed. To compress that decay phenomenon, still to be discovered, of course, into something of the shortness of our lifetime, implies a very large amount of material and the total elimination of spurious backgrounds, which can obviously hide the effect. The most important background is cosmic rays. Cosmic rays, in fact, are producing signals at a very high rate with respect to the proton decay and they have to be separated. I am working in an experiment, called ICARUS at the Gran Sasso. If we succeed, it would be an absolutely fundamental discovery.

The second subject is the search for the so-called "dark matter". The existence of additional, invisible "dark matter" is today almost a necessity in cosmology. Ordinary matter has been created by the so-called Relic Nucleosynthesis, about three minutes after the Big Bang. However, it is insufficient to produce the whole amount of matter in the Universe. Therefore, there is room for another large contribution, something like a factor of 10 more, of matter that does not come from nuclear synthesis. This matter is not hadronic, it is not — like you and me are made — out of protons, neutrons and electrons. The Standard Model covers only such small fractions due to hadronic matter. Maybe another tremendous discovery is ahead of us, finding what constitutes the rest of the matter of the Universe. Four centuries ago Kepler showed that the Earth is not the center of the Universe and that the Earth is just one little object among millions of different things. Later Darwin came along and said that man is not different, but just the last step of a long evolutive process. These ideas have fundamentally changed the way we are perceiving the world around us. Now imagine: someone may come and say that the most important element of the Universe is not ordinary matter but something else. So we will discover that the matter of which we are made of is not the most basic constituent of the Universe. That would be, in my view, a result which will extend well beyond the field of physics and have fundamental cultural consequences.

Now, as you probably know, supersymmetry is a possible way for these new particles. Many people keep saying that you might produce the first supersymmetric particles with the LHC. If this will happen, we will have to explain, while at least one of these particles is sufficiently stable to live for the over 12 billion years from the Big Bang without decay.

But one can take another, complementary point of view. Since relic "dark matter" particles contribute to about 10 times more total mass than ordinary particles, it should also be possible to detect them directly in nature. We are working on an experiment in the Gran Sasso, looking for very tiny recoils of these, so-called wimps. Wimp is a funny name that is how we call these particles. These particles could collide with ordinary matter, producing tiny and very rare recoils. The flux traversing the earth is extremely large, of the order of one million of such particles crossing every second on each centimeter squared, but notwithstanding many orders of magnitude less than solar neutrinos, which have been already very hard to detect. Many options are open: supersymmetry may not be the right thing, but wimps may exist, or supersymmetry may exist, but may not be stable enough to survive in space.

In my opinion, there are two kinds of experiments that can be done: one that brings supersymmetry in through the LHC (I would say, the Goliath of the field) and another, the much smaller underground experiments (the David of the field). You may say that they are the two faces of the same coin, and there is an interesting competition between the two methods. I personally hope that we have enough time to take a good look at these things even before LHC is operational. But that is not important. In fact, we need both methods. We need to know that these particles exist and LHC will certainly do something in this respect. But we also have to demonstrate that these particles have been stable for 15 billion years and for that you have to go to natural sources, naturally emitted particles and that you can only find with underground experiments.

These are fantastic things.

Yes. You know that all the stars we can see, all these billions of stars that make the sky so beautiful, are only half per cent of the total matter of the universe. The matter of which we are made of, namely hadrons are about 5% of the matter of the universe. About 95% of what constitutes the Universe is unknown!

It is rather incomprehensible.

Yes, it is. But let me also discuss another topic on which I am active: this is the question of energy. As you know, energy is a physical quantity. Yet, there is a tremendous amount of political, humanitarian, financial aspects related to it. There is a time limit to fossils which mankind can burn. And how the world population, by then of 10 to 12 billions people, may survive without plenty of abundant energy?

Everybody will agree on the fact that the future progress of mankind will be impossible without a very substantial and continuing energy supply, namely that "Energy is necessary for mankind and always will be so!" The main underlying issue is obviously the explosive population growth, which is now of about 90 million new human beings born each year (10,000 people/hour), mostly from developing countries. In order to live decently they all presumably will expect plenty of additional energy. There is a huge correlation between lack of energy and poverty: 1.6 billion people — a quarter of the current world's population — are without electricity, and about 2.4 billion people rely almost exclusively on traditional biomass as their principal energy source.

At the present consumption level, known reserves for coal, oil, gas and nuclear energy correspond to a duration of the order of 230, 45, 63 and 54 years. The actual longevity of the necessarily limited fossil's era will be affected on the one hand by the discovery of new, exploitable resources, strongly dependent on price, and on the other by the inevitable growth of the world's population and of their standard of living. Even if these factors are hard to assess, taking into account the long lead-time for the massive development of some new energy sources, the end of cheap and abundant fossils, with the exception of coal, may be at sight. The consumption of fossils, especially of coal, may indeed be prematurely curbed by unacceptable, greenhouse related, environmental disruptions. The climatic effect of the combustion of a given amount of fossil fuel produces one hundred times greater energy capture due to the incremental trapped solar radiation. Doubling of pre-industrial concentration will occur after *roughly* the extraction of 1000 billion ton of fossil carbon. We are presently heading

After signing a collaboration agreement between CERN and ENEA (Institute for New Technologies, Energy and the Environment) by Luciano Maiani of CERN and Carlo Rubbia of ENEA, 2002.

for a greenhouse dominated carbon dioxide doubling within *roughly* 50–75 years. The shear breathing of so many people represents as much as 10% of the global emissions of carbon dioxide.

It should be said, however, that the Kyoto prescriptions, even if universally applied, are largely insufficient. For instance they will introduce only about a 7-year delay in the doubling of carbon dioxide. Since the carbon dioxide remains in the atmosphere for many centuries, a slowing down of the emission rates only delays the reaching of a given greenhouse concentration, without preventing it. It is generally believed (IPCC) that only a major technology change can modify drastically the present traditional energy pattern. New dominant sources without greenhouse emission are needed in order to reconcile the huge energy demand, growing rapidly especially in the Developing Countries, with an acceptable climatic impact due to the induced warming up of the Earth.

The time has come to seriously consider and strongly develop appropriate R&D for other, new sources of primary energy, without which mankind may be heading for a disaster. Only two natural resources have the potential capability of adequate, long term, alternatives: nuclear and solar energies. But solar energy is not today's solar energy and nuclear energy is not today's nuclear energy. You have to realize that in many countries, solar energy is something that has great potentials. I am working on an innovative technology, which is called solar thermodynamics. What is it? It copies the idea of Archimedes. He used mirrors and with these mirrors he concentrated solar power into a "spot" and this way he created so much heat that he could burn the ships of the Romans, who tried to occupy Greece.

So I divide my time between these two projects, one is creating innovative technologies in energy production. The other is such fundamental issues as proton decay and dark matter. I feel I am still active, I am very happy and I keep doing these things and I find them very exciting. Whether we will get anywhere with either of them, is not the point in this respect.

Where do you live?

I still live here, in Geneva. I still have a small group here and CERN allows me to have an office and a secretary. Formally I am a visitor from the University of Pavia because I have a professorship there now. But my family lives here, my children and my children's children, all live here. This is a nice place to live. But, because of my other activities, I travel — just as I did most of my life.

Please, tell me something about your family.

What do you want to know?

Something about your wife, children ...

My wife is a perfectly normal person, just as I think I am. She used to be a teacher but she is retired now. We have a son and a daughter. The daughter has one child, a boy, and my son two girls and a baby on the way that is supposed to come in September. My son is a physicist at ETH in Zurich and my daughter is a medical doctor, but also has a Ph.D. in biology. She is at the Hôpital Cantonal in Geneva and she works in the field of liver transplants.

I always admired Italians because they seem to have a gift of knowing how to enjoy life. Are you a typical Italian?

I am not quite sure. I think that I am happy wherever I get and I belong to this Mediterranean class, to which, by the way, you also belong. Hungarians, in my view, are very similar to Italians.

Are you a public figure?

I try not to be. Sometimes I have to attend public ceremonies, celebrations, but the demand is often too pressing and I do not have any interest in that or in any political role.

You received the Nobel Prize many years ago. Has it changed your life?

I think the Nobel Prize increases your visibility, people listen to you more, but in my view the Nobel Prize is a challenge to try to do better and more effectively what you do. In other words, the Nobel Prize is a promise that you better do something good rather than becoming a self-indulging person about what you had already done. For me the Prize has always been a reason to do better things; it has been a challenge for the future.

With that visibility that the Nobel Prize gives you, have you ever felt a responsibility to try to make science more accessible for the general public?

Obviously, you have to use that visibility for such purposes. But at the same time, you also have to refrain yourself from taking the attitude of

Carlo Rubbia during the centennial celebrations of the Nobel Prize, Stockholm, 2001 (photograph by I. Hargittai).

always being right. I read a study that showed that after receiving the Nobel Prize the scientific productivity of the laureates drops seriously. I think I understand why. When you are an "ordinary" person, you can allow yourself to make all the mistakes you want and so you can take up all the crazy ideas that come to your mind. After being recognized with the Nobel Prize you have to be more careful and sometimes it slows down your ability to come up with new ideas; you tend to be more conservative, much more careful because if you make something stupid — which you always do in science — people will say, hey, a Nobel laureate has done something stupid! I try not to be affected by this feeling and try not to be locked up in a structure that is binding you more than necessary. So far I more or less succeeded, but it is a borderline issue. Another issue is that you have to be very careful about giving your opinion on something that is outside your field. Often people assume that a Nobel Prize winner knows everything about everything, which is obviously not true! You have to be prepared for that.

Can you mention any mentors who helped you in your career?

I was very fortunate in my work. I have always had strong contact with very valuable and smart people and that is extremely important in our field. Today we are talking about teamwork. We have a very large amount of cross-fertilization which develops all the time. I could list you hundreds of people who were essential in my work everywhere. I cannot give you any specific name, but I have had cooperation and friendship with very many people, each one giving me something different.

You mentioned teamwork. Looking at some papers coming out of accelerator works, sometimes there are hundreds of authors on a paper. Where is the room for individual creativity there?

Well, there is plenty of room for individual creativity because without that you would not make any progress. I do believe that you need a lot of collaborators. In the largest experiment that I participated in, the UA1 experiment, there were about a hundred physicists, which was justified due to the multitudes of facets of the work. Today the LHC experiments have a few thousand people, so this explosion of connectivity is growing. If you look at this from outside, from your side, you could think that every person could be replaced by someone else. But this is not true. Every person has a specific facet. Let me give you an example. For instance, when they built in the Renaissance these big cathedrals, every little part of them was an individual contribution and I don't think that any of the sculptors or painters felt limited because there was another sculptor or painter nearby. Everybody had his role and brought about his own contribution, which was unique — the same is true for the accelerator experiments.

What are the ingredients of being successful in science?

A lot of work, a lot of patience, and never give up.

For me the most fascinating topic we talked about today was the experiment at Gran Sasso in which you are looking for dark matter.

You see, it is absolutely fascinating!

I have always wondered, we are made of matter since nature seems to favor matter over anti-matter...

Be careful, be careful! The point is that this is already a problem in itself and is connected with proton decay and with CP violation. You start at the beginning of everything, there is nothing yet but energy. Then comes the Big Bang and you have all kinds of situation; there is mass density, energy density, incredible numbers. Then rather early you find that the symmetry between matter and anti-matter is no longer exact. If today I make a particle with an accelerator, say a proton, I also produce an antiproton with it. If I make an electron I also make a positron with it. All we do we are reproducing ordinary matter. When we make protons at CERN we turn energy into mass using accelerators and these protons are exactly the same, absolutely identical with the kind of protons that were produced in the Big Bang. There is a great similarity between the two situations. However, when we produce a particle at CERN we also produce an anti-particle. This is the way in which van der Meer collected anti-protons. All these anti-protons, as well as the protons produced at the same time were new particles. However, there is a big difference between the accelerator and the Big Bang. Then it started with a very large energy, much larger than the one we can produce in an accelerator. But at the beginning only particles were produced; in fact, there are very few anti-particles in the Universe even now. Thus, apparently, matter dominates over anti-matter. Now how is this possible? The answer comes from theory through CP violation. CP violation that we can see only as a tiny sign with the K_{long}–K_{short} is a big signal if we consider the Big Bang. But even so, it is not enough to separate particles from anti-particles, we also need non-static, evolutionary conditions. Evolution, decay, expansion, associated with CP violation, can produce matter and anti-matter, but then we have to be able to say which is which because there has to be only one type. The way matter was produced at the Big Bang, is fundamentally different from the way we produce it at CERN, therefore, we need a different mechanism. That mechanism almost inevitably tells you that some of the matter that was produced then and winning over anti-matter, would eventually decay and transform back to energy again. This would be the proton decay we already talked about, and it would be an incredible discovery! This is what we are looking for. The problem is that we don't know what the number is. We do know that such a number exists and that it is non-zero for many reasons, but whether it is 10^{33} years or 10^{44} or 10^{60} years, we just don't know. We may never know because the magnitude of this instability is so small that the magnitude of the available experiments is not sufficient to do this. But it also may happen that Nature will be kind to us and give a little signal, a little spark. That spark would be a fantastic element because studying it we would be able to understand better what happened at the beginning. So it would be the beginning of a completely new story. The other new story is that the Universe is full of energy.

Would it be possible that there is a part of the Universe that is made of anti-matter?

It is not so simple. But you should talk with Sam Ting about this because anti-matter is what he is interested in.

I am going to talk with him tomorrow. When we talk about symmetry breaking there is also the amazing fact that living matter is built up from only left-handed amino acids and right-handed nucleic acids and never the other way around.

This is also something that I have been working on. It is a surprising fact that there are about 10^{22} or 10^{25} DNA on Earth and they all have the right screw. That might be an indication that life evolved in a very narrow road, that there was essentially one elementary case that started everything. It also is possible that one of the two different chiral forms was eating up the other. This is one of the most fascinating issues in biology.

Is there anything I didn't ask and you would like to comment on? Any message?

I think we had a very long and nice conversation. Let's keep something for the next time! Thank you.

Simon van der Meer, 2004 (photograph by M. Hargittai).

33

SIMON VAN DER MEER

Simon van der Meer (b. 1925, in The Hague, the Netherlands) received the Nobel Prize in Physics in 1984, together with Carlo Rubbia, "for their decisive contributions to the large project that led to the discovery of the field particles W and Z, communicators of the weak interaction". He received an engineering degree from the University of Technology, Delft, the Netherlands, in 1952, and worked for the Philips Research Laboratory in Eindhoven till 1956. That year he moved to the European Organization for Nuclear Research (CERN) where he worked until 1990 on the design and construction of particle accelerators.

He received honorary degrees from Geneva University (1983), Amsterdam University (1984) and Genoa University (1985). He is a foreign honorary member of the American Academy of Arts and Sciences (1983) and of the Accademia Nazionale dei Lincei (1987), and correspondent of the Royal Netherlands Academy of Sciences (1984).

We recorded our conversation in his office at CERN on April 6, 2004.[*]

Please, tell me something about your family background.

My father was a schoolteacher and my mother came from a teacher's family. They wanted the best for their children's education, sending us to the gymnasium and university. I studied at Delft, at the Technical University.

Since I was a small boy I was always interested in building things. This was still the heroic age for electronics with tubes; transistors did not exist

---

[*]Magdolna Hargittai conducted the interview.

yet. You made everything yourself and you could understand how things worked in detail — this is hardly possible nowadays. I studied physics and finished the university as a "physical engineer" in 1952. I went to work for Philips, at Eindhoven, the Netherlands, on high voltage equipment and electron microscopes, for about four years. Then I went to CERN.

Why?

That seemed a very interesting opportunity. CERN had just started a few years before and they offered positions for physicists and engineers. The salaries were good, but most importantly it was of great technical interest to me. I thought it would be a challenge, not in the least because of the international atmosphere. This was a very interesting time, CERN was much smaller than now, you knew nearly everybody. There was a lot of development going on, the laboratory was growing continuously and you were free to do things that are not easy nowadays, now that CERN is big and somewhat bureaucratic.

How did it work then? If you were interested in technical development you must have had an interaction with the particle physicists to know what they were planning to study.

Not at that time. In the beginning of my stay at CERN I only worked on the 30 GeV accelerator and I had relatively little contact with the few particle physicists that were around then. What happened was that CERN defined what specialists were needed; particle physicists or machine builders, and the latter formed the largest group in the beginning.

Did you have to learn a lot?

Of course. I knew nothing. I knew very little about magnets. We learnt from books and publications. During the Second World War little accelerator work had been done in Europe in comparison with the U.S.A.

How long did it take to build the accelerator?

About four years. There were different groups; one that made the magnets, another made the vacuum, again another the radio-frequency equipment, and so on. And, of course, there were a few people at the top who designed it. I worked in the magnet group.

I am jumping a little ahead. I read the Nobel presentation by Gösta Ekspong in which he said, "Van der Meer made it possible, Rubbia made it happen." Would you mind telling us your side of this story?

We wanted to find the W and Z particles; that was Rubbia's main goal. The idea was to have the protons and anti-protons circulate in one ring. Since they have opposite charge they can use the same ring, one vacuum chamber, and the protons and anti-protons will follow the same orbits but in opposite direction. Then they collide at certain places and you have interactions — but it is very difficult for the protons and anti-protons to collide because they are so small and there are not enough of them. Protons are easy to get, but anti-protons are rare. Our 30 GeV proton accelerator produces a pulse of protons every few seconds. These are used to produce anti-protons by shooting them against a metal target. But even when accumulating many pulses of these anti-protons in an accumulator ring (for instance during a day) we have far from enough. Moreover, it is not easy to inject and store many pulses in the accumulator ring because of the momentum spread of the beam.

Do you have to get rid of the other particles that are also produced?

No, that is not a problem. Protons have the wrong charge and are not accepted by the anti-proton accumulator ring. All the unstable particles will rapidly decay. The only remaining stable particles are electrons, but they are light and therefore lose energy by radiation. Only anti-protons make it to the accumulator ring.

The problem now is that the particle beam is too wide both horizontally and vertically, and its longitudinal momentum spread is too large. All this must be reduced to get enough anti-protons accepted by the accumulator ring. This process is called cooling, because the temperature is the same as the momentum spread (horizontal, vertical and longitudinal) of the beam.

The momentum accepted by the accumulator ring is only a few times the beam's momentum spread. Therefore, without cooling, we could only inject a few pulses of anti-protons. By the cooling process we reduce the momentum spread of the beam and make it possible to get a lot of pulses accepted.

Now there exists a very general theorem that says it is impossible to increase the particle density in phase space, where phase space is a 6-dimensional space with 3 spatial and 3 momentum components. This ("Liouville theorem") is valid if only so-called conservative forces act on the particles. And we use electrical and magnetic forces (which are conservative) to influence the particles. Therefore it would seem that we cannot increase the particle density in phase space. Cooling seems impossible. If, for instance, you deflect the particles so that the beam-width decreases, then the transverse momentum spread of the beam will increase.

The trick we use to get around this problem is called "stochastic cooling". We use the fact that in phase space the particle beam is a collection of points; each particle is represented by a point. In between the points is empty space. Now if we can observe the individual particles, we can push them towards the center of the distribution with electrical forces that obey the Liouville theorem. The empty space in between will then automatically be pushed outwards, but the average particle density will be increased. The trick is looking at each individual particle with a so-called "pick-up" whose electrical signal is used to push it in the desired direction. The same signal will also push the other particles, but that is a random push and therefore less important on the long run than the small systematic push on each particle. This looks simple, but there are lots of details, of course, that you have to get right.

How did you think of this?

I invented the stochastic cooling in 1968. That was long before it was used, by which time it had been long forgotten. Then Carlo Rubbia came with his proposal to make *W*'s and *Z*'s, for which you obviously needed cooling. There was another cooling method, developed by the Russians, the technique called electron cooling, so Carlo's original proposal used that. But it was more difficult than stochastic cooling for this application. Carlo came with his proposal in about 1976. Then we did a lot of thinking. First we did an experiment with a small ring to test stochastic cooling; that took a year or so. Then the actual ring, the anti-proton accumulator, was built in two years. After some developments, in 1983, the *W* and *Z* were found. And we got the Nobel Prize right away.

What was it that you originally proposed it for?

It was for increasing the number of interactions in CERN's Intersecting Storage Rings (ISR), a proton-proton colliding beam machine. There it was important to reduce the size of the beams to get more interactions. We tried this, but it took a long time (hours) to cool the beams, because of the large number of particles in the ISR machine. The cooling made only a small improvement. But for Carlo's proposal, with less intense beams, the cooling was far more important.

You actually achieved an enormous concentration of particles compared to what could be done earlier?

Oh yes. The density in six-dimensional phase space was increased by a factor of 10^9.

I was wondering about the experiment. You have the collisions between the protons and anti-protons. How many are relevant? It seems like looking for a needle in a haystack.

Of course, most of the collisions are not interesting, only a very few are relevant. To find the W and Z particles, you have to use their properties. They decay almost immediately because their lifetime is very short. As a result of the collision you see all kinds of secondary effects. For W and Z particles there is a very specific effect: they decay into muons and electrons that can be distinguished from the background caused by the uninteresting collisions.

So basically it is just the interpretation of these millions of results?

Yes.

Do you consider this work as your most important achievement?

Yes, definitely.

Are there other experiences that you remember with pleasure?

Yes, certainly. There were experiments with neutrinos. As you know, they do not interact very much. You make neutrinos by taking a proton beam and shoot it at some metal. Many particles come out. Among these are pions and these pions decay into muons and neutrinos. These neutrinos can penetrate all kinds of material. All other particles are stopped by steel, for example, but not the neutrinos; they get through. To increase the number of neutrinos we have to make their parent beam more intense. The pions, when they come out of the target, have to be focused so that the neutrinos that come out of the decay more or less fly in the same direction.

To focus the pions, since they have different momenta, we cannot use a normal magnetic lens that would work properly only at a particular momentum. My invention was an aluminum cone with a high current flowing through it (a "horn") that increased the beam intensity by a large factor.

I also remember with much pleasure the method I developed to measure the luminosity (that is a measure for the probability that the particles in

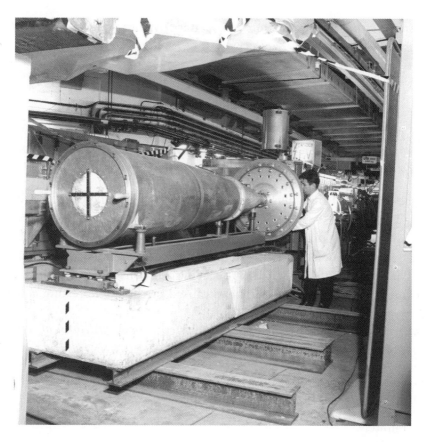

Neutrino horn (focusing device invented by S. van der Meer) in the neutrino beam line, 1963 (courtesy of CERN).

two colliding beams will interact) of the proton-proton collider ISR. In this machine the intersecting beams cross each other in the horizontal plane. To calculate the probability of collision you have to know many things about these beams. It depends on the beams' height but not on their width. The particle density is a function of height. It is difficult to measure the vertical distribution without destroying the beam. In the method that I invented the two beams were displaced vertically with respect to each other and from the interaction rate versus displacement it was possible to calculate the luminosity.

How did this work? Did experimental physicists come to you saying that they would like to measure a particular thing and asking you if you had any idea how to do that?

No. For instance, in the case of the luminosity measurement, there was a workshop where people discussed different machinery to measure the vertical distribution of the beams. After the workshop I thought of my method, which is much easier.

You have been here, at CERN, from almost the beginning. Is there any interaction between experimental physicists, theorists, and engineers?

Not much with theorists; but yes, we meet with experimental physicists, but not very much, because CERN is big and the accelerator is one thing and the experiments are another. Also, many of the experimental physicists are not permanently here, they just come for one experiment and then you lose contact.

I am interested in what really brings particle physics forward. There has to be an interaction between different people. Or is it just conferences and publications where you hear about others?

It is both. For example, when I started thinking about stochastic cooling, I had not been involved in its use; I was making power supplies, an entirely different kind of job. But I always liked to do things as a sideline. I realized that people very much needed something to help them, so I started to think about it. And I found out how to do it.

This also means that people here, at CERN, have a certain freedom to do what they would like to do, right?

Yes, that's right, at least as far as thinking is concerned.

You retired about 13 years ago. What do you do since then?

Nothing special. One thing I do not do is think about accelerators. That is finished. The main reason is that you cannot do it alone. Accelerators are built by large groups of people. You can start thinking about improvements alone but even there you soon run into a wall because you need the interaction with colleagues and with the machines. Another thing with accelerators is that there is little really new there anymore. They are just getting bigger, like this Large Hadron Collider (LHC) that they are building now.

As I understand LHC will reach an energy higher than ever before reached by an accelerator.

Yes. But to go further than that becomes very, very difficult. Except for the superconducting magnets there is little new in that machine, it is just big and expensive. It becomes more and more difficult to make improvements.

How do you then see the future of particle physics?

As for accelerators, linear colliders will probably be the next development. Electrons, when deflected, will lose too much energy by radiation. Linear colliders will not suffer from that, but there the beams cross only once and it is difficult to get sufficient energy and luminosity. Many groups are now working on this and there are still some interesting developments. But it will be a very expensive project and whether it will be done in the States or in Japan, or in Europe, nobody knows. It will probably not be done in more than one place because of the cost.

Concerning these huge expenses, don't you ever get the question whether society should build these large accelerators when there are so many obvious problems in the world?

Of course, this question comes up frequently. And, of course, you can reply that the cost of such an accelerator, say a billion Swiss francs, means only some 3 francs per person considering that about 300 million people live in the countries involved. That is not so terrible for getting such interesting answers to so many important questions. This is very fundamental work.

Do you have hobbies?

Less nowadays. I have always made things with my hands and I still like to do that. I read a lot. I used to walk a lot but I do not do that anymore.

I would like to ask you about your family.

I have two children. The oldest is a girl who studied econometry, lives in Rotterdam and works with an institute in the field of monitoring various development programs. The younger one is a son; he is an engineer/physicist, much interested in practical things.

Is your wife also Dutch?

Yes.

What does she do?

She is a physiotherapist, though not practicing any more.

What language do you speak at home?

Dutch. Both our children also learned that language in addition to the local French, so they are bilingual.

You received the Nobel Prize as an engineer. There are not many engineers who received it.

Yes, but there are some.

How did you feel when you received the prize?

It was wonderful. Of course, it was not totally unexpected because people were already talking about it. The *W* and *Z* particles are very important in physics and it was foreseen that finding them would bring a Nobel Prize. Of course, the prize might have gone to the people who made the detectors or the experiment as a whole, but I probably got part of it because my stochastic cooling made it possible.

Has the Nobel Prize changed your life?

A little bit, yes. Something somehow is sticking to you and you are always treated somewhat differently. The prize is the ultimate recognition.

Carlo Rubbia (left) and Simon van der Meer in 1984 after the Nobel Prize announcement (courtesy of CERN).

Simon van der Meer in 1984 (courtesy of CERN).

Did you become a public figure after the Prize?

No, not at all. For a few weeks after the announcement there was much publicity, but basically nothing has changed.

Considering the size of Holland, there are many famous physicists there. Are the schools so good there?

There was a golden time of physics in the Netherlands, around 1900–1910. It is still good, but perhaps not quite what it was at that time. Of course, there are Veltman and 't Hooft, and many others.

Are you religious?

No.

Do you have heroes?

No, I do not think so. There are people whom I admire, but I do not want to pinpoint anyone.

Can you mention any mentors who meant a lot to you?

I had a physics teacher in the gymnasium who was a very special character. He taught everything by means of experiments and made everybody enthusiastic about it.

What was the greatest challenge in your life?

I think it was probably the anti-proton accumulation project. It had a lot of stochastic cooling in it and we had calculated how it should work, but it had never been done before. Some people even said it would not work. There were things that could have gone wrong and then the whole set-up would have been useless. It was a project of 50 million Swiss francs, so it was a responsibility.

Knowing all that you know today, would you choose the same career?

I think so. I would still be an engineer and would still be interested in physics. My work on accelerators happened to coincide with a most interesting period in their development. That was difficult to foresee.

What do you find the most interesting thing to do in physics nowadays?

I think it is biological work that is the most interesting nowadays. There are many possibilities for physicists there.

Do you like to live in Geneva?

Yes, we like it here. We have many friends, also our son lives here.

Is there anything you would like to add?

They often ask me what I suggest for young physicists. I always say that no matter how crazy ideas you have, always follow them up because sooner or later you will find something that works. And, if not, you will in any case learn a lot.

But nowadays there are fewer and fewer places where you have the opportunity to work out your crazy ideas...

Well, I am not so sure. There are plenty of crazy ideas at CERN. I am sure that there are also plenty of crazy ideas in molecular biology, for instance. And the ideas that you get are usually connected with the job you are supposed to do.

Douglas D. Osheroff, 2004 (photograph by M. Hargittai).

34

Douglas D. Osheroff

Douglas D. Osheroff (b. 1945, in Aberdeen, Washington) is Professor of Physics and Applied Physics at Stanford University in Palo Alto, California. He received his Ph.D. from Cornell University (1973). He spent 15 years at AT&T Bell Laboratories in New Jersey from 1972 till 1987. He received the Nobel Prize in Physics in 1996 jointly with David M. Lee and Robert C. Richardson "for their discovery of superfluidity in helium-3". He has received numerous other prizes such as the Simon Memorial Prize (1976), the Oliver E. Buckley Prize of the American Physical Society (1981), the MacArthur Prize (1981) and the Walter J. Groves award for teaching (1991). He is a member of the National Academy of Sciences (1987). We recorded our conversation in his office at Stanford University on February 12, 2004.*

There have been quite a few Nobel Prizes given for low temperature physics, starting from 1913 or so. Why is that so?

I think it's probably because there are interesting phenomena occurring at low temperatures that do not occur at higher temperatures. For whatever reason the Nobel Prize seems to be given for fundamental discoveries and new ideas, especially in physics.

I also noticed that there were Nobel Prizes before for studies of helium, or at least liquid helium was especially mentioned; the very first one,

---

*Magdolna Hargittai conducted the interview.

in 1913, to Kamerlingh Onnes and later in 1962 to Landau. When did you start to get involved with this field?

I suppose it was when I was a graduate student at Caltech. First in my junior year there I started to work in an infrared astrophysics group that obviously had nothing to do with helium. It took me roughly a year to realize that astrophysicists seldom if ever do experiments. They build instruments and make observations. I wanted to be more involved with the things that I studied, be able to probe them by looking at their responses to various stimuli. It was while I was thinking about this that I had the opportunity to work in a low temperature group at Caltech. The professor, David Goodstein, was on sabbatical but there were two guys there on their sabbatical, one from UC Riverside, the other from Pomona College. They just filled my mind with all of the marvelous and counterintuitive things that happened in helium-4 at low temperatures. I saw the seeds there.

Then I went to Cornell and in the first semester there I heard two talks. One on Pomeranchuk cooling, which was based on a proposal by Isaac Pomeranchuk in 1950, which was just one year after the first publication on the liquifaction of helium-3. Pomeranchuk proposed this very unusual cooling technique that we ultimately used, where, below roughly 0.3 kelvin the liquid is more highly ordered than the solid. This is a very unusual situation in Nature. If you form solid helium-3 by compressing the liquid and if you do this adiabatically and reversibly, the solid will rob entropy from the liquid and thus the liquid will cool. The prediction was that one would be able to cool helium down to temperatures below two thousandths of a degree and this seemed such a bizarre thing that I thought that this was very exciting. The other was the development of the helium-3–helium-4 dilution refrigerator. Pomeranchuk cooling is not used any more but dilution refrigerators are used through a good fraction of all condensed matter physics; essentially all mesoscopic physics is done with dilution refrigerators.

During the second semester of my first-year graduate studies I built most of the dilution refrigerator. It is rare for a first-year graduate student to do that. So this is basically how I got involved in ^{3}He physics. It was because I felt that these new cooling technologies gave promise of allowing people (and in particular me) to look at Nature in a new and hopefully very exciting way, and that turned out to be the case.

What was so special about helium-3? Is it that the liquid is more highly ordered than the solid?

Yes. In fact, it is also true for ^{4}He but only in a very, very small amount at low temperatures. Helium-4, of course, becomes a superfluid, so below the superfluid transition temperature the liquid does become very highly ordered. But the solid is quite highly ordered as well. The disorder in the solid is contained within the phonon spectrum and the entropy associated with it is not very high. In helium-3 the nuclei have a spin and a magnetic moment. So if the interactions between the nuclei are weak, which they are, then you end up having an entropy — there are two states available up and down for the spin, that gives you an entropy of $R \log 2$, which is a pretty substantial entropy. That gives you a remarkably high cooling capacity. In helium-3 the reason that the liquid is so highly ordered is that it is a degenerate Fermi fluid at low temperatures. Helium-3 atoms are Fermi particles just like electrons are Fermi particles and, as you know, no

Osheroff with his experimental set-up in April of 1972 just after they discovered superfluidity in helium-3 (all photographs are courtesy of D. Osheroff unless indicated otherwise).

two Fermi particles can occupy the same quantum state and that leads to this phenomenon. This was known for a fairly long time, certainly before Landau worked out his Fermi liquid theory. There you have two particles in each momentum state, one with its spin up and one with its spin down and then you start filling up the states and at absolute zero there is a sharp cutoff between the unfilled empty states and the filled states. That gives you an entropy, which changes linearly with the temperature. It isn't that the liquid has such a low entropy rather that the solid has such a high entropy.

How do you get helium-3?

We buy it. Helium in the atmosphere is only a few particles in a million, and helium-3 is only a few particles of a million of that. So you can't get ^{3}He from the Earth's atmosphere. The helium that comes out of natural gas wells has a little bit of helium-3 but it's only about a part in 10^7. Thus people were never be able to get enough helium-3 to liquefy by extracting it from liquid helium-4. Although Bill Fairbank here at Stanford was trying to do that. When ^{3}He was first liquefied at Los Alamos National Laboratory by a group lead by Ed Hammel, that helium-3 came from the decay of tritium. Los Alamos was working on the hydrogen bomb and the hydrogen bomb uses tritium as part of the trigger device to create copious quantities of neutrons. Tritium decays to helium-3 with a half-life of about 12 years and I think it is still true that the tritium is usually contained as a compound in hydrogen bombs. Of course, helium-3 as the result of that decay does not form any compounds, so the pressure starts building up and you have to milk the helium-3 out of the hydrogen bombs. I find it ironic that the states that I discovered at very low temperatures, these very fragile states resulted from material that was born inside a hydrogen bomb.

So generally they get helium-3 from tritium. Of course, not all tritium ends up in a hydrogen bomb — but a good fraction does. It's been estimated that there is about 64 kg helium-3 worldwide. The reason it's a relevant number is because, say, for me to use a big sample would be 3 grams. But helium-3 has potential applications in medicine now, which will require lots of ^{3}He.

What is it?

It turned out that atomic physicists have learned how to spin-polarize the ^{3}He nuclei. It's a fairly complicated process. You polarize the electrons in an alkali metal, usually rubidium. Then you have collisions of the rubidium atoms with helium-3 atoms and in a certain fraction of the time they will transfer the spin polarization of the outermost electrons of the rubidium atoms to the helium-3 nuclei. Originally this was developed by atomic physicists to produce polarized targets for high energy physics experiments. It is interesting how this thing works. I think it was Will Happer at Princeton who had this brilliant idea. It turns out that the spin-polarized gas, if in the right container, can remain polarized for many hours. If you look at the polarization, it is roughly 50%. If you look at the polarization of protons in the human body in a 1 Tesla magnetic field, it is about 3×10^{-6}. This is a very small polarization and therefore the signal to noise ratio is not very good. As a result people have never been able to do very good MRI imaging of the lungs because there is not much material there. Well, Bill Happer had this idea that if you have the patient inhale spin polarized helium-3 and hold his breath, because the polarization is so high, even though the density is down by three orders of magnitude, the polarization is up by almost six orders of magnitude and so you could do a wonderful job of MRI imaging of the human lungs in as little as fifteen seconds.

Do they use it now?

They developed this about 7 years ago, I don't know how common the practice is now. The problem is that they feel that if the medical people started using it, the price of ^{3}He would go up very much. Now for low-purity ^{3}He the cost is about 100 dollars per liter of ^{3}He gas. That's peanuts for MRI and they could use it all up. Will Happer, who used to work for the Department of Energy, estimated that the cost of creating helium-3 by first getting tritium, would be about 6000 dollars per liter. I surely can't afford that for my research. So let me tell the end of the story. It's been hypothesized that there should be a large amount of helium-3 on the Moon and now people at NASA Ames Research Center here at the Bay area are interested in what other uses there might be for helium-3, if they went to the Moon mining helium-3. Of course, for my research it does not make any sense. In principle, it can be used for inertial-confinement fusion, so that would be one scenario and it's still many years away if we'll make it a viable energy source. However, the application of helium-3 in medicine is imminent.

Maybe we could turn to the research that led to the Nobel Prize.

That was a serendipitous discovery. We should go back to 1957 when Bardeen, Cooper and Schrieffer published their BCS theory, which explained the origin of superconductivity in terms of the formation of correlated pairs, called Cooper pairs. These correlated pairs would be much like Bose-particles and you could imagine some kind of a Bose–Einstein condensation of these correlated pairs. Although this theory was based on an attractive interaction to form the correlated pairs that is not relevant to helium-3, within about two years or so people began to speculate that other degenerate Fermi fluids at low temperatures, well below the Fermi temperature, might form a similar ordered state. The only two examples that one could come up with were the neutrons in neutron stars that are difficult systems to study experimentally and liquid helium-3. There was a worldwide effort to discover superfluid helium-3. It started about 1959 and ended about 1966, which is just one year before I went to graduate school.

The person who did the most impressive work was John Wheatley. He was originally at the University of Illinois at Urbana Champaign but then he moved to UC San Diego and he was there when the last part of this work was done. Originally I think, it was Phil Anderson, who made the prediction that Tc would be about 100 millidegrees. It did not take experimentalists too long to cool helium-3 below 100 millidegrees but they didn't find any superfluidity. Then the theorists started to get worried. Eventually experimentalists had cooled helium-3 to 2 millidegrees without seeing any superfluidity. Then Phil Anderson said that we don't really understand the pairing interaction and that Tc's would likely be below 50 microdegrees. At that point most people just regarded this as a dream that theorists had but something that wasn't connected with reality.

There were, however, two theorists who continued working on this. It had early been understood that helium-3 atoms liked to surround themselves with atoms that had nuclear spins parallel to their own because these atoms cannot get too close to each other, since they have to obey the Pauli exclusion principle. This is a way for them to avoid the core-repulsive interactions. It was shown that the ferromagnetic spin fluctuations in the liquid would inhibit the formation of Cooper pairs with anti-parallel spins. Therefore, people originally were thinking about not *s*-way pairing but *p*-way or *d*-way but this argument suggested that *d*-way was unlikely. There were two other theorists, around 1965, Layzer and Fay, who took the opposite approach saying that in fact, if you formed *p*-wave BCS states, where the two spins would be parallel, the presence of the ferromagnetic spin fluctuations would enhance Tc for these states. They predicted Tc at 4 millidegrees. This work was published in an obscure place and neither I nor the two professors I was working with were aware of this article. Unfortunately, they never got credit for what they suggested, it was an article that did not have an impact.

In 1967 when I got to my graduate studies people weren't looking for helium-3 superfluidity anymore because the urban legend was that this idea was an aberration of the theorists. But Pomeranchuk proposed this novel cooling technique, in 1950, and proposed that probably you can get down to below a millionth of a degree with this technique. Pomeranchuk believed that the dominant interaction between nuclear spins in solid ^{3}He would be the direct magnetic interaction and the strength of that is μ^2/a^3 where μ is the magnetic moment and a is the lattice spacing. That ends up being about ten millionth of a degree. It turns out that by the time I got to graduate school, people realized that what actually happens is that the particles can trade places with their neighbors, at melting pressures 40 million times in a second. This atom-atom exchange is very similar to the electron exchange between sites in electronic magnetic materials. This would lead to nuclear spin ordering at much higher temperature.

There was a graduate student of Peter Kapitza, in Moscow, who actually tried this Pomeranchuk cooling process. He didn't have very good equipment then, so he was only able to go from above a 100 millidegrees down to about 20 millidegrees. By that time we had dilution refrigerators, which would get to even lower temperatures. There were a few people in the United States, including my thesis advisor, Dave Lee and John Wheatley, who believed that this would allow us to cool solids into a nuclear spin-ordered state. This is what I was trying to do for my thesis work.

The trouble was that none of the thermometers stayed in thermal equilibrium with either the liquid or the solid down at these temperatures so it was not clear how you could even study these thermodynamically. In the summer of 1971 Lee gave me a preprint from John Wheatley's group in which they measured the melting pressure as the function of the applied magnetic field and temperature. They found that there was a very large suppression of the melting pressure with relatively low magnetic fields, suggesting that the relatively low magnetic fields substantially change the entropy of solid helium-3. I decided to look into this. First I had to reproduce their results, and I found that the actual effect was a very tiny effect rather than a huge effect. I studied this for about 3 months using the lab's only NMR magnet, which was an iron-core electromagnet, and eventually two other students argued that I should give up the magnet so that they could do their own experiment. I did give up the magnet, but I kept my cryostat cold, hoping that their cryostat may open up a leak or something so that

Douglas Osheroff soldering at one of his cryostats in 1997.

their experiment would fail, in which case I would get the magnet back. In the meantime, since I had to keep consuming liquid helium, I had to do some other experiment and I decided to try testing the limits of cooling of this Pomeranchuk process.

The helium-3 melting curve had been measured by John Wheatley and myself down to about 3 millidegrees, and I thought that below that temperature we could make a gradual extrapolation down to 2 millidegrees and see if this process would take us down low enough to observe nuclear spin ordering in solid helium-3. The first time I tried this experiment I saw this very strange kink in the curve of pressure versus time, which I interpreted as being due to heating. That was the day before Thanksgiving in 1971 and then I pre-cooled over the whole four-day holiday weekend and the next Monday I started at a much lower temperature and cooled down but saw exactly the same kink with much less solid in the cell. The pressure reproduced itself to one part in 50 thousand. It therefore seemed very unlikely that this could be just a statistical fluke that it would be so reproducible, thus it had to be the result of some highly reproducible phase-transition occurring in this mixture of liquid and solid helium-3. This kink was hard to miss but I think this impressed the Nobel Prize Committee that I pursued this unexplained kink rather than just ignoring it.

It wasn't the transition in solid helium-3 that we expected. We did believe that it was a transition in solid helium-3, because another product of this early work of mine was to show that there was much more entropy in the solid at very low temperatures than what the thermodynamic expansion theories suggested there would be. So we said that there was something very strange in solid helium-3 probably due to this phase transition. We published this. All the data were correct, probably still some of the best there is, but the explanation was all wrong. Ironically, it sailed through the publication process and appeared in *Physical Review Letters*.

There were two people then who suggested that this change in slope was consistent with a jump in the heat capacity expected at the superfluid-transition temperature. One was a colleague of Peshkov in the Soviet Union. Peshkov had claimed to have observed superfluidity in liquid helium-3 at 5.5 millidegrees using much less powerful refrigeration techniques. Then John Wheatley had cooled down all the way to 2 millidegrees and didn't see any superfluidity — there was a huge fight between these two. Pomeranchuk certainly had not observed superfluidity earlier. There was also a second person to make the same comment, John Goodkind at UC San Diego, who had spent a few years of his time searching for superfluidity in helium-3. So this idea that there was an alternative explanation was certainly plausible and it really bothered me and I thought that somehow I have to prove that this transition happened in the solid and not in the liquid. I then put a new NMR coil in the cell, which would measure the susceptibility of the helium-3 that was inside. The trouble is that you are looking at a mixture of liquid and solid. We had found not one but evidence for two phase transitions in our pressurization studies. At the first, higher temperature transition we would see a kink in the magnetization. The kink was in the wrong direction because it appeared as if the susceptibility of the ordered phase was higher rather than lower. It should have been lower if it was an anti-ferromagnetic phase as expected. But we already thought that this was an unusual phase transition. At the low temperature transition there was about a 2% instantaneous drop in the magnetization. It seemed like there was strong evidence that the phase transitions were in the solid.

Then I started to think. We were focusing on the high-temperature transition, we really didn't understand the other one. In helium-4 when you go through the superfluid transition the thermal conductivity of the liquid increases by many orders of magnitude. If that happened in helium-3 as well it might be that the solid would form in a different place inside the cell. It is the solid that has the very high magnetization, which is temperature dependent. So what I saw was consistent with saying that suddenly we were growing more solid inside the NMR coil. I didn't like that, of course, and decided that I better do a better experiment. The trouble was that we didn't know where the solid formed. What happened at this point — quite fortunately for me — was that there was a power interruption at Cornell caused by a squirrel running along the high-voltage power line, so we were without power for a couple of hours. When the power came back on I found that air had leaked backward through one of my pumps and formed a solid air block inside the liquid helium Dewar in a line that pumped on a pot of helium-4; that was the first stage of refrigeration below 4.2 degrees. We had to warm it up and when we cooled down again a leak opened up in the cryostat, which was very bad. We had to fix this but in the meanwhile I had time to think of this other problem and I went to see Michael Fisher, whose office was at the Chemistry Department but actually he was a member of the Chemistry, Physics, and Math Departments; a very bright guy. I asked him about the Pomeranchuk cell. He said that you have to understand the effect of the surface tension between the liquid and the solid. Physicists never learn anything about surface tension. He talked about the fact that first-order phase transitions are inhibited because as you grow a very tiny amount of a solid the surface energy dominates so it actually will increase the energy to form a little bit of solid, even though it may be the low-energy state. He predicted that we would form solid only at a few places, maybe 2 or 3 and I started to think how I could differentiate between those. I created an early form of magnetic resonance imaging. In fact, Paul Lauterbur, who recently shared the Nobel Prize for MRI, actually read our article describing how I did this. The static magnetic field in our experiment pointed horizontally. We then created a magnetic field gradient in the vertical direction as well. Then we would sweep the NMR frequency, and the resonant layer would go from the top to the bottom of a long NMR coil. We could then differentiate in one direction where the liquid and solid were. Sure enough, we found that all of the solid peaks in this NMR spectrum showed this funny little drop in magnetization at the low temperature transition.

Then, on the night of April 20th I was taking care of the cryostat, a process that takes about an hour, and I started looking back over some recent data that while we got this 1–2% drop in magnetization of the solid at lower temperature (B) phase transition, the liquid magnetization dropped by a factor of 2. I knew right away that this was it, so I wrote in my lab-book, "2:40 a.m.: Have discovered the BCS transition in liquid helium-3 tonight." I didn't get any sleep that night at all. It's interesting because the phone call from Stockholm came about 2:30 in the morning — needless to say I didn't go back to sleep that night either.

You mentioned that there were many people looking for superfluidity in helium-3, but some of them gave up. Then the Nobel Prize was given to three. Were they the right ones?

They were the people who were directly involved with this discovery. John Wheatley was already dead by this time. It was certainly my feeling at the time of the discovery and I believe also that of Richardson and Lee as well that if Wheatley had made this discovery, because he has done so much of the pioneering work in liquid helium-3 physics, proving that the Landau theory was correct and others as well, then he would have gotten the Nobel Prize for the discovery. But I felt that we wouldn't; we were just spoilers. Dave Lee was my thesis advisor; he was responsible for the decision to develop Pomeranchuk cooling; and maybe more importantly, he was the one who gave me complete freedom to do whatever I wanted. Bob Richardson just became assistant professor when I joined the low temperature group. He was in the lab all the time and he was largely responsible for what I knew about low temperature research.

How many years did it take to get the Nobel Prize?

It was 24 years.

Did you expect to get it?

In 1976 Richardson, Lee and I were given the Simon Memorial Prize in Britain. It was a year after that that someone told me that he had nominated me for the Nobel Prize. For many years after that people kept telling me that — although they are not supposed to do so. For years as October came about I would get nervous. There were two possibilities; one was that my life is suddenly thrown into a state of utter chaos; the other is that I would be mildly disappointed again.

Mildly?

Yeah. First of all we certainly didn't expect the Nobel Prize to begin with. But as time went by the number of groups studying superfluidity started to diminish and it didn't seem like such an important field any more. There was something else. I basically was born with a silver spoon in my mouth as a physicist. This was regarded as an important experiment regardless whether it was a Nobel Prize winning experiment or not. The first informal offer I got for a tenured faculty position was when I was less than six months out of graduate school. After that for about 20 years each year there was someone who asked if I was interested in joining their university. I was one of the first 21 MacArthur fellows, which was quite a bit of money; it was 184,000 dollars and not taxed, so that was a very similar amount of money I got for the Nobel Prize, after taxes. It came to about 192,000 dollars. I won other prizes as well. But as time went on it was less and less likely that I would win the Nobel Prize. I don't remember what happened to the MacArthur money. But I know that I told myself, if I ever get another prize I would buy myself a really good camera — I am an avid photographer. But that never happened. So eventually I decided I should enjoy myself prizes or not and I went out and bought a very nice camera, a 7000-dollar camera — one year before the Nobel Prize.

What happened after the Prize? Has your life really become so chaotic?

It did. The morning after the announcement, there was a press conference at Stanford. The first question they asked was: why did it take so long? In fact, the average for physics is about 17 years; they want to make sure that the discovery has been important beyond just its own field. I am not sure how important helium-3 superfluidity was in understanding unconventional BCS states, but the heavy fermion systems and of course, the high-Tc

Douglas Osheroff with Princess Victoria of Sweden and then mayor of San Francisco, William Brown, at a celebration of the centennial anniversary of the Nobel Prize, in San Francisco, 2001.

superconductors are all unconventional BCS states. I guess that's why finally we were given the Prize. The year when it came I was totally unaware of the day and that the prizes were already being given out, so it came completely out of the blue. When that early-morning call came I was very irritated because I was convinced that it was the wrong number waking me up in the middle of the night — even after that person asked me by name.

Anthony Leggett received recently the Nobel Prize for work connected with helium-3. Was that work done before or after yours?

After. He'd done some work on p-wave BCS states and in 1965 he worked out the Fermi liquid corrections to the magnetic susceptibility of those states. But I believe that the theory he got the Nobel Prize for was the theory of what he called "spontaneously broken spin-orbit symmetry" and the consequences of that in these very strange superfluid phases. He is a wonderful person and he definitely deserved the Nobel Prize. He also did very important work on the physics of macroscopic quantum tunneling. He would have ultimately gotten the prize, probably similar to the de Gennes prize, which was given for a "volume of work", although they are not supposed to give Nobel Prizes for that but definitely that's what happened in the case of de Gennes.

Can we say that superfluidity is the appearance of a quantum mechanical effect on a macroscopic scale?

Definitely, there is no question about that. In superfluid helium-4 what has been known for a long time is that circulation around a closed loop is quantized. The quantum of circulation, h/m, has been confirmed quite well. In the case of helium-3 you see the same thing but in fact the quantum of circulation is not h/m but $h/2m$ because the entities are the Cooper pairs. But if you look at the change in phase as you go around the loop it would still be 2π, 4π, 6π, etc. There are, in fact, 8 vortex structures in helium-3 and as you go around the loop once some of them have phase changes of 4π rather than 2π even though it is just one vortex. The complexity of superfluid helium-3 is one of the remarkable things and there is an enormous richness in the behavior that one observes as the result of that complexity.

We already talked about the possible practical applications of helium-3. How about superfluidity in general in this respect?

For example, in high energy accelerators they utilize the fact that superfluid helium-4 has a very high thermal conductivity in cooling the superconducting magnets. There are lots of other things you can do with helium-4. Not that many applications, mostly because the temperature is so low. In fact, with helium-4 it is not that low, it's just below 2.17 degrees. But there hasn't been much tendency in industry to maintain those temperatures. Hospitals do because superconducting magnets are used in MRI but that is not in the superfluid phase. Richard Packard at Berkeley first showed that using superfluid helium-3 one can make possibly the world's most sensitive gyroscope, a detector of rotation, but now he has found he can also do that with superfluid ^{4}He.

I would like to switch topics and ask you about your family background.

I was the second of five children of a medical doctor; my mother had been a nurse originally. I grew up in Aberdeen, in Washington State, which

The Osheroff family about 1953: Douglas is the child on the left.

is a logging town. I thought it was a great place to grow up. A relatively small fraction of the students from my high school went to college there, but there was no friction between those who were college-bound and those who were not. My father was a physician and he came from a medical family. I always fainted at the sight of blood so I had no interest in that. However, from my early age I was interested in how things worked, I tore apart my toys to get the electric motors out. My father gave me the camera that he'd used as a child and I tore that apart. He was, I think, fascinated by my fascination and nurtured that. That is probably why I went to science. Unfortunately, he'd died in 1977 before I got any of my prizes but I am sure he'd have been very happy to know that I won the Nobel Prize. My mother had a really debilitating stroke in 1988 and she'd lived another 8 years totally bedridden. Eventually she just got tired of living. They woke her up in the morning when the Nobel Prize was announced and told her that her son, Douglas, just received the Nobel Prize. She told them: we all knew he was going to get that; you didn't have to wake me up. She died about a month later.

Is your name's origin Russian?

I didn't know. My grandparents came to New York from Russia in 1906, I guess. There was a failed coup attempt and the czar came down very heavily on Jewish people living in Russia. They were only allowed to live in the Belarus and the Ukraine. The Belarus Academy of Sciences eventually contacted me by email and asked whether I had any roots in Belarus. I said I had no idea. My mother was the daughter of a Lutheran minister so we had a mixed family. I myself married a Chinese girl and when we got engaged I called up my parents and my mother started lecturing me about the necessary attention that one has to spend in inter-racial marriages, and that was the first time I realized that she regarded her marriage as an inter-racial marriage. But the bottom line is that I think my grandfather was from Belarus and my grandmother was from Russia just on the other side. But it could have been the other way around.

Was your family religious?

We all grew up in the Lutheran church. When my father came into this marriage he did not bring with him any of the traditional culture of the Jewish people with him. We always celebrated Christmas, for example. My father was a student of all religions. We would go to Sunday school and

Douglas Osheroff and his wife, Phyllis Liu Osheroff on their wedding day, 1970.

on Sunday afternoons he would tell us about other religions, but he never told us about Judaism. He himself had no religion but found religion as such interesting. I was confirmed in the Lutheran church and we all were given the option to continue in the church or not; it was our decision. I did not. I guess the reason was that in church you had to recite these things that always started with "I believe in …" and then there was a long litany of things that I could not possibly believe in and I felt very uncomfortable with that. In some sense it seemed that lying in church is the worst place to lie. I guess at some emotional level I accept the idea of God but I have no idea how God would manifest itself.

About your wife.

She is of Chinese origin. We have no children. She is a biochemist and I am a physicist and the hours between 5 p.m. and 9 p.m. were very important for us and still are, I suspect.

Do you have heroes?

There are people whom I admire, but I would not call them heroes. Scientists, like Albert Einstein, are certainly heroes of mine. Actually Bill Clinton is also a hero of mine. I think that this country has not treated him well, he was one of the best presidents that we've had. He created a wonderful foreign policy and gave the U.S. a presence of leadership, a presence in the world community and that has all been destroyed by now. I think that his sexual transgressions should have remained private as they had for every president up to that time. There are a few things that really stand out in my mind about Clinton. One is his ability to focus on foreign policy even while being impeached. I found that absolutely amazing. The other thing is his decision to get involved in Bosnia. There was something that went on there what was fairly close to genocide and no one in Europe was willing to touch that issue. I can tell you that at the big banquet in Stockholm on December 10th I was sitting next to the Swedish foreign minister and I lectured her about the ethical issues involved in this and after a while I said, my God, what am I doing; lecturing a foreign minister of a country but she said, "That's all right, sometimes it's good for us to be lectured to!"

What are you working on these days?

I have to say that my mind is growing away from my research in some sense. I was a member of the board that investigated the Columbia shuttle

Reconstruction Team members discuss debris with Columbia Accident Investigation Board (CAIB) board member Douglas Osheroff, 2003 (CAIB photo by Rick Stiles 2003).

accident and that ended up taking away much more of my time and took much longer than I ever expected. I was being rather outspoken and I have not hesitated to explain the issues that I thought were relevant in this accident. It was also important that Nobel laureates are generally treated well by the press. When President Bush announced, after two controlled press leaks, that NASA should go to Mars I did not think — and I don't think now — that that is the smart thing to do. I don't think that we will send people to Mars for probably 30 years. It was estimated back in 1989, when George Bush senior proposed going to Mars, that the costs would be about 400 billion dollars. That would be about 700 billion dollars now, and NASA is not known to overestimate the cost of its programs. If you do that over 30 years that would mean that we would be spending three times as much money to send humans to Mars as we would be spending in all other scientific research save biomedical research. I think it's crazy. My guess is that what's going to happen is that NASA will adjust its program to be able to do whatever it has to do to go to Mars and that will cause a substantial cut in the amount of science that NASA does, space science in particular. I think that the decision to abandon the Hubble Space Telescope is fundamentally a reaction to the cost and the scheduling problems associated with the completion of the international space station. They called it a safety issue because it's difficult to argue with that, but I believe it is basically a matter of cost and schedule.

You were relatively young when you received the Nobel Prize. Does that spoil someone?

I wasn't that young, I was 51. What's happened was that I ended up doing lots of other things and they are usually done at the expense of my research. Of course, the research continues because it's done by postdocs and graduate students but I am overseeing this research less well than before. I daresay that until I made the discovery that I made, Dave Lee certainly did not oversee what I was doing basically at all. He would give me a broad direction and then would leave me alone for several months. He would be aware of the general progress I was making but would never tell me what to do. So sometimes it's not bad to leave people alone. But I realize that I am less involved than I used to be.

I came to academia when I was 41 years old and I was very much interested in training young professionals in science, in physics in particular, of course. I think that's still true, I got a university prize for undergraduate education and I am university fellow for undergraduate education. I was even sometimes highlighted at football games, as a Nobel laureate.

Do you enjoy all those activities that came, so to speak, with the Nobel Prize?

Yes, I do. My only problem is that I am not very good at saying no and the things that I would say no to, are some of the things that I would enjoy most. Look; Stanford treats me as a professor. Maybe occasionally I get some advantage because of the Nobel Prize but there are three Nobel laureates in the Physics Department, three more at SLAC, 17 Nobel laureates total at Stanford University. There is no reason that Stanford should treat Nobel laureates as prima donnas. But when I go somewhere, to give a series of talks, they treat me very well and it is fun. When in September my wife and I went to Beijing, I gave the C. N. Yang Nobel Lecture, C. N. Yang showed up for my talk; very regrettably his wife was dying of cancer at that time. They put us up in a hotel suite, they gave us a vacation, we spent four days in the Yellow Mountain, one of the most beautiful places in the world, and I met many young students. Recently I gave a talk at Halifax in Nova Scotia; teachers drove their students from three different provinces to attend my lecture! These are great things and you know that you are stimulating these kids. In fact I started doing those sorts of things well before the Nobel Prize, when I was still at Bell Labs in New Jersey. The idea was that I'd realized that I had been stimulated by various people; first of all my father. Then there was the General Motors Parade of Progress that came through my town. It was like a circus carnival but it was all about science and technology — that was the first time that I saw liquid oxygen. Then there was a chemistry teacher in high school, who had been in graduate school and took the time to explain to the students what research was all about. I was a tinker, I didn't do research but his explanation would not be lost on me, in fact it had a profound impact on me. You have to support those sorts of activities.

Do you expect another such breakthrough in your scientific career as helium-3 superfluidity?

It's funny. I will give a talk in less than two weeks from now in Old Dominion University in Virginia, called the Nature of Discovery in Physics. I start that lecture by first talking about Kamerlingh Onnes first liquefying helium and then discovering superconductivity two years later. Then I point out that Kamerlingh Onnes failed to recognize that his liquid cryogen, helium-4 was undergoing a superfluid transition, which was about as profound as superconductivity. Then I say that no one really got the Nobel Prize for discovering superfluidity in ^{4}He, but Peter Kapitza, probably he came the closest. Kapitza shared the Nobel Prize with Penzias and Wilson for their discovery of cosmic microwave background radiation, which was, of course, a completely serendipitous discovery as well. Then I talk about four discoveries that I had made and only one of them won any prizes, superfluid helium-3. But I and collaborators also basically discovered antiferromagnetic resonance in solid helium-3, and with two theorists we were able to determine, due to the complex NMR spectra, the symmetry of the ordered state and actually guess the exact state, which was later confirmed by neutron scattering experiments. Another scientist at Bell Labs named Jerry Dolan and I were the fist people to observe "weak localization" but that occurred while Phil Anderson and his famous "gang of four" were actually writing their theory of weak localization. The fourth one is the observation, here at Stanford, of a dipolar gap due to interactions between tunneling two-level systems in structural glasses at low temperatures. So you see, for me each one of these is almost equally exciting because Nature suddenly tells you that there is more to her behavior than anyone had guessed. Doing an experiment is asking Nature a series of questions and

Norman Ramsey, Douglas Osheroff, Charles Townes, and Magdolna Hargittai at the Lindau meeting, 2005 (photograph by I. Hargittai).

the theories end up hopefully constraining Nature's answers so that you can figure out what's actually going on. Of course, Nature only gives you very obscure answers to your questions. That's what it's all about.

What are your future plans?

I don't know. I soon will be 59 years old. I doubt that I will continue being active beyond 70. I am certainly thinking about a change by the age of 65. The question is what do I do? I just don't have any personal time; I do lots of exciting things; but my wife is only working part-time. We love traveling. We do a lot of it but perhaps not enough. There is not much time for taking pictures and enjoying things. The last time we went hiking in the mountains was one month before the announcement of the Nobel Prize. So I don't know. People are in fact suggesting things and ask me if I would be interested in doing things but they are usually things that would force me to leave Stanford and probably get rid of my research program completely and stop teaching and these are all things that I enjoy doing. I haven't agreed to any of these things so far. I don't know. In some sense a university is a wonderful, maybe the most wonderful place to be. It makes it possible to do things that have value and that you have some control over and I don't know if life could be much better than that.

Jack Steinberger, 2004 (photograph by M. Hargittai).

35

JACK STEINBERGER

Jack Steinberger (b. 1921, in Bad Kissingen, Germany) is at the European Organization for Nuclear Research (CERN). He received his Ph.D. in 1948 at the University of Chicago. He spent a year at the Institute of Advanced Study at Princeton (1948), and at the University of California at Berkeley (1949). He was Professor of Physics at Columbia University from 1950 till 1967 and in 1967 he joined CERN, where he stayed ever since.

Dr. Steinberger was awarded the Nobel Prize in Physics in 1988, together with Leon Lederman[1] of Fermi National Laboratory and Melvin Schwartz of Digital Pathways, Inc., "for the neutrino beam method and the demonstration of the doublet structure of the leptons through the discovery of the muon neutrino".

We recorded our conversation in Dr. Steinberger's office at CERN on May 15, 2004.*

I would like to ask you first about your family background.

My father was cantor in the Jewish community in Bad Kissingen in Germany where I was born in 1921. He had this job from the time he was eighteen years old until he left Germany at the age of sixty-three, in 1937. My mother had a university education, which was not common in those days, and gave English lessons and French lessons. It was a modest existence; half of my father's salary was paid by the Jewish community and the other half by the State. This was because religious education was obligatory at

---

*Magdolna Hargittai conducted the interview.

Jack Steinberger's parents (all photographs courtesy of Jack Steinberger unless indicated otherwise).

the time in Germany. This continued even in Nazi times. Hitler came to power in 1933, and I left Germany in 1934 when I was thirteen years old. I still experienced the virulent anti-Semitic propaganda during the political campaigns and the marching of the SA, the *Sturm Abteilung*. After a while there was a putsch within the Nazi party; the SA was eliminated; and the SS, *Schutzstaffel*, took over. When I was there, the SA still held marches during the night at torchlight, saying "wenn's Juden Blut vom Messer fliesst, dann geht's noch mal so gut" (when the blood of the Jews is flowing from our knives, then things are really going well).

How did you leave Germany in 1934?

It was made possible by an offer of American Jewish charities to find foster homes for children. My parents decided that it would be better to send two of their three children away. My older brother and I went together to the United States where we went to two different foster homes in the Chicago area, which were close to each other, and we went to the same school. Of course, it was not completely easy for someone to take a child into his

Jack Steinberger at fifteen with Barnet Faroll in Winnetka, Illinois.

family, especially a teenager from a different background. In my particular case, my foster home was in Winnetka, a very rich neighborhood; the man was also very rich; his wife died soon after he took me in; nevertheless, he kept me. This man had a cook and a chauffeur. He was a grain broker; Chicago was a center of grain commerce in America. By then, I had already developed a socialist leaning and he was a very capitalist gentleman; he hated Roosevelt; and I was in favor of Roosevelt.

Did they have their own children?

They had no children.

Did you keep contact with him after you left his home?

A little bit. Our relationship was friendly, but not very close. He was very kind and helpful, but he had never been a father before, so it was not a well-defined relationship. He died rather soon after I had left.

When did your interest in science begin?

I was not aware of any such interest. I had the privilege of going to a famous high school in this rich suburb of Chicago on the North Shore, the New Trier Township High School. I wanted to become a doctor, but we had no money for that. My parents in the meantime had joined us in America with the help of my foster father, and when I finished high school, I left my foster home, and from then on I lived again with my parents. We were very poor; we had a little grocery store in Chicago. Instead of medical school, I went to the Armour Institute of Technology, which today is the Illinois Institute of Technology.

However, after two years I could not continue in the engineering school because I did not get a scholarship, and we had no money to pay my tuition. I got a job and continued my studies in night school. Engineering did not have night school, but the University of Chicago had a night school, so I went there. There I switched to chemistry from chemical engineering, and after a year, the University of Chicago gave me a scholarship, and I could become a full-time student again. This is how I got my Bachelor's degree in chemistry. In the meantime, Pearl Harbor happened, and we all had to go in to the army. There was a program at the University of Chicago, which trained people in some basic physics to make them useful to serve in the Signal Corps where they had new radar devices. I attended this course at the end of 1942, which took three months, and this course turned me to physics.

When I completed the course, rather than joining the Signal Corps, they sent me to MIT, where there was a laboratory to develop radar sets for bombing German civilians at night, and this is what I did during the war. I could also take some courses in basic physics at MIT. So it was by chance that I entered physics.

What happened after the war?

You see, when you interview people like me, and I suppose you pick us because we are successful, these are people, who had, naturally, a lot of luck. We may differ from others in various aspects, but most of all, in that we had more luck. Such luck was that during my stay at MIT during the war, I met some people, among them one of Robert Oppenheimer's former students. Oppenheimer was the person who, before the war, brought theoretical physics to the United States. Before Oppenheimer, there was no school of theoretical physics in America, and he was *the* man, who taught courses and the younger generation of theoretical physics were all his students. This man told me that when the war was over, I should go back to school, and I should go to Pasadena, where Oppenheimer was a professor. So I applied to become a graduate student of Caltech in Pasadena, but I needed money. Caltech could not offer me a scholarship, so I didn't go there, but had I gone there, Oppenheimer wouldn't have been there anyway. After the war he stayed on the East Coast. So I returned to the University of Chicago where I could get some support from the University, and my parents still lived in Chicago. After the war, there was a completely new faculty, which was absolutely extraordinary. I had the great luck of having Enrico Fermi as my teacher for my doctor's degree. Moreover, Fermi brought some outstanding students with him from Los Alamos; others came to study with him from China. They knew Fermi's name, whereas I had never heard of him. Marshall Rosenbluth, Lee and Yang, Marvin Goldberger, Richard Garwin, and Owen Chamberlain were among the students. Among the other professors, there was Edward Teller, Maria Goeppert-Mayer and her husband, Joseph Mayer, there was also Harold Urey, and there was James Franck.

For my thesis work, I wanted to do some theoretical work, but it's not always easy to find some theoretical project, so Fermi suggested to me an experimental topic, which was in cosmic rays, and I did that. Fermi was a magnificent teacher, and I didn't think much of this at the time, but later I appreciated it a lot. He devoted a huge amount of his energy and time to help his students and other students as well. He did everything he could to help people. I was the only one who did an experimental thesis with him; and he merely suggested the topic, and let me then do what I wanted. When I needed help, he helped me, in finding resources, for instance. In cosmic rays, you get more of them at the summit of the mountain than at its bottom, so after a month of running my experiment, he suggested to drive my equipment to a mountaintop and found a driver for me because I couldn't drive. He did not interfere, however, in designing the experiment and building it. I always found the right people who could teach me to do things what I needed. I have full gratitude and admiration for Fermi. It took me half a year to finish my project. My daughter is working on her doctorate in physics at MIT, and has already spent five years on it. In my time it was easier to find projects and it was faster to complete them.

Where did you go after Chicago?

I went to Princeton for a year to the Institute for Advanced Study in 1948. Just because Oppenheimer was there, I wanted to get back to theory, which was not easy, but finally I found something and did some work.

Hideki and Mrs. Yukava at the Institute for Advanced Study in Princeton, New Jersey.

It was a special year when Feynman and Schwinger had finished their work on learning how to do renormalization theory and figured out how to do higher-order radiative correction in electrodynamics, which was *the* thing to do in theoretical physics, but nobody before them knew how to do it. They presented their findings in a way that was difficult to understand; each had his way of explaining things. There was then a very young guy in Princeton, Freeman Dyson, who explained it to the world, and extended the theory. I had a wonderful year as a friend of Freeman Dyson; it was sheer pleasure. At the end of my year at Princeton, Gian Carlo Wick visited the Institute from Berkeley, and he offered me to become his assistant, which I accepted and joined him at the University of California at Berkeley. That was the only place at the time where they had accelerators, where it was possible to work with the newly-discovered π-mesons. So I chose to do experiments there. I was very successful in my experimental work there, and did a number of things in just one year, including the photoproduction of pions, the discovery of neutral pions, and measured the pion's mean lifetime, all in collaboration with others. The Lab was run more or less by Luis Alvarez, who took a very solid disliking to me, and made it impossible for me to stay at Berkeley.

Was your refusal of signing the loyalty oath a factor in your departure?

It didn't really matter, because I did not have a job, only a postdoctoral fellowship. Thus for me it was not a real sacrifice not signing the oath, but it was a different story for Gian Carlo Wick, whom I admired a great deal, who had just become a professor there, and Berkeley was *the* most advantageous place to be in physics at that time. Wick was a man of principle, he considered it an infringement on his political freedom, he did not sign the loyalty oath, and he had to leave. Most people though signed it because it was not easy not to sign it. There were also people who signed it, and yet left sooner or later because the atmosphere was so much poisoned. Wolfgang Panofsky was an example and Robert Serber was another of those who signed, but then left. Some others just did not sign it, and had to leave immediately.

Why was Alvarez against you?

I had no idea, but about fifteen years ago, when Serber was still alive, I wrote him a letter asking him whether he knew what Alvarez's reason was, and Serber wrote me that it was jealousy. What kind of jealousy, I don't know, but I was truly very successful during the year I spent at Berkeley.

The Nobel citation mentioned the neutrino beam method and the demonstration of the double structure of the leptons through the discovery of the muon neutrino. How did it come about?

This happened at Columbia University, and it wasn't my favorite experiment; I was not the key person in this experiment; the key person was Melvin Schwartz. He had the idea, and it's interesting to remember how the idea came to him. Until that time, everything about neutrinos had been learned from particle decays. The neutrino was discovered in beta-decay, and afterwards it was understood in other decays. It had been seen by Frederick Reines and Clyde Cowan in 1956–1957 in connection with nuclear reactors. T. D. Lee, who was a key person at Columbia, asked the question during the regular Friday afternoon coffee times — I was not present, but know the story from Mel's description — how could we learn something about neutrinos at higher energies rather than in the decays of particles? The idea came to Mel that it would be possible to do it in accelerators. It is also interesting to note that he couldn't have had this idea about these neutrino experiments five years before because the accelerators that existed five years before did not have enough energy to give hope for this kind of an experiment. When T. D. Lee posed the question in 1959, they were just building a new machine at Brookhaven, called the AGS machine, which made it possible for the first time to imagine such an experiment. It was Mel's idea, Mel had been my Ph.D. student at the time, and he and I and Leon Lederman together did the experiment.

What was Lederman's role?

You don't ask me this question; I don't remember what he did. Don't ask me too much about what I did either, but I did some things.

Why did you leave Columbia and join CERN?

There were all sorts of peculiar reasons. I had had some pleasure in returning to Europe, starting in the 1950s. As early as 1954, I was at a summer school in Varenna, Italy, together with Fermi. In 1956, I spent a sabbatical in Bologna and Rome and visited CERN in the same year; there is a nice picture of Heisenberg and me at CERN from that visit. I often came to CERN in the summers because I'd become interested in mountaineering, and I found someone here to go into the Alps with. I came here again for my sabbatical in 1965, first for half a year only, but I asked

Jack Steinberger and Werner Heisenberg at CERN in 1956. M. Patrascu of Romania, from the Dubna Joint Institute for Nuclear Research is in the background.

Columbia for an extension for another year without pay. However, Columbia said, no, they wanted me to return, which I didn't do, so they told me that if I didn't go back, I would lose my job there. I still didn't go back and started looking for a job here. Nonetheless, when after a year I wanted to go back to Columbia, they took me back. In the meantime, I had several job offers, including one at CERN. I also remarried and my two sons from my first marriage were old enough to take care of themselves, the youngest being fifteen or sixteen, living with my first wife. I liked it here, and I decided to settle here.

Yesterday I talked with Dr. van der Meer, and he told me that he contributed to your experiment here with the idea of the so-called horn.

Van der Meer is the man whom I admire more than anybody else here at CERN. He had this idea about the horn, but it came a bit later, not for the first experiments. His idea about the horn was a completely original idea; nobody had this idea before. It was used in many subsequent experiments. However, to make such a horn was not a trivial thing. In Brookhaven, Mel Schwartz and Lederman continued to make improvements in the experiment that we had started together, and I was no longer involved in the work. They tried to make a horn and did not succeed. Here, van der Meer not only had the idea, but made the horn. Later he had the idea of stochastic cooling, which involved very far from trivial mathematics and basic physics.

As I understand, you came to CERN for good in 1968.

I simply preferred to be here. There were some benefits and there was some price. I regret two things in coming here about which I didn't think much when I decided to leave the United States. One is that I left my children in America, and my contacts with them have diminished. The other is that there is no teaching here, and I miss the contact with students. On the other hand, in order to do the experiments in America, with Fermilab superceding Brookhaven, I would have had to commute to Chicago from New York for my experiments, all the time.

Would you mention any of your students from your Columbia days?

One of them was my wife; she was very bright. Melvin Schwartz was also my student. He had been in my class and one day he came to me and said that he would like me to be his thesis advisor. I tried to shift his interest toward theoretical work instead, but he insisted. It was a great pleasure working with Mel Schwartz. The third student in my class — I forget his name — had come from Harvard, but I didn't know that at the time when he was my student in the graduate class. When I looked at the answers he gave in the written examinations, I realized that he understood everything better than I did; he was strikingly brilliant. I talked with my friend T. D. Lee, who had this student also in his class, and he had the same impression. Later I learned that while he was at Harvard, he solved an outstanding problem in mathematics, which had been around for decades waiting for somebody to solve it; I'm not sure what the problem was. Afterwards he came to CERN sometimes. He had received his degree and he played the harpsichord. We played some chamber music in our home with me playing the flute, which is not very good, to say the least. He played the harpsichord well even when the instrument was falling apart. He was a special person, but not in very good shape psychologically. He had spent some time in psychiatric institutions, and never really did physics on his own. He worked with T. D. Lee, and solved difficult problems, but never developed his own direction in physics. It's a very interesting case how we can do certain things and can't do certain other things. This was my answer to your question about students.

Would you care to tell me something about your family?

Maybe it was because of the circumstances of my youth, but I did not have any experience with ladies until I entered the course for the Signal Corps at the University of Chicago. There was a very peculiar character there, someone by the name of Joe Heller, and he made a date for me. Joe Heller is also an author, and I sometimes wonder if these two people might be the same. But Joe played an important role in my life. He came and stayed with us in a small house that several of us rented together. He signed up for the same course and made a date with a lady who a few months later, at the end of 1942, became my first wife. Joan was a simple girl from Northern Wisconsin, who had a secretarial job with the Signal Corps, where she met Joe Heller. In a couple of years we had our first child; he was not especially wanted; in those days we didn't have abortions. So at 23 years I had a boy, who is an interesting character, and has his second wife now, who is 35 years younger than he is, as far as I know. Then came a second boy, who did not finish college, but has become very well known in the music world as a designer and builder of musical instruments. He's certainly much better known than his father. Joan didn't like my style of life and eventually found her own way; in the late 1950s, she studied art in New York; started painting, and found a different social group. Finally Joan left me in 1960. We divorced in 1962, and I immediately married the next girl around, who is my present wife, Cynthia, who was my student. We have two very nice children. Cynthia and I did some research together for another few years, but later she switched to biology.

As I understand, the experiment for which you received the Nobel Prize was not your favorite work; which was that?

The experiment that I think was most useful didn't involve any deep idea or important invention; it was just something I thought had to be done. It was looking for the π^0. It was part of the experiments when we were looking for the π-mesons at Berkeley. The same is true when I came to Columbia University in 1950, and did the experiments on π-mesons in the bubble chamber. Again, they were obvious experiments, we made some progress in how to use bubble chambers, but we did not invent the bubble chamber. What I remember with especially great pleasure were some experiments that were not terribly important, but there I could use some ideas of mine; the experiments were related to the resonances that people discovered in the 1950s and 1960s. There were particles produced, but they decayed before having been observed due to their extremely short lifetime, and for which the decay products were observed. So the experiments were about these decay products.

You mentioned that you have a photograph picturing you and Heisenberg together. I wonder if you could share with us what you think about his role in the German atomic bomb project and what you think about his behavior during World War II, and what is your impression about the play Copenhagen?

First I would like to comment on what the American physicists have against Heisenberg. He visited the United States just before the war and his visit included Columbia University, but it was before my time there. He then returned to Germany, and stayed there during the war, and the Americans did not forgive him for that. I, on the other hand, understand him and I understand why he went back. Nobody ever accused Heisenberg of being anti-Semitic and he was never a member of the Nazi party. Apparently, his mother was a friend of Himmler's mother, but I can't accuse Heisenberg of that. I can understand Heisenberg because I was brought up in Germany, and it was taken for granted by myself and by everybody that in case of war you owe your life to your country, and you don't question your country. Even if he was not an admirer of Hitler, Germany was at war, it was his duty to do what he could for his Vaterland. This is how I view Heisenberg's return to Germany. Then, this project came up, the possibility of making a chain reaction and creating a bomb; the basic discovery was made in Germany, by Otto Hahn, but the resources they had were inadequate. Nobody knew in advance — either in the United States, or in Germany — how difficult it would be to create the atomic bomb. As it turned out, they had huge resources in America and the Europeans, especially Szilard and Fermi made the first reactor, Szilard being the key person. I knew him a little and I wish I knew him a little better. They had nothing comparable in Germany, and even if they had had the people, they still would've lacked the technical resources. It took a huge effort in America to finally build the bomb at Los Alamos, utilizing the input of Chicago and Oak Ridge, and the rest. In Germany, they never even got close to understanding the enormity of the efforts they would have needed to build an atomic bomb. The Nazi party had a meeting with the people who were involved with building the bomb in 1942. They had to make up their mind what projects to support, and at the end they supported the Penemünde project for the rockets to attack the British rather than the atomic bomb.

Heisenberg visited Bohr because Bohr was the key person for all the physicists at that time, and Heisenberg had spent some time with him years before, when Copenhagen was one of the top centers of physics in Europe. By the time of Heisenberg's visit in 1941, Denmark was under German occupation, and Niels Bohr was not very happy that Germany was occupying Denmark. The purpose of Heisenberg's visit was something technical, and he just used the opportunity to see Bohr. Their personal relationship was independent of the circumstances under which the visit was taking place. In my opinion Heisenberg's project did not play a role in his visit. Heisenberg didn't know enough nuclear physics to have been very serious about it. In his visit I merely see a former pupil visiting his old master and friend. There is then the famous letter that Bohr wrote afterwards but never sent. I read it, and I just couldn't follow what he wanted to say with his letter.

Why do you think Melvin Schwartz left academia?

Everything is complicated. I may not know exactly his reasons, but it may have played into his actions that he liked gambling. I would never go to Las Vegas, but he liked to go there. He took a job at Stanford, and he was not very successful there. Even if you are very bright, sometimes things don't work out. Also, he had problems in having his colleagues get along with him. He didn't always get the support he thought he should get, and he became frustrated with his situation. Then one day someone came to his lecture in short pants and asked Mel whether he could sit in his class, just auditing it? The person was one of the future founders of Apple Computers, which shortly became tremendously successful. From this encounter, Mel got the notion that he would be happier doing something in the informatics business. He had an idea; he established a company, it managed to survive, but it was always marginal. At some point Mel wanted to get out, but he didn't know how to get out. It was the Nobel Prize, which finally provided a way for him to get out, and he went back to academia. He went to Brookhaven and became some sort of an advisor to the management, then he went back to Columbia University, but he became quite sick, and he retired, and has lived in the mountains in Iowa ever since. He was not terribly lucky in his later years.

Did the Nobel Prize change your life?

Of course, I sit here in my office; people don't dare to throw me out. Had I not won the Prize, they would've thrown me out. And you are here. If I hadn't won the Nobel Prize, I could do some honest work instead of talking with you. It certainly has changed my life.

How do you view the Nobel Prize as an institution?

I don't think the Nobel Prize is a good thing. The whole business of prizes in science does not have much sense. Maybe it did at one time — and even that I doubt — but it certainly has no sense now. Einstein would be Einstein for me with or without that Nobel Prize. You need a lot of luck to stumble into something. The question of merit is also ill defined. It takes so many people for science to go forward. In my field a thousand people may be working on an experiment, and the individual contribution may lose its meaning. Van der Meer is an exceptional guy, but there are others who may be also bright, but did not get the Prize. The fact that van der Meer got

Jack Steinberger in Stockholm, 2001 (photograph by I. Hargittai).

it was an accident. They got the Nobel Prize for the experimental discovery of the *W* and *Z* particles, but the idea came from three guys, Rubbia and two others, Cline was one and the other was a young guy, and I suspect that the original idea was that young guy's, who wasn't there when the experiment was finally done. To whom the merit belongs is not that obvious in our business. Then, the Nobel Prize has this unfortunate feature that it provides notoriety, and people give credit to Nobel laureates' opinions, which they shouldn't. When Mr. Charpak voices his opinion on an issue, people take note of it in France regardless whether it is in the scope of his expertise or not. The merit of his opinion has usually nothing to do with the work for which he got the Nobel Prize. The notoriety of the Nobel Prize, and especially that of the Nobel Peace Prize, gives a platform to the laureates. Some of them make such a use of this notoriety that I regret. I consider prizes evil because they make unjustifiable distinctions between people. The real privilege is doing science; society supports scientists to do what they like to do. Why do you need a prize on top of it?

You have mentioned luck as a necessary element to be successful in science. What else would you single out as necessary for success?

I was lucky; I also had a character that made it possible for me to concentrate on what I was doing. I enjoyed having results. I wanted to have some style of living, but it was compatible with doing science. Other things were obviously missing from my life. In my productive years, I always felt that the next step followed from the previous one in a natural order, without much reflection about what should be done next. Now, it's different; now, it's not obvious what I should do. I didn't face this problem in my productive years, but it may be that I might have been better off if I had reflected more, if I had questioned doing what seemed so obvious to do then.

What was the greatest challenge in your life?

[Long silence.] I don't know. It wasn't easy to learn quantum mechanics. Now I'm trying to learn cosmology; there is a lot of physics, which enters these studies, and I'm finding it very hard. I have a lot of frustration; part of this may be my deteriorating brains; part of this may be that the field has become very technical with many different people doing many different things. It's more than I can manage.

How do you see the future of particle physics?

I wish you hadn't asked that. Clearly, there are many unanswered questions and riddles: supersymmetry, dark matter particles, cosmology, basic questions like why are the masses, the interaction-strengths what they are. The problems are very difficult, and for me it's impossible to make a prognosis of how one might hope to make progress. There is good reason to expect from the Large Hadron Collider, to find supersymmetry and maybe even find the Higgs. That would mean enormous progress.

Building these big machines requires sacrifices from society. What if such support halts? What would then happen to particle physics?

I'll volunteer something to answer this impossible question. What basic physics can provide is interesting to me, and I would like to know the answers to the outstanding questions just for my pleasure. If we have some kind of commitment to human culture, if we ask the question about the contributions to human culture in the twentieth century, particle physics figures heavily in it. Relativity was invented, so was quantum mechanics, and we understood something about particles. To me — and I am biased — this is the greatest achievement of the twentieth century. For society, direct benefits mean technical progress in informatics, transportation, medical progress in extending human life, the fact that I am still alive, and so on, but I am not in agreement with such views. To me the chief effects are that we're destroying our ecology in a massive way; the people at the time of my parents were better off because society was more stable; they didn't have to worry about their jobs from one year to the next; I'm very unhappy about the instability of what we have today. There is general economic instability and there is increasing rather than decreasing tension in particular, between the rich and the poor. My opinion is that a lot of these problems have been created by the progress of science. So it's not obvious to me that we are helping society all that much. The immediate problem of the future is how to resolve these problems. Karl Marx tried to find a way, but did not succeed too well, but at least he had some ideas that weren't so bad. Some of his things were not thought through; for example, what he meant by the dictatorship of the proletariat was not obvious. For me the chief problem today is how humanity with all these technical resources can find a social organization which allows people to live together; for America not to invade Iraq; for Africans not to live much more miserable than Americans do, and so on. These are the social aspects.

Where does terrorism come into your equation?

Terrorism is a very particular thing; to me it consists of two basic components. One is religion; and it's not restricted to one particular religion. But if we take Islam for example, they used to be on top of society in the tenth and eleventh centuries, and they're now nowhere, and this breeds frustration. The cultural frustration is then combined with economic frustration and political frustration. They want to offer something to society, and my guess is that they are more egalitarian, and poor people can find themselves more comfortable in that society than in some others, but I'm not about to become a Muslim. I have visited Toledo in Spain — where I would've not gone before Franco had died, but I went afterwards — and I saw the synagogues in Toledo, which the Muslims allowed there; Maimonides is also an important part of Jewish heritage, and all this comes from the tolerance of that people and from that time. How Islam has further developed, I don't know.

I would like to tell you about one experience. It happened two weeks after September 11. We celebrated the one hundredth anniversary of Fermi's birth at the University of Chicago, which I attended, and then I went on to visit my youngest son, who at that time was studying mathematics at Waterloo University in Canada. It is not very far from Chicago, but I had to fly to Toronto, and I took a limousine from Toronto airport to Waterloo. I was alone with the driver; he was black, and I asked him about his life. He was born in Eritrea; he studied some kind of engineering in Yugoslavia; he got a job in Somalia; but with the war between Somalia and Eritrea, he found himself out of his job; and he was now a cab driver in Canada. I asked him about his religion, and he was Copt — Christian Orthodox. He told me that the dominant religion in his region is Christian Orthodox, and Catholics and Muslims were in the minority. I asked him how they got along, and here comes the answer, which was absolutely vitally interesting for me. He told me, "When I was a child, we did get along fine; I had Muslim friends and when I went to church, they played in front of the church till I got out. When they were in mosque, I played in front of the mosque till they got out. But now things are different and the different religious groups don't get along." To me it was interesting to see that we can't look at these things as static; they are dynamic; they evolve with time. The Islamic fundamentalism is the product of the last decades and it's still growing.

Are you religious?

I'm not although I have some appreciation of religion. I'm now a bit anti-Jewish since my last visit to the synagogue, but my atheism does not necessarily reject religion. My last visit made me very hostile toward the repetition of the phrase, which I remember from my youth, when I was a religious Jew. The phrase is that "The Jews are God's chosen people," and I can't accept that. If there's a god, he didn't choose any people.

Where do you live now?

We used to live in France, but we moved to Switzerland when our children reached school age; we thought that Switzerland offered better schools.

How do you feel about the behavior of Switzerland towards the refugees during World War II?

There was a conflict between the right and the left in Switzerland before the war, and by the time the war came, the right prevailed. Then, the Swiss wanted to stay out of the war at all cost, but they were very close to Germany, and this motivated their policies. Besides, I'm sure there were many Swiss who were perfectly happy with Hitler. Concerning immigration policies though, my parents couldn't have entered America if my foster father hadn't sent them an affidavit about his taking responsibility for any financial burden in connection with my parents' immigration.

You mentioned Fermi as your mentor to whom you looked up. Did you have other mentors, people you looked up to?

I had one at MIT, Laszlo Tisza. He was my first real physics teacher.

I know him.

Fantastic. I like him very much, and am very much indebted to him. We still correspond from time to time. Then, there was Gian Carlo Wick whom I've mentioned before. Then, there were Lee and Yang, with whom we were friends all our lives although Lee is no longer my friend for reasons I don't know. Then there was George Placzek. Oppenheimer had some weaknesses, but he also had some strengths. One of the pleasures of my life has been to be able to get to know people whom I could admire.

May I ask you about Bruno Pontecorvo?

I knew him in 1947 when I was working on my thesis at Chicago. He worked in Canada, and came to visit Fermi in Chicago. We were doing very similar physics, but he was already an established scientist, and I was a student. He was very generous. He was one of the most original physicists with brilliant — sometimes silly — ideas, but always remarkable. He then went to the Soviet Union, where he was limited by the resources of Dubna, which could not compete with America. In his later years, in the Gorbachev days he was able to come out for visits to Italy, and he came here, and I introduced him when he gave a seminar. Pontecorvo was the first one, before Schwartz, to suggest not that it was possible to do neutrino physics, but that one should look for that second neutrino, the experiment that we eventually got the Nobel Prize for. He had this idea a year before, but he could not do it in Dubna because the facilities didn't exist there. Looking for neutrinos from the Sun was also his idea, which played a big role in recent years. I respect very much that Pontecorvo went to the Soviet Union; he had certain ideals; he stuck to them regardless of the cost in personal sacrifice.

Was there anything you missed during you career; something you regret you didn't do?

[Long silence.] There have been lots of things done by others that were much more interesting than what I've done. Others had much luck, but so did I; I can't be unhappy about my life.

Reference

1. An interview with Leon Lederman appeared in Hargittai, M.; Hargittai, I. *Candid Science IV: Conversations with Famous Physicists.* Imperial College Press, London, 2004, pp. 142–159.

Masatoshi Koshiba, 2005 (photograph by M. Hargittai).

36

MASATOSHI KOSHIBA

Masatoshi Koshiba (b. 1926, in Toyohashi city, Aichi Prefecture, Japan) is Councillor at the International Center for Elementary Particle Physics at the University of Tokyo. He studied physics at the University of Tokyo and received his Ph.D. from the University of Rochester, New York in 1955. He was Research Associate at the University of Chicago, 1955–1958, and Associate Professor, later Professor at the University of Tokyo from 1958 till 1987. He spent three years as acting Director at the Laboratory of High Energy Physics and Cosmic Radiation at the University of Chicago between 1959 and 1962. He was Professor at Tokai University from 1987 to 1997.

He shared half of the Nobel Prize in Physics in 2002 with Raymond Davis, Jr., "for pioneering contributions to astrophysics, in particular for the detection of cosmic neutrinos". He received many awards, among them the "Grosse Verdienstskreuz" from the President of Germany (1985), the Order of Culture from the Japanese Government (1988), the Bruno Rossi Award from the American Physical Society (1989), the Alexander von Humboldt Prize from the Humboldt Foundation (1997), the Order of Cultural Merit from the Emperor of Japan (1997), the Wolf Prize in Physics (Israel, 2000) and the Panofsky Prize from the American Physical Society (2002). We recorded our conversation on May 6, 2005, in Lindau, Germany.*

---

*Magdolna Hargittai conducted the interview.

I noticed that you did not submit an autobiography to the Nobel site, only a curriculum vitae. Therefore, I would like to start with your family background.

Sure. My father was an army officer and when I was about four years old, my mother died. Then my father remarried and his new wife already had two sons, so I have a younger and older sister and two younger brothers. In those days there was a country called Manjuria and as an army officer, my father had to go to Manjuria for a while. He took all of the family with him except me because, as a first son, he wanted me to become an army officer. So he told me to remain in Japan and take the entrance exam to the junior army college. Thus, I remained in Japan and stayed in my uncle's house. When I was in my first year of high school, I was stricken by polio and my arms and legs became numb. But somehow with a lot of training, my two legs and my left arm became usable again. My right arm is still numb. So this was the condition I had just before the end of the war.

When the war ended I was in the first year of the junior college at the University of Tokyo. My father was detained in China as a prisoner of war and came back to Japan one year later. But the American occupying forces forbade the former army officers of Japan to take jobs, so we had no income. Therefore, my elder sister and myself had to do almost anything to earn money for our living. This was still the situation when I graduated from the Department of Physics at the University of Tokyo, and I didn't really know whether I should stay with physics or not. Then in the second year of graduate school, I encountered an experiment, using what is called nuclear emulsion, which was invented by Cecil Powell. This little experiment made me an elementary particle experimentalist.

If you were working on nuclear emulsions the best place to go, in the whole world, was either Bristol University or some of the American universities. I was lucky enough to be recommended by Professor Sin-Itiro Tomonaga, who is a Nobel laureate, to go to the graduate school of Rochester University. I got my Ph.D. degree from Rochester and then moved to the University of Chicago. After about 3 years there, I was offered an assistant professorship in Japan, so I went back to Japan. Then two years later I received a letter from a professor at the University of Chicago, who was my boss earlier. The letter said that he started a very large international collaboration project with a big balloon and a big emulsion block. He wanted me to help him there. So I took a leave of absence from the University of Tokyo and joined the group at the University of Chicago. Four months later this professor, who asked me to go there, got a heart attack and died; probably because of this big project; he had a one million dollar research grant and a big emulsion block to be exposed by big-balloons as a high-altitude cosmic rays experiment. Then the University of Chicago appointed me to lead that big international collaboration team and somehow I did that.

After about 3 years of this international collaboration work I came back to Tokyo again and started teaching in the graduate school. Then I was thinking: what is a good project for a young student to work on and then at that time I was approached by a Russian physicist, his name was Budker, a very famous physicist. There is an institute of nuclear studies in Novosibirsk and he was building an accelerator there, in which electrons and positrons were colliding. This e^+–e^- experiment was very new those days, not many people worked on that. Budker asked me to come to Siberia and do the joint experiment there and I thought that this would be a very good opportunity because this field certainly has a bright future. Unfortunately, however, two years later Budker also got a heart attack. So in order to do e^+–e^- experiment, I came to Europe and looked around. I found a German e^+–e^- colliding machine in Hamburg, so I joined one of their experiments. Our best collaborator was a Heidelberg group headed by Professor Heinze and through this collaboration with the German physicists, I received a nice medal from the Bundespresident von Weizsäcker. That was the first medal I received from a foreign country. That was about the time when I started the experiments at Kamiokande.

Was it your idea to start the Kamiokande experiment?

Yes.

As I read originally your idea was to study proton decay.

That's correct. I tell you a little about that. As you pointed out, the Kamiokande experiment was originally built to search for proton decay in such a way that when the protons decay into positrons and neutral pi-mesons, pi-mesons decay to two gammas and these high-energy gammas initiate a, what is called, "cascade". High-energy positrons also initiate a cascade. Therefore, we expected to have one high-energy cascade in one direction and two cascades in the opposite direction. Those cascades produce a special light, called Cherenkov-light in water, thus one should be able to see one Cherenkov-light on one side and two Cherenkov-lights on the other. This is a very simple detection, and therefore, if we have a large amount of water, surrounded by light-sensitive devices, the detection should be easy. So I designed three thousand tons of water surrounded by photomultipliers, located one thousand meters underground. It has to be so deep so that we can exclude most of the cosmic ray particles. With the same design, that is, a large amount of water deep underground, surrounded by photomultipliers, but about ten times more research money and about seven times more water, an experiment is being prepared in the United States now. As you can easily imagine that if protons actually decay, then the big American experiment will find it first. Indeed, they found one event after the other and after about five events, we also found one. This is reasonable. But I had to think very seriously, how to compete with this big American experiment. At this stage, I could not expect to have more research fund. The amount of research money is fixed. That means that the number of photomultipliers is fixed. The only way to compete with the American experiment is to make individual photomultipliers that are much more sensitive than the American ones. So this is what we did; we developed the largest photomultiplier. This development was successful and when we installed this big phototube in our Kamiokande, it worked very nicely. I immediately noticed the possibility that with this good sensitivity we can observe solar neutrinos. When they hit electrons in the water, these electrons will move in the direction of the neutrino and then stop. We can measure these electrons by means of the Cherenkov-light that they emit. By the observation of electrons we can determine when this electron started in this water, in which direction and with what energy. All these things can be measured. That means that we can get the information about the neutrino that hit the electron. We can learn when it came, in which direction, in what energy spectrum. Thus we can make astrophysical observation of a neutrino!

So I decided to do this seriously and decided to make the necessary improvements at Kamiokande, which took about a year and a half. At the very beginning of 1987 we were ready to start taking the data on solar neutrinos. Luckily, within less than two months, a supernova occurred in the southeast and then immediately we observed about 12 neutrino reactions in the detector, coming from the supernova. And that is the story.

I understand that your experiment had an advantage over the American and Italian underground detectors in that you could measure the direction of the neutrino and also that you could detect them in real time.

Yes, that is right.

Later you built the so-called Superkamiokande experiment as well. It probably costs a lot of money. Who finances it?

It is the Ministry of Education of Japan.

So it is a Japanese operation alone, not international.

Yes. But it was very difficult to get that much money because it's cost was a hundred million dollars. Fortunately, the Ministry of Education was very happy because we discovered the supernova neutrinos and this was reported by all of the world's mass communication systems. Not only that, we, of course, published our results of the astrophysical observation of the solar neutrinos. Not only that, but we also discovered that neutrinos oscillate, which was a completely new discovery. Therefore, the Ministry of Education was very happy and with a little push we could get the hundred million dollars for Superkamiokande.

These experiments are very expensive. But the other possibility for experimental particle physics is the big accelerators that are even more expensive. How do you see then the future of particle physics?

There is one project, which is already a global project; a very high-energy electron-positron linear collider experiment. Instead of a circular ring it uses a linear arrangement. This is considered seriously as a future project for all elementary particle experiments globally. Now people are discussing where to build it and how to finance it. This will cost about 6 billion dollars.

There is also another question related to this. Although the information we learn from these experiments is extremely interesting, considering the costs, isn't it too much of a luxury to do them, when there are so many more imminent problems that cannot be solved due to lack of resources.

You look at how much money is spent for what purposes. The amount of money spent for basic science research is not a large fraction of the national resources. Even in those countries where the support of the natural

Masatoshi Koshiba during the interview, 2005 (photograph by M. Hargittai).

sciences is favored, it is less than 10%. When you are talking about hunger and poverty it is a matter of much, much bigger budget sizes. It should not be compared directly.

I have a question about the Nobel Prize. You shared it with Ray Davis, what do you think of the involvement of John Bahcall? He is the one who did the theoretical calculations.

I don't think that I am justified to make a statement on that matter because this is exclusively in the hand of the Nobel committee. The experiment of Ray Davis was a very difficult experiment and after about 10 years of most difficult times he could produce his amazing results. Although the flux was only one third of what was expected. This has been the shock to every physicist. I think he deserves the Nobel Prize. But as to Bahcall, I have nothing to say.

You mentioned that your father was in the military in World War II. What is your opinion about the recent disturbances in China concerning a revisionist textbook published in Japan?

It is true that there is some difference in understanding the history between the two countries and unfortunately there is a National Shrine, the Yasukuni Shrine in Tokyo that also honors about 12 or 20 war criminals. This is a reason for the Chinese attack. I feel that it is a silly thing that the Yasukuni shrine incorporated those generals together with the millions of soldiers who died in the past. But the Chinese government is also very neurotic in the sense that if another country's minister pays a visit to such a shrine, this is not their affair, it is that other country's affair. This is the opinion of the majority of the Japanese people.

Do young Japanese people today know about the history of Japan in World War II?

Of course, they learn that the Japanese army did many bad things during this period. But if we are talking about soldiers doing very bad things, that happened to every country's soldiers, like the American occupation army did so many bad things in Japan. Which can be criminalized but they kept quiet.

Did your father eventually get a job?

Yes, but a very modest job. I had to support him.

Did he live to witness your successes?

He was happy when I became full professor.

I would like to ask you about women in science in Japan.

In the field of chemistry, the number of lady chemists is larger percentage-wise than the number of lady physicists. We do have a number of good ladies in physics, but the number is not very large.

Will it ever change?

It depends on how the ladies feel. Most ladies are interested in fashions and other things.

Tell me something about your present family.

I have a son, he is professor of engineering, in one of the Japanese universities. I also have one daughter who is married to a dentist and she also has a daughter.

Masatoshi Koshiba with his wife and surrounded by students at the Lindau meeting of Nobel laureates, 2005 (photograph by M. Hargittai).

How about your wife?

She is a housewife.

If you have to single out one of your achievements, which one would you say was the most exciting?

The ones that I aimed at and accomplished. It is the astrophysical observation of the solar neutrinos. In doing this experiment I got two bonuses. One is the supernova neutrino and the other is the neutrino switch.

What are your future plans?

I am an old man, I don't think about the future anymore.

What was the greatest challenge you had to face in your life?

Let me see. It may be the time when I tried to get funding for the e^+–e^- experiment in Germany because the amount of money we needed was very large and it was not easy to get it. But I finally succeeded.

Do you have heroes?

Who could that be? I know one whom I admired very much, Professor Tomonaga.

Any mentors?

No.

Is there anything you have missed in your career?

I am satisfied with what I have got.

Riccardo Giacconi, 2005 (photograph by M. Hargittai).

37

RICCARDO GIACCONI

Riccardo Giacconi (b. 1931, in Genoa, Italy) is University Professor at the Johns Hopkins University in Baltimore. He received his Ph.D. at the University of Milan in 1954 and was an assistant professor there till 1956. He spent a year as a Fulbright Fellow at Indiana University and two years at Princeton University. He worked for the American Science and Engineering Company in Cambridge, Massachusetts from 1959 till 1973; was Professor of Astronomy at Harvard University (1973–1981); served as principal investigator at several NASA programs; was Director of the Space Telescope Science Institute (Baltimore, MD, 1981–1992); Director General of the European Southern Observatory (1993–1999), and President of Associated Universities, Inc. (1999–2004). He received half of the Nobel Prize in Physics in 2002, "for his pioneering contributions to astrophysics, which have led to the discovery of cosmic X-ray sources". He is a Fellow of the American Academy of Arts and Sciences (1971); member of the National Academy of Sciences of the U.S.A. (1971); of the American Philosophical Society (2001); and other academies. He has several honorary degrees and received many awards, among them the Wolf Prize in Physics (Israel, 1987), and the U.S. National Medal of Science (2003).

Dr Giacconi founded what is called X-ray astronomy and is often called the "father of X-ray astronomy". He built the first X-ray detectors and launched them on rockets in 1962. X-rays can only be investigated with instruments placed in space because they are found towards the short-wavelength end of the electromagnetic spectrum and are thus absorbed by the Earth's atmosphere. He was the first to detect X-ray sources outside of our Solar System and also the first who showed the existence of X-ray background radiation in the Universe. It was also he and his group that proved for the first time the existence of black holes — as sources of X-ray radiation. The first X-ray satellite, named Uhuru, was also built and launched under his direction. Dr. Giacconi was the leader of the team that proposed the Chandra X-ray Observatory, the most powerful instrument of its kind. Chandra can detect X-rays emitted by particles up to the last second before they fall into a black hole and the light from some of its sources has traveled through space for ten billion years.

We recorded our conversation on May 6, 2005 in Lindau, Germany.*

You were a child in a turbulent time in Italy. How did it affect you?

I don't know the answer to that question. I am writing a book now in which I ask myself the same question, but I don't know the answer. Another thing I can tell you is, that when it is the Fourth of July and everybody launches fireworks in the United States, I feel like I would like to go to a shelter. I think that life was more accelerated. In the European education system they try to cram you full of information. When you come out of that at high-school graduation, you do know a lot. In the United States you catch up later. Then, of course, you had to be somewhat responsible. You had to help to find food for your family, for example. So, in a way, I felt that we were deprived of a part of our childhood because we grew up faster.

I also read that your parents divorced when you were quite young. Who did you stay with?

With my mother. But I kept in contact with my father as well.

How did your interest in physics start?

I have a story that people don't believe, but I'll tell you anyway. I did one year in elementary school, then the teachers decided that I didn't need to do the second year, so I went to third year. Then my mother was not happy with my grades, so I did the third again. By the time I got close to the end of high school, I was tired of school. But there was a way in Italy in which I could take an exam and not have the last year in high school. The requirement was that you had to have at least 80% high grades in each subject and then you could take the exam. So

---

*Magdolna Hargittai conducted the interview.

I did that. I took some exams in June, others in October, and I passed. The worse grade I got was in physics.

At that point I had to decide how to continue. I had many interests, architecture most of all. But I guess I was an ambitious child so when I was thinking of architecture I was thinking of at least becoming as important as being able to change people's mind about shapes as, say, Corbusier, or others did, and I was worried that I would not be creative enough. Then there was philosophy that I was very much interested in, but there the prospect was of becoming either a politician or a teacher and I didn't want to become either. My mother was trying to push me towards engineering as she thought I had a mind for it and also in Italy engineering was a respected profession. But I didn't like that because I thought that it was boring and intellectually not challenging. So I chose physics because I thought that even if I was mediocre in it I could still do interesting things; nuclear power was coming out about that time, for example. But if I was lucky and I happened to be gifted, I could have an interesting and intellectually challenging life.

You studied at Milan and worked with cloud chambers. Why did you go to the United States?

I got a Fulbright fellowship. There was one person in the U.S. I wanted to work with, Robert Thompson, at Indiana University. He was doing research on cosmic rays in a manner I liked. When I arrived there he seemed to be in some kind of trouble and never published again. I stayed there for a year and after that I got an offer to go to Princeton to finish my three-year fellowship. I went there because I felt I would learn more there.

Then my visa was expiring but I didn't want to go back to Italy or Europe because I didn't see any prospects there. I was looking for a job and finally I got an offer from a little company, American Science & Engineering, Inc. in Cambridge, Massachusetts, which, however, had very good people. The chairman of the board was Bruno Rossi who was also professor at MIT, a very good physicist. He also happened to be the teacher of my mentor, Giuseppe Occhialini, who was also a world-class physicist. He worked with Patrick Blackett and they together discovered pair production, and he also worked with Cecil Powell when he discovered the pion; I don't know why Occhialini never received the Nobel Prize but at least he got the Wolf Prize. This company offered to help me with my visa and my job was to start research in space. That was in 1959, the Sputnik just happened, and it was a wide-open field and it was a wonderful opportunity for a young man.

So I started to work there. I was a physicist and I got exposed to space physics. Space physics has no meaning. What does have a meaning is the size and the enormity of what you can do in space. You can do all kinds of research; you can look at the Sun, you can look at the planets or the stars, many different things. So I started to look at different things and eventually narrowed down because Bruno Rossi suggested to me, "Why don't you do X-ray astronomy?" I looked into it and I realized that my background as a physicist allowed me to design instruments, it allowed me to understand the problem and set it as a physical problem, which was to make an estimate about what to expect if Nature is like I expect it, estimate about what kinds of instruments I need, what kind of technology must I invent that is not there. All of this happened at once. Of course, I also used instruments that were already available, such as Geiger counters, anti-coincidence systems, etc. I felt that using all these I could make a new instrument a hundred times better than any of the previous ones. But I had an idea for another one that was a million times better. By now it has become a billion times better. So there was opportunity at this little company.

Nature was also very nice, very abundant. I sent up the rockets. The first one failed, with the second one the door didn't open, so there were small problems. But I had a very good group of people around me and we built the sensitive instruments and then put them on the rocket provided by NASA, it's just like putting something on a car. We soon got results, showing that there was a type of star that was not known to exist and a process in Nature that was not known to take place. In fact, it was known to exist but was not believed to be important for astronomy. It was that if you have a very high temperature gas, about a hundred million degrees, the emission of that gas occurs mostly in the X-rays.

You mentioned in your talk that you give lectures on how to organize big projects. What is the biggest project you organized?

One, in which I organized every part, was a very large telescope which was a billion German marks, so about half a billion dollars. That was for the European Southern Observatory. Then, a project where I organized only a piece, was the Hubble Space Telescope Science Institute, where I was the scientific head of the project. That project was huge, about 1.7 billion dollars, but there I didn't have direct management control. I developed this idea that you develop by magic, that is, to manage without the power to manage, only having the power of persuasion. And that was very interesting! The last one that I was involved with was a billion dollar project. That was while I was Director

General of the European Southern Observatory (ESO) and it involved the cooperation of ESO, the United States, and Canada to develop and build a big observatory being put in place in the Atacama Desert in Northern Chile.

How did you learn the skills for this, this really isn't just physics.

Well, by doing smaller programs and then larger and larger ones. I was given the responsibility more and more to manage things and it was clear that I could lead people and persuade people to do things. Within the corporation, when I started in 1959 I was alone, maybe a couple of people to help me, but by the time I left, ten years later, there were 500 people in my group alone.

What are the possibilities in the future for X-ray astronomy?

At the moment with the X-ray telescopes we are reaching a sensitivity equal to that which you can reach in any wavelength. We can see X-ray sources that are very weak, very far-away, most of which we can see with the Hubble telescope but there are some others that the Hubble cannot see. There we try with the ground-based telescopes and if they cannot see them, we try with the Spitzer telescope that is an infrared telescope and then we can see them. With X-ray telescopes we can see objects as far away as you can with any wavelength. The thing that is peculiar about X-rays is that, for example, I had an experiment with Chandra, looking at the region of the sky, and we saw these very, very faint sources and 90% of them are supermassive black holes. That is one thing we do. The other one is the discovery that in clusters of galaxies the space between the galaxies is filled with gas that has a very high temperature. That sounds curious but the point is that the mass in the gas exceeds by a factor of ten all the mass of the stars and everything that is in the galaxies. And this gas has been heated by the collapse of these galaxies together. So you can see very far back and you can ask questions about the formation and evolution of the Universe, as this is the largest body of matter congregation in the Universe. This gives us a tool to study the formation and evolution of the Universe. There is also the possibility to test general relativity in strong field situations such as the black hole.

Which of your work are you the most proud of?

The one that I did while I was quite young, the discovery of the first X-ray star, the invention of the first X-ray telescope, and the first satellite that I did in 1970. That's what was the most rewarding.

What was the greatest challenge that you faced?

Human beings. Organizations, my colleagues, scientists, skepticism, things like that.

Are there any mentors of yours whom you would like to mention?

I am thankful to Giuseppe Occhialini, to Robert Thompson. I have mixed feelings about Bruno Rossi. Obviously, I am very grateful to him for suggesting the problem of X-ray astronomy, but later on we didn't get along very well.

Any heroes?

Fermi.

Perhaps this is obvious because he was also Italian.

No, it is not that! It is that I hate the division between theory and experiment that people often do. Fermi was a student of Nature who asked himself questions, then did experiments to find the answer, then came up with new questions, and so forth, so he was a complete man and that is my idea of a physicist. It starts off from Galileo who said, "To try and try again." What he meant by this was that you make a prediction, you try the model, you learn something from it, then you develop a second model and you test it again. It is a continuum of finding things, developing concepts and models and testing them again. This was, I think, what characterized Fermi's work as well. That's why he is my hero.

Did you also manage to be a theoretician and an experimentalist at the same time?

I am an experimentalist. I am not as good as Fermi was. But the time when this question came up most was when we had this little satellite and it was so much more powerful than anything else before, it was like a new world. We saw one of those binary sources. The first time we saw it we asked what it was and how we could look at it. Then we decided what to do next and we slowed down the spacecraft to be able to look at it better. Then we tried to fit a function, which didn't work. Then somebody came up with another idea and we tried that, too. We had meetings every week with huge debates about what to do next, so it was a constant interaction of theory and experiment. Finally, in about two years, we had a full model and knew what we were looking at.

Has the Nobel Prize changed your life?

Well, many pleasant things happened. There is also some feeling of obligation towards the community. I go to places to give lectures, I sit down with you …

Please, tell me something about your family. I liked very much the way you wrote about your wife in your Nobel autobiography.

Thank you; it is true. We have two children, two daughters. We also had a son who died in an automobile accident and parents can never get over that. We had a very difficult time. Our daughters are not scientists, both are in banking and making a good life. I have two grandchildren, they live in Chicago. I wish I spent more time with my family when we were young; now it is too late.

What are your future plans?

I am writing a book. I retired from the job I had and am now a University Professor at Hopkins at the Physics and Astronomy Department with no teaching duties. The book I am writing is an autobiography in the following

Riccardo Giacconi during the interview in Lindau, 2005 (photograph by M. Hargittai).

sense: it is about discoveries in X-ray astronomy and the methodology that I applied to astronomy has changed the field. So I would like to get it across why we were doing things the way we did.

Were there any failures in your career that you might regret?

Well, you know, rockets fail. I also made mistakes. OK, there is one thing. I think the biggest failure has been in trying to create institutions or trying to change institutions. I think my failures were in my hope for what could be done in science and about the constraints. My biggest success, in my opinion, was the way I was able to start from zero and organize the Hubble Space Science Institute and achieved both human and scientific results. The Space Science Institute was an institute in which much attention was given to personal, intellectual, and professional growth of people, male and female, by the way, it became a very important example of a workplace where women were treated properly and were given opportunity. There was a kind of rationality and mutual trust and openness, and hopefully meritocracy. My colleagues used to make fun of me saying that yes, yes, we have democracy here but you have 51% of the vote — because I was the director. But basically the approach we took was that the best scientists are the ones who can give the best service to the community. It turns out that in ten years this new institution was the fifth largest publishing astronomical institution in the world. And we did a great deal to bring the Hubble data both to the scientific community and also to the public. From the scientific point of view, we created a software system that was able to digest the incredible amount of data that we received and then archive it so that it could be available any time. All this requires a very disciplined planned approach, which was not common in optical astronomy.

Then it got transported from the Hubble telescope to the very large new telescope and it could start with this enormous amount of background knowledge. In this last one I was too far away from the actual work because I was the president of the university consortium that directed the observatory. But the distressing thing is that you create something with the hope that it will become a permanent system — but in the United States nothing is permanent so I don't know how long it will survive. It also is very much people-dependent. You never know what the next director would do. So I think that trying to set up permanent institutions is probably my biggest failure.

What is your opinion about the level of science in Italy?

Italy is a very strange country. In my opinion it has one of the worst educational systems in the world. Particularly higher education. The universities at the moment are overburdened by this idea of universal education. They don't have the means. It is free, it is open to everybody, and the system simply cannot work. It's like giving free health coverage but you don't have enough hospitals. That is one side. Then, they are doing very poorly in industrial research having to do with the way the capitalist system is working in Italy, which is very strange with ownerships in the hands of powerful families and the state. It is a mixture of the socialistic approach, failing totally due to the well-known reasons, but at the same time, it is not favorable to innovations because of these powerful families. I mean, Mr. Fiat, for example, has succeeded in driving his corporation completely out. There are problems with metallurgical industries and power industries; they also have a very strong green movement. They succeeded in killing nuclear power production — for a country that almost invented it! This is insane. Particularly, because what happens now is that all the nuclear power plants are in France, and Italy is totally dependent on France buying 90% of its power.

Notwithstanding this, for reasons that I cannot quite comprehend, Italy continues to produce very good scientists. To make it short, there are two things: one is that every Italian scientist who got the Nobel Prize did that for work that he or she did abroad. None of them were working in Italy when they got the prize. They simply did not have the means in Italy for their research; they lack the opportunity there. Then, Italian students, postdocs are very bright, basically in spite of the system that is so bad that they have to survive on their own. So there is a certain survival value in this.

I think Natta received the Nobel Prize for work done in Italy.

Oh, but that was long ago! In fact, what is truly frightening is that the best Italian school in physics with Fermi, Amaldi, Segrè, Majorana, was set up in Rome under the fascist regime by this very illuminated minister, Corbino. Many of the people were Jews who then left and they have not reconstituted that kind of school in Italy. After the war, the best advice a teacher could give to his bright student was: Go west, young man! And that, of course, is very sad. This is why I push for excellence.

soon after the discovery of the so-called BCS [Bardeen–Cooper–Schrieffer] theory. I wanted to understand how superconductivity works and some time in 1961, I came upon the concept of phase coherence. Flux quantization had just been discovered experimentally, which you could explain by maintaining phase around the superconducting ring. There were various theoretical approaches and I was interested in the question whether that phase had any physical significance. To cut a long story short, I was led to consider two systems for which the phase difference between them could be important. I was able to calculate what the current for the superconducting tunnel junction would be as a function of phase, trying to explain the results of Ivar Giaever. The calculations showed that there should be a current even at zero voltage, with the current depending on the phase difference. Then I worked out a number of phenomena that would be important, for example, there would be an AC super-current of a frequency depending on the voltage difference. That would depend on Planck's Law relating energy and frequency. There would be a strong magnetic field sensitivity and so on. Roughly speaking, the superconductor would be like a laser with coherent oscillation. You have two separate blocks of superconductor, both oscillating and the phase difference has physical manifestations, which you find when you're trying to couple the two systems.

Then I moved in the direction towards understanding the mind and the brain. I found this a more challenging problem than physics. I got also interested in things, which orthodox science seems to be prone to reject, like psychokinesis and other paranormal phenomena. I became interested in consciousness, which by now has become part of orthodox science. The origin of my interest can be traced back to discussions with two people in my College. I was also impressed by some parallels between paranormal things and quantum theory. Things were strange in the same way. There were things that are understood today but of which people were not aware of then, like two identical particles behaving like slightly different particles. There were specific contacts with people that moved me to consider that may be current science wasn't including everything. I moved completely away from physics but I'm now back in physics, working on the relationships between physics and those phenomena.

It was a slow process to understand the brain as I was trying to put ideas together, studying things like neuropsychology, and trying to work out my own theories. I also got interested in the descriptions of the nature of intelligence. There were some very intuitive ideas of what intelligence is. It is something that gives direction to developments. It's a set of general statements, may be like the laws of thermodynamics. I wrote a computer program, which collected all the ideas that I found meaningful.

People talk about language instincts, which we have, which cause us to develop in a certain way. The implication is that development, as it were, is intelligently directed. It's not so much a trial-and-error process, as people sometimes tend to assume. There's a clash nowadays between people who work in neural networks who don't like to think in terms of innate mechanisms. People, like Stephen Pinker, say that there are processes that direct our development and make things happen more efficiently. The question is whether intelligence is there initially in an innate form or whether it is acquired by experience. If it is there in an innate form, it should have a genetic source, which would cause the brain an appropriate structure. To some extent you could simulate the networks by training. This is an argument, which the constructivists produced. They say you can train your network to become an expert of particular things. This is like learning a skill, a sport, for example. By learning, you acquire certain processes and do them very easily. Your training is restructuring your nervous system so it can do things that it couldn't do before. However, it's more difficult to do for what we call mental intelligence and you can't train people to think deeply if they haven't got that kind of brain. We have a particular mechanism for learning that causes us to learn from experience and that's part of what intelligence is. There are components of intelligence that are unacceptable to talk about today for political correctness.

I agree with Pinker's views that we have instincts, we act instinctively with things like language and enhance linguistic skills more quickly than general learning because we have the right circuits, which may appear as instincts. You can either say a neural circuit makes you process language in the right way or you have the instinct to do it. These are just different descriptions of the same thing.

In 1971 I had a sabbatical at Cornell and I was thinking along more orthodox lines in terms of the brain. In 1976 I did a collaboration with Herman Hauser. We wrote down some ideas as to how development might take place. At that time it was out of the orthodox line but I don't know what people would think of it today. I actually submitted it to an artificial intelligence (AI) conference and it was not accepted in the program. Even my Nobel Prize did not do the trick, but I was allowed a ten-minute talk in the evening. It contained ideas, which were not accepted by the AI people, that you might just train your neural network to learn things like motor skills, rather than having an algorithm that you learn. People in

AI were just not thinking in terms of such models and didn't like that kind of explanation. The idea was basically that you build up developmental schemes step by step. Having learned one skill, you could incorporate the knowledge, and not just the knowledge but the system that you trained for a particular skill, into a more advanced system of skills. We saw a logical connection between the steps, which an innate developmental strategy could capitalize on. It was published in *Kybernetes*; apparently the editors were pleased to have a Nobel laureate author. However, the paper was ignored, as tends to happen if you don't work in the field. This publication contains the basis of our approach I'm working on today, which I hope will get more attention when I'm going to read it at the end of this month at a big conference called JCIS 2000, which is the Joint Conference on Information Sciences.

Brian Josephson lecturing at a meeting commemorating the centennial of the discovery of the electron, Cambridge, England, 1997 (photograph by I. Hargittai).

Do you have support for your work?

I don't have any grants. The only grants I have is the remains of what I got from the Research Corporation in 1969. Having a Nobel Prize did not help me in getting grants. Prejudice against my interests in research has proved stronger then any previous recognition. It's true though that I haven't tried too hard because it takes a lot of time, applying. I have small expenses and the Laboratory gives staff allocation and my College pays some as well. That's enough to get me a computer and travel, and so on. I don't have experiments to pay for.

Would you care to give an example of what shocks people?

Just stating that there's good evidence for telepathy seems to shock people. But as in other fields, there are carefully done experiments, which, when you analyze them, indicate, that phenomena like telepathy and psychokinesis occur. There is a vicious circle in that the results of such experiments very rarely get published in the ordinary journals and people then say that because they are not published, there's no evidence for it, and therefore they don't exist.

Such experiments must be designed very carefully. There is a sender and receiver in a telepathy experiment. The sender sees one of four pictures and which of the four is chosen is decided by a random number generator. There is no personal choice involved. The receiver writes down impressions. The environment is carefully controlled; there is a uniform light field and white noise coming through headphones to encourage a state where there's nothing much coming through the ordinary senses, so people are more sensitive to what might come in elsewhere. They record any visual and other impressions they have. If telepathy does not exist and you've made sure there isn't enough sensory connection, there can be no correlation between what impressions the receiving person gets and which was the choice of target. Then the information is given to a judge; the judge sees all four targets and doesn't know which one was chosen and examines all the recorded impressions. If telepathy doesn't exist, there could be only chance correlation and the target closest to the records would be 25 per cent all the time. Of course, you have to get enough statistics to make it significant but they come to something like 33 per cent, which is very significant if you have enough observations. This is the approach you have in a scientific experiment. You don't rely on things like somebody saying, "I always think of my friend just before he telephones." You have the whole set-up under control and just by seeing if there's a correlation or not, you can tell if there's an effect. You take certain precautions, which make sure there aren't any artifacts.

In psychokinesis you have a random number generator, which people try to influence in some objective way of recording what the effect might be, to see if there is correlation between what people are trying to do and what the random number generator does. The experiments point strongly to there being effects but the scientific community always takes no notice.

As a physicist you must ask the question, "What may be the medium, the carrier of correlation?"

Yes, except, quantum theory indicates that that may be the wrong way of looking at it. I think it's more likely to be a question of organization of nature at, may be, sub-quantum levels. The way I see it, there is some kind of system, which is more strongly interconnected than the systems we usually study. This could come out from conventional physics.

You mentioned that lately you are returning to physics.

Brian Josephson at the Cavendish Laboratory in Cambridge, England, 2000 (photograph by I. Hargittai).

I'm particularly interested in complexity because that is bringing up counter-intuitive situations that may be the basis of quantum mechanics, for example. I got interested in mind partly because I felt this would help me to understand unusual kinds of organization. The thing about complexity is that you can't reduce the system to ordinary type treatment because of the unpredictability. You get some chance fluctuation, which has significant effects. In a complex system there are very many things that might happen and might fit some general plan and you can't say which it is. There's a sort of hidden organization, which will become manifest when one of these transitions occurs. What appears to happen is that certain relationships develop, at least in the biological context, and this has strong effects on what phenomena you get. A new aspect of a scientific theory emerges as people study complex systems.

Somebody tried to handle complexity through category theory but I'm not so sure whether it is good for biological systems, for example. Category theory presumes very high degree of regularity, which may be OK for fundamental physics but may not be good for biology. I think relationships do come into it. The work I'm doing now is discussing how the possibilities of the mind, as it develops from one level to another, are because it takes into account new forms of relationships. The example is that the level of action involves the only relationships that are very concrete, that are already there. As you get to more abstract levels, you deal with relationships between how things are now and how they might develop. You can sort of create a related situation.

In language, for example, you can represent certain more complicated relationships, and then use these relationships to construct the kinds of possibilities that you are interested in. Language itself capitalizes on relationships that are possible and is getting control of them. I am working on this. It's fairly descriptive but you can understand how things like language could work by going through some of the details and some of the structures of how you built it up. Some people here working on condensed matter are becoming interested in this problem. There are also people in biology who may be getting interested. But it seems difficult to get people interested in psychology but, perhaps, once word gets about these approaches, people may change their orientation.

Do you sense any change now in how your interest and work are being viewed by others? You have been now for close to 30 years in this area.

Yes. People are more accepting what I'm doing in the Department, partly because I've arranged a number of talks by people involved in this work.

I have students but not very many because it's difficult to get support for work in this field.

Have you had a lot of frustration during these 30 years?

Yes.

But you don't mind.

It's not that I don't mind; I'm detached from it and believe it won't be very long before it's more accepted. The world direction of science is changing and it is changing towards accommodating these things. Developments like quantum information and quantum holography are going on. When people are trying to understand what really happens to quantum systems when you go into the more microscopic level and you can influence individual systems, you find new kinds of theory. For example, people who are involved in superstring theory find that space and time need to be treated differently. Science is unconsciously going in this direction.

Did you come across Arthur Koestler's work, did it have any impact on you?

Not very much. There was a conference on beyond reductionism, which I attended. I have also some contact with the people in Edinburgh who work in the chair that Koestler had endowed but I'm not directly involved. He did not have a great impact on my interests.

What is the relationship between your work and religion?

One of my papers discusses the synthesis of physics and spirituality. This relationship with religion concerns more my interest in meditation and all the states of consciousness. It does open the possibility of there being deeper levels of reality, which could include spiritual levels. Science may also come to some kind of accommodation with religion by allowing background systems, which have an influence. These may eventually be models that would include God, to take an extreme view of things.

There are people in mainstream science, like Francis Crick and Gerald Edelman who study consciousness. Is there a connection?

Both of them are too materialistic as far as I'm concerned. Crick is concerned with just a pack of neurons. The notion that mind is not local could not be made fit into that conventional picture. Edelman probably thinks similar to Crick although I never managed to really understand him.

Addressing a broad audience of scientists that I hope our readership will be, what would be your message?

The science establishment has quite a number of misguided beliefs. They have a rather restricted view of what kind of things might be a case, which lead to their rejection of telepathy and others. Another example is, perhaps, complementary medicine. I don't say everything in complementary medicine is valid but a lot of things are, such as healing and homeopathy. In both cases it doesn't fit the usual belief system and it's just dismissed without evidence. I think it won't be too long before these conventional believes are shown to be wrong. They have after all had to be changed about things like acupuncture, which is now accepted to have some effects whereas previously it was believed to be totally nonsense. This is one example. Then there is psycho-neuro-immunology, something, which was just dismissed at the beginning. I anticipate rather dramatic changes following quite an embarrassment to science. A lot of things will be put on solid scientific foundation before too long.

I wouldn't like to project you just as a Nobel laureate physicist turned telepathy freak. I'd like to bring you also into human proximity in this conversation.

I hope it will turn out to be better then the *Scientific American* interview. What they published in April 1998, they had several questions that they slighted and they didn't show me what they were to print before it was published. They said I was uncertain about what I said whereas it was just unusual being interviewed.

I would like to ask you about your family background.

I was born in Cardiff in Wales and my schooling was in Cardiff. My father taught French at one of the local schools. My mother did some journalism and also some poetry. I was an only child. My mother was the artistic side in our family. My father would've preferred mathematics to French but he came back in the middle of the academic year from the Great War (WWI) and French he could take right away. He was interested in mathematics and science.

My mother was born in Swansee and my father south of Manchester. They moved to Cardiff when my father got a job there. My grandparents came from Eastern Europe, Jewish. We weren't observant and went to the synagogue on Rosh Hashana and such occasions. We changed to a Reform Synagogue when it appeared because it was more congenial. I see religion primarily as a matter of conscious experience whereas it is commonly presented as a belief system and faith. I was not inclined to accept it on faith. To me it's in a way an experimental matter.

I was always interested in mathematics and was something of a mathematical prodigy. I got to grammar school (high school) at the age 11 and they let me work on my own. I did sixth form mathematics right after I got to grammar school. The physics Master gave me a theoretical physics book to read. I was obviously interested in science. I got a scholarship to Cambridge when I was 15 but they suggested not to come up immediately, so I came up when I was 17.

Who are your heroes?

David Bohm. He had lots of opposition, outside science as well. Then, John Bell. Earlier on, Einstein, then Feynman. I met him on one remarkable occasion when I had Feynman on one side and Gell-Mann on the other side. This was when I visited Caltech.

Brian Josephson in Lindau, Germany, 2005 (photograph by I. Hargittai).

Where do you think your interest is the closest with what you call orthodox science?

The question of what music is got me interested at one point as a possible entry point into things being along with the orthodox point of view. I found it in my meditation on consciousness, and I started to perceive music in a different way. I suppose it is really the way musicians hear music. I discussed the meaning of music with a musicologist Fellow of my College. These is a great discrepancy between what psychologists think and what musicians think. The psychologists have a view, which just looks abstractly at structure and tries to find generative rules and so on, whereas musicians tend to feel some elements of meaning in the things, which are being developed. We looked at the question whether there was any way to explore this dimension meaningfully. The question is really whether the future of music is in its meaningfulness, which goes beyond the psychologist's structural models.

You may compare music to DNA, which is clearly meaningful. You can't talk about DNA in purely syntactic terms. It's possible that vitamins and hormones would be even better examples, which more directly influence the functioning in healthy ways. There have been suggestions to explore the idea of music as food. This is giving us a semantic dimension to music, which is rather denied by the orthodox theories. This is one direction I've been interested in. If you ask where this comes from, it seems hard to explain it in orthodox terms, to say, for example, that it's just pattern recognition. This is why we're interested in music. I asked my collaborator whether there was anything one could say definitively on this and he came up with the obvious idea that themes can be translated from one piece to another, a composer may utilize a theme of one composition in another composition in a different way. There's something fundamental about a theme, which is not any obvious way relatable to structure. There's something of a criterion for good music, which is a sort of permanence in that you can always see new meaning in it, rather than just learning it, whereas more ordinary music you know what it is and it loses its interest. The interest to use it is a criterion for differentiating different levels of significance in music. There's a whole new world there; and there is difference between the musician's perception of music and the scientist's perception.

Music may be one case where you can objectify different levels of consciousness, in a musical score. There seems to be something there, which doesn't fit in with ordinary science. I've left that on the side but now I've got ideas of how mind developed more and I may get back to it. I think it is a better argument for a Platonic world than Penrose's arguments for mathematical understanding because you might explain mathematical understanding because it's all the way logical and say that mathematical truth is connected, ultimately, with specific structures. Music seems to be different. There seems to be no logical dimension to music apart from what the psychologists produced, something additional to music. The followers of Indian philosophy talk about fundamental connection between sound and form, which applies to music as well as to ordinary sound, making it a basic property of nature. Then, someone has recently produced a new model of music, different from the psychologist's model. He built a program, which absorbed the content of a particular composer, saying that the composer has discovered important new ideas so let's put all these in. His computer program then will generate new compositions and the person running the program may apply some aesthetic judgment of his own to make some choices. This would stress some mechanical aspects of music. But I think this is not the case because you're really using the composer's sense of meaning as input to a program. Of course, this is all extremely speculative. We hope though to eventually get a proper theory of this, which then I hope we can connect with my models of mind.

Do you have any favorite composers?

I didn't get interested in classical music until the age 40. I like Mozart, Beethoven, and so. I also like Tanner because I find deep spirituality in his music.

What does your wife do?

She was trained as a nurse but she gave it up when our daughter was small. Now my wife, Carol, helps in the local school. Our daughter, Miranda is a vet student in Liverpool.

Anything you would like to add?

I'd like to add something about the unfortunate situation of scientific censorship. There are examples of it on my web site. As a consequence, there's very little research on paranormal things. People no longer send their papers to *Nature* because they know, they will be rejected. There was a book review, which criticized parapsychology on grounds that turned out to be misunderstanding. But it was a very long time before the editor would publish a correction. There's also a tendency against speculative papers, which would make it impossible today to publish the Einstein–Podolsky paper. I have to resort to publishing my papers in journals that ordinary scientists don't read.

go to the United States. The pay was also better in the U.S. So I came to Schenectady.

We are getting close to the time when you did the work that eventually led you to the Nobel Prize. Would you mind saying something about that?

First when I came to the United States I had various assignments with famous people in General Electric. For example, I worked with a Hungarian here, called Gabe Horvay. He was very helpful to me. I thought I would do applied mathematics there but then I worked with a German mathematician, Hans Brückner, and recognized that I could never be as good as he was. I worked in mathematics for a while with various people and then I had an assignment with Gabe Horvay at General Electric Research and Development Center and I recognized that that's where I wanted to work. So I applied for a job there and I got the most wonderful recommendation from Gabe. I was only told later that when they asked him if I was any good he said

Ivar Giaever, Walter Harrison, Charles Bean and John Fisher at General Electric Corporation, around 1966 (all photographs courtesy of Ivar Giaever unless indicated otherwise).

that he didn't really know but he did know that I was better than Gabe Horvay. So I got a job there.

I had a mentor named John Fisher. He is still a good friend of mine; he is in his early eighties. He said that we were going to work on thin films. I didn't know what it meant; I thought it was about photography. Thus when soon after that we went to Norway for a holiday, I studied photography because I thought that was what I was to work on. But it was not that at all. When I got back I started to work with John, and he had the idea of doing tunneling work and that's how I started to work on tunneling. He was a theoretician and I was doing the experiments.

The idea of tunneling came up very soon after quantum mechanics was discovered but was there any experimental indication of it before you started to work on it?

Not really in solid state; but tunneling was well known in nuclear reactions, for example, the emission of alfa particles happens through tunneling. So in nuclear physics tunneling was well accepted. But as far as I know there was no experimental evidence of tunneling of electrons in solid state. Actually, there was Leo Esaki who did tunneling in semiconductors before I did but I never heard of Leo Esaki. John Fisher, my mentor, probably have, but I hadn't.

Ivar Giaever with Leo Esaki, about 1985.

Would you care to tell us the story of your discovery?

First of all, I am a skeptical person. You have to be if you want to be a good scientist; you have to be skeptical and optimistic at the same time. When I first heard of tunneling from John Fisher, I didn't believe in it. I didn't believe in it because I didn't know quantum mechanics. I could not imagine something going some place where it didn't have enough energy to be. So I thought may be I could show that tunneling doesn't exist. At the same time, I started to study quantum mechanics at Rensselear Polytechnic Institute and after I learned quantum mechanics I understood that tunneling really has to be. It took me about a year to get convinced that tunneling is possible. The experiments showing that tunneling really existed took about a year or a year and a half. We passed electrons through a thin oxide layer between two metals, that's how we managed to do this. It was difficult to figure out how to make this thin layer. John Fisher called my irreproducible experiments "miracles" because miracles happen only ones. But of course, science is not that; you cannot publish miracles; you have to have reproducible experiments. But finally we managed to do it. Then I had to give a talk on this and I didn't even have a doctorate degree so I got nervous. About a hundred people came and there were many questions and I could tell that people were very skeptical. They were very

Ivar Giaever with the evaporator, around 1973.

nice but still, you could tell. There were many other possibilities, these materials could be semiconducting through this thin layer, there could be a hole in the layer; there could be all sorts of things. So I decided that I needed to find a way to absolutely prove for certain that what I observed was tunneling. By that time I believed in tunneling.

I thought of many different ways but they didn't work very well. In the meanwhile I took courses at Rensselaer, by a professor named Huntington, and he talked about superconductivity and that was the key for me. He said that in a superconductor there is an energy gap and in his class I recognized that that's the experiment that I'd been looking for. I knew that if I could do the tunneling into a superconductor I could see this gap. He also said that there was a new theory that he didn't know much about and nobody really knew how big the energy gap was. I went back to the lab and talked to the people there and they didn't know it either. But then they came up with the estimate that the gap was a few millivolts. And I knew that that was just the right kind of gap for me to measure.

This was a nice bit of luck that everything came together. There were the superconductors, then the right energy gap — actually another person also got this by another type of method, which was not very direct — and then I came along. I remember, I told Charles Bean, a very good friend of mine who helped me a lot, that look, my gap by my method didn't agree with what other people measured and I was nervous about that, but he said you don't have to worry; they have to agree with you, your method is superior.

When finally it was clear that you have shown tunneling in superconductivity, did you know the importance of the work that it would be of Nobel caliber?

No, I did not. I was shocked, when relatively early, I received the Buckley Prize in solid state physics. I was more shocked by that than by the Nobel Prize.

When the prize came to Bardeen, Cooper, and Schrieffer in 1972, did you think that you might also have a chance or just the other way around?

I thought the other way around. There was a very famous conference in Copenhagen about tunneling. A professor, Eli Bernstein, who helped me a lot, organized this conference, and then I heard rumors that I had been from the government and that's how they started this physics of geological processes. Now they have established a board, which is going to be between this PGP group and the University and the people who give the money. I am the chairman of this board. This is a rather complicated thing; we try to isolate this PGP from the University but the University of Oslo is very bureaucratic and they don't like to give up any powers and that's why this board was created. So this is a very good thing. About a month ago I went with this group for an expedition up to Spitzborgen, which is an island way north from Norway, practically at the North Pole. That island is supposed to be similar to Mars in many ways and that's why this expedition was arranged. That was great fun for me to go there; really in the wilderness and see what these geologists do. Of course, I am not a geologist and my main role on that board is to make sure that the money is well spent.

Which of your work are you most proud of?

For me the most exciting thing was when I developed a DC transformer, which has to do with superconductivity as well. The fact is that you can send DC voltage in and you can amplify the DC voltage and get a higher DC voltage out; that was an invention. It never became practical but anything with superconductivity can't really become practical. But it was a nice piece of work.

You received your undergraduate degree as an engineer, you got your Ph.D. in theoretical physics, and then later on you became a biologist. What do you consider yourself foremost?

I don't really think about that. But you should see the reason: when I started in physics I was 30 years old; I didn't know anything about physics but was interested. I told my friend, John Fisher, that I didn't think I could do much there because I was too old for that. But he said that the time you do good work is when you are learning new things. He said that the most ideal thing would be to change one's career every 5–10 years and learn new things all the time. So I tried to do that. Nature doesn't distinguish between physics, chemistry or biology, nature is just nature. I try to look at science as one field, science.

You make it sound easy but in our world when your research needs support and for that you need to publish in that particular field and show some results, this interdisciplinary attitude may be difficult to keep up.

You are right; the most difficult part of it is funding.

Do you have heroes?

Yes. Richard Feynman is one of my heroes; he had the kind of personality I like. He looked at everything and I also like to look at everything. There are many kinds of scientists, but most of them are very dull. They do a thesis and then they keep refining that thesis for the next 40 years, digging and digging deeper and deeper. I am a broader person. I am more like an inventor and not a deep thinker in that sense. When there is a problem I like to try solving it; you have to try to get a whole coherent picture and if there is something that sticks out and doesn't fit you have to try to figure out why it doesn't. Or sometimes I get an interesting idea and then try to figure out how to make it work. I try to tell my students that if you were really a scientist you would know that because if you lay awake at night you wouldn't think about your sweetheart, you would think about science. Then you are a real scientist. Sometimes you think about it for a year and then suddenly an idea comes up, how to solve it. Somebody said that a good idea only comes to a prepared mind; that is absolutely true; if you don't think about it the ideas won't come; you have to work for good ideas.

Talking about the mind, what do you think about the way your co-laureate's, Brian Josephson's career turned?

In my view, unfortunately, he has stepped over the bounds. That is very unfortunate. I don't agree with what he thinks about what he does. I know him reasonably well and when we see each other, I always tell him that; I told him many times but he doesn't listen. He definitely stepped over the line, in my opinion he is dealing with fiction.

Has the Nobel Prize changed your life?

Of course! It does that in a way, which is not too good. The Nobel Prize has a great prestige and somehow people think that if you received the Nobel Prize, no matter in what field, you should be able to solve the crisis in the Middle East. It has this great prestige but I think that know that they are superior. I think it is a big mistake of Norway not to join the European Union. Norway is a rich country. Still the European Union spends twice as much money on science percentage-wise then they do in Norway. This is very unfortunate. Being a professor in Norway, for example, is not a prestigious or important position. A professor in Germany is an important man; a professor in Norway is not an important man. Science in Norway has no stature. I am trying to change that as much as I can but I have not been very successful.

Is this a new phenomenon or has it always been like this?

I think it has always been so. Norway used to be a farm country and science has never been important in Norway. For example, there was Lars Onsager, the great chemist, Nobel Prize winner and he was Norwegian. They have sculptures of sports figures and everybody knows about them but nobody there knows of Onsager. Whenever I go there I always try to tell them you should remember him and be proud of him.

I am distressed that people don't get enough money for science in Norway. The problem is that the government is not interested. A politician doesn't have to have an education, a politician just becomes a politician and then they don't appreciate science. Norway doesn't have a good industry. They have the oil industry now and they have the fish but they don't make anything in Norway.

What do they do with their money?

A large amount of that is put into funds; they buy stocks all over the world and they save the money for the time when the oil will run out. But they don't invest it in Norway which they should do. For example, the road system in Norway is terrible. There are so many good things that they could use the money for but they just don't do it.

Please, tell me something about your family.

I met my wife when we were about 15 years old and we have been together more or less ever since. We've had a wonderful marriage; we have four children. The oldest one will be fifty this December. We have seven grandchildren.

Any scientists among your children?

Ivar and Inger Giaever after their engagement, 1951.

Yes; my third child (second daughter) is a biophysicist at Stanford. My son is an engineer — I tried to talk all my kids to become engineers — he works in Seattle. My first daughter is a schoolteacher and my youngest daughter is an artist. I told her that she cannot possibly make a living as an artist but she wanted to study that. Eventually she became a painter and in fact she makes a very good living because she became a graphic artist at one of the big New York City newspapers. So it worked out very well for her. They all live in the United States; all of them were born there, except my son. But they all speak Norwegian.

Do you have any hobbies?

I like outdoor activities. My wife and I play tennis together; we are at about the same level, we also ski. With a group of people that I went to college with in Norway — amazingly all of them live in the United States — we bought a timeshare condominium together and in the last week in February we all get together and go skiing in Utah. Even though we are getting older we still manage very well with the ski.

How do you define success in science?

You have to be curious. When I watch television I want to know how it works. I can't imagine not understanding all things that are around us. Most people are not interested at all, even when they turn on the light, they have no idea how it happens. These are all the things that have been made; but there are also the things that are there; such as the blue sky. I want to understand why it is blue. If you want to become a good scientist you have to wonder about things. Children are wonderful that way. Unfortunately, the way we treat the kids somehow we take that away from them, rather then encourage them.

Were you a curious child?

I think so; I was. I was wondering about everything. Of course, I didn't understand how things worked. My parents didn't have a good education; although my father was a pharmacist, that time in Norway all you had to do for that was ten years of schooling. But he was very interested in things and my mother was very ambitious, too. I had a very nice upbringing. I lived on a farm in a house my parents rented at a very small village in Norway and I had no idea that you could become a scientist, I didn't know that science existed.

Ivar and Inger Giaever in Lindau, Germany, 2005 (photograph by M. Hargittai).

Were your parents alive when you received the Nobel Prize?

My mother was still alive. It was very fortunate because she happened to be visiting with us in the United States when it happened, so she could experience it all. She also came with us to Stockholm.

Is there anything you would like to add?

No, I think we have covered a lot.

Vitaly L. Ginzburg, 2004 (photograph by I. Hargittai).

40

VITALY L. GINZBURG

Vitaly Lazarevich Ginzburg (b. 1916, in Moscow) is a Member of and Advisor to the Russian Academy of Sciences (RAN) and Associate of the P. N. Lebedev Institute of Physics of RAN, known by the abbreviation of FIAN (Fizicheskii Institut Akademii Nauk) in Moscow. He graduated from the Faculty of Physics, Moscow State University in 1938, and received his Candidate of Science (Ph.D. equivalent) and D.Sc. degrees, both in physics, in 1940 and 1942, respectively. He has worked at the Lebedev Institute since 1940, where he was Head of the I. E. Tamm Department of Theoretical Physics between 1971 and 1988. He was elected corresponding member and full member of the Academy of Sciences of the U.S.S.R. in 1953 and 1966, respectively.

Professor Ginzburg was co-recipient of the Nobel Prize in Physics for 2003 jointly with Alexei A. Abrikosov[1] (b. 1928, currently at the Argonne National Laboratory, Argonne, Illinois) and Anthony J. Leggett (b. 1938, University of Illinois at Urbana) "for pioneering contributions to the theory of superconductors and superfluids". He has been much decorated in the Soviet Union and later in Russia with high awards, orders, and prizes, including the Mandelstam Prize (1947), various state prizes, the Lomonosov Prize (1962), the Vavilov Gold Medal (1995), the Lomonosov Big Gold Medal of the Russian Academy of Sciences (1995), and the Triumph Prize (2002). He was elected as foreign member of numerous science academies, including the American Academy of Arts and Sciences (1971), the National Academy of Sciences of the U.S.A. (1981), and the Royal Society (London, 1987). He has received other expressions of recognition, including the Wolf Prize (Israel, 1994/1995).

in the creation of the Soviet hydrogen bomb in the period of 1948–1953. At that time it never occurred to me that the Soviet Union might use such a weapon as a means of aggression. I am sure of the same for Tamm and others with whom I had occasions to discuss this question candidly. I must admit that we did not understand Stalin's real aspirations. It was only recently that I learned about a Soviet physicist who worked on the bomb while he understood Stalin's aims, and this physicist was acting out of fear. He kept silent about it at that time and I cannot condemn him now, and would not identify him either.[10] At this time I fully understand that Stalin was an arch bandit who would have employed even the most terrible weapons without hesitation if he had thought he would need them in accomplishing his goals, and could get away with such an action. It is the luck of humankind that Stalin and Hitler did not possess atomic bombs first. Of course, I understood only much later the danger of placing such terrible weapons into Stalin's hands.

The responsibility of the scientist participating in the creation of such terrible weapons depends on many factors. This responsibility depends on the goals for which such weapons are being developed. I certainly sanction the creation of weapons serving the protection of one's country from aggressors and terrorists. I don't want to leave here any doubt that I specifically mean Israel in this context.

My question had a specific aspect concerning any debate or discussion of the general morality of creating a weapon like the hydrogen bomb that might endanger the survival of life on our planet, as we know it.

We did not have such discussions. I do not remember any such discussion at that time.

Were you aware at all of any discussions that were going on in America at that time concerning the development of the hydrogen bomb?

What do you think? Of course, not. You may not be able to imagine the isolation in which we were living at that time. We could not even have communications with our western colleagues about purely scientific matters. The only exception was David Shoenberg[11] who subscribed to some of our Russian-language journals and helped us in disseminating some of our results in the West. Our paper with Landau was published in Russian and it was Shoenberg who — just as a personal initiative — translated it into

English and sent the English translation to some people; otherwise it might have remained completely unknown. At that time we did not have any possibility to publish our results in English; the foreign-language scientific journals that had been published in the Soviet Union, had been closed down.

But you did not think about the dangers of the hydrogen bomb in more general aspects, regardless whether it was in the hands of a democracy or a dictator?

I think that even asking such a question would have been a stupid act. We have to be realistic about life. When nuclear fission was discovered, it gave the possibility of creating atomic weapons and sooner or later many understood this possibility. May be there might have been some religious people with some general moral considerations, but who would have paid attention to them? You have to be realistic about life.

I hope you understand that I am just trying to understand the atmosphere in which you were living at the time of your working on the hydrogen bomb.

Of course, I understand, and I have no idea what some members of the so-called intelligentsia were thinking. By the way, the Americans were fully justified in starting the development of the atomic bomb, as they had no idea what was going on in Germany.

This is for the atomic bomb. How about the hydrogen bomb?

Our spies got hold of some vague notions about the development of the hydrogen bomb without getting any information how it should really be made.

At that time, were you aware of those spies?

Not at all. Only Kurchatov and Khariton knew about them. I can tell you even about a humorous aspect of these dealings that I heard about; I was never present at these meetings. Everybody considered Kurchatov to be a super-clever man. When there were discussions concerning several alternatives to solve a problem, Kurchatov made the choice in favor of one of the several alternatives. He was the only one who possessed the information supplied by our spies. You cannot imagine the level of secrecy under which we were operating. I personally did not know anything. Of course, I was trusted even less than anybody else. My wife was in exile so I could not be trusted. There was a theoretical group with Yakov Borisovich Zel'dovich in charge at the Institute of Chemical Physics. Zel'dovich was in the secret location, but came periodically to Moscow. Sakharov went to the secret location in 1950.

There were problems and Tamm was invited to join the project. Tamm used to be a Menshevik, that is, a social democrat. He was proud of an event that happened at the first congress of the Soviets in 1917. He was sitting amidst his Menshevik comrades and the Bolsheviks were sitting on the side. This has been immortalized even on paintings. There was an issue being debated and then there was a vote and Tamm in the middle of the Mensheviks voted together with the Bolsheviks. It was so conspicuous that Lenin shouted at him, saying something like, "Bravo, Tamm." This is probably what later saved him because after the victory of the Bolsheviks, the Mensheviks were destroyed. Tamm eventually ended all his political activities and restricted himself to physics. His brother was shot in the late 1930s. Their father was folk-German and their mother was Russian. Initially Tamm was not involved in the nuclear bomb project because he was not very much trusted politically and they did not think that he would be needed. Eventually, however, they realized that his participation would be useful. I still do not quite understand how they could entrust me with participation in the nuclear weapons program; my wife was in exile for counter-revolutionary activities; apparently my rating in usefulness was very high. I never considered myself a great physicist. I was elected corresponding member of the Soviet Academy of Sciences in 1953. However, not very long before that I was expelled from the Scientific Board of FIAN under the pretense of strengthening the Board. This is a contradiction and my explanation is that Kurchatov appreciated my contribution.

There was a man, Vannikov, a well-known official, who was imprisoned and beaten terribly. Then he was freed because they needed people after World War II had started and he was brought before Stalin, who appointed him to be Minister of Ordnance or something like that. He was also made deputy of Beriya and responsible for the nuclear weapons project. Vannikov invited Sakharov and Tamm and told them that for the work on the hydrogen bomb ["vodorodka"][12] they will have to move to Arzamas. It was probably Tamm, who told me the story. Sakharov and Tamm did not want to go because they had their families in Moscow; they assured Vannikov that they could do all work in Moscow with periodic visits to Arzamas. At that point the telephone rang; Vannikov answered it, obviously it was somebody important because Vannikov talked with great respect, and when he replaced the receiver, he told Tamm and Sakharov that Beriya advised them strongly to accept the assignment. Tamm and Sakharov thus left for Arzamas and I stayed behind. I did not do much because I was poor in calculations, so I was happy when they suggested to me to work on the TOKAMAK, which is the Russian acronym for the toroidal magnetic chamber.

Then, one day in 1951, as I was coming to work, they refused to give me my notes in the 1st Division. The 1st Division in every organization belonged to the secret police. We had to deposit our notes there every evening and receive them every morning. I only found out recently what did really happen. A little while ago, our journal *Uspekhi Fizicheskikh Nauk (Advances in Physics)* published a historical paper, including some documents. Included was a letter, which was written at that time by Kurchatov, Sakharov, and someone else, to Comrade Beriya. They suggested developing the thermonuclear project about which I first thought that they meant the production of energy. It was soon clear, however, that the hydrogen bomb was meant. They described the suggested technique and requested the necessary resources for it. Among their requests was that I be added to the project. Beriya made a note on the letter; he expressed agreement but gave instructions to check the people for reliability. I was duly investigated and it was established that I should not be given access. This greatly disappointed me at the time because I found the problems I was working on being of great interest, and they did not let me use my own notes.

Do you have any comment on Edward Teller?

I met him twice, but only superficially. I had hardly any personal experience with him, but I know that he was a good physicist, no doubt about that. The fact that he was unpopular among many of his colleagues was a consequence of his testimony in the Matter of Oppenheimer, and I found it unfair. I don't think that Teller was incorrect. Oppenheimer was an important physicist, but did not understand the Soviet threat and Teller understood it.

On August 29, 1949, the first Soviet atomic bomb was exploded.

I can tell you about it. It was a bomb and it was a copy of the American atomic bomb. In fact, our scientists built a better bomb than the American one but the Soviet scientists were afraid to explode first something different from the American construction. They thought if anything went wrong with an improved construction, they would all be sent to prison.

Then, on August 12, 1953, the first Soviet hydrogen bomb was exploded …

It was our construction, Sakharov's and mine. It was based on the idea of a layer structure. I am not familiar with the details, but can explain it to you in general terms. We had to burn tritium and deuterium. In any case, it was important to compress the reaction mixture because the reaction was proportional to the square of density. For compression, Sakharov suggested

Vitaly Ginzburg standing in front of Andrei Sakharov's memorial plaque at the Physics Institute of the Russian Academy of Sciences (FIAN), Moscow, 2004 (photograph by I. Hargittai).

to employ layers of deuterium and tritium and to surround them by layers of a heavy element. He suggested uranium, but this was not a question of principle, lead might just as well have been used. If uranium is burnt to complete evaporation, the pressure will increase 92-fold. According to Sakharov, a conventional atomic bomb would evaporate uranium and thus increase the pressure 92-fold. The substance that should be compressed was deuterium with tritium. Deuterium with deuterium would be less effective, but tritium was not available. Then came my suggestion according to which a layer should be made from lithium deuteride, in which lithium would provide tritium upon burning:

$$^6Li \text{ (in LiD)} + \text{neutron} \rightarrow {}^3H + {}^4He + 4.6 \text{ MeV}.$$

Here D is deuterium (or ^2H) and ^3H is tritium (or T). In further work I did not participate and I did not even know whether my suggestion had been followed up or not. Only many years later did I ask Sakharov about it; we were not especially close to each other, but had normal interactions. So I asked him, "Andrei Dimitryevich, did they employ your layer structure?" His answer was that it was not utilized. I was taken aback and only many years later I learned about the reason when these things were declassified. In what I read about, the explanation was that the utilization of the layer structure, which was employed in the first two Soviet hydrogen bombs, was able to produce a bomb merely about twenty times more powerful than the Americans' Hiroshima bomb. The Soviet leaders, these bandits, wanted more. This led to a third idea, which was the same as the Teller–Ulam idea, that is, compression by irradiation.

Did they know about the Teller–Ulam suggestion?

I have no idea. I can only advance my hypothesis, but this is not evidence. What I suppose is that our people must have studied the American explosions. I don't know whether they knew anything about the Teller–Ulam suggestion or they thought of it themselves. It was not Sakharov's idea although undoubtedly he contributed to it. The main theoretician was Yakov Zel'dovich. However, he was Jewish so he was not even elected to be an academician in 1953. He was very upset about it. In any case, they were looking for a solution and found a way to do the compression by irradiation. Taking such an approach, there was practically no limit to make stronger and stronger bombs and Khrushchev was very much taken by making larger and larger bombs. They ended up with a bomb of half a million tons of TNT equivalent. When it was exploded, many people died. It was absolutely unnecessary.

Do you think that the arms race and in particular what is known as Star Wars contributed to the dissolution of the Soviet Union? Many think that the Soviet system would have collapsed in any case, sooner or later.

I think so too and I will answer your question with Pope John Paul II's words. He said that Communism collapsed under the weights of its own crimes. It is also attributed to him that the medicine (that is, Communism) was more harmful than the illness it was supposed to cure.

In that sense, it might have lasted, say, fifty years longer.

Of course, and I am not a political scientist. Before telling you about my personal view, I would like to tell you that I used to be a party member. The only excuse for this I have is that I joined the Communist Party in 1942 when the Germans reached the river Volga. I hardly understood anything in politics.

The story of your (second) marriage is very romantic.

I see nothing romantic in it. We loved each other and got married. I did not worry about anything, except about my little daughter from my first marriage. She was nine years old at that time. Returning to the question about the collapse of the Soviet system, I did not understand anything for a very long time, but now I understand.

At the time, in 1917, many believed in the ideas of the revolution. I am convinced that the majority of the Bolsheviks were decent people. But totalitarianism, totalitarianism of any kind, leads to systems that were Stalin's and Hitler's systems. I like what Churchill said about them that Stalin and Hitler differed only in their mustaches. Democracy is better in spite of its many deficiencies. Suppose, for example, that I would be given unlimited power; although I consider myself to be a decent person, I am certain that I would also misuse such power. What I am trying to say is that nobody should be given unlimited power. If anybody is given unlimited power, if there is no democracy, if there are no appropriate laws, anything may happen. Stalin was a real criminal, but the people who followed him in the summit of the Soviet system, there was not a single honorable person among them. Gorbachev was better or may be he was forced by the situation he found himself in; in any case he took a different course. Currently, Russia is in a very bad shape and it is not at all clear yet what will happen. Nonetheless, I am an optimist.

We always learned that the victory of communism was final. Do you think that the victory of democracy is final in Russia?

Of course, not. However, it would be easier for me to tell you about superconductivity. In politics, I am an average person.

But you write a lot about political questions and your writings strike me as candid and even brave.

There was a famous singer in Russia by the name of A. Galich, who then emigrated; his original surname was Ginzburg, but no relation. His words were something like "More than anybody else, be afraid of the man who will tell you how things should be done."[13] Great words and he meant that Lenin, Stalin, and Hitler preached that they possessed the only right answers to everything. I do know that I don't, but I have my opinion, and here it is: totalitarianism is worse than anything else. The only acceptable way to go is democracy. Also, I know that in Russia there are liberties that were non-existent in Soviet times. I agree with the great Churchill that democracy has many problems, but it is still the best form of government. I disagree with those who can complain only and sing the praise of the old regime. There used to be censorship and there is no censorship today, well, there is only partial censorship today, especially for television. There was no freedom to travel; today, there is freedom to travel. Although I am an atheist, I appreciate today the freedom of worship.

You spoke today about it at the Presidium of the Russian Academy of Sciences protesting the introduction of religious instruction in the schools. Then one of your colleagues said that everything has been in flux for centuries in Russia, except for the Eastern Orthodox Church, which has provided permanence and stability.

I have been an atheist all my life and have not been interested in such debates. Recently I got involved in such a debate by accident. There was an article in our literature magazine, *Literaturnaya Gazeta*, declaring that there are hardly any atheists left in Russia, more or less suggesting placing all remaining atheists in quarantine. My friend, Evgenii Feinberg and I were so upset that we responded to this article rejecting the suggestion in no

Old headquarters of the Russian Academy of Sciences, Moscow, 2004 (photograph by I. Hargittai).

uncertain terms. From this, everything then started and I have written copiously about this topic ever since. At the same time I condemn in the strongest terms the crimes the Bolsheviks committed against those serving various churches or simply being religious.

At this time, the suggestion is to offer religious instructions as an elective subject in school. However, what does "elective" mean? It means to take it or not to take it; I find it unacceptable and I am determined to fight it. My wife tries to constrain me; she tells me that at my age — I am 88 years old — this should not be my concern.

Something else. You have given a rather long list of directions in physics that you consider as most exciting and most promising today. If you were, say, 25 years old today, what would be your personal choice?

I am against making such choices. Rather, I am for having it on a broad base. I have myself been engaged in numerous areas and I have listed them all in my scientific autobiography.[14] Generally speaking, our writings in physics become obsolete, most of it anyway, as science progresses. Whatever represents value finds its way into the textbooks. There are exceptions. For example,

I found it very useful that Zel'dovich published his collected works (in two volumes),[15] for the following reason. Zel'dovich was a Foreign Member of the Royal Society (London); they publish biographical memoirs of their Fellows, and they asked me — I am also a Foreign Member of the Royal Society — to compile such a memoir.[16] I could not have compiled this biographical memoir without his collected works. He was also a good organizer and made various people write commentaries and prepare other contributions for his two volumes. When he gave me these two volumes as a gift, he told me, "You will soon be 70 years old." In fact, I liked his idea although I did not count myself as outstanding as Zel'dovich, and I did not want to take up the huge task of organizing such a compilation as he did. Besides, I have published an enormous amount of papers. What I decided to do was to compile my scientific autobiography, giving emphasis to those works that I considered most important.

Answering your question, I have worked in many areas and especially in theoretical physics. I often did something one day and something very different the next day. If starting anew today, I'm sure I would again become a theoretical physicist. I have a prayer and although I usually do not explain it to people, I am going to explain it to you. As you know, Jewish men have such a prayer in which they thank God that he did not make them into women. In my prayer, I am thanking God for having made me into a theoretical physicist. This does not mean that I have anything against experimental physicists. In my eyes they have the most difficult job possible. They have to sit at some apparatus all their lives. The great luck of a theoretical physicist is that he can easily change his topics all the time.

In my Nobel lecture, I raised this question, why it took me so long to generalize the London equation. I understood that it needed generalization as early as 1943 and yet we published this generalization only in 1950.[17] The reason was that I was busy with many other things. I always dealt with many problems. This is why I cannot give you a specific response to your question about what would be my choice today if I were 25 years old again. Problems in theoretical physics would be sufficient to keep busy a thousand Ginzburgs. If then taking a closer look at theoretical physics, I do have some fixed ideas. From 1964, I have been interested in high-temperature superconductivity. Today, the question is about room-temperature superconductivity. Could we make a superconductor that would be possible to utilize at room temperature for which, for example, water-cooling would suffice? This is what I find to be a most interesting problem. It may not be the greatest challenge in physics today, but I would probably select this one to pursue further if I could suddenly become young again.

Do you have pupils in the direct sense of this word?

There may be quite a few of them but I am against me choosing whom to call my pupils. I would like to call my pupils those who consider me their teacher. In a formal sense, I was Landsberg's doctoral student ["aspirant", equivalent to a Ph.D. student]. Then when I was working on my higher doctorate [similar to a British D.Sc. degree or the German habilitation] Igor Evgen'evich Tamm was my supervisor. So in a way, in an organizational sense, I was connected with them although the connections were very loose. However, even such loose organizational connection did not exist between Lev Davidovich Landau and me and I did not even take his examinations, the so-called Theor-Minimum — which I regret to this day because it would have strengthened my background. Nonetheless, I consider Landau as my teacher along with Tamm.

Vitaly Ginzburg standing in front of Igor Tamm's memorial plaque at FIAN, Moscow, 2004 (photograph by I. Hargittai).

I am trying to answer your question about my pupils. I conducted my seminar for many years and I had a lot of Ph.D. students, but I cannot decide whom I should be calling my pupils.

Did you attend Landau's seminars?

Of course.

Did he come to your seminars?

Never. I used to go to his seminars and we had plenty of interactions anyway. Also, the Ginzburg seminars started some time towards the end of the 1950s and he stopped his activities after his terrible accident in 1962. Besides, the Ginzburg seminars were more of an educational character attended by a broad circle of people. In any case, Landau did not attend any other seminars except his own. Many people attended my weekly seminar and on the 1700th session, I closed my seminar. This was in 2001.

Why did you not ask someone else to continue?

A good seminar should have a very active person in charge with a broad vision in physics. There are many good physicists around but nobody has taken up bringing a seminar together. When I closed my seminar, I left open the possibility for others to continue. What was easy for me, what

Vitaly Ginzburg lecturing (courtesy of V. Ginzburg).

I did virtually effortlessly, appear difficult to others. I do not consider myself to be great but it appears that there are few such people, and there are even fewer like Landau. It is hard to explain what makes the emergence of special talent possible.

This is an interesting question, which has been posed repeatedly in connection with the appearance of several exceptional physicists coming from Budapest at the beginning of the twentieth century, Szilard, Wigner, Neumann, Teller, and others.

It was a most intriguing phenomenon. Coming back to your question about pupils, I would not like to single out people by name in that respect. Lacking close associates is regretful to me because I cannot even ask anybody for doing something. I might have some ideas and would like to have some people to try out some ideas but I have nobody to ask. Did you read my Nobel lecture?

I even listened to it on the Internet.

The one I gave in Stockholm in English lasted only 45 minutes. Then I gave one in Russian in Moscow, which lasted two hours. My lecture was unusual in the sense that I spoke mostly about what I did not succeed in doing. I had some ideas about thermoelectric phenomena as early as 1944, I published them, and to this day I could not induce anybody to develop those ideas. I never had the ability to make people do things. In this sense I am a poor teacher. A good teacher can make people do things in the best sense of this word. I do not mean exploitation rather I mean joint work.

All your grandchildren live abroad ...

My daughter (from my first marriage) is often visiting her daughter who lives in Princeton. My granddaughter has two children, twins. I am alone with my wife, with whom I have been married for 58 years. We did not have children in this second marriage and there are fewer and fewer friends around, so we feel lonely.

What did your wife do?

First she graduated from a polytechnic; later she studied foreign languages. Landau helped her find an occupation. In a way those who had experienced

Vitaly and Mrs. Ginzburg (courtesy of V. Ginzburg).

Vitaly Ginzburg and István Hargittai in Ginzburg's office at FIAN, Moscow, 2004.

prison, repression, tend to stick together; they find a common language. So she worked in physics of low temperatures. She defended her Ph.D. dissertation; she is very gifted. Later she was engaged in making translations.

Anything to add?

I enjoyed our conversation because I found that we speak a common language.

References

1. An interview with Alexei Abrikosov appeared in Hargittai, B.; Hargittai, I. *Candid Science V: Conversations with Famous Scientists*. Imperial College Press, London, 2005, pp. 176–197.
2. I was invited to this meeting because it was also a festive ceremony of my receiving the Diploma of Dr. honoris causa from the Russian Academy of Sciences.
3. Hargittai, B.; Hargittai, I. *Candid Science V: Conversations with Famous Scientists*. Imperial College Press, London, 2005, pp. 176–197.
4. Ginzburg, V. L. "Why Soviet scientists did not always receive the Nobel Prizes they deserved?", *About Science, Myself, and Others* (in Russian, *O nauke, o sebe, i o drugikh*). Third edition, augmented. Fizmatlit, Moskva, 2003, pp. 408–418. There will be an English translation of this book to be published soon.
5. Solov'ev, Yu. I. *Herald of the Russian Academy of Science*. **1997**, *7*, 627.
6. www.tamm.lpi.ru/staff/ginzburg.html
7. Ginzburg, V. L. "Superconductivity and superfluidity (what I have and have not managed to do)", *About Science, Myself, and Others* (in Russian, *O nauke, o sebe, i o drugikh*). Third edition, augmented. Fizmatlit, Moskva, 2003, pp. 198–265.
8. Ginzburg, V. L. "Responses to some questions", *About Science, Myself, and Others* (in Russian, *O nauke, o sebe, i o drugikh*). Third edition, augmented. Fizmatlit, Moskva, 2003, pp. 513–517.
9. *Physics Today* **2000**, *June*, 38.
10. I happen to know from other sources that Lev D. Landau was such a scientist.
11. An interview with David Shoenberg appeared in Hargittai, M.; Hargittai, I. *Candid Science IV: Conversations with Famous Physicists*. Imperial College Press, London, 2004, pp. 688–697.
12. Professor Ginzburg is referring to the hydrogen bomb by a nickname as in Russian people often use such names. It is not always easy to find the equivalent of such nicknames in English, especially when the subject being marked by such a name is a monstrous thing like the hydrogen bomb. The Russian word is "vodorodka" from the word vodorod, meaning hydrogen. The nickname means something like "this little hydrogeneous thing".

13. In Russian, "Bol'she vsego boisya togo kto skazhet chto znaet kak nado."
14. Ginzburg, V. L. "Experience of scientific autobiography", *About Physics and Astrophysics* (in Russian, *O fizike i astrofizike*). Third edition, revised and augmented. Buro Kvantum, 1995, pp. 312–349.
15. These two volumes are available in English translation: *Selected Works of Yakov Borisovich Zel'dovich. Volume I. Chemical Physics and Hydrodynamics.* Princeton University Press, 1992; *Selected Works of Yakov Borisovich Zel'dovich. Volume II. Particles, Nuclei, and the Universe.* Princeton University Press, 1993.
16. Ginzburg, V. L. *Biographical Memoirs of Fellows of the Royal Society,* Volume 40, The Royal Society, London, 1994, pp. 431–441,
17. Ginzburg, V. L.; Landau, L. D. *Zh. Eksp. Teor. Fiz.* **1950**, *20*, 1064.

David J. Gross, 2005 (photograph by M. Hargittai).

David J. Gross

David J. Gross (b. 1941, in Washington, DC) is Frederick W. Gluck Professor of Theoretical Physics and Director of the Kavli Institute for Theoretical Physics at the University of California, Santa Barbara. He received his B.Sc. degree from the Hebrew University of Jerusalem in 1962 and his Ph.D. degree from the University of California at Berkeley in 1966. For three years he was Junior Fellow of the Harvard Society of Fellows and taught at Princeton University from 1969 till 1997, serving as the Eugene Higgins Professor of Physics and then as Thomas Jones Professor of Mathematical Physics during the last decade of his association with Princeton. He has been at Santa Barbara since 1997.

His highest recognition was the Nobel Prize in Physics for 2004, which he shared with H. David Politzer of the California Institute of Technology and Frank Wilczek of the Massachusetts Institute of Technology "for the discovery of asymptotic freedom in the theory of the strong interaction".

He has been a Fellow of the American Academy of Arts and Sciences (1985); a Member of the National Academy of Sciences of the U.S.A. (1986); among his many awards, there is the Dirac Medal of the International Center for Theoretical Physics (1988); the Oskar Klein Medal of the Royal Swedish Academy of Sciences (2000); the Harvey Prize of the Technion — the Israel Institute of Technology (2000); and the Grande Medaille D'or de l'Academie des Science (France, 2004).

We conducted our conversation with David Gross in Lindau, Germany; on June 28, 2005.*

---

*István Hargittai and Magdolna Hargittai conducted the interview.

and started describing these experiments and I drew the Feynman diagrams that picture these collisions and I described what the experiments indicated. He didn't show any reaction, just listened very politely, and I went on for forty minutes. Finally, I got to the end and asked him what he thought. He said, it's very interesting, you really take these pictures — and he pointed to the Feynman diagrams I drew on the blackboard — very seriously.

He was famous for saying that it was very interesting when he didn't like what he heard.

After the war Wigner, unfortunately, decoupled — like many people do — from the frontiers of physics. He continued to do things here and there that were of interest to him, but they were not mainstream. When I came to Princeton, he was the great old man and all the theoretical physicists used to have weekly lunches with him, but he wasn't really into what was happening at the time and he stopped following recent developments. If you stop following recent developments in an area of science that's rapidly developing, you are out. He was involved in so many other things.

You wrote about him beautifully.

I was asked to. I'd read everything he wrote about symmetry. He understood an incredible amount, which I understood much better after reading his papers, about the nature of symmetry. This was an article in praise of him, so I didn't criticize him. But he did not understand, for example, gauge symmetry. Wigner, in the articles I referred to, talks about gauge invariance, gauge symmetry, as not a real thing, not as a physical symmetry. He just didn't pay attention to the modern developments in gauge theory. When he thought about physics, he focused on very fundamental issues, and due to his incredible stubbornness a lot of his criticism was misdirected. For example, Wigner could never accept the theory of superconductivity, and would have many arguments with Philip Anderson at teatime. Anderson would just say, you're being stupid, Eugene.

Did he use the word stupid?

Probably not. But Wigner was an absolutist, which, again, is a typical mathematician's approach. Wigner would say — correctly — that there couldn't be a state of superconductivity except when the number of particles goes to infinity. However, $N = 10^{23}$ for a physicist is infinity. The degree of instability of the system is ignorable when there are 10^{23} atoms present.

This is just one mole and Wigner was also a chemist.

He was an engineer and a chemist, but when he thought about physics, he was an absolutist, like a mathematician. I would have intense arguments with Wigner at teatime about quarks. He could not accept the principle that you could base fundamental physics on anything but observable particles. One of his greatest contributions to mathematical physics was his mathematical description of particles as irreducible representations of the Lorenz group. That gave him an operational definition of an elementary particle. Actually it did not apply to the neutron, which is unstable, but that didn't bother him. However, for Wigner it was nonsense to base a theory on quarks; particles that could never be pulled out of the nucleon. He would have vicious arguments with me about this. He was very stubborn and you couldn't convince him. It was the same with his political views. He was very stubborn, a very determined man.

You went to the Hebrew University for your undergraduate studies. How did this happen?

I was living in Israel. We went there when I was 13 years old. My father was a refugee from the Eisenhower Administration. He had worked for the U.S. Government for twenty years under democratic administrations. When Eisenhower was elected, my father was out of a job. He joined a group of advisors to the Israeli Government when the United States sent the first big aid package to Israel. My father was not particularly a Zionist although he was pro-Israel. It was more of an adventure than a job. The whole family went for two years and then he stayed on and started an academic career at Hebrew University. I went to high school there and then to university. My father returned to the United States with the rest of the family at about that time. He had an undergraduate degree in English literature, then went into the government, and ended up as a self-made and self-taught political scientist-economist-general social theoretician. He wrote a very well-received book on political science and he was in public administration. For the last 25 years of his life he had a career in academia in American universities. He was passionately interested in lots of issues but not in science.

You have mentioned that you were very secular. So what defines your Jewishness?

As for all of us, the Nazis defined it. You know how it is.

I have my views but this is your interview.

Somehow you can't avoid it and I don't want to avoid it, I'm proud of it. It's partly different in my case because I did go to Israel and there you get quite an indoctrination of Jewish heritage. When I was in Israel, it was extraordinarily secular, but this has changed since. I'm not religious at all. My children are like me, except they go to the synagogue once in a while and I never do. It must be very different in Hungary because Hungary has not come to grips with its past yet and it hasn't dealt with its collective guilt either.

Where did your ancestors come from?

They could've come from Hungary. Our name used to be Grosz rather than Gross. However, following the family name is less important once you go back a few generations. I don't know. This is on my father's side. My mother's family came from the Ukraine, from near Kiev. My father's family was American for generations whereas my mother was born in Kiev in 1913. Her father had already left for the United States before she was born. The rest of the family was stuck in Russia when the war and the revolution came. My grandmother and her children managed to move to America only years later. We were lucky since all those who stayed were killed.

The luckiest got out.

The luckiest and the smartest, those who were ambitious and had courage, got out.

Do you have any comment about the large proportion of Nobel laureates who are Jewish?

I can think of two reasons. One is obvious, being an immigrant; recent immigrants have had an enormous impact over the last one hundred years in the United States. Currently other immigrant groups, the Chinese and others from Southeast Asia are replacing the Jews. Fifty years ago Jews would dominate competitions in science; now Chinese and Indians and others from Southeast Asia are replacing them. There is then another reason. Jews used to be excluded from many professions and science and education was one way they could make their mark. Scholarship was very highly respected in Judaism and in Jewish culture and that also played a big role. Now that Jews are no longer excluded from any profession in the United States, they go into many other things. The first generation would go into science and the second generation goes into business; they're making money. There's then a third theory, which is more genetic, and I have no idea whether there is any truth to it, but with the genome, it should be possible to find out. The European Catholic Church had this very stupid policy of encouraging the smartest people to become priests and at least most of the priests had no children. In the Jewish tradition, the smart Yeshiva students were supported by the community and encouraged to marry the rich man's daughter and have a large family. If that is a five per cent effect, you multiply it by fifty generations, and it could be quite substantial.

You received the Nobel Prize with your former graduate student. Is there any way to delineate your contributions? How did your interaction work?

Frank was one of my first two graduate students and he was much better than the other one. I always worked very closely with my students. I was 31 years old when we started this work. I essentially worked with him like a colleague although he didn't know as much, which might have been an advantage in this case. I had the idea and gave the overall direction. The particular work that was cited by the Nobel Prize, the discovery of asymptotic freedom, was in one sense a "eureka moment". But in another sense it was

David Gross and his wife in Lindau, Germany, 2005 (photograph by M. Hargittai).

part of a long program that I had been working on for over five years. After many attempts of finding an explanation for the basic phenomena that had been revealed at the Stanford experiment, I tried to prove that it couldn't be done in quantum field theory. The work with Frank was aimed at closing the last hole in that argument. That development led to QCD, quantum chromodynamics. When Frank came along, I told him what I was interested in doing and we started this project. I had other projects going on, other parts of the program with other people. In general, I like very much to collaborate with smart people.

Did you have a mentor?

Not really although I had people whom I learned a lot from and people who were very helpful to me, but not mentors. My teachers in high school and at the university weren't that good.

You have moved around quite a bit.

Compared with most people in the United States, I haven't moved around so much. I was at Princeton for 27 years. That's a long time. I had no reason to leave Princeton, but the place I went to, the Institute for Theoretical Physics in Santa Barbara is very special and the job I got is very interesting. They tried to get me as director of this Institute ten years before. It wasn't as attractive back then as it became later and I wasn't at the same point in my life either. There is always a temptation for scientists with leadership qualities to take over some administrative position. It's a very dangerous thing and most of my life I tried to avoid that and stay off as many committees as I could. However, among the various kinds of administrative jobs, the one I hold is the best in the world because my responsibilities are scientific rather than administrative.

There are four Nobel laureate physicists currently at Santa Barbara although two of them received the chemistry prize.

Three of them are in the Physics Department and Kroemer is in Engineering. Walter Kohn was the first director of the Institute and he came to Santa Barbara about twenty six years ago. Alan Heeger is an experimentalist, who also came to Santa Barbara about that time. They are not members of the Institute. The Institute has only a few permanent members and it is a group of young theorists.

You said in your Nobel Banquet speech and you repeated this here in Lindau that today's questions in physics are more profound and more interesting than they were years ago. You don't think that the big questions had already been asked?

I don't think so. Of course, many big questions have been asked and many big questions have been answered, but the questions only get more interesting. They also seem to get harder, but that's always a temporary difficulty. The questions I was discussing in my lecture today about space and time are just as profound and in time will become more profound as some time ago the questions that led to quantum mechanics were.

You mentioned the start as the Big Bang, but what was before it?

Well, what is the Big Bang? I think the answer to what happened before the Big Bang, or what is the Big Bang, is that we still do not know how to formulate the question. What I really meant is that you have to know a lot to ask intelligent questions. People ask questions about the origin of the Universe and they have been asking these questions for centuries, millennia, who knows? The more we know the more able we are to make those questions precise and answerable. There are still questions that for most scientists — except for Johnny Wheeler — are still ill-formulated, philosophical, and not so interesting. People ask, why is there something as opposed to nothing? But this is a very ill-defined, bad question. It is unapproachable, unanswerable and untestable at the moment. But a question like: how did the Universe begin? has become a scientific issue and it is one by the way that we have been concentrating on for the last twenty years. We think that we now have ways of measuring, by observation of the microwave background radiation, that takes us back to a time close to the Big Bang. The inflationary model is a direct way of probing that era.

How do you assess George Gamow's contribution?

Gamow was an extremely important physicist. Had he lived longer he would have certainly deserved a Nobel Prize. He was extraordinarily imaginative and with respect to understanding the implications of the Big Bang, he had enormous foresight. It was Gamow and his two colleagues, who are much less known.

Ralph Alpher and Robert Herman.

Yes.

István Hargittai and David Gross in Lindau, Germany, 2005 (photograph by M. Hargittai).

is very time-dependent. It's almost inevitable at the beginning of a scientific understanding that things appear quite complicated, but later appear to be beautiful and simple. I can't imagine that the laws of Nature could be anything but beautiful and simple — but, of course, there is no definition of beautiful and simple. Simple is what our minds appreciate as making sense. But where does our mind come from? Our minds have evolved to understand the real world, by the process of natural selection. One is more likely to survive if one understands the world. Natural processes created our minds and they evolved to regard and appreciate things that can exist in nature as beautiful, as making sense, as simple. So to me it seems somewhat tautological to say that Nature is simple. It is simply that our minds that have evolved to regard what exists in the real world as simple and beautiful.

How will we know whether there is supersymmetry or not?

The large hadron collider will observe and prove supersymmetry or rule out low energy supersymmetry. It will resolve the story one way or another.

The recent realization that neutrinos have mass doesn't fit the Standard Model, does it?

No, but it requires only a minor modification of the Standard Model. There was always an understanding in the Standard Model that neutrinos would be very light. But the discovery of the neutrino mass was not a surprise at all. There is no principle — although people searched for it — according to which the neutrino should have no mass. There are many features of the Standard Model, or in the case of neutrinos a simple extension of the Standard Model, for which we don't understand the underlying principles and, accordingly, cannot calculate the respective parameters. We can't calculate the masses of the quarks. This is all part of the mysteries that are not explained by the Standard Model and that require a much better theory.

My impression is that people get emotional about string theory. They either love it or hate it. Why is this so?

People always get emotional about big things. String theory is very ambitious. The people who love string theory, like me, love it because it's beautiful and it has so much promise. It has so much richness and beauty. Also it explains certain puzzles, like how quantum mechanics and gravity can be consistent. This has been a puzzle since quantum mechanics was first formulated. With string theory the quantization of gravity is tame and under control. That's a great success. String theory also allows us to understand paradoxes that arise when one considers quantum gravity such as black hole physics. Stephen Hawking made a lot of noise with his analysis of black holes that concluded that the laws of quantum mechanics would have to be violated. He felt that information would be lost when black holes were formed which would then radiate away and disappear, and this would violate fundamental tenets of quantum mechanics. String theory finally resolved that this information loss doesn't happen and Hawking has conceded and given up. He publicly announced that he now sees that information is not lost, based on arguments that originated in string theory.

People who hate string theory do so for a variety of reasons. The best reason is, of course, that they're rightly worried about the enormous extrapolation to physics at a very high energy scale, very far removed from present-day observation, where we don't directly have any experimental evidence. That is indeed a dangerous way to do physics. They're concerned that the best graduate students, who often dictate the direction of science by voting with their feet, will go into string theory. In the United States there are dozens and dozens of jobs for string theorists in departments around the country, which never hired string theorists before partly because of this opposition. The reason is not that the older professors have suddenly become convinced that string theory is so interesting or promising, but because their graduate students say, "I want to do string theory and I won't come to your school unless I can do string theory." Young people move into fields that they regard as exciting and that moves science. The older people are suspicious — and it's good to be suspicious and careful because this is a dangerous game to play. But there is another reason. To express this less kindly, if you say that string is crazy, too ambitious, and too speculative, then that gives you the excuse not to have to study it, which is hard work. Many senior scientists like to have a good reason not to have to keep up with what's going on in string theory.

Are you waiting for the new hadron collider?

The whole field is.

What will be then left for the field after that?

The field of experimental particle physics faces great danger; it's a very scary situation. I'm now involved in studies in the United States about the

David Gross surrounded by students on a boat excursion on Lake Constance, 2005 (photograph by I. Hargittai).

next linear collider, the International Linear Collider that might be a complementary machine to the large hadron collider. It's not such a big step; it's just a complementary electron-positron collider. But it will also cost about ten billion dollars and whether the world can get together to build it is not at all clear. We face a real danger that the experimental effort of probing the fundamental laws of physics might get to a point where it will become too big and too costly, and too difficult to carry out. This is the first time in science that such a thing would happen. The other area where something similar is happening is astrophysics, especially cosmology. They're building these incredible detectors and observatories and satellites and they're getting to the point that some of their desired instruments have become too big and too expensive. This is an unusual situation for science to find itself. The big questions might remain open and we will not have the will or the means to address them. It also could be however that the experimentalists will find new ways to do cheaper and better experiments and get around these problems, but nobody can see the way out of this problem at the moment.

Looking at it from a utilitarian point of view, this kind of research is driven purely by curiosity whereas molecular biology, for example, carries the promise of new drugs against deadly diseases.

I don't think that it's a question of competition, say between fundamental physics and biology, because there is room for everybody. It's not that you can save money by not doing high energy physics and then spend it on biology. The amount of money being spent in biology is at least by a factor of ten more than all the physical sciences put together. So it's not a question of money. I don't know what will happen if we get to the point where there are fascinating questions, which interest everyone, that we cannot afford to study. Also, it's hard to predict what technology will make use of fifty years from now. In the past we have found applications for developments in basic physics, which, at the beginning, seemed totally removed from everyday life, like electricity. More recently we have the example of quantum mechanics — pure science in the 1920s, today the foundation of modern technology. So who knows what will happen. I have a feeling that it would be a momentous and tragic moment in human history if we get to the point when we aren't able to answer deep questions like the origin of the Universe because they are too big and too expensive to attack. Actually, I'm — for no good reason — very optimistic that we'll find solutions for these problems that today may seem hopeless.

My last question concerns the laws of Nature. There are different levels of its laws; there is classical physics and there is now quantum mechanics, they apply to different levels of Nature. Maybe quantum mechanics is not the last frontier, maybe there is a deeper level.

We are in the process of trying to go deeper. That's what we are up to in theoretical physics at the moment. The real revolution that's occurring is to try to reconcile quantum mechanics with general relativity, with the dynamics of space and time. We've learned a lot about the strange things that happen in the quantum theory of relativity through string theory. But string theory is not so far a real break with our present theoretical framework by any means, as much as quantum mechanics was a break with classical physics. We have many ways of describing string theory in different backgrounds, in different circumstances. Some of these descriptions are equivalent to ordinary gauge theories, such as we use for the Standard Model. String theory seems to be part of the same theoretical framework; we really haven't broken with the past. We have hints that it will be necessary to break with the past and that the break has something to do with the nature of space and time. We are in a situation similar to that that occurred with respect to quantum mechanics in the early 1920s. At that time they had the old quantum theory, they had a bit of quantum mechanics but they haven't made the break. That break, or revolution happened in 1924 with Heisenberg's work. In quantum gravity, or string theory, I believe that the real revolution is yet to come. It will change our understanding of space and time and will, most likely also change the way we view quantum mechanics, since quantum mechanics itself, indeed the whole framework of physics, is embedded in our traditional notions of notion of space and time. I think that the framework must be modified. We have hints of the many kinds of changes that might be needed, coming from our study of quantum gravity, string theory, and gauge theory. However, I can't see the answer yet, just as in 1920 nobody could see the uncertainty principle or matrix mechanics. The emergence of quantum mechanics required a real revolution, a discontinuous change of concepts that's very hard to come by.

Do you think that the human mind is capable for this deeper level of understanding?

This is a question that I often discuss in talks; it's about whether there is a limit to our capability to understand the universe and its laws. I have a feeling that there is no such limit. I like dogs, but I know very well that I can't teach my dog quantum mechanics. Not only can't I teach it quantum mechanics, I can't even teach it classical mechanics. There's a limit to what a dog can understand. Is there a limit to what the human mind can comprehend? Does the Universe contain levels that are beyond human understanding?

I believe strongly that the answer is no; we are capable of understanding everything. I believe this for a variety of reasons. One is that language has an infinite capacity. Noam Chomsky made the famous remark that one of the most important characteristics of human language is that even a baby can formulate an infinite number of sentences once it acquires a small vocabulary. In other words, a baby has an innate ability to be able to formulate new sentences that have never been formulated before. There's something about language, and the higher form of language called mathematics, that suggests an infinite capability. I see no evidence in the structure of knowledge — that has developed over the centuries — of reaching boundaries, of facing problems that we would not be capable of understanding.

Finally, I have empirical evidence. If it was true that the world, beyond some point, is impossible for us to understand, if there was a limit to what we could understand, then surely it would've been the case that it would take longer and longer to educate young people to get to the point where they could start making contributions to science. But that's not the case. It takes about the same amount of time for the brightest minds to get to the point when they can shake the world as it did five hundred years ago. Newton made his contributions in his twenties. Today we're dealing with much more complex things, but it takes young people no longer to get to the frontiers of physics. I don't see any sign that we're getting anywhere near a limit to our ability, and until there is evidence to the contrary I will continue to believe that there is no such a limit.

References

1. *Physics Today* **1995**, *December*, pp. 46–50.
2. Wigner, E. P. "The probability of the existence of a self-reproducing unit." Originally published in *The Logic of Personal Knowledge: Essays in Honor of Michael Polanyi*, Chapter 19, Routledge and Kegan Paul, London, 1961; reprinted in Wigner, E. P. *Symmetries and Reflections*, Indiana University Press, Bloomington and London, 1967, pp. 200–208.

Frank Wilczek, 2005 (photograph by I. Hargittai).

42

FRANK WILCZEK

Frank Wilczek (b. 1951, in New York City) is Herman Feshbach Professor of Physics at the Massachusetts Institute of Technology (MIT). He was awarded the Nobel Prize in Physics in 2004, jointly with David J. Gross of the University of California at Santa Barbara and H. David Politzer of the California Institute of Technology "for the discovery of asymptotic freedom in the theory of the strong interaction". Frank Wilczek received his B.S. degree in Mathematics from the University of Chicago in 1970 and his M.A. degree in Mathematics and his Ph.D. degree in Physics from Princeton University in 1972 and 1974, respectively. He first stayed at Princeton University and became a Professor of Physics there. Then in 1980–1988 he was Professor of Physics at the University of California at Santa Barbara and in 1989–2000 he was Professor at the Institute for Advanced Study. He has been at MIT since 2000.

He has received many honors. He has been a Member of the National Academy of Sciences of the U.S.A. (1990), a Member of the American Academy of Arts and Sciences (1993), the American Philosophical Society (2005), and many other learned societies. He was awarded the 1994 Dirac Medal by the International Center for Theoretical Physics (Trieste), received the Lorentz Medal of the Royal Netherlands Academy of Arts and Science (2002), the King Faisal International Prize for Science, and numerous other recognitions. We recorded our conversation in Lindau, Germany, on June 27, 2005.*

---

*István Hargittai conducted the interview.

Trying to list them from the most senior ones, Jeffrey Goldstone, Roman Jackiw, Bob Jaffe, Alan Guth, John Negele, Barton Zwiebach, Eddie Farhi, and a couple of younger people, Washington Taylor and Krishna Rajagopal, these are the tenured people.

So you are the only Nobel laureate among them.

At present, yes.

Does this make you feel nervous or them intimidated?

You have to ask them, but I try to wear it lightly. Also, I haven't been there very much since last October. There are several Nobel laureates in experimental physics at MIT, Wolfgang Ketterle, Jerry Friedman, and Sam Ting, although I am the only one in theoretical physics. Cliff Shull was there also in experimental physics; he passed away recently. There are then other Nobel laureates in biology and also in economics. At MIT, it's not a singularity. It's very nice and I get different treatment. For example, I'm not teaching now and I won't be teaching next year. I got an especially nice office; they gave me basically anything I asked for.

What did you ask for?

I asked for some things. I always wanted to set up a little workshop at home and now I was able to get money for that.

At MIT?

At home.

What kind?

Mechanical and electronic.

How did you justify it?

This is the nice thing that I didn't have to justify it. I wanted it. It's fun.

You were awarded the Nobel Prize for your graduate work. Your Nobel lecture was a little more general than your former supervisor's Nobel lecture.

Right.

Have you done anything of comparable importance since your graduate work?

Comparable, yes, although I wouldn't say more important. A couple of things might be comparable although they haven't had yet the wide ramifications that that breakthrough had. I did work on the axions, which is a proposed new particle that is a dark matter candidate that would solve some profound problems of particle physics. People are looking for it and it would have a big effect in cosmology. If that were discovered, I would have a serious chance for a second Nobel Prize. It's out there.

Did you do it alone?

I did it alone although there were other people on the track. Then I did work on fractional quantum numbers in a special fractional quantum statistics. You have heard of bosons and fermions. It turns out that in two-spatial-dimensional quantum mechanical systems there are other consistent possibilities than bosons and fermions. I discovered that and they are called anyons. The name was chosen to indicate that instead of a plus sign or a minus sign they can get any phase at all. Those have appeared now in real physical systems of the quantum Hall effect. The theory is completely well defined and in that frame you can show that the particles have this property. Direct experimental demonstration that leaps to the eye is much harder, but that's happening right now.

Frank Wilczek and István Hargittai in Lindau, Germany, 2005 (photograph by M. Hargittai).

Last November, just before I went to Stockholm, I got a paper from an experimental group at Stony Brook, by Vladimir Goldman and others, entitled "Direct Evidence for Fractional Statistics". I'm not sure that the paper quite lives up to that description, but it's close. I think it's coming. You must have heard of the Josephson effect, which is a kind of complication on top of the superconductivity theory. We have one superconductor interfacing with another with a boundary in between them. Very interesting phenomena happen. Similarly in quantum Hall effects when you have one quantum Hall bordering with another there can be very interesting effects. It's kind of generalized Josephson effect. That's where the fractionalized statistics play a crucial role. This situation is about to become qualitatively more fruitful than it has been up till now. Even as it is now, that's a really nice thing.

The citation for our Nobel Prize was for the discovery for asymptotic freedom, but it really recognized a larger body of work having to do with the strong interactions. The original breakthrough wasn't the end of the story by any means, and I've done other things in the strong interactions that are quite significant. They include things about the high-temperature QCD, how you use these things to describe matter in extreme conditions at very high temperature and at very high density. These are interesting and beautiful theories. These are some highlights so you can see that I haven't retired since my graduate studies.

Back in Princeton, did you come across Eugene Wigner?

By the time I came to Princeton, Wigner was no longer a young man. In his later years he developed Alzheimer's or something that looked like Alzheimer's. When I was there he was already having problems with his memory. I was at colloquia where he asked good and sharp questions, especially in the early days. I saw his style. He was very courtly. I could see as a student that he would ask these questions in an almost apologetic way, saying that "I'm sure this is my fault, but I don't understand…" in his almost whispering voice. Then he would ask some questions that showed that he had really penetrated the ideas and gone beyond. He was a remarkable person.

Was he still relevant to physics when you were there?

I didn't think he was. This was the early-1970s.

How about Freeman Dyson?

Freeman I only got to know when I came to the Institute for Advanced Study. I knew that he was a historically important physicist when I was a student at the University. He is also a very original writer. When I got to the Institute, I ran into him all the time. He's probably the smartest person I've ever met in terms of raw talent. He's very soft-spoken and modest, but he's made some comments that really impressed me. For instance, I was asked to write a review of quantum field theory for the hundredth anniversary of the American Physical Society and I was having a hard time thinking about what to write. It's a big subject and I was wondering what I could say that was new. Finally, I hit on the idea of focusing on what does quantum field theory tell you that quantum mechanics and classical field theory don't tell you separately. I thought about that and finally I realized that the most important thing, the most profound thing that quantum field theory explains is why you can have many, many electrons in the Universe and they're all identical. They're all aspects of one field. It's easy to say that, but I was proud of myself for having come to that realization. I don't think it's in the literature anywhere. I asked theoretical physicists at group gatherings about the most important thing in quantum field theory and they gave all kinds of answers that weren't as good. The only correct answer is the one I gave you. I've asked hundreds of people in different settings and only one person got it and that was Freeman. He answered immediately.

He's missing from the roster of Nobel laureates.

I don't understand that. In the history of QED he was a major figure. The rule that the Nobel Prize can be awarded only to three people comes into it.

Was he the fourth person?

For sure.

You mean he couldn't have been the third, the second, or the first?

Oh, oh, I see what you mean. That I don't know well enough. You could say that his contribution was the second generation contribution in the sense that Feynman and Schwinger did calculations at the one-loop level — and I'm not familiar with what Tomonaga did. They showed how to interpret the simplest kinds of divergences in a sensible way that was adequate to describe the experiments. What Dyson did was, he gave a general proof that this procedure could be used in all orders in perturbation theory and principles. It was a second generation contribution, but it was absolutely fundamental for the developments of the field and other fields. What I would've done if I were giving out the Nobel Prizes, is that when it came to 't Hooft and Veltman, who clearly merited a Nobel Prize for their contributions to renormalization in gauge theories, it would've been very natural to include Dyson in that. His work was very much connected to the same kind of problem and they used some of his techniques.

Did you expect to receive the Nobel Prize?

Oh, yes. I expected it for a long time. I thought it was a serious possibility ever since the data became solid, which was by the early 1980s, certainly by the mid-1980s. So I thought it was possible. However, I thought that there was a big problem, which was that since our breakthrough work was clearly based on the techniques that 't Hooft and Veltman had developed, which was based on non-Abelian gauge theories which they pioneered, I thought that it was rather unlikely that we would get the Nobel Prize before they would get theirs. Not impossible, but unlikely.

So you were expecting them to receive the Nobel Prize as well.

Oh, yes. The only thing why I wasn't sure of their Prize was because their work was essentially mathematical. I didn't think so but I could see how some conservative experimentalist could say that their work was essentially mathematical and didn't directly describe a physical phenomenon. We could have received the Prize without them receiving it, but I thought it was more likely to happen what did happen. Once they got it, I became very nervous. At that point I didn't see any barrier. I thought it was imminent.

Was it unambiguous that the three of you should get it?

I thought it was pretty clear-cut.

Although you participated in the discovery as a graduate student, you never had any doubt that you would be included.

No, I never had any doubt about that.

Was there any fourth person whom you were very sorry to see losing out?

No. Once 't Hooft got his Prize for other things, he was taken care of, otherwise, if he hadn't, he would've been a natural candidate for this also.

't Hooft and Veltman?

Just 't Hooft in that case. But I think they did the right thing. Only, they could've done it a lot sooner. It has taken them a while to sort it out.

Did I understand you correctly that from now on you will be living in anticipation of a second Prize?

I think it's possible if the axions are discovered and they would have the properties that I'd predicted, or if the fractional statistics develops in a very fruitful way. Another thing that I didn't even mention, but could also be important, very important, as a matter of fact, is that we — in this case we meaning Savas Dimopoulos and Stuart Raby and I — applied these ideas of renormalization group and running up the couplings to supersymmetric unified theories and found that they made predictions about unifications that are different from non-supersymmetric unified theories. It subsequently turned out that they agreed much better with experiment. To this day that's the best quantitative indication both for supersymmetry and for unification. If low-energy supersymmetry is in fact discovered that would indicate that. I don't know if that quite rises to the level of a Nobel Prize but it's not ridiculous. But for the next two years I'm not nervous. Ten years from now, may be.

I would like to ask you about your parents. Did they go to Stockholm?

They didn't go to Stockholm. They're almost eighty now and they don't get around so well. They're also very self-contained. They don't like to do things differently from what they've been doing. They now live on Long Island. When I was growing up we lived just inside the city limit of New York, in Queens. Now they moved a little bit and live just outside the city limit. They were born in America, but my grandparents were not. They came from the region of Galicia of the Austro-Hungarian Empire, now Poland on my father's side. I know that my paternal grandmother came from Galicia and my paternal grandfather may have come from near Warsaw, but he is kind of a shadowy figure and I don't know much about him. My maternal grandparents were from Italy, near Naples. My paternal grandfather was Jewish, but my other grandparents were Catholic. I was brought up as a Catholic.

John von Neumann's ancestors came from Galicia.

Of course, I heard about him a lot in Princeton. He was a real giant. He is legendary just in terms of pure intellect. Everyone who knew him said that he was like another species. These were people who knew Einstein, too.

What did your parents do?

his course because I was somewhat interested in physics. He taught this course where he showed SU(3) and SU(6) and how they were used in physics and suddenly I saw that they weren't just abstractions, had not just aesthetic value, but you could actually make contact with the physical world, and it wasn't ossified at all, it wasn't all done. That really planted the seeds and although I went to Graduate School in Princeton in mathematics, I wasn't at all sure of what I wanted to do and I kept in touch with physics, and that's what led to my career. I switched from mathematics to physics when I was two years in. It was in the early-1970s and very exciting things were going on in physics. That was the point when I started to work with David Gross. I had sat in a course by David and then I joined him.

Was there a division of labor in your prize-winning work?

There was a dynamic to it. Now it's a touchy thing and I don't want to cause any tension and touch on a delicate situation. We worked very intensely together. There were some parts to which I contributed more and David contributed more to other parts.

Did he initiate the project?

In a sense. But no, I wouldn't say it that way. Ken Wilson had given a series of lectures at Princeton about renormalization group that impressed a lot of people and I sat on those and so did David. Wilson's lectures later became a book. Then there were dramatic developments in what now is called the Standard Model of electroweak interactions. David was coming more from the point of view of trying to understand the strong interactions. I was coming from the point of view of gauge theory and the weak interactions. I was interested in studying the high-energy behavior of weak interactions. I wanted to figure out whether the new theories avoided the famous problem of the Landau ghost that electrodynamics had, which is the question whether there is asymptotic freedom in a different language. I brought some expertise in gauge theories to it and David brought much more maturity and expertise in the strong interactions. He used renormalization to make concrete predictions. We both brought things together. I was a graduate student looking for a problem and discussed many possibilities. We agreed that this was a terrific thing to work on.

Is the Nobel Prize changing your life?

It's definitely changing my life. My life has been very different for the last few months in terms of busyness, but it's extremely gratifying. People

Frank Wilczek surrounded by students in Lindau, Germany, 2005 (photograph by I. Hargittai).

have started treating me differently. But this is not sustainable, I can't be traveling this much. It has been very disruptive and some of my ongoing projects have been put on the back burner. I'm eager to take them up again. I'm very excited about the new developments in fractional statistics, but I had to drop them because I haven't had the time to deal with them.

Will you get back to your routine?

In two weeks. (I've been saying that for months.) I'm trying to reach a new equilibrium. It won't be the same, but I do definitely want to return to full-time research or close to full-time research. But I realize that it's a duty of the Nobel laureates to reach out to the public so I am probably going to write up some of my lectures as books. I will probably work on more ambitious and riskier projects than in the past. I can afford it. I'm no longer worried about producing results that would ensure my getting tenure.

There are people who feel they have to worry about their reputation more when they are Nobel laureates.

I've thought about that and I think it's a very unhealthy attitude. I've seen people destroyed by that. I'm not going to name names, but people who've taken that attitude have not prospered. I had time to think about this. I said to myself years ago that as soon as I get the Nobel Prize I'm going to write some not-so-important papers just to make sure that I don't get intimidated. I'm not going to write deliberately bad papers, but I'm not going to set some unrealistic standards. That would be the way to sterility. I feel very strongly about that.

Name Index

Abeddesa, J. P.   641
Abelson, P.   460
Abrikosov, A. A.   809, 812, 813, 816–821, 836
Adams, M.   120
Agmon, I.   396
Agnew, H. M.   488, 511
Akhmatova, A.   357
Alberts, B.   112, 334, 342
Alferov, Z.   567, 574–576, 585
Alonso, C.   473
Alpher, R.   847
Alt, F.   344
Alvarez, L. W.   507, 509, 511, 537, 538, 604, 606, 609, 623, 738, 739
Amaldi, E.   771
Anderson, C.   603
Anderson, P.   716, 730, 842
Andrew, R.   223, 224, 226
Apeloig, Y.   276
Arbatov, G.   514
Arber, W.   **152–163**
Archimedes   15, 692
Aristotle   15, 57
Armstrong, L.   214

Astbury, A.   686
Avery, O.   158
Bachmair, A.   349, 358, 359
Baelz, E. von   383
Bahcall, J.   758
Baisden, P.   473
Baltimore, D.   **164–181**, 192, 193, 195, 301, 319, 327, 331–334
Barak, A.   399
Barber, W.   639
Bardeen, J.   629, 715, 793, 794, 819
Barnett, L.   111, 867
Barrell, B.   103
Bar-Tana, J.   270
Bartel, B.   350, 359
Bashan, A.   396
Bean, C.   790, 793, 797, 799
Becker   568, 569
Beethoven, L. van   784
Bell, J.   683, 782
Ben-Gurion, D.   270, 271, 295
Benzer, S.   **114–133**, 356, 357, 526
Berg, P.   123, 327
Berger, H.   380
Bergius, F.   816

Page numbers in bold refer to interviews

Beriya, L.   824, 825
Berkleman, K.   594
Bernadotte, B.   220
Bernal, J. D.   14, 16, 24
Bernstein, E.   793
Bethe, H.   490, 493, 496, 511, 611, 623
Bijvoet, J. M.   9
Bishop, J. M.   **182–199**, 201, 208, 209, 213
Bitter, F.   634–636, 639
Bjorken, J. D.   637, 639
Blackett, P.   765
Bloch, F.   609, 633
Bloch, K.   376, 377
Blumberg, B.   633
Bohm, D.   782, 840
Bohn, M.   341
Bohr, N.   16, 120, 583, 744
Borden, G.   112
Bosch, C.   816
Botstein, D.   343
Bovari, T. H.   104
Boyarsky, L.   119
Boyer, H.   194
Bradbury, N.   466, 511
Bragg, W. L.   17
Brattain, W.   118
Bray, R.   119
Brecht, B.   557
Brenner, S.   9, 10, 12, **20–39**, 46, 69, 83, 94–96, 103, 105, 115, 121–123, 544, 545
Bretschler, M.   16
Brickwedde, F.   490
Bridgman, P.   356
Brockway, L.   406, 413, 427
Brower, C.   355
Brown, G.   287, 333–335
Brown, H. C.   616
Brown, W.   722
Brückner, H.   790
Buchsbaum, S.   510
Buck, L. B.   286
Budenstein, Z.   306
Budker, G. I.   685, 755
Buerger, M.   414, 429

Bulmahn, F.   584
Busch, H.   241, 345
Bush, G. H. W.   296, 728
Bush, G. W.   296, 619, 728
Byers, B.   350, 359
Cairns, J.   544
Calvin, M.   13
Canuto, V. M.   513
Carlsson, A.   128
Carter, J.   332, 514, 628
Chamberlain, O.   737
Chandrasekhar, S.   866
Chapeville, F.   123
Chargaff, E.   4, 5
Charpak, G.   505–507, 516, 534, 658, 685, 686, 746
Chau, V.   350, 359
Cherenkov, P. A.   817, 818
Chevalier, H.   623
Chiang, K.-S.   449, 450
Choibalsan, H.   814
Chomsky, N.   855
Chu, S.   818
Churchill, W.   829
Ciechanover, A.   111, 239, 240, 242, 247, 249, 257, **258–303**, 305–307, 311, 346, 347, 353, 354, 358
Cline, D.   746
Clinton, W. J.   205, 413, 461, 628, 629, 726, 727
Cocconi, G.   657
Cohen, J.   74
Cohen, M.   36
Cohen, S.   194
Cohen-Tannoudji, C.   818
Cohn   560
Collins, M. K. L.   110
Condon, E. U.   607
Cool, R.   589
Cooper, L. N   532, 715, 793, 794
Corbino, O. M.   771
Cori, C.   375
Cori, G.   375
Cork, B.   589
Cotton, F. A.   391
Cowan, C.   739

Cozzarelli, N.   342
Crick, F.   **2–19**, 22, 23, 26, 32, 35, 36,
    42, 43, 46, 59, 83, 94–97, 103–105,
    111, 115, 119, 123, 140, 189, 191,
    197, 198, 287, 321, 330, 333, 393,
    532, 545, 780, 781, 840, 848
Crick, O.   4, 8, 17
Cronin, J. W.   **586–599**, 663, 683, 684
Crothers, D.   344
Curie, E.   464
Curie, M.   464, 670
da Vinci, L.   309, 356
Daly, J.   416
Damadian, R.   222, 226, 227, 233–235,
    501
Darwin, C.   16, 83, 689
David (King)   271
Davis, R. Jr.   753, 758
Dayhoff, M.   241, 345
de Gennes, P.-G.   723
de Hoffman, F.   498
Delbrück, M.   24, 45, 47, 54, 59, 115,
    119–121, 159, 521, 522
Deutsch, M.   485, 634, 635
Diebold, R.   594
Dimopoulos, S.   865
Dobson, G. N. P.   559
Doctorow, E. L.   320
Dolan, J.   730
Doty, P.   512
Drell, S.   594, 637
Dulbecco, R.   47, 121, 165, 171
DuMond, J. W. M.   603, 604
Dunitz, J.   59
Duve, C. de   261
Dyson, F.   49, 580, 738, 797, 862–864
Dyste, J.   355
Ecker, D.   359
Eckert, W. J.   492
Edelman, G.   780, 781
Einstein, A.   8, 16, 167, 278, 315, 321,
    544, 560, 670, 677, 726, 745, 782,
    821, 859, 865, 867
Eisenhower, D. D.   514, 612, 628, 629
Ekspong, G.   685, 701
Elgin, S.   341

Elon, A.   278
Elson, D.   393
Engelhardt, V. A.   318, 323–325, 332
Engels, F.   453
Erdös, P.   116, 124
Erez, E.   338, 339
Ernhardt, F.   556
Ernst, R.   227, 441
Esaki, L.   773, 787, 791, 795
Etlinger, J. D.   307
Euripides   57
Evgenev, M.   329, 330
Eyring, H.   444
Fairbank, B.   714
Faraday, M.   28, 665
Farhi, E.   860
Farley, F. J. M.   506
Faroll, B.   735
Fay, D.   716
Feder, J.   799
Feinberg, E.   817, 829
Feldman, G.   673
Fermi, E.   407, 484–487, 493, 494, 496,
    536, 560, 588, 603, 628, 651, 677,
    737, 740, 744, 748, 750, 768, 771
Feynman, R.   511, 544, 633, 637, 639,
    651, 657, 738, 782, 801, 819, 863,
    867
Fidecaro, G.   505
Finbak, Chr.   416
Finley, D.   111, 269, 288, 345, 347, 349,
    350, 358, 359
Fisher, J.   790–792, 800
Fisher, M.   720
Fitch, V. L.   587, 594, 599, 663, 683,
    684, 686
Fok, V.   320
Forsgren, K.   110
Franck, J.   737
Franco, F.   748
Frank, E.   511
Frank, I. M.   817, 818
Frank, R. M.   511
Franklin, B.   309
Franklin, R.   5, 15, 17, 104
Fraser, S.   478

Freund, P.   867
Frey, M.   263, 264, 269
Friedman, F.   635, 639
Friedman, J.   644, 860
Frieman, E.   510
Frisch, D.   560, 635
Frost, R.   357
Fuchs, K.   495
Fukui, K.   385
Gagarin, Y.   419
Galant   45
Galich, A.   829
Galileo   13, 15, 768
Gamow, G.   12, 13, 15, 23, 42, 43, 123, 494, 847, 848
Gamow, I.   13
Garfield, E.   820
Garwin, Jeffrey   483
Garwin, Joe   484
Garwin, L.   505, 511, 516
Garwin, R.   482
Garwin, R. L.   **480–517**, 737
Garwin, T.   483
Gatrousis, C.   473
Gelfand, I.   328
Gellert, M.   342
Gell-Mann, M.   588, 594, 782
Georgiev, G.   319, 320, 323, 324, 332, 340
Gergely, J.   250
Gerhart, J.   102
Gerlach, W.   117
Giacconi, R.   **762–771**
Giaever, Inger   805, 806
Giaever, Ivar   773, 774, **786–807**
Gilbert, W.   316, 526
Gilcrease, F. W.   426
Gilman, A.   27
Ginzburg, V. L.   **808–837**
Ginzton, E. L.   609, 610, 613
Gittelman, B.   639
Glaser, D. A.   **518–553**
Glaser, L.   547, 548, 550
Glashow, S. L.   500, 688
Glotzer, M.   347
Gödel, K.   31

Godknopf, I.   345
Goebl, M. G.   359
Goeppert-Mayer, M.   737
Goldanskii, V.   810
Goldberg, A. L.   307
Goldberger, M.   513, 588, 737, 840
Goldhaber, G.   594, 640, 643
Goldhaber, M.   594
Goldman, V.   862
Goldstein, G.   240, 287, 345
Goldstone, J.   860
Gonda, D. K.   351, 359
Goodkind, J.   719
Goodstein, D.   712
Gor'kov, L. P.   812, 817, 818
Gorbachev, M. S.   453, 514, 515, 583, 828
Gore, A.   413, 629
Graciet, E.   355
Graham, D. O.   510
Grannell, P.   218
Green, D. E.   372–374
Green, H.   334
Greengard, P.   93
Greenstein, J.   167
Gregorich, K.   474
Gross, D. J.   286, **838–855**, 857, 858, 868
Guntaka, R.   213
Gustaf, C., XVI (King of Sweden)   802
Guth, A.   860
Gyurcsány, F.   250
Haas, A. L.   257, 264, 266, 284, 285, 307, 346
Haber, F.   557
Hahn, I.   500
Hahn, O.   460, 744
Hall, H.   475, 820
Hammel, E.   714
Hansen, W. W.   609
Hanson, C.   410
Hanson, J.   410
Hanson, L.   410
Happer, W.   715
Harker, D.   9
Harrison, W.   790

Hartwell, L. H.   63, 68, 69, 89, 93, 350
Hauptman, H.   412, 414, 423, 430, 431, 437
Hauser, H.   775
Hawking, S.   851
Hayaishi, M.   363
Hayaishi, O.   **360–387**
Hayaishi, T.   386
Hedberg, K.   408
Hedberg, L.   408
Heeger, A.   577, 846
Heine, H.   278
Heinze   755
Heisenberg, W.   740, 743, 744, 821, 854
Heller, H.   257
Heller, J.   742
Hellwege, P.   569
Helmholtz, H.   528, 545
Herman, R.   847
Herold, E.   570, 574
Herrmann, G.   467, 474
Herschbach, D. R.   439, 440, 444, 445
Hershey, A.   120
Hershey, B.   337
Hershko, A.   111, 112, **238–257**, 259–263, 265–270, 282–289, 291, 292, 298–301, 305–307, 311, 344, 346, 347, 351, 353, 354
Hershko, J.   248, 254
Hershko, M.   252
Hertz, G.   557
Herzl, T.   271
Herzog, C.   381
Hicke, L.   350
Hilbert, D.   321
Hill, A. V.   14, 16
Himmler, H.   744
Hinshelwood, C.   15, 26
Hitler, A.   51, 91, 148, 558, 602, 734, 744, 750, 822, 828, 829
Hochstrasser, M.   348, 351, 358
Hod, Y.   257, 264
Hodgkin, D.   14
Hoffman, D.   477
Hoffman, D. C.   **458–479**
Hoffman, M.   465–467

Hoffman, M. R.   477
Hoffman, R.   328–334, 356
Hofstadter, R.   594, 609, 610, 633, 637, 639, 642, 647
Hogness, D.   260
Holdren, J.   513
Holm, C.   343
Holmes, F. L.   41, 43, 124
Homer   57
Hooft, G. 't   686, 863, 864
Horvay, G.   790, 791
Horvitz, H. R.   21, 334
Hu, C.   355
Hu, R. G.   358
Hughes, R.   594
Hulse, R.   508
Hunt, R. T.   63, 68, 69, **88–113**, 246, 347
Hunter, T.   101, 168
Huntington, H. B.   793
Hutchins, R. M.   57
Hutchinson, G. E. L.   60
Hutchison, C.   486
Huxley, H.   37
Hwang, C.-S.   355
Ilyin, Y.   340
Imanishi-Kari, T.   178
Ingram, V.   16
Ipat'ev, V. N.   816
Ishimori, K.   379, 380
Jackiw, R.   860
Jackson, D.   559
Jackson, R.   101
Jacob, F.   10, 23, 56, 113, 115, 137, 316
Jaffe, B.   860
Jeans, J.   425
Jentsch, S.   349, 359
John Paul, II   368, 513, 828
Johnson, E. S.   351, 359
Johnson, L. B.   50
Johnston, H.   497
Jones, L.   672
Jones, R. V.   90
Josephson, B. D.   **772–785**, 787, 801
Josephson, C.   784
Journe, V.   516

Jouvet, M. 380
Judson, H. 10
Kandel, E. 131
Kapitsa, S. 624
Kapitsa/Kapitza, P. 624, 717, 730, 819
Karle, I. L. **402–421**, 423, 427–429, 431–434, 437
Karle, Jean 410
Karle, Jerome 404, 406, 410, 412–414, 416, 417, 420, **422–437**, 441
Kármán, T. von 167, 494
Karplus, M. 59
Katzav, M. 399
Katzir, E. 391
Kavenoff, R. 330
Kay, L. 24
Keats, J. 321
Keese, C. 797
Kellenberger, E. 159
Kendall, H. 644
Kendrew, J. 9, 14, 16, 35, 37
Kennedy, J. F. 172, 498, 551, 612, 628
Kenrick, F. B. 553
Kepler, J. 689
Kerman, A. 686
Ketterle, W. 860
Keyworth, G. A. 499, 510
Khariton, Yu. B. 823
Khorana, H. G. 23, 527
Khrushchev, N. 56, 314, 318, 552, 827
Khvol'son, O. D. 816
Kirschner, B. 288
Kirschner, M. 347
Kissinger, H. 49–51, 514
Kistiakowsky, G. 56, 629
Kleitman, N. 380
Klug, A. 12, 17, 33, 37, 86, 108, 109, 323, 324, 393
Koch, C. 7, 12
Koestler, A. 780
Kohn, W. 846
König 568
Konopka, R. 121, 130
Konrad, G. 295
Kornberg, A. 246, 374, 375, 387, 527
Kornberg, R. 334

Koshiba, M. **752–761**
Koshland, D. 343, 548
Kotake, Y. 364, 371
Koussevitzky, D. 293
Koussevitzky, M. 293
Kraus, K. 195, 196
Kroemer, H. **566–585**, 846
Kumar, T. 528
Kurchatov, I. V. 495, 823–825
Kurti, G. 555
Kurti, N. **554–565**
Kusch, E. 505
Kusch, P. 505, 671
Kwartin, Z. 293
Laden, O. bin 52, 516
Land, E. 508
Landau, L. D. 320, 444, 576, 712, 714, 812, 816–822, 832–834, 836, 837, 868
Landsberg, G. S. 815, 816, 832
Langevin, P. 557
Lark-Horovitz, K. 117, 118, 128
Larsen, P. L. 340, 358
Laue, M. von 557
Lauterbur, P. 217, 218, 222, 227, 228, 233–235, 501, 720
Lawn, F. 50
Lawrence, E. O. 608, 623
Lawrence, F. O. 461
Lawrence, P. 96, 98, 103
Layzer, A. 716
le Carré, J. 329
Lederberg, J. 159
Lederman, L. 203, 504–506, 594, 657, 682, 733, 740, 741, 751
Lee, D. M. 711, 717, 721, 728
Lee, E. 676
Lee, T. D. 502, 504, 591, 657, 682, 737, 739, 742, 750, 866
Lee, Y. T. **438–457**
Leggett, A. J. 723, 809, 812, 816
Lehn, J.-M. 564
Lenin, V. I. 829
Letokhov, V. S. 818
Leviant, I. 529
Levine, J. 507

Levine, M.   160
Levinger, L.   345
Levinson, W.   209
Levinthal, C.   333
Levintow, L.   209
Lewin, B.   341
Lewis, E. B.   122, 128, 135, 138, 169, 209
Lewis, H.   132
Lewis, J.   112
Libby, W.   588
Lifshits, E. M.   819
Lindeman, F. A.   558, 559
Lipmann, F.   123, 246
Lipscomb, W.   391, 392
Liu, P.   726
Livnat, L.   399
Livshitz, E.   390
Livshitz, H.   390
Lodge, C.   51
Lodish, H.   247, 268, 287
London, I.   101, 112
Longo, M.   672
Lorenz, K.   129, 136
Lowry, O.   266–268
Luria, S.   119, 120, 159
Lwoff, A.   115
Lysenko, T.   129, 318
Mackay, A.   24
Maddox, B.   17
Magasanik, B.   335
Magdoff, B.   12
Mager, J.   244, 245
Mahan, B.   444
Maiani, L.   686, 691
Maimonides   748
Majorana, E.   771
Malavskys   293
Mandelstam, J. von   815, 816
Mansfield, G.   230
Mansfield, P.   **216–237**
Mansfield, S.   230
Mantsch, P.   598
Margolis, L.   328
Mariott, D.   359
Mark, J. C.   491, 511

Marshall, L.   484
Martin, D. S., Jr.   465
Martin, S.   95, 185
Marx, K.   453, 748
Mason, H.   376
Masui, Y.   67–69, 71
Maudsley, A.   223
Mayer, J.   737
McCarthy, J.   55, 495
McDaniel, B.   594
McFarlane, R.   510
McGrath, J. P.   349, 359
McMillan, E.   460, 607
Meinhardt, H.   141, 142
Mendel, G.   137, 545
Mendelsohn, F.   278
Mendelsohn, M.   278
Mengele, J.   149
Merritt, F.   594
Meselson, M.   10, **40–61**, 160
Metropolis, N.   511
Mewherter, J. L.   461, 462
Miller, C.   132
Miller, S.   574
Millikan, R. A.   132, 167, 556, 603, 676
Milosevic, S.   53
Milstein, C.   37, 441
Minsky, M.   337
Mirick, G. S.   371
Mitchison, G.   12
Mitchison, M.   75
Mizutani, S.   171, 181, 193
Moncada, S.   284
Monod, J.   23, 44, 94, 95, 115, 137, 260, 316
Moore, P.   396
Morgan, T. H.   132, 138
Morris, P.   229
Mössbauer, R.   584
Moult, J.   392
Mozart, W. A.   784
Müller-Hill, B.   149, 316
Mulliken, R. S.   486
Mullis, K.   527, 553
Murray, A.   98, 347
Naipaul, V. S.   82

Nambu, Y.   866
Nasmyth, K.   108, 343
Nasser, De L.   132
Nathans, D.   153, 159, 160, 163
Natta, G.   771
Naylor, N.   463
Neal, K.   74
Neddermeyer, S.   499
Negele, J.   860
Nernst, W.   557
Neumann, J. von   25, 28, 489, 556, 811, 834, 865
Newton, I.   544, 677
Neylan, J. F.   608
Nirenberg, M.   23, 65, 123, 245, 441
Nixon, R. M.   50, 172, 296, 510, 616
Nordheim, L.   496
Nun, S.   620
Nurse, P. M.   **62–87**, 89, 93, 107, 180
Nüsslein-Volhard, C.   **134–151**, 209
O'Neill, G. K.   638, 639
Occhialini, G.   765, 768
Ochoa, S.   123, 306
Ohkuma, S.   303
Olby, R.   16
Onnes, K.   712, 729, 730
Onsager, L.   804
Oppenheimer, R.   116, 499, 604, 607, 610, 621–625, 652, 736, 737, 750, 825
Orgel, L.   12
Osborne, L.   636
Osheroff, D. D.   **710–731**
Özkaynak, E.   349, 358
Packard, R.   724
Pais, A.   113
Palmer, R.   594
Panofsky, E.   602
Panofsky, W. K. H.   513, **600–629**, 637–641, 647, 672, 739
Parker, R.   213
Patrascu, M.   740
Paul, W.   568
Pauling, L.   8, 24, 41, 44–46, 49, 56, 58, 59, 167

Peebles, J.   594
Peerce, J.   293
Peierls, R. E.   511, 559, 560
Pelham, H.   98, 100, 104
Penrose, R.   784
Penzias, A. A.   13, 730
Perl, M. L.   617, 640, 642, 657, **668–679**
Perl, T.   678
Perutz, M.   9, 14, 16, 34, 37, 94, 95, 104, 109, 123
Peshkov, V. P.   719
Pfefferkorn, E.   197
Phillips, D.   392
Phillips, M.   607
Phillips, W. D.   818
Piatkov, K.   355
Pickart, C. M.   307, 348, 358
Piéron, H.   380
Pine, A.   456
Pinker, S.   775
Pinochet, A.   54
Pintér, J.   556
Piroué, P.   500
Placzek, G.   750
Planck, M.   557
Poggio, T.   532
Polanyi, J. C.   439, 440
Polanyi, M.   48, 440, 444, 557
Politzer, H. D.   286, 839, 857
Pomeranchuk, I.   712, 717, 719
Pontecorvo, B.   750
Poole, B.   261, 303
Powell, C.   754, 765
Powles, J.   220
Prescott, E. C.   286
Press, F.   332
Primakov, E.   514
Prusiner, S.   93, 276
Ptashne, M.   98, 316
Purcell, E. M.   502
Putin, V. V.   619
Qi, X.   358
Rabi, H.   511
Rabi, I. I.   493, 511, 629, 659, 671

Raby, S.   865
Radda, G.   205, 206
Raff, M.   112
Rajagopal, K.   860
Raman, C. V.   816
Ramsey, N. F.   502, 563, 730
Ranganathan, D.   418
Rao, H.   343
Ravel, M.   392
Ray, B.   123
Reagan, R.   235, 439, 499, 509, 510, 513, 622
Reardon, P.   594
Rees, J.   641
Reifhart, W.   528
Reines, F.   669, 739
Reshevsky, N.   58
Revzin, A.   344
Rich, A.   13, 45, 319, 333, 334, 393
Richardson, R. C.   711, 721
Richter, B.   82, 594, 617, **630–653**, 655, 658, 672, 673, 676, 677
Riezman, H.   350
Ritson, D.   640
Robb, W.   797
Roberts, K.   112
Roberts, R.   38
Robertson, J. M.   9, 104
Rocap, G.   359
Roentgen, W. C.   803
Rojansky, V.   670
Roosevelt, F. D.   167, 560, 735
Rose, I. A.   239, 241, 242, 245, 257, 259, 262, 264–266, 283–287, 289, 291, 301, **304–309**, 354
Rosenblatt, Y.   293
Rosenbluth, M.   492, 737
Rossi, B.   765, 766, 768
Rotblat, J.   512
Rourke, F. M.   461, 462
Rous, P.   184
Rubbia, C.   658, **680–697**, 699, 701, 702, 707, 746
Russell, B.   336, 867
Sabin, A.   194

Sagane, R.   604
Sagdeev, R.   514
Sakharov, A. D.   496, 591, 624, 663, 684, 821, 824–827
Salk, J.   194
Samson   270
Samuel, D.   375
Sander, K.   141
Sanger, F.   8, 34, 35, 37–39, 94, 103, 122
Sauli, F.   506
Sauter, F.   569
Savonarola   335
Scherrer, P.   159
Schiff, L.   609
Schimke, R.   344
Schmid, B.   173
Schnell, W.   685
Schoenheimer, R.   260
Schopper, E.   687
Schottky, W.   573
Schrieffer, J. R.   715, 793, 794
Schrödinger, E. T.   27, 28, 109, 119, 127, 557
Schwann, T.   64
Schwartz, L.   482
Schwartz, M.   657, 682, 733, 739–742, 745, 750
Schwarz, A.   267
Schwinger, J.   637, 657, 738, 863
Scorcese, M.   337
Scott, W.   341
Seaborg, G. T.   471–474, 629
Seaborg, H.   472, 473
Seethaler, G.   99
Segrè, E.   492, 771
Sens, H.   506
Serber, R.   739
Sewell, D.   473
Shakespeare, W.   84
Shankland, R.   484
Shapiro, B.   270
Shapiro, V.   821
Sharon, A.   399
Sharpless, B.   93

Shatt, R.   534
Sheldrick, G.   431
Sheng, J.   355, 358
Shockley, W.   570, 573
Shoemaker, C.   408
Shoemaker, D.   408
Shoenberg, D.   822, 836
Shoolery, J.   622
Shrader, T. E.   359
Shubnikov, L. V.   819
Sigler, P.   392
Simon, F. E.   555, 557–560
Simpson, M.   260, 307
Singer, M.   337
Skidmore, L.   410
Skinner, B. F.   129
Skrinsky, A.   648
Slichter, C.   221
Smith, H. O.   153, 159, 160
Smith, M.   594
Smith, S.   594, 658
Solomon, M. J.   340, 358
Solov'ev, Y. I.   836
Sophocles   57
Spector, D.   213
Spence, R.   466
Sperry, R.   115, 127, 128
Stahl, F.   43, 46–48, 160
Stalin, J.   313, 314, 318, 814, 819, 822, 824, 828, 829
Stear, E.   579, 580
Stehelin, D.   213
Steinbeck, J.   9
Steinberger, J.   485, 607, 657, 682, 683, **732–751**
Steitz, T.   396
Stent, G.   13, 15, 16, 32, 59, 60, 121, 159
Stern, O.   117
Sternglanz, R.   343
Stevenson, A.   197
Strassmann, F.   460
Strauss, F.   344, 358
Sugihara, T.   473
Sulak, L.   594
Sulston, J. E.   21, 39, 83

Sundin, O.   341, 358
Svedberg, T.   47
Szentpéter, J.   556
Szilard, L.   35, 36, 54–56, 116, 117, 132, 488, 498, 499, 548, 549, 557, 560, 671, 744, 811, 834
Takahashi, T. T.   358
Tamm, I. E.   320, 625, 817, 818, 821, 824, 825, 832
Taniguchi, T.   364, 370, 371
Tanner   784
Tawney, B.   410
Tawney, Madeleine   410
Tawney, Michael   410
Tawney, N.   410
Taylor, D.   617
Taylor, J.   508
Taylor, R.   82, 644, 677
Taylor, T.   491
Taylor, W.   860
Telegdi, V.   484, 505, 506
Teller, E.   54, 116, 452, 487–490, 493–495, 498–500, 510, 548, 549, 556, 560, 588, 611, 612, 620–625, 652, 737, 825, 834
Temin, H. M.   165, 171, 173, 181, 191–193, 195
Thompson, G. P.   560
Thompson, R.   765, 768
Tigner, M.   594, 648
Ting, S. C. C.   82, 631, 642–644, **654–667**, 860
Tisza, L.   750
Tlsty, T.   344
Tobias, J. W.   359
Tomkins, G.   112, 240, 242, 243, 245, 261
Tomonaga, S.-I.   637, 657, 754, 761, 863
Tonegawa, S.   384
Townes, C.   93, 647, 730
Traub, W.   391
Treisman, R.   98
Trilling, G.   643
Truman, H. S.   488, 493, 583
Tsibur, S.   293
Tucker, R.   293

Tukey, J. 508
Turner 60
Uhlmann, F. 343
Ulam, S. 488–490, 499
Ulitskaya, L. 329, 330
Urban, M. 264–266, 284, 285, 346
Urey, H. 588, 737
Vámbéry, Á. 564, 565
van der Meer, S. 681, 684, 685, **698–709**, 741, 745
Vannikov, B. L. 824, 825
Varmus, H. E. 180, 183, 184, 189, 190, **200–215**
Varshavsky, A. 111, 128, 171, 239, 247, 259, 267, 269, 287–289, 291, 292, 298–301, **310–359**
Varshavsky, J. 7, 49, 312, 317, 319
Varshavsky, V. 337, 338, 355, 356
Vavilov, S. I. 817
Velikhov, E. 514
Veltman, M. 82, 93, 863, 864
Victoria (Princess of Sweden) 722
Vinograd, J. 47
Vizi, S. E. 250
Vlosky, D. 117
von Verschuer, O. 149
von Weizsäcker, C. F. 755
Walker, G. 335
Wang, J. 342
Warburton, M. 303
Ware, D. 221
Watkins. J. D. 510
Watson, J. D. 3, 5, 6, 9, 11–14, 16, 17, 23, 37, 43, 45, 59, 75, 94, 97, 104, 112, 121, 123, 126, 140, 189, 287, 317, 393, 840, 848
Waugh, J. 219, 221, 222, 228
Weber, J. 469, 507
Weigle, J. 121, 159, 160
Weinberg, A. 629
Weinberg, S. 682
Weiner, J. 116, 124
Weisblum, B. 123
Weismann, A. 60
Wellstone, P. 205
Wenzel, W. 589
Weyl, H. 867
Wheatley, J. 716–719, 721
Wheeler, J. A. 494, 623, 625, 847
Wick, G. C. 738, 739, 750
Wieland, H. 374, 378
Wieschaus, E. F. 135, 138, 143, 209
Wiesel, E. 268, 276
Wiesel, M. 276
Wiesner, J. 498, 508, 629
Wigner, E. P. 13, 25, 256, 494, 522, 556, 557, 560, 834, 840–843, 855, 862
Wigzell, H. 70
Wilczek, F. 286, 839, 846, **856–869**
Wilhelmy, J. B. 469
Wilkins, M. 3, 5, 9
Wilkinson, K. 264–266, 284, 285, 346
Wilson, J. 112
Wilson, K. 868
Wilson, R. W. 594, 730
Winter, K. 686
Witkop, B. 416
Wittgenstein, L. 336
Wittmann, H. G. 392, 401
Wojcicki, S. 594
Wolf, M. 632
Wollman, E. 121
Wooldridge, D. 127
Wu, B. 445, 446
Wu, C.-S. 341, 503, 504, 657, 666, 682
Wul'ff, F. D. 813
Wünning, I. 350, 359
Wüthrich, K. 83
Xia, Z. 355
Xu, J. 355
Xu, Z. 358
Yamada, M. 287, 346, 347
Yamamoto, S. 377
Yanagida, M. 108, 343
Yang, C. N. 502, 504, 591, 729, 737, 750, 866
Yanofsky, C. 122
Yochem, J. 359
Yonath, A. **388–401**
York, H. 494, 495
Yuan, B. 160

Yukava, H.   738
Zacharias, J.   508
Zamir, A.   393
Zeitlin, M.   312
Zel'dovich, Y. B.   320, 824, 827, 831
Zernike, F.   569

Zetterberg, A.   93, 94
Zewail, A.   170
Zhou, J.   355
Zichichi, A.   505–507, 673
Zwiebach, B.   860

Cumulative Index of Interviewees
Candid Science I–VI

Abrikosov, A. A.   V/176
Agnew, H.   V/300
Alferov, Zh. I.   IV/602
Altman, S.   II/338
Alvarez, L. W.   V/198
Anderson, P. W.   IV/586
Arber, W.   VI/152
Bader, A.   III/146
Bahcall, J. N.   IV/232
Bahcall, N. A.   V/266
Balazs, E. A.   III/120
Baltimore, D.   VI/164
Bartell, L. S.   III/58
Bartlett, N.   III/28
Barton, D. H. R.   I/148
Barton, J. K.   III/158
Bax, A.   III/168
Bell, J. B.   IV/638
Benzer, S.   VI/114
Berg, P.   II/154
Bergström, K. S. D.   II/542
Berry, R. S.   I/422

Bishop, J. M.   VI/182
Black, J. W.   II/524
Blobel, G.   II/252
Blumberg, B. S.   V/578
Bodánszky, M.   V/366
Boyer, P. D.   III/268
Bréchignac, C.   IV/570
Brenner, S.   VI/20
Brown, H. C.   I/250
Calvin, M.   V/378
Carlsson, A.   V/588
Chamberlain, O.   IV/298
Chargaff, E.   I/14
Chulabhorn, M.   V/332
Ciechanover, A.   VI/258
Cohn, M.   III/250
Conway, J. H.   V/16
Cooper, L. N   V/164
Cornforth, J. W.   I/122
Cotton, F. A.   I/230
Coxeter, H. S. M.   V/2
Cram, D. J.   III/178

Crick, F.   VI/2
Cronin, J. W.   VI/586
Crutzen, P. J.   III/460
Curl, R. F.   I/374
Deisenhofer, J.   III/342
Dewar, M. J. S.   I/164
Djerassi, C.   I/72
Dresselhaus, M. S.   IV/546
Dulbecco, R.   V/550
Dunitz, J. D.   III/318
Dyson, F. J.   IV/440
Eaton, P. E.   I/416
Edelman, G. M.   II/196
Eigen, M.   III/368
Elion, G. B.   I/54
Ernst, R. R.   I/294
Ernster, L.   II/376
Eschenmoser, A.   III/96
Finch, J. T.   II/330
Fitch, V. L.   IV/192
Fowler, W. A.   V/228
Friedman, J. I.   IV/64
Fukui, K.   I/210
Furchgott, R. F.   II/578
Furka, A.   III/220
Gajdusek, D. C.   II/442
Gal'pern, E. G.   I/322
Garwin, R. L.   VI/480
Giacconi, R.   VI/762
Giaever, Ivar   VI/786
Gilbert, W.   II/98
Gillespie, R. J.   III/48
Gilman, A. G.   II/238
Ginzburg, V. L.   VI/808
Glaser, D. A.   VI/518
Goldhaber, M.   IV/214
Greengard, P.   V/648
Gross, D. J.   VI/838
Hassel, O.   I/158
Hauptman, H. A.   III/292
Hayaishi, O.   VI/360
Heeger, A. J.   V/410
Henderson, R.   II/296
Herschbach, D. R.   III/392
Hershko, A.   VI/238
Hewish, A.   IV/626

Hoffman, D. C.   VI/458
Hoffmann, R.   I/190
Hooft, G. 't   IV/110
Hornykiewicz, O.   V/618
Huber, R.   III/354
Huffman, D. R.   V/390
Hulse, R. A.   IV/670
Hunt, R. T.   VI/88
Jacob, F.   II/84
Josephson, B. D.   VI/772
Kandel, E. R.   V/666
Karle, I. L.   VI/402
Karle, Jerome   VI/422
Ketterle, W.   IV/368
Klein, G.   II/416
Klug, A.   II/306
Kornberg, A.   II/50
Koshiba, M.   VI/752
Krätschmer, W.   I/388
Kroemer, H.   VI/566
Kroto, H. W.   I/332
Kuroda, R.   III/466
Kurti, N.   VI/554
Larson, C. E.   V/316
Laurent, T. C.   II/396
Lauterbur, P. C.   V/454
Lederberg, J.   II/32
Lederman, L. M.   IV/142
Lee, Y. T.   VI/438
Lehn, J.-M.   III/198
Leonard, N. J.   V/324
Levi-Montalcini, R.   II/364
Lewis, E. B.   II/350
Lipscomb, W. N.   III/18
MacDiarmid, A. G.   V/400
Mackay, A. L.   V/56
Mandelbrot, B. B.   IV/496
Mansfield, P.   VI/216
Marcus, R. A.   III/414
Mason, S.   III/472
McCarty, M.   II/16
Merrifield, B.   III/206
Meselson, M.   VI/40
Michel, H.   III/332
Milstein, C.   II/220
Moncada, S.   II/564

Mößbauer, R.   IV/260
Müller-Hill, B.   II/114
Mullis, K. B.   II/182
Nathans, D.   II/142
Ne'eman, Y.   IV/32
Nirenberg, M. W.   II/130
Nurse, P. M.   VI/62
Nüsslein-Volhard, C.   VI/134
Olah, G. A.   I/270
Oliphant, M. L. E.   IV/304
Orchin, M.   I/222
Osawa, E.   I/308
Osheroff, D. D.   VI/710
Ourisson, G.   III/230
Panofsky, W. K. H.   VI/600
Pauling, L.   I/2, V/340
Peierls, R. E.   V/282
Penrose, R.   V/36
Penzias, A. A.   IV/272
Perl, M. L.   VI/668
Perutz, M. F.   II/280
Pickering, W. H.   V/218
Pitzer, K. S.   I/438
Polanyi, J. C.   III/378
Polkinghorne, J. C.   IV/478
Pople, J. A.   I/178
Porter, G.   I/476
Prelog, V.   I/138
Prigogine, I.   III/422
Pritchard, D. E.   IV/344
Radda, G. K.   II/266
Ramsey, N. F.   IV/316
Richter, B.   VI/630
Robbins, F. C.   II/498
Roberts, J. D.   I/284
Rose, I.   VI/304
Rowland, F. S.   I/448
Rubbia, C.   VI/680
Rubin, V.   V/246
Sanger, F.   II/72
Schawlow, A. L.   V/138
Scheuer, P. J.   I/92
Schleyer, P. v. R.   III/80

Seaborg, G. T.   III/2
Segrè, E. G.   V/290
Semenov, N. N.   I/466
Shechtman, D.   V/76
Shoenberg, D.   IV/688
Skou, J. C.   V/428
Smalley, R. E.   I/362
Stankevich, I. V.   I/322
Steinberger, J.   VI/732
Stent, G. S.   V/480
Stork, G.   III/108
Sulston, J. E.   V/528
Taube, H.   III/400
Taylor, J. H.   IV/656
Telegdi, V. L.   IV/160
Teller, E.   IV/404
Ting, S. C. C.   VI/654
Tisza, L.   IV/390
Townes, C. H.   V/94
Tsui, D. C.   IV/620
Ulubelen, A.   I/114
van der Meer, S.   VI/698
Vane, J. R.   II/548
Varmus, H. E.   VI/200
Varshavsky, A.   VI/310
Veltman, M. J. G.   IV/80
Walker, J. E.   III/280
Watson, J. D.   II/2
Weinberg, S.   IV/20
Weissmann, C.   II/466
Westheimer, F. H.   I/38
Wheeler, J. A.   IV/424
Whetten, R. L.   I/404
Wigner, E. P.   IV/2
Wilczek, F.   VI/856
Wilson, K. G.   IV/524
Wilson, R. W.   IV/286
Yalow, R.   II/518
Yonath, A.   VI/388
Zare, R. N.   III/448
Zewail, A. H.   I/488
Zhabotinsky, A. M.   III/432

Book Shelves
Q141 .H264 2006
Hargittai, Istvan
Candid science VI : more
conversations with famous
scientists